Silicon-Containing Dendritic Polymers

ADVANCES IN SILICON SCIENCE

VOLUME 2

Series Editor:

JANIS MATISONS

School of Chemistry, Physics and Earth Sciences, Flinders University, South Australia.

Advances in Silicon Science is a book series which presents reviews of the present and future trends in silicon science and will benefit those in chemistry, physics, biomedical engineering, and materials science. It is aimed at all scientists at universities and in industry who wish to keep abreast of advances in the topics covered.

Series Editor

Professor Janis Matisons
Nanomaterials Group
Chair of Nanotechnology
School of Chemistry, Physics and Earth Sciences
Flinders University
GPO Box 2100,
Adelaide 5001
SOUTH AUSTRALIA

Volume 2

Silicon-Containing Dendritic Polymers

Volume Editors

Dr. Petar R. Dvornic
Michigan Molecular Institute
1910 W. St. Andrews Rd.
Midland, MI 48640
USA

Dr. Michael J. Owen
Michigan Molecular Institute
1910 W. St. Andrews Rd.
Midland, MI 48640
USA

For other titles published in this series, go to http://www.springer.com/series/7926

Petar R. Dvornic • Michael J. Owen
Editors

Silicon-Containing Dendritic Polymers

Springer

Editors
Petar R. Dvornic
Michigan Molecular Institute
1910 W. St. Andrews Rd.
Midland, MI 48640
USA
dvornic@mmi.org

Michael J. Owen
Michigan Molecular Institute
1910 W. St. Andrews Rd.
Midland, MI 48640
USA
michaelowen01@chartermi.net

ISBN: 978-1-4020-8173-6 e-ISBN: 978-1-4020-8174-3
DOI: 10.1007/978-1-4020-8174-3

Library of Congress Control Number: 2008936828

Printed on acid-free paper

springer.com

Preface

In this book, we have attempted to present a coherent picture of the field of silicon-containing dendritic polymers which has attracted outstanding research attention during the last 20 years. This interest resulted from an almost explosive development of dendritic polymers in general, including primarily dendrimers and hyperbranched polymers, which have become one of the fastest growing areas of polymer science, and from a unique combination of properties of silicon-containing polymers that is not found in any class of "purely" organic materials.

The almost contagious fascination of many scientists from different disciplines with dendritic polymers came mostly from their unprecedented molecular architecture, unique resulting properties and the realization that they represent ideal building blocks for "soft", chemical, bottom-up nanotechnology. As a consequence, several hundred different dendritic polymer families have been reported, among which the silicon-containing ones (similar to the situation in other fields of polymer science, including linear and crosslinked polymers) have attracted most of the attention among the non-fully-organic members of this polymer family (i.e., those containing element(s) other than C, H, N and O).

Appropriately, siloxane dendrimers were the first heteroatom dendritic polymers to be reported as early as 1989 by the founders of the field, Evgenij Rebrov and Aziz Muzafarov of the Institute of Synthetic Polymer Materials of the Russian Academy of Sciences in Moscow. Since then, many other types of related materials have been described including carbosilane, silane and silazane dendrimers, as well as their copolymeric counterparts. In all of these, silicon has been used as an ideal branching junction with the rare ability to provide controllable branching functionality of either two or three, resulting in a plethora of new materials with properties that are not attainable in any other way.

It is a special feature of this book, which is the first on this subject matter, that it compiles the developments in each of the major areas of the field of silicon-containing dendritic polymers as seen through the eyes of those who were either the pioneers who initially opened each area or who played one of the major roles in their development, or both. Hence, the list of chapter authors truly is a "who's who" in the field of silicon-containing dendritic polymers, and we are extremely grateful to all of them for accepting our invitation to participate in this project and for delivering their masterpieces in such an excellent form and timely manner. As a result, all chapters represent the state of the art in each individual area, meet the highest scientific standards and contain a wealth of information and a variety of inspiring suggestions for future developments.

This book consists of two main parts: Chapters 2–11 are devoted to the silicon-containing dendrimers and Chapters 12–16 are devoted to the silicon-containing hyperbranched polymers. In Chapter 1, we briefly overview the entire field of dendritic polymers and emphasize the role that silicon plays within it. Chapters 2–7 describe some of the most important classes of silicon-containing dendrimers, including those primarily built on siloxane, carbosilane, silane, silazane, silyl ether and polyhedral oligomeric silsesquioxane compositions. Chapters 8–10 present some of the so far most successful applications of silicon-containing dendrimers, including those in electrochemistry (of metallo-silicon dendrimers), catalysis and liquid crystalline materials based on dendrimers with mesogenic side groups. Chapter 11 bridges the worlds of silicon and purely organic structures in dendrimer chemistry and it highlights unique radially layered copolymeric PAMAMOS dendrimers which are the first commercially available silicon-containing dendrimers. Chapters 12–15 describe the state of the art in the field of silicon-containing hyperbranched polymers, wherein Chapter 15 addresses the very important issue of intramolecular cyclization. Finally, Chapter 16 describes a rather unconventional method of preparing the hyperbranched polymers by the so-called bimolecular nonlinear polymerization derived from the classic polymer crosslinking technologies and receiving quite an interest lately.

For all these reasons, we are sure that this book will serve as a useful guide and a source of reference for experienced scientists interested in this and related fields, as well as for advanced graduate students either as a source of creative inspiration or as a textbook for appropriate courses. We also believe that it will find its readers in a variety of different interdisciplinary fields, including those of synthetic chemistry, polymer and materials science, silicone chemistry and technology, and particularly nanotechnology.

Finally, we wish to point out that, of course, we are entirely responsible for any conceptual errors or omissions, and in cases where we had to make choices we apologize to those few who we could not invite to contribute because of space limitations. We also wish to thank Springer, our publishing company, and particularly Dr. Sonia Ojo our publishing editor for expert help, understanding, and cooperation during the preparation of the manuscript, and superb realization of the final text during the production process.

Petar R. Dvornic
In Midland, August 2008 Michael J. Owen

Contents

Contributors

Natalia Boiko
Chemistry Department, Moscow State University, Moscow, Russia

Isabel Cuadrado
Departamento de Química Inorgánica, Universidad Autónoma de Madrid, Madrid, Spain

Jianfu Ding
Institute for Chemical Process and Environmental Technology, National Research Council, Ottawa, Ontario, Canada

Petar R. Dvornic
Michigan Molecular Institute, Midland, MI, USA

Holger Frey
Johannes Gutenberg-University Mainz, Mainz, Germany

Daniel Graiver
Department of Chemical Engineering and Material Science, Michigan State University, East Lansing, MI, USA

Katherine J. Haxton
School of Physical and Geographical Sciences, Keele University, Staffordshire, UK

Leonard Interrante
Department of Chemistry and Chemical Biology, Rensselaer Polytechnic Institute, Troy, NY, USA

Chungkyun Kim
Dong-A University, Busan, Korea

Robertus J.M. Klein Gebbink
Chemical Biology & Organic Chemistry, Debye Institute for Nanomaterials Science, Faculty of Science, Utrecht University, Utrecht, The Netherlands

Dale J. Meier
Michigan Molecular Institute, Midland, MI, USA

Russell E. Morris
EaStChem School of Chemistry, University of St. Andrews, St. Andrews, UK

Aziz Muzafarov
Laboratory of Organoelement Polymers Synthesis, Enikolopov Institute of Synthetic Polymer Materials of Academy of Sciences of Russia, Moscow, Russia

Masato Nanjo
Department of Chemistry and Biotechnology, Graduate School of Engineering, Tottori University, Koyamacho-minami Tottori, Japan

Michael J. Owen
Michigan Molecular Institute, Midland, MI, USA

Evgenij Rebrov
Laboratory of Organoelement Polymers Synthesis, Enikolopov Institute of Synthetic Polymer Materials of Academy of Sciences of Russia, Moscow, Russia

Jacques Roovers
Formerly of National Research Council of Canada, Canada

Rakesh Sachdeva
Michigan Molecular Institute, Midland, MI, USA

Hanna Schüle
Institute of Organic Chemistry, Organic and Macromolecular Chemistry, Mainz, Germany

Akira Sekiguchi
Department of Chemistry, Graduate School of Pure and Applied Sciences, University of Tsukuba, Tsukuba, Ibaraki, Japan

Qionghua Shen
Starfire Systems, Inc., Malta, NY, USA

Valery Shibaev
Chemistry Department, Moscow State University, Moscow, Russia

David Y. Son
Department of Chemistry, Southern Methodist University, Dallas, TX, USA

Gerard van Koten
Chemical Biology & Organic Chemistry, Debye Institute for Nanomaterials Science, Faculty of Science, Utrecht University, Utrecht, The Netherlands

Maaike Wander
Chemical Biology & Organic Chemistry, Debye Institute for Nanomaterials Science, Faculty of Science, Utrecht University, Utrecht, The Netherlands

Chapter 1
The Role of Silicon in Dendritic Polymer Chemistry

Petar R. Dvornic and Michael J. Owen

1.1 Introduction

During the last two decades, dendritic polymers, particularly dendrimers and hyperbranched polymers, have become one of the fastest growing areas of interest in polymer science [1]. This can be easily seen from the impressive growth in the number of publications on these unique polymers, which soared from less than a dozen in the 1970s, to over 10,000 (in scientific journals and patent literature) by the end of 2007 [2]. At present, new publications continue to appear regularly, and only the future will tell how much further this trend will continue.

Of many different reasons that may have caused such a great interest in dendritic polymers, the following seem especially important. First, the natural beauty and symmetry of dendritic, particularly dendrimer, structures is hard to resist and it has certainly inspired many scientists to design novel chemical compositions, architectural arrangements and artistic presentations of these unique molecules. Regardless of whether they are shown as simple schematics, or as elaborate computer-generated 3D images, dendritic structures have great aesthetic appeal, and are very inspirational for creative thinking, further modifications or potential applications (see Fig. 1.1).

Second, it has been widely accepted that after traditional linear, crosslinked and lightly branched polymers, dendritic polymers comprise a new class of macromolecular architecture (see Fig. 1.2) which, analogously to the three traditional classes, exhibits its own, unique, architecture-specific properties that are not found in any of the other classes [3, 4]. Among others, such properties include: often unusually low polydispersities and Newtonian flow of dendrimers even at very high molecular weights, encapsulation abilities in host–guest interactions, much smaller hydrodynamic volumes and much more compact molecules than their corresponding linear counterparts, much higher solubilities and much lower viscosities than the corresponding linear polymers of the comparable composition

P.R. Dvornic and M.J. Owen
Michigan Molecular Institute, 1910 W. St. Andrews Rd., Midland, MI, 48640, USA
E-mails: dvornic@mmi.org; michaelowen01@chartermi.net

P.R. Dvornic and M.J. Owen (eds.), *Silicon-Containing Dendritic Polymers*,
Advances in Silicon Science 2, © Springer Science+Business Media B.V. 2009

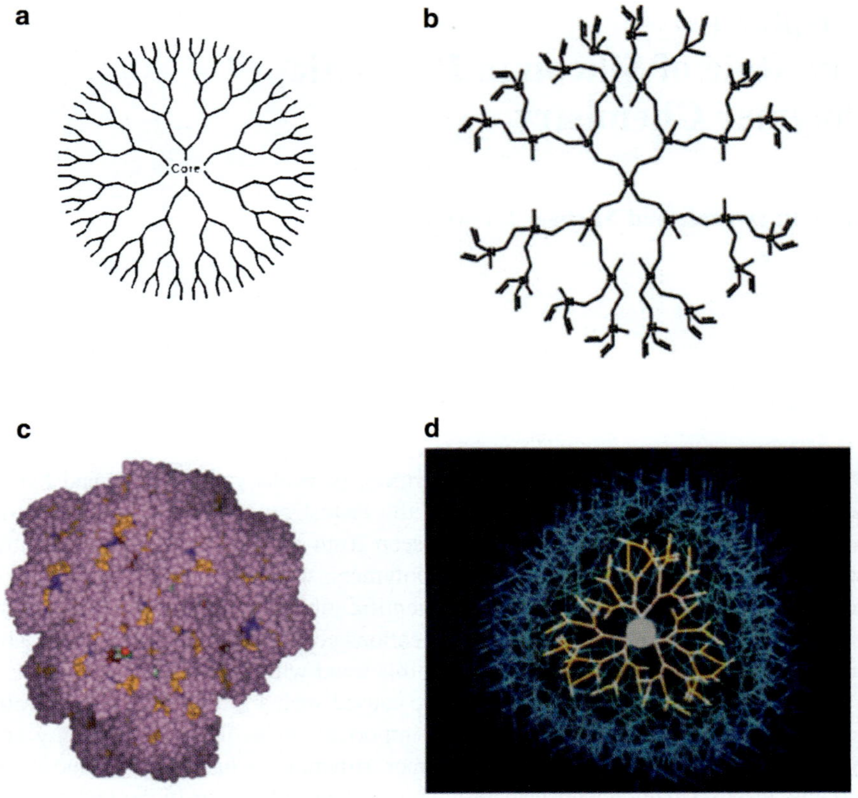

Fig. 1.1 Some of the different ways used to present planar projections of ideal dendrimer structures. **a**: A simplistic, sketchy presentation of a generation three tetradendron dendrimer contour; **b**: a structural formula of the third generation of a tetradendron vinyl-terminated polyethylenesilane; **c**: a CPK or space filling model of a fourth generation tetradendron PAMAMOS dendrimer (see Chapter 11) with trimethylsilyl end-groups; **d**: an idealized representation of a high generation tridendron dendrimer with a sense of three-dimensionality added to the structure.

and molecular weights. Consequently, it is not surprising that because of the attractiveness of these properties, many investigations have been directed at assembling new polymer architectures from well-known, commercially available monomers in a desire to achieve new material properties avoiding traditional, often painstaking, approaches based on the syntheses of compositionally new monomers followed by their polymerizations.

Third, together with fullerenes [5], polyhedral oligosilsesquioxanes (POSS) [6], and carbon nanotubes [7], dendrimers [8] and hyperbranched polymers [9] are among the few known chemical building blocks of truly nanoscopic size (i.e., ranging between 1 and 10 nm in the longest molecular dimension, see Fig. 1.3) that can be used for the directed synthesis of more complex nano-structured materials by controlled bottom-up approaches. As a consequence, with the spectacular rise of nanotechnology in the 1990s and 2000s, this, "chemical" approach to "soft"

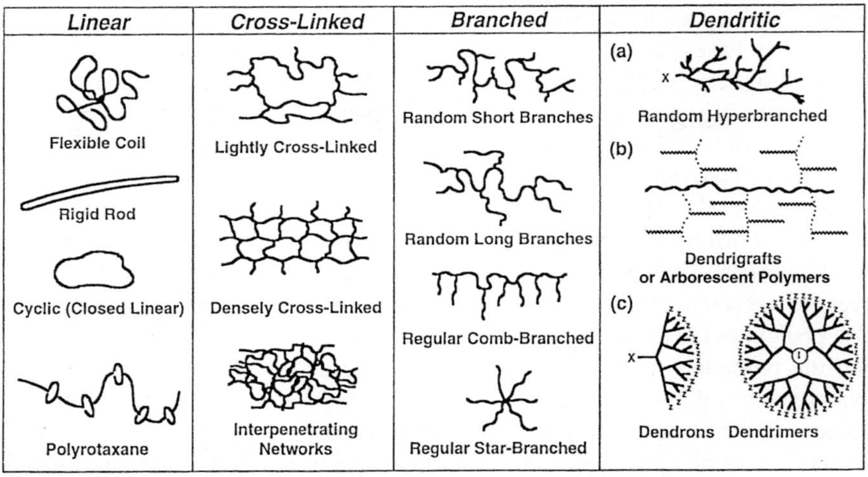

Fig. 1.2 Dendritic polymers as one of the four major classes of macromolecular architecture known today. The columns indicate representative examples of several subclasses that can be identified within each class.

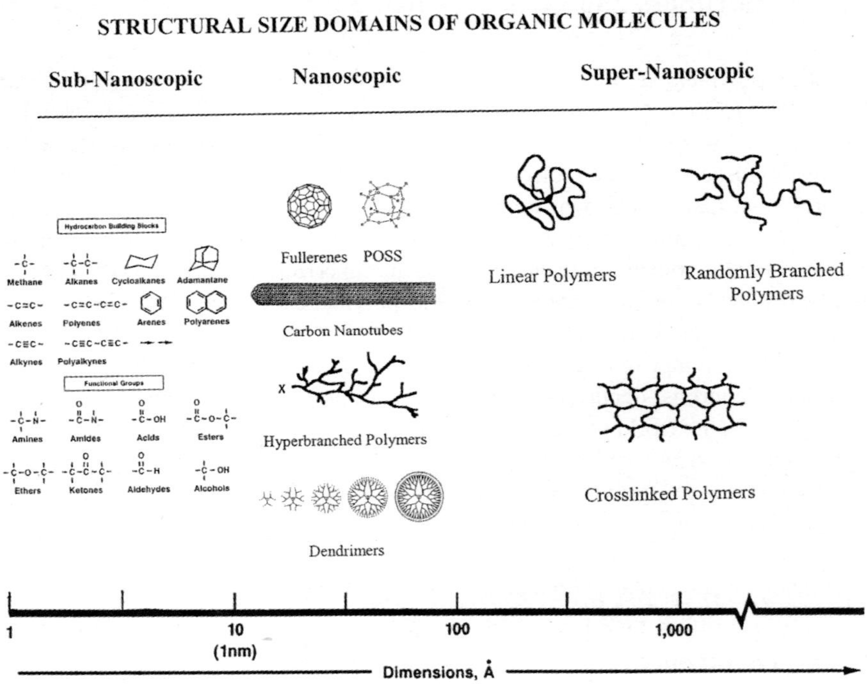

Fig. 1.3 Dimensional hierarchy of organic molecules. Some typical functional groups and carbon nanotubes are also included as important building blocks for more complex material structures. Note that the sizes are not exactly to scale but are in relative order of increased complexity. The sub-nanoscopic size domain "belongs" to traditional organic chemistry, while the super-nanoscopic size domain includes molecules of polymer science.

nanotechnology gained outstanding importance and popularity [10], and within it dendrimers have played a very important and unique role. Namely, while both fullerene spheres and POSS cubes are very precise molecules in both their sizes and shapes, they are also limited in dimensions to a diameter of about 1.5 nm and a diagonal of about 0.5 nm, respectively. In contrast, a generational series of dendrimers (see Fig. 1.5) provides not only preciseness of the molecular diameter of each of its members, but also the ability to select at will the molecular size of the building block by choosing the appropriate dendrimer generation. In essence, this enables selection of the size domain at which the synthetic chemist will exercise control over the structure being built by using dendrimers as building blocks, which, in turn also predefines the resulting properties of the obtained material, specific for the size and the properties of the dendrimer generation used. On the other hand, while hyperbranched polymers cannot compete with dendrimers in molecular uniformity because of their broader distributions of shapes and sizes [9], they can still provide a useful level of size control by the proper monitoring and adjustment of their molecular weights and molecular weight distributions in synthetic strategies that are generally easier and more amenable to large scale preparations.

1.2 Dendrimers and Hyperbranched Polymers

Dendritic polymers are highly branched macromolecules that are based on characteristic "branch-upon-branch-upon-branch" or "cascade-branched" structural motifs (see column 4 in Fig. 1.2) [4, 11]. These motifs represent the single structural feature that distinguishes dendritic molecules from all other types of macromolecular architectures, including traditional branched polymers, where branching is generally occasional (i.e., primary branches, rarely secondary branches and very rarely tertiary branches) and is sometimes even present as an imperfection rather than the defining and repeating structural feature. Dendritic polymers are also almost exclusively synthetic in origin, with a rare exception of naturally occurring amylopectin [12]. In principle, they comprise five major subgroups, including dendrimers, hyperbranched polymers, dendrons, dendrigrafts or arborescent polymers, and dendronized polymers [3, 4]. In the world of silicon polymers only the first two groups have played a significant role so far (see Chapters 2–11 and 12–16 for silicon-containing dendrimers and hyperbranched polymers, respectively). In addition to these, examples of dendronized silicon-containing polymers comprising of a linear polymer backbone with more or less regularly appended dendrons have also been reported as potential candidates for molecular cylinders or nano-wires (see Chapters 3 and 8).

1.2.1 Dendrimers

Dendrimers [8] (see Figs. 1.1 and 1.4 which represent two dimensional projections of the respective three dimensional models) are globular macromolecules that consist of two or more tree-like dendrons emanating from either a single central atom or

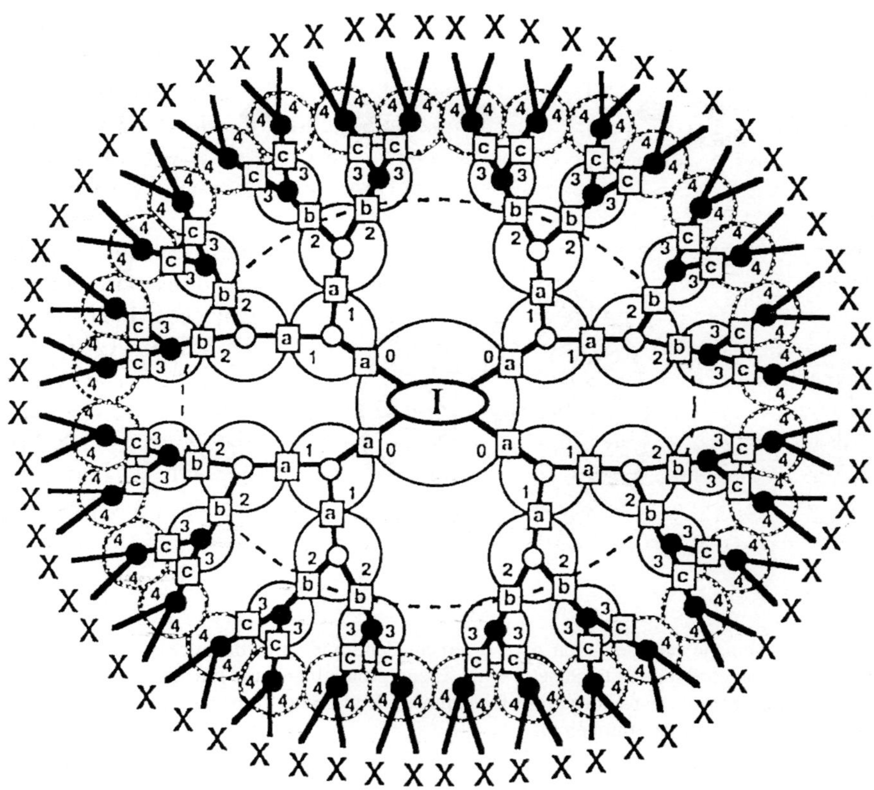

Fig. 1.4 A generalized representation of a tetradendron dendrimer architecture and its constitutive building blocks. An ellipse with an "I" in the center denotes the core. The big circle in the center with four small zeros denotes a generation zero part of the structure. Medium size circles denote dendritic branch cells with Arabic numbers indicating their generation level and small letters in squares denote connecting bonds between the adjacent cells. Small circles (both filled and unfilled) in the center of each cell represent branch junctures which in this case are all of 1→2 branching functionality. Symbols X represent end-groups, which may be chemically reactive or unreactive.

atomic group called the core (denoted I in Fig. 1.4). The main building blocks of dendrimer molecular architecture are the branch cells (represented as large circles with Arabic numerals in Fig. 1.4), which can be considered as three-dimensional dendritic repeat units that always contain at least one branch juncture (represented as small filled or unfilled circles in Fig. 1.4). In an idealized case of complete and perfect connectivity (i.e., in an ideal dendrimer structure) these branch cells are organized in a series of regular, radially concentric layers (called generations and denoted by Arabic numbers in Fig. 1.4) around the core. Each of these layers contains a mathematically precise number of branch cells which increases in a geometrically progressive manner from the core to the dendrimer exterior (often also referred to as the dendrimer surface). The outermost branches end with the end-groups (denoted X in Fig. 1.4), which can be either chemically inert or reactive. Their number depends on the particular dendrimer composition (i.e., the branching

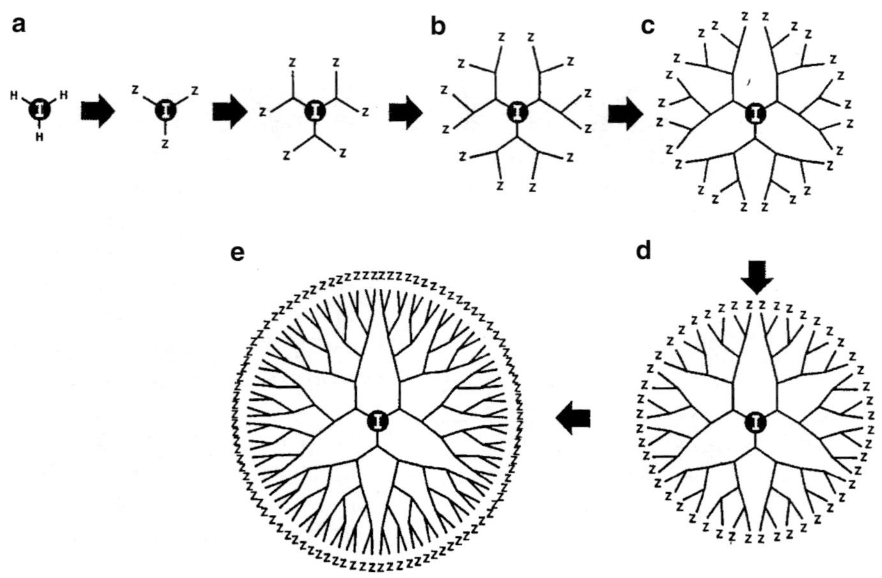

Fig. 1.5 Schematic representation of a divergent dendrimer growth process starting from a three-functional core such as trisilane. I: core; Z: end-group; branching functionality: 1→2. Note the geometrically progressive increase in the number of end-groups from generation to generation. **a** through **e**: generations 1 through 5.

functionality at each generational level) and generation, and may range from only a few (i.e., three or four functional groups of common cores) to several hundreds or even thousands at very high generations [11]. Because of this, high generation dendrimers may be viewed as globular, reactive, nanoscopic building blocks with some of the highest density of functionality known to chemistry.

Dendrimers are synthesized by the repetitive addition of branch cells in the radial molecular direction, one layer or generation (G) at a time, either from the core to the outer surface (divergent method), or from the end-groups to the core (convergent method) [8]. Divergent synthesis [11] (see Fig. 1.5) provides outward dendrimer growth from the core (I) to the "outer molecular surface" and it can be realized by reiterative utilization of two orthogonal chemical reactions as represented by the following generalized scheme:

$$A_x + x\,B\text{-}C_y \;\rightarrow\; A\text{-}(B\text{-}C_y)_x \qquad (1.1)$$

$$A\text{-}(B\text{-}C_y)_x + xy\,D\text{-}A \;\rightarrow\; A\text{-}(B\text{-}C\text{-}D\text{-}A_y)_x \qquad (1.2)$$

reiteration of 1.2 and 1.3

$$A\text{-}(B\text{-}C\text{-}D\text{-}A_y)_x \rightarrow\rightarrow \ldots\ldots\ldots\ldots \rightarrow\rightarrow \text{Dendron (if } x=1\text{) or Dendrimer (if } x \geq 2\text{)} \qquad (1.3)$$

The scheme involves reiteration of two basic reactions, 1.2 and 1.3, in which, under the reaction conditions used, functional groups A can react only with functional groups B and not with functional groups C or D, while functional groups C can react only with functional groups D and not with functional groups A or B. Clearly, this is a work-intensive process since after each completed reaction step the obtained reaction product must be isolated and thoroughly purified before it can be further used in the next reaction (see Reaction 1.3) to continue growth to higher generation products. A classic example of an excellent set of orthogonal reactions for divergent dendrimer synthesis is the Roovers-van Leeuwen combination of hydrosilylation and Grignard addition, which leads to high generation polycarbosilane dendrimers, treated in detail in Chapter 3. Graphically, a divergent dendrimer synthesis can be represented as shown in Fig. 1.5.

Convergent dendrimer synthesis [8b] consists of two main stages. In the first stage, individual dendrons are grown by a process which begins from what will later in the synthesis become the "dendrimer surface" (i.e., end-groups, E) and progresses inward to the focal point (A), as shown in Reaction Schemes 1.4–1.6. This process usually involves a protect–deprotect strategy and it can be represented as follows:

$$y\ E_x\text{-A} + B_y\text{-C} \rightarrow (E_x\text{-A-B})_y\text{-C} \tag{1.4}$$

$$(E_x\text{-A-B})_y\text{-C} + \text{D-A} \rightarrow (E_x\text{-A-B})_y\text{-(C-D)-A} \tag{1.5}$$

reiteration of 1.4 and 1.5

$$(E_x\text{-A-B})_y\text{-(C-D)-A} \rightarrow\rightarrow\ldots\ldots \rightarrow\rightarrow \; |\{[(E_x\text{-A-B})_y\text{-(C-D)-A-B}]_y\text{-(C-D)-A-B}\}_y\text{-}\ldots.|_y\text{-(C-D)-A} \tag{1.6}$$

In the first reaction, a compound carrying the desired dendrimer end-groups and having a reactive "handle" A is reacted with a branching agent B_y-C which contains groups B that react with A and a group C which does not react with A under the A + B reaction conditions and where y is most often equal to 2 or 3. In the next step (Reaction 1.5) group C is converted back into A and the resulting product is then used in another reiteration of Reactions 1.4 and 1.5 to build the next generation dendrons (Reaction Scheme 1.6). Once the desired size of the dendrons is attained these dendrons are coupled together into the dendrimer molecules by using a polyfunctional core or anchoring reagent in what is usually referred to as the anchoring reaction:

$$|\{[(E_x\text{-A-B})_y\text{-(C-D)-A-B}]_y\text{-(C-D)-A-B}\}_y\text{-}\ldots.|_y\text{-(C-D)-A+}B_z\text{-W} \rightarrow \; \langle|\{[(E_x\text{-A-B})_y\text{-(C-D)-A-B}]_y\text{-(C-D)-A-B}\}_y\ldots.|_y\text{-(C-D)-A-}B_z\text{-W} \tag{1.7}$$

The anchoring reaction is an important difference between the divergent and the convergent synthesis. It constitutes an additional synthetic step, in which the anchoring reagent is to become the core W of the newly formed dendrimers, and it must be carried out to completion with respect to the anchor functionality if ideal dendrimers are desired. For that reason the reacting dendrons are usually used in slight excess, but the crude products can be easily purified due to the huge difference in sizes (i.e., molecular weights) of dendrimers and dendrons. Graphically, a convergent dendrimer synthesis can be represented as shown in Fig. 1.6.

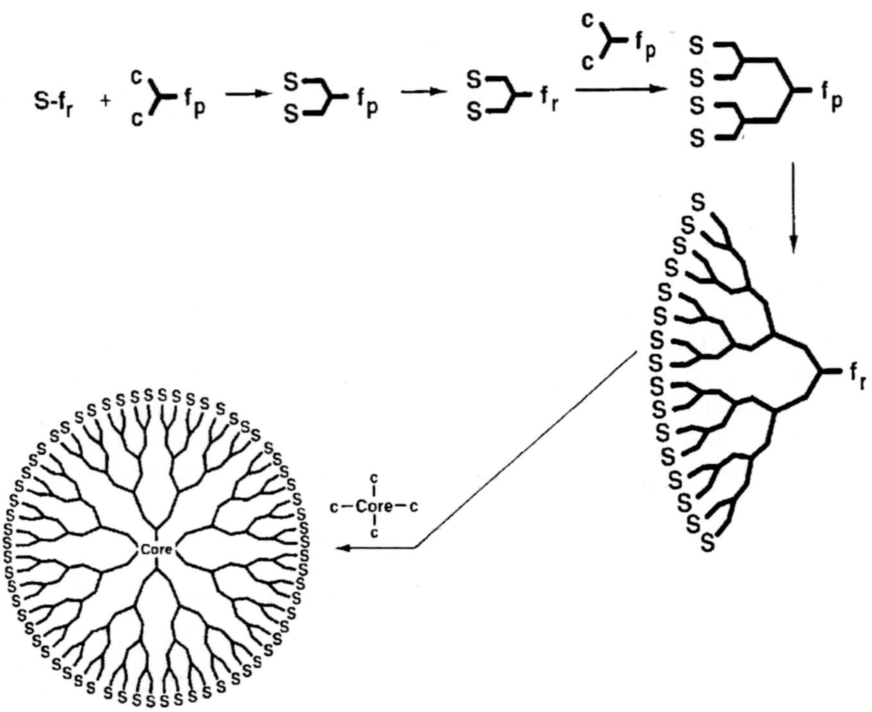

Fig. 1.6 Schematic representation of a convergent dendrimer growth process to a tetradendron dendrimer of the third generation. s: "surface" group; f_r: reactive focal group; f_p: protected focal group; c: functional group of the branching and/or anchoring reagent which reacts with the reactive focal group. Note that the last reaction step represents the anchoring reaction which involves coupling of four preformed dendrons through their reactive focal groups with the anchoring reagent $(Core)c_4$.

It can be seen from Reactions Schemes 1.1–1.7 and Figs. 1.5 and 1.6 that two major differences between the divergent and the convergent dendrimer syntheses include: (a) the direction in which dendritic molecules grow, and (b) the necessity of an anchoring reaction in the convergent synthesis. Thus, while in divergent synthesis the growth process progresses from the core to the outer dendrimer surface (see Fig. 1.5), in the convergent case it occurs in the opposite direction, i.e., from the surface to the focal point (see Fig. 1.6). As a consequence, in a convergent synthesis, the number of chemical events occurring within each growing dendritic molecule is the same in all reiterative steps as long as the same branching reagent is used throughout the entire synthesis. It is thus independent of the generation and equal to the branching functionality of the selected branch cell reagent (y in Reaction Scheme 1.4). In contrast, for the divergent growth process, this number increases geometrically with generation, since it is equal to the number of reactive surface groups of the dendritic reactant (i.e., Z of Fig. 1.5).

Because of this, a convergent growth process should exhibit a constant probability of structural defects formation at all generations (where the term "defect" includes any omitted basic reaction event which results in the local deviation of the produced dendritic structure from ideality [13]), while divergent growth should show this probability geometrically increasing with generation expansion. The most prominent defects normally include the so called "missing arms" which result from incomplete A + B or C + D reactions (as in Reactions 1.1. and 1.4, and Reactions 1.2 and 1.5, respectively) and lead to unreacted sites that remain entrapped within the formed dendritic structures with consequently reduced degrees of branching. While the divergent synthesis should result in a relatively large number of small defects which propagate through the dendrimer structure progressively through generations [13], in the convergent synthesis a relatively small number of such defects may occur at any generation stage leading to "instantaneous" large defects (e.g., as large as half a dendron) at higher generations. This difference may be viewed as a potential advantage for the convergent synthesis, since it generally results in a relatively large molecular weight difference between the ideal and defective dendrimer products which, in turn, should provide for easier separation (e.g., by liquid chromatography) of ideal structures from their defective mutants.

A serious limitation of the convergent dendrimer synthesis, however, is the possibility of reduced "reactivity" of the focal points in higher generation dendrons. This arises from the fact that these molecules may experience severe steric hindrance at their focal points which become "buried" within the bulk of the large developing dendrons. This may significantly reduce the probability of successful collisions between such focal points and the reactive groups of either the branch cell reagents in the growth stage or the anchoring reagent in the anchoring reaction, disrupting the ideality of the growth process and further increasing the probability of defect formation. In both cases (i.e., divergent as well as convergent synthesis), however, the formation of structural defects is always present. It is a rule rather than an exception, and for this reason the beautiful, symmetrical and complete structures by which dendrimers are usually represented in presentations and publications reflect idealization rather than reality. As a consequence, dendrimers are *not ideally monodisperse* macromolecules, although they may come close, often exhibiting very high degrees of monodispersity which in some cases are even higher than those found in the products of the best living anionic polymerization systems.

Another important difference between the divergent and convergent synthesis is that the latter yields monodendrons which opens up some attractive synthetic possibilities for preparation of unique dendritic products. For example, dendrimers with different end-groups in region-specific "sectors" of "hemispheres" of the "outer surfaces" can be prepared by co-anchoring dendrons with different end-groups. Similarly, compositionally (i.e., segmentally copolymeric) or generationally asymmetric dendrimers can be produced by co-anchoring dendrons of different compositions or sizes (i.e., generations). In addition, if the anchoring reagent is a multifunctional linear polymer, this approach also leads to dendronized polymers (for examples see Chapters 3 and 8). Copolymeric dendrimers can also be prepared by divergent synthesis if compositionally different branch cell layers are constructed at different generational levels which leads to the radially layered copolymeric dendrimers, such as those discussed in Chapters 2 and 11. Interestingly, almost all of the silicon-containing dendrimers reported to date are prepared divergently (for some notable exceptions, see Chapters 4 and 8).

With rigorous process control and purification procedures after each iterative growth step, dendrimers can be obtained in very high purities and monodispersities of shapes and sizes, even at truly "polymeric" high molecular weights, which can reach into several hundreds of thousands, or even a million [11]. Their sizes increase with generations in a manner that leads to the crowding of the end-groups, which in turn, forces dendrimers to adopt spheroidal or globular shapes at sufficiently high generations (for a detailed analysis of dendrimer shapes and symmetries see Chapter 1 in Reference 8a). This crowding is particularly important for the divergent synthesis where it leads to a critical growth stage at which the available "surface" area of the dendrimer is smaller than the total area required to pack together all end-groups that need to be introduced into the structure at that generational stage. As a consequence, no further ideal growth is sterically possible at and after this stage which is usually referred to as the de Gennes dense-packed stage as he was the first to realize the necessity of its existence [14].

For most compositions, dendrimer diameters increase in a regular manner with generations in increments of up to 1 nm per generation, from about 1 to about 10 nm [11]. As a consequence, dendrimers represent a truly unique, homologous series of compositionally identical, chemically reactive, spheroidal building blocks that increase in size in a stepwise, controllable manner, and span the very interesting lower domain of the nanoscopic size range (see Figs. 1.3 and 1.5). Because of this, they provide an unprecedented opportunity to manipulate this fundamentally important size domain during the bottom-up synthesis of more complex nanostructures and thus achieve control of the structural organization of matter at the nano-scale, with precision that was not possible before.

1.2.2 Hyperbranched Polymers

Similar to dendrimers, hyperbranched polymers (see Fig. 1.7), also consist of tree-like macromolecules with extensive intramolecular branching and large numbers of end-groups [9]. In silicon chemistry, they have been prepared by two main synthetic approaches: a "traditional" one, based on the polymerization of "branched" monomers of the general type AB_x (see Chapters 5 and 12–14), and a recently "rediscovered" bimolecular (or even multimolecular) nonlinear polymerization, BMNLP, of at least two different monomers of the type A_x and B_y (see Chapter 16). In this notation, in both cases, A and B represent functional groups that can react with each other (i.e., A + B → –A–B–) but do not react with themselves (i.e., A + A → no reaction *and* B + B → no reaction), while x and y are integers such that $x \geq 2$ and $y \geq 3$. For the simplest case of an AB_2 monomer, the traditional, AB_x-type polymerization can be represented as shown in Fig. 1.7 or in Reaction Scheme 1.8:

$$n\,AB_2 \rightarrow \text{-}[A\text{-}B]_n < \qquad (1.8)$$

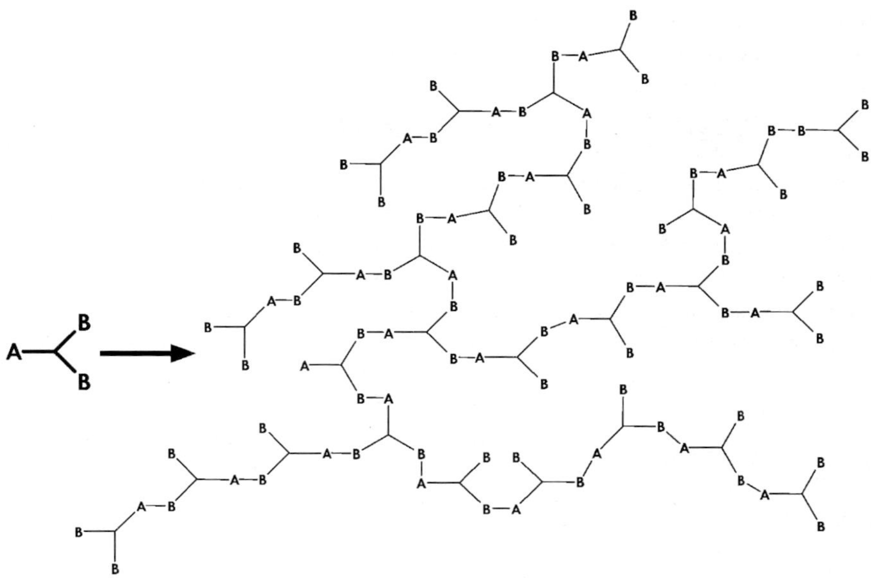

Fig. 1.7 A generalized representation of the formation of a hyperbranched polymer molecule by the coupling of many AB_2 molecules. Note that this is a typical step-growth polymerization reaction in which there is no possibility for polymer gelation, i.e., formation of an infinite network. Also note that in the ideal case each polymer molecule will have many B end-groups and only one A focal group. In reality, however, A almost always reacts with a neighboring B group to create a cyclic structure (for details on cyclization reactions in such systems see Chapter 15).

This is a typical step-growth polymerization process in which growing polymer molecules are composed of –A–B < repeat units which contain 1→2 (or, in a general case 1→x) branch junctures (i.e., branching multiplicity). However, since this process is governed by the statistical occurrence of individual reaction steps, and because of steric hindrance associated with increasing polymer molecular sizes (see Fig. 1.7), it results in different lengths of branches within the polymer molecules, leading to generally broad distributions of molecular weights, sizes and shapes in the resulting polymer products (i.e., polydispersities which can reach into the $M_w/M_n > 10$ range). In addition to this, the usual way of representing such hyperbranched structures (e.g., as used in Fig. 1.7) is fundamentally misleading, since it assumes only intermolecular growth reactions and completely ignores the occurrence of intramolecular cyclization ones (see Chapter 15). In reality, the latter occur almost always and often already at low degrees of conversion, leading to complete disappearance of the A groups, thus eliminating the reactive focal points of the growing polymer molecules and limiting their growth (i.e., molecular weight) [9]. In order to eliminate cyclization and remedy the problems it creates, specific techniques for experimental performance of these polymerization reactions have been developed [9b]. Some of these include the use of sterically relieved monomers (i.e., monomers with longer segments between the reacting functional groups) to reduce the probability of successful ring closure, and the so-called core-seeded polymerizations and the slow monomer addition protocols (see Chapter 13 and 15) both of which enhance the consumption of A groups in the growth reactions rather than in cyclization. In essence, all of these methods aim at enabling preparation of truly high molecular weight products. For silicon-containing hyperbranched polymers, several different chemical reactions have been used in AB_x-type polymerizations (see Chapters 5 and 12–15), but the most successful ones have been the nucleophilic substitution reactions with Grignard or organo-lithium reagents (see Chapter 12) and hydrosilylation (see Chapter 13).

The BMNLP method for the preparation of hyperbranched polymers is discussed in detail in Chapter 16. In principle, for a generalized $A_x + B_y$ system it can be represented as shown in Reaction Scheme 1.9, or for the simplest and the most often used one: $A_2 + B_3$, as shown in Fig. 16.1.

$$a\,A_x + b\,B_y \rightarrow \text{products} \tag{1.9}$$

For a useful reaction chemistry A + B (i.e., one that leads to very high, almost quantitative, functional group conversions, generally higher than 98%), coefficients a, b, x and y must be chosen very carefully, in order to avoid possible gelation and enable preparation of soluble hyperbranched polymer products. Overall, this method has many practical advantages (see Chapter 16 for details), including the large assortment of available monomers, ability to produce polymers of identical composition but with different end-groups (i.e., $[A–B]_n$–type compositions with either A or B end-groups, as opposed to only B end-groups being possible from AB_x systems, see Fig. 1.7), and no monomer shelf-life problems. However, there is an ever-present danger of reaction mixture gelation if the synthesis is not carefully and precisely controlled at all times, and it generally leads to lower molecular weight products than the AB_x approaches.

As a result of the different ways by which dendrimers and hyperbranched polymers are prepared, they closely resemble each other in their basic "branch-upon-branch-upon-branch" structural motifs, their highly branched, tree-like molecular topologies and high molecular density of functionality (compare Figs. 1.7 and 1.5 or 1.6), but they also differ in polydispersity of molecular shapes and sizes, where dendrimers may be considered as highly isomolecular compounds, with molecules of almost perfect sphericity (particularly at higher generations) and very precise sizes, while their hyperbranched relatives are often broadly polydisperse with individual molecules spanning a range of shapes and sizes. Expectedly, these important similarities and differences express themselves in related properties of these dendritic polymer materials and in their preferred applications. For example, since dendrimers are ideal nanoscopic building blocks for the bottom-up preparation of very precise, more complex nano-structures (see Chapter 11) they have attracted outstanding interest for nanotechnology in areas such as electronics, photonics, catalysis and biomedicine (see for example Chapters 3 and 7–11). On the other hand, hyperbranched polymers are significantly more economical and easier to produce, and have, therefore, inspired research as potential substitutes for dendrimers particularly in applications that can tolerate their polydispersity [9c]. Among others, such applications may include topical drug delivery, biotechnical reactor-based processes, various engineering applications including special functional and protective coatings, sensors, decontamination and antifouling surfaces, and construction of biomimetic materials (see Chapters 12–14 and 16). At the moment, however, dendrimers seem to have no competition in the areas that require exceptionally high precision at the nanoscale, including nanolithography, templating, drug delivery, gene therapy, diagnostics, quantum dots preparation, etc.

1.3 A Brief Historical Overview of the Main Developments in Dendrimers and Hyperbranched Polymers

The concept of a dendritic polymer was first described not for dendrimers but for hyperbranched polymers by Paul Flory in the 1950s [15]. In a sense, this was a by-product of his work on gelation theory where he pointed out that a step-growth polymerization of an AB_2 monomer (or in general an AB_x monomer where $x \geq 2$) will yield a soluble high molecular weight product, which we now call a hyperbranched polymer, if two conditions are satisfied: (1) reactive functional groups A and B will react only with one another and not each with itself, and (2) there are no intramolecular cyclization reactions occurring in the system (see Fig. 1.7).

It should be noted, however, that the earliest realization that transient branched molecules are formed during the polymerization of a difunctional and a trifunctional monomer (e.g., phthalic anhydride and glycerol) can be attributed to Kienle and co-workers from Rutgers University in 1929–1939 [16]. These authors did not call the resulting structures hyperbranched (the name caught on only in the late 1980s and early 1990s), nor polymers (in those days Staudinger was still fighting for his "macromolecular hypothesis"), but rather they termed them "complex three-dimensional molecules". However, they clearly understood their branched character consisting of mono-connected (what we call today end-groups), di-connected (linear), and tri-connected (intramolecularly branched and/or looped), repeat units [16b]. They even determined molecular weights of these transient species and showed that once they reached about 1,100 the viscosity of the reaction mixture containing stoichiometric amounts of the reacting functionalities rapidly increased toward gelation (see also Chapter 16).

Following Flory, the next major contribution to the field of dendritic polymers came in 1978 with the development of the so-called cascade synthesis by Fritz Vögtle and his coworkers at the University of Berlin [17]. Although these authors focused on constructing "molecular cavities" rather than dendritic structures *per se*, they conceived a synthetic strategy [18a, b] that later paved the way for the development of poly(propylene imine) dendrimers by E. W. (Bert) Meijer and his group at the Eindhoven University of Technology and DSM Corporation [18c].

The discovery of dendrimers, however, can be attributed to Robert Denkwalter and his coworkers at Allied Corporation, who in two patents (filed in 1979 and 1981, and granted in 1981 and 1983, respectively) described the synthesis of the first homologous series of dendrimers, the polylysines, via a protect–deprotect divergent strategy [19], up to generation 8 [20]. A few years later, in 1985, George Newcome and his colleagues at the University of South Florida [21] and a group at the Dow Chemical Corporation led by Donald Tomalia [22] reported the synthesis of arborols and polyamidoamine (PAMAM) dendrimers, respectively. Of these, the PAMAMs later became the most studied and the best characterized dendrimers of all, because they were commercially introduced by Dendritech Inc., a company spun off in 1992 from the Michigan Molecular Institute in Midland, Michigan, as the first (and still the largest) producer of dendrimers in the world [23].

In 1988, Kim and Webster of Du Pont de Nemours [24], and in 1990, Miller and Neenan of AT&T Bell Laboratories [25], inspired by the fast growing scientific popularity of dendrimers, revisited Flory's early concepts of the 1950s and reported on the first successful syntheses of hyperbranched polyarylates and aromatic polyamides, respectively, from AB_2 monomers (see Fig. 1.7). At first, this breakthrough raised hopes that hyperbranched polymers would provide a "poor man's" alternative to dendrimers. However, it was soon realized that although hyperbranched polymers could clearly replace dendrimers in some potential applications, there were still many areas where dendrimer perfection resulted in unprecedented properties that are not matched by any other type of molecular architecture, including the hyperbranched one. Some of these include biomedical applications (e.g., diagnostics, gene therapy, drug delivery, etc.) [26], chemical applications such as catalysis (see Chapter 9), bottom-up synthesis in nanotechnology (see Chapters 10 and 11), MEMS, rheology [27], and advanced engineering areas such as electronics and photonics.

The next breakthrough in the field was made by Hawker and Fréchet, who in 1990 developed a convergent synthesis of polybenzylether dendrimers [28]. This new approach enabled preparation of dendrimers with compositionally and/or generationally different segments emanating from the same core as well as precise control of the type and density of end-group functionality, where even dendrimers with a single functional group different from all others could be prepared. The same authors also introduced the concept of copolymeric dendrimers having more than one compositional type of branch cell layers in a single molecule [29]. Following this, in 1998, Russo and Boulares reported on the first synthesis of a hyperbranched polymer, an aromatic polyamide, by a bimolecular nonlinear polymerization (BMNLP) from an A_2 and a B_3 or, in general, $A_x + B_y$ monomer (in their case di- and tri-functional amines and acids, respectively) [30]. This approach (see Chapter 16) offers the potential to greatly expand on Flory's original concept and avoids the need for rare and expensive AB_x monomers, which often have limited shelf lives. In addition, because of the abundance of commercially available A_x and B_y monomers, this approach also offers the possibility of practically unlimited polymer compositions, a versatility restricted only by the synthetic chemist's imagination much in the same way that the copolymerization principle brought to the field of polymer synthesis. Most of all, it provides a real possibility to make dendritic polymers economically viable for large scale engineering applications because it can utilize very cheap commercial monomers in economical one-shift processes.

1.4 Silicon in Dendritic Polymers

In analogy with its widespread use in polymer chemistry of traditional linear, crosslinked and lightly branched macromolecules, silicon has also assumed a dominant role in the field of heteroatom-based dendritic polymers. Many excellent reviews describe earlier efforts in various areas of this field [31]. Such interest in silicon-containing dendritic polymers resulted from a variety of reasons of which the following appear to be the most prominent ones. First, it is certainly the high level of maturity that organosilicon chemistry has reached during the last century that provided an abundance of commercially available organic and inorganic silicon compounds and monomers that are suitable for use in the preparation of novel silicon-containing dendritic polymers. For example, it was estimated that in the year 2000 the total production of silicone polymers alone reached over 10^6 t/year [32] yielding a variety of different materials with unique combinations of exceptionally high elasticity, high thermal, thermo-oxidative and electrical stability, and very low surface energy that is not common to any other class of synthetic or natural materials. Together with the fact that after oxygen, silicon is the second most abundant element of the Earth's crust, this acquired knowledge and the unique properties of silicon compounds provide strong reasons to expect that their production and use, and with them also the interest in silicon chemistry, will only continue to grow in the future.

Second, silicon chemistry provides an abundance of very high yield synthetic reactions, without which the construction of architecturally perfect dendrimer molecules is not possible (see Section 1.2). This is required by the generational nature of dendritic growth processes, where formation of higher generation dendrimers necessitates complete transformation of a large number of end-groups per molecule, which condition in silicon chemistry is met by a number of well-controlled and reliable reactions, such as hydrosilylation, Grignard reaction, controlled condensation of silanols, and aminolysis of chlorosilanes.

Third, of particular relevance for dendritic polymer chemistry is the fact that silicon can play a role of either a di- ($1\rightarrow2$) or a tri-($1\rightarrow3$) functional branch juncture and in each case its functionality can be easily tailored by appropriate selection of the reagent used in the synthesis (see Chapter 3). This enables controlled preparation of dendritic polymers with very similar compositions but different branching densities (one branch in – two branches out: –Si(R) < or one branch in – three branches out: $-Si\equiv$), which can result in substantial differences in the resulting polymer properties, including the hydrodynamic diameters and the maximum molecular size attainable (e.g., the de Gennes dense packed stage), encapsulation ability (i.e., intramolecular density and available free volume), molecular end-group density and ease of end-groups conversion to other functionalities.

Fourth, most silicon-containing dendritic polymers exhibit the same or a very similar unique combination of highly desirable physical properties for which their traditional linear counterparts are very well known. These properties primarily include: unusually pronounced low temperature flexibility (manifested through some of the lowest glass temperatures known to polymer science), high thermal and thermo-oxidative stability, very low surface energies (second only to fluorocarbon polymers) and small viscosity-temperature coefficients. And when to this list one adds pronounced chemical inertness of some of the most prominent silicon scaffolds (such as polycarbosilanes or polysiloxanes for example) and a variety of other extremely useful properties contributed by the highly branched dendritic architecture, including Newtonian flow of both solutions and bulk, spheroidal shape, nanoscopic size, a variety of different end-groups available for further modification (such as: Si–H, Si–Cl, Si–OH, Si–OR, Si–CH=CH_2, Si–CH_2–CH=CH_2, etc.), extremely high density of molecular functionality at high generations, etc., it is not hard to envision a variety of potential applications that are opening up for these new polymers in areas such as nanotechnology, surface science, bio-medicine, catalysis, electrochemistry, liquid crystallinity, new ceramics materials, organo-inorganic hybrids and nanocomposites, etc. Many of these will clearly play dominant roles in the development of science in the twenty first century, and we have no doubt that silicon will play a prominent role in all of them as well.

The first silicon-containing dendritic polymer was reported in 1989 by Evgenij Rebrov and Aziz Muzafarov and their co-workers from the Institute of Synthetic Polymer Materials of the Russian Academy of Sciences in Moscow [33]. They synthesized four generations of polymethylsiloxane dendrimers with trifunctional silicon branch junctures using a diethoxysodiummethylsilanolate-based divergent growth strategy. Interestingly, this dendrimer analogue of polydimethylsiloxane (PDMS), the most important silicon-containing polymer of them all, was the first silicon-containing dendritic polymer reported, although the quest for the best synthesis of polysiloxane dendrimers does not yet seem to be over (see Chapter 2 for details). Following this, the same group continued to report on a variety of different siloxane-carbosilane copolymeric dendrimers, but their 1989 paper remains the seminal contribution that started the silicon-containing dendritic polymers era and did so in the early days of the development of the overall field of dendritic polymers. In fact, this paper appeared only 4 years after the first dendrimer papers in scientific journals [21, 22] and 1 year before the first report on the convergent dendrimer synthesis [28]. In Chapter 2 of this book Dr. Rebrov and Professor Muzafarov describe the developments in polysiloxane dendrimers from their early efforts until the present day.

The most utilized, and the most versatile synthetic scheme for the preparation of silicon-containing dendrimers so far is the reiterative series of hydrosilylation and Grignard reactions that leads to carbosilane dendrimers [34]. Impressive developments in this most prolific area of silicon-containing dendrimers are presented in Chapter 3 by one of its pioneers, Professor Roovers, and his colleague Dr. Ding. Note that because of: (a) the high level of reproducibility and synthetic control to which the chemistry of polycarbosilane dendrimers has been brought to date, (b) the ability to extend their synthesis to truly high generations (e.g., generation five and higher) with several tens or hundreds of different end-groups per molecule, (c) the high solubility of these dendrimers in a variety of useful solvents and relative ease of their characterization, and (d) the inertness of their scaffolds to a variety of environments and conditions, polycarbosilane dendrimers have opened the doors to some of the most investigated applications of silicon-containing dendrimers. Among others, their most prominent applications to date, in electrochemistry, catalysis and liquid crystalline systems are described in detail in Chapters 8–10, respectively.

Following polycarbosilanes, a series of more "exotic" silicon-containing dendrimer structures are presented in the next four chapters. First, two other main classes: polysilane [35] and polysilazane dendrimers [36] are discussed in Chapters 4 and 5, by three of their respective originators: Professors Nanjo and Sekiguchi of the Tottori University and University of Tsukuba in Japan (polysilanes), and Professor Son of the Southern Methodist University, Texas (polysilazanes), respectively. Next, in Chapters 6 and 7, respectively, dendrimers with cyclic siloxane cores [37], and POSS-containing dendrimers [38], are described by their respective pioneers, Professor Kim of the Dong-A University in Pusan, Korea, and Professor Morris of the University of St. Andrews in St. Andrews, Scotland and his colleague Dr. Haxton.

Some of the most promising applications of silicon-containing dendrimers to date seem to be in the areas of electrode coatings in analytical electrochemistry [39], in catalysis [40] and in a completely new type of architecturally copolymeric liquid crystalline materials the molecules of which are composed of globular (!) and linear parts [41]. While the first mentioned one is based on various metal-containing silicon dendrimers (most prominently their ferrocenylene derivatives), and is described in Chapter 8 by the pioneer in this field, Professor Isabel Cuadrado of the Autonomous University of Madrid in Spain, dendrimers in catalysis are treated in Chapter 9 by a group of authors including Professor Klein Gebbink and Dr. Wander led by the "father of dendrimer catalysts", Professor Gerard van Koten of the University of Utrecht, Holland. In our opinion, dendrimer-based catalysts will eventually lead to the realization of the ultimate dream of catalysis chemists to combine the best features of homogeneous and heterogeneous catalysis in one and thus achieve ideal catalysis showing the best of both worlds [42]. In Chapter 10, liquid crystalline, mesogen-modified silicon-containing dendrimers are discussed by Professor Shibaev and Dr. Boiko of the Moscow State University, Russia, and in Chapter 11 we conclude the dendrimer part of this book by presenting dendrimers which contain silicon in otherwise organic compositions and the first radially layered copolymeric silicon-containing dendrimers known as poly(amidoamine-organosilicon)s (PAMAMOS) [43], which are also the first and still the only silicon-containing dendrimers that are commercially available since 2003 [23].

The second part of this book is devoted to the silicon-containing hyperbranched polymers. Similar to the situation in silicon-containing dendrimers, this field has been heavily dominated by the hyperbranched polycarbosilanes and, therefore, this part of the book begins with two chapters on these hyperbranched polymers. The first of these describes their preparation by nucleophilic substitution reactions (Chapter 12) [44], presented by the pioneer in this area and the scientist who described the very first silicon-containing hyperbranched polymer, Professor Leonard Interrante of the Rensselaer Polytechnic Institute in Troy, New York, and Dr. Shen, while the second treats their preparation by hydrosilylation (Chapter 13) [45] presented by Professor Frey and Dr. Schülle of the University of Mainz, Germany. Interrante's technology has led to the first commercial production of silicon-containing hyperbranched polymers [46] which is one of only a few commercial productions of hyperbranched polymers available worldwide today, while Frey's work on hyperbranched polycarbosilanes by hydrosilylation has led not only to new methods of making hyperbranched polymers in general with an unprecedented level of synthetic control and degrees of branching, but also to the better theoretical understanding of the structure of hyperbranched polymers and relationships that govern the development of that structure.

Following this, in Chapter 14 Dr. Graiver of Michigan State University, East Lansing, Michigan, describes a very interesting chemistry of rearranging hyperbranched poly(silyl-ethers) [47], and finally, this book concludes with two chapters that are fundamentally important for both the theory and practice of hyperbranched polymer chemistry in general, including the cyclization issues in hyperbranching (Chapter 15) and bimolecular non-linear polymerization (BMNLP) (Chapter 16) [48]. The former is another contribution to this book by Professor Son, while the latter presents the technology developed during the last 8 years for the preparation of silicon-containing hyperbranched polymers by this alternative approach to hyperbranching at the Michigan Molecular Institute in Midland, Michigan. With respect to BMNLP, it is interesting to note that this highly versatile "new" method for dendritic polymers preparation is actually older than all other ones, because it draws its origins from the initial Flory's theory of gelation dating back to the mid 1950s. Its "rediscovery" in modern days, however, offers serious promise that economical production of a variety of hyperbranched polymers of vastly different chemical compositions based on a broad assortment of easily available commercial monomers may indeed be possible. If this turns out to be so, the entire field of dendritic polymers, including that of silicon-containing ones, will undoubtedly receive a considerable boost that could trigger another burst of interest and further development. Only the future will tell.

References

1. In principle, the class of dendritic polymers also includes dendrons, arborescent polymers (sometimes also referred to as dendrigrafts) and dendronized polymers, but (and especially in the silicon-containing dendritic polymers field) the interest in dendrimers and hyperbranched polymers has been disproportionally more pronounced.
2. A SciFinder search of the field in November of 2007 revealed 11,844 references containing the concept of dendrimers and 13,769 references containing the concept of hyperbranched polymers.
3. Dvornic PR, Tomalia DA (1996) Curr Opin Colloid Interface Sci 1:221.
4. Dvornic PR, Tomalia DA (1996) Sci Spectra 5:36.
5. See for example: Kadish KM, Ruoff RS (2000) Fullerenes: Chemistry, physics, and technology. Wiley-Interscience, New York.
6. See for example: (a) Voronkov MG, Lavrent'yev VI (1982) Top Curr Chem 102:199. (b) Feher FJ (2000) A survey of properties and chemistry: Polyhedral oligosilsesquioxanes and heterosilsesquioxanes, Gelest Inc. Catalog, Tullytown, PA, pp. 43–59. (c) Lichtenhahn JD (1996) Silsesquioxane-based polymers. In: Salamone JC (ed) Polymeric materials encyclopedia. CRC Press, New York, Vol. 10, pp. 7768–7778.
7. Ijima S (1991) Nature 354:56.
8. See for example: (a) Newkome GR, Moorefield CN, Vögtle F (2001) Dendrimers and dendrons. Concepts, synthesis, applications, Wiley-VCH Verlag, Weinheim, Germany. (b) Fréchet JMJ, Tomalia DA (eds) (2001) Dendrimers and other dendritic polymers. Wiley, Chichester.

9. See for example: (a) Mishra MK, Kobayashi S (1999) Star and hyperbranched polymers. Marcel Dekker, New York. (b) Sunder A, Heinemann J, Frey H (2000) Chem Eur J 6(14):2499. (c) Hult A (2003) Hyperbranched polymers. In: Mark HF (ed) Encyclopedia of polymer science and technology, 3rd edn. Wiley-Interscience, New York, Vol. 2, pp. 722–743. (d) Gao C, Yan D (2004) Prog Polym Sci 29(3):183.
10. Descriptor "soft" is used here to distinguish nano-structured materials and products of synthetic organic chemistry, usually prepared by controlled bottom-up synthesis from sub-nanoscopic small molecules, from "hard" counterparts of inorganic chemistry and metallurgy prepared either by the bottom-up manipulations of single atoms or by the top-down grinding of macroscopic samples.
11. Tomalia DA, Dvornic PR (1996) Dendritic polymers, divergent synthesis (Starburst polyamidoamine dendrimers). In: Salamone JC (ed) Polymeric materials encyclopedia. CRC Press, Boca Raton, FL, Vol. 3, pp. 1814–1830.
12. Tomalia DA, Hedstrand DM, Wilson LR (1990) Dendritic polymers. In: Mark HF, Bikales NM, Overberger CG, Menges G (eds) Encyclopedia of polymer science and engineering. Wiley-Interscience, Vol. Index Volume, pp. 46–92.
13. Dvornic PR, Tomalia DA (1995) Macromol Symp 98:403.
14. de Gennes PG, Hervet H (1983) Phys Lett 44:L351.
15. (a) Flory PJ (1952) J Am Chem Soc 74:2718. (b) Flory PJ (1953) Ann NY Acad Sci 57(4):327.
16. (a) Kienle RH, Hovey AG (1929) J Am Chem Soc 51:509. (b) Kienle RH, van der Meulen PA, Petke FE (1939) J Am Chem Soc 61:2258.
17. Buhleier E, Wehner W, Vögtle F (1978) Synthesis 2:155.
18. (a) Moors R, Vögtle F (1993) Chem Ber 126:2133. (b) Wörner C, Mülhaupt R (1993) Angew Chem Int Ed Engl 32(9):1306. (c) de Brabander-van der Berg EMM, Meijer EW (1993) Angew Chem Int Ed Engl 32(9):1308.
19. (a) Denkewalter RG, Kolc JF, Lukasavage WJ (1981) US Patent 4,289,872. (b) Denkewalter RG, Kolc JF, Lukasavage WJ (1983) US Patent 4,410,688.
20. (a) Aharoni SM, Crosby III CR, Walsh EK (1982) Macromolecules 15:1093. (b) Aharoni SM, Murthy NS (1983) Polym Commun 24:132.
21. Newkome GR, Yao Z, Baker GR, Gupta VK (1985) J Org Chem 50:2003.
22. (a) Tomalia DA, Dewald JR (1985) US Patent 4,507,466. (b) Tomalia DA, Baker H, Dewald JR, Hall M, Kallos G, Martin S, Roeck J, Ryder J, Smith P (1985) Polym J (Tokyo) 17(1):117.
23. http://www.dendritech.com
24. Kim YH, Webster OW (1988) Polym Preprints 29(2):310.
25. Miller TM, Neenan TX (1990) Chem Mat 2(4):346.
26. Majoros I, Baker JR, Jr (eds) (2008) Dendrimer based nanomedicine. Pan Stanford, Singapore.
27. Dvornic PR, Uppuluri SE (2001) Rheology and solution properties of dendrimers, Chapter 14 in [8b], pp. 331–360.
28. (a) Hawker CJ, Fréchet JMJ (1990) J Am Chem Soc 112:7638. (b) Hawker CJ, Fréchet JMJ (1990) J Chem Soc Chem Commun 1010.
29. Hawker CJ, Fréchet JMJ (1995) Three-dimensional dendritic macromolecules: Design, synthesis and properties. In: Ebdon JR, Eastmond GC (eds) New methods of polymer synthesis. Blackie Academic and Professional, London, UK, Vol. 2, pp. 290–330.
30. (a) Russo S, Boulares A (1998) Macromol Symp 128:13. (b) Russo S, Boulares A, da Rin A (1999) Macromol Symp 143:309.
31. For excellent earlier reviews in silicon-containing dendritic polymers see for example: (a) Neumann D, Matisons JG (2003) Multiple roles of silicon in dendritic chemistry; Chapter 8 in Nalwa HS (ed) Handbook of organic–inorganic hybrid materials and nanocomposites. American Scientific Publishers, Stevenson Ranch, CA, Vol. 2, pp. 295–330. (b) Lukevicz E, Arsenyan P, Pudova O (2002) Main Group Met Chem 25:135. (c) Lang H, Lühmann B (2001) Adv Mater 13:1523. (d) Frey H, Schlenk C (2000) Top Curr Chem 210:69. (e) Krska SW, Son D, Seyferth D (2000) Organosilicon dendrimers, Chapter 23 in Jones RG et al. (eds) Silicon-containing polymers. Kluwer, Amsterdam, pp. 615–641. (f) Son DY (2000) Main Group Chem News 7:16. (g) Majoral J-P, Caminade A-M (1999) Chem Rev 99:845.

32. Brook MA (2000) Silicon in organic, organometallic and polymer chemistry. Wiley, New York, p. 4.
33. Rebrov EA, Muzafarov AM, Papkov VS, Zhdanov AA (1989) Dokl Akad Nauk SSSR 309:376.
34. (a) van der Made AW, van Leeuwen PWNM (1992) J Chem Soc Chem Commun 1400. (b) Roovers J, Toporowski PM, Zhou L-L (1992) Polym Preprints 33(1):182.
35. (a) Sekiguchi A, Nanjo M, Kabuto C, Sakurai H (1995) J Am Chem Soc 117:4195. (b) Lambert JB, Pflug JL, Stern CL (1995) Angew Chem Int Ed Engl 34(1):98. (c) Suzuki H, Kimata Y, Satoh S, Kuriyama A (1995) Chem Lett 293.
36. Hu J, Son DY (1998) Macromolecules 31:8644.
37. (a) Kim C, Choi S-K (1997) Main Group Met Chem 20(3):143. (b) Kim C, An K (1997) Bull Korean Chem Soc 18(2):164.
38. Jaffrès P-A, Morris RE (1998) J Chem Soc Dalton Trans 2767.
39. Alonso B, Cuadrado I, Morán M, Losada J (1994) J Chem Soc Chem Commun 2575.
40. Knapen JWJ, van der Made AW, de Wilde JC, van Leeuwen PWNM, Wijkens P, Grove DM, van Koten G (1994) Nature 372:659.
41. Ponomarenko SA, Rebrov EA, Boiko NI, Vasilenko NV, Muzafarov AM, Freidzon YAS, Shibaev VP (1994) Polym Sci A36:896.
42. Tomalia DA, Dvornic PR (1994) Nature 372:617.
43. (a) de Leuze-Jallouli AM, Swanson DR, Perz SV, Owen MJ, Dvornic PR (1997) Polym Mat Sci Ed 77:67. (b) Dvornic PR, de Leuze-Jallouli AM, Swanson D, Owen MJ, Perz SV (1998) US Patent 5,739,218. (c) Dvornic PR, de Leuze-Jallouli AM, Owen MJ, Perz SV (2000) Macromolecules 33:5366.
44. Whitmarsh CW, Interrante LV (1991) Organometallics 10:1336.
45. (a) Mathias LJ, Carothers TW (1991) J Am Chem Soc 113:4043. (b) Lach C, Frey H (1998) Macromolecules 31:2381.
46. http://www.starfiresystems.com
47. Decker GT, Graiver D, Tselepis AJ (1999) US Patent 6,001,945.
48. (a) Dvornic PR, Hu J, Meier DJ, Nowak RM (2002) US Patent 6,384,172 B1. (b) Dvornic PR, Hu J, Meier DJ, Nowak RM (2003) US Patent 6,646,089 B2.

Chapter 2
Polysiloxane and Siloxane-Based Dendrimers

Aziz Muzafarov and Evgenij Rebrov

2.1 Introduction: Historical Background

Three different original protocols for the synthesis of siloxane dendrimers have been described. However, after appearing in the literature at much the same time they did not result in any follow-up, unlike many other strategies proposed simultaneously or even later. To appreciate the suddenness of the introduction of siloxane dendrimers and the reasons for their subsequent neglect we need to make an excursion into organosiloxane chemistry. Silicones are based on a relatively small number of starting functional monomers. Moreover, their combination into new structures is restricted by the reversibility of most reactions and as a consequence by the statistical character of the main methods for the synthesis of siloxane polymers [1–3]. Although the existence of some selectivity results in the formation of useful polycyclics, the possibilities of structural control of linear and especially of branched oligomers and polymers are seriously limited. The nonequilibrium polymerization of strained cyclosiloxanes [4] is a rare exception to this rule which emphasizes the hopelessness of attempts to overcome the "statistical ill fate" that haunts the siloxane structures. All efforts to apply heterofunctional processes to achieve structural diversity are limited by this specificity of the siloxane bond and by a scanty number of appropriate reagents.

Because of this, it is not surprising that after the discovery of sodiumoxyorganoalkoxysilanes, compounds unique in their synthetic versatility and often called Rebrov's salts [5], the branched functional oligomers and acyclic methylsilsesquioxanes came into the focus of our research attention. Furthermore, most of the oligomers obtained using this approach became the first generations of the corresponding dendrimers, but only after the finding of an appropriate method for quantitative transformation of alkoxy groups into the chlorosilyl ones, and a way to reiteratively use them in pairs did it become evident that possibilities of the new approach for dendrimer synthesis were enormous (see Chapter 1). By that time, dendrimers had become well known in the field of organic polymers, largely due to the efforts of D. Tomalia [6, 7].

A. Muzafarov and E. Rebrov
Laboratory of Organoelement Polymers Synthesis, Enikolopov Institute of Synthetic Polymer Materials of Russian Academy of Sciences, Profsoyuznaya Street, 70, Moscow 117393, Russia
E-mail: aziz@ispm.ru

P.R. Dvornic and M.J. Owen (eds.), *Silicon-Containing Dendritic Polymers*,
Advances in Silicon Science 2, © Springer Science+Business Media B.V. 2009

In 1989, the abstract of our second paper entitled "*Organosiloxane dendrimers*" appeared in Chemical Abstracts and provided the keyword for introduction of silicon into the dendrimer community [8]. Aiming at the synthesis of acyclic polymethylsilsesquioxanes we wanted to emphasize the appearance of a new level of control over the siloxane polymer structure. Soon thereafter, two other protocols for the preparation of siloxane dendrimers were also proposed, probably inspired by one of Tomalia's reports at a conference in Japan [9, 10]. This, almost simultaneous, appearance of three original methods for the siloxane dendrimer synthesis signaled the great potential of this new field at the time of the emergence of dendrimers in general (see Chapter 1).

2.2 Chemistry of Siloxane Dendrimers

The first synthetic route to siloxane dendrimers was characterized by its simplicity and allowed formation of one of the densest dendrimer structures reported [8]. This synthesis of methylsilsesquioxane dendrimers is shown in Reaction Scheme 2.1.

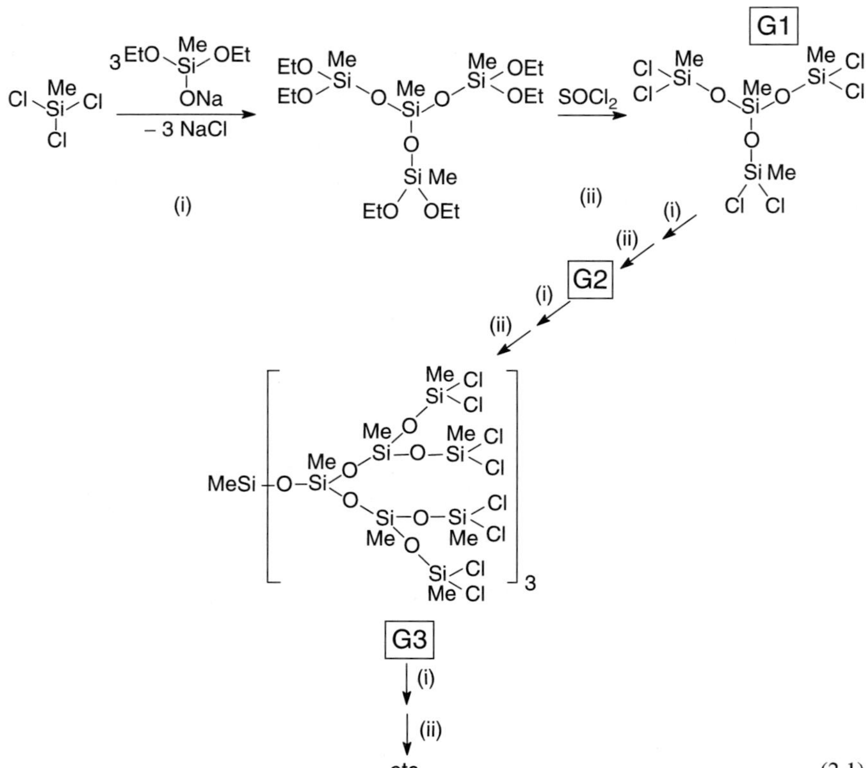

(2.1)

Unlike the common sodium silanolates, the main reagent used in this synthesis, sodiumoxymethyldiethoxysilane, was readily soluble in the majority of organic solvents, including nonpolar alkanes and aromatic hydrocarbons, which greatly facilitated its use and preparation. The main problem with the strategy, however, was that both the resulting siloxane dendrimer skeleton and its chlorosilane end-groups were sensitive to nucleophilic reagents. As a consequence, the major goal of our further study became to find conditions that would exploit the difference in reactivity of the chlorosilane and siloxane bonds towards the sodiumoxy groups of the branching reagent.

However, this reaction scheme was published without synthetic details and for this reason it was not always taken seriously although it was solidly supported by the results of structural characterization of at least three dendrimer generations which were isolated, purified by rectification in vacuum and individually characterized. Furthermore, the absence of a reliable blocking method made purification of higher generations difficult, so that the reported data for the fourth and higher generations were those of raw reaction products.

In a later work [11], we focused on the development of combinatorial possibilities of our protocol, and a synthetic platform that included a set of reagents and synthetic approaches that enabled fine regulation of dendrimer structure to eventually emerge. Unfortunately, we did not go beyond the simplest models, although development of the proposed methods could have given an opportunity for a wide comparison of structure–property relationships for this type of dendrimers. Today, we can only hope that the recent publication of the details of the synthesis of sodiumoxy-derivatives and the study of their properties [12] may stimulate further use of this route to siloxane dendrimers.

The second protocol for the synthesis of siloxane dendrimers [13–15] was based on the use of telomeric dimethylsiloxane oligomers and a methylchlorosilane branching agent, as shown in Reaction Scheme 2.2.

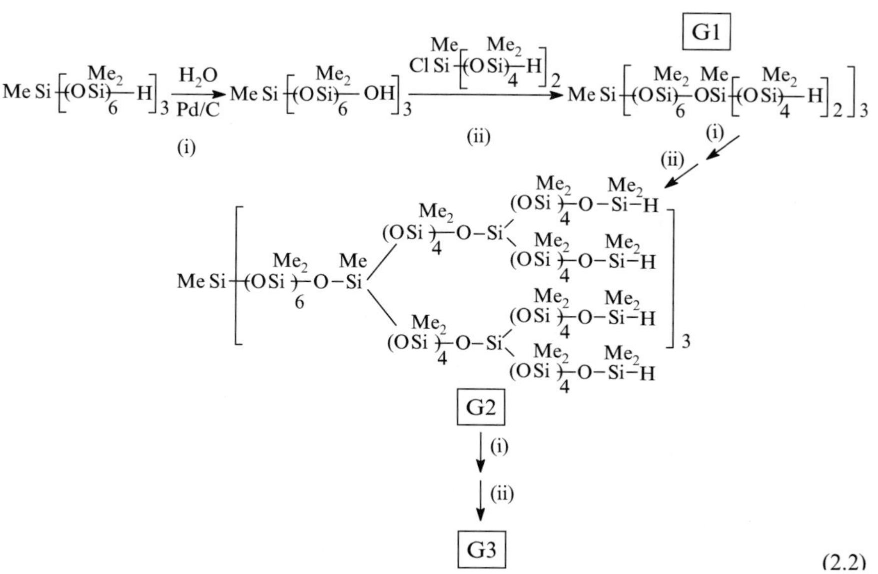

(2.2)

Unlike "dense" methylsilsesquioxane dendrimers obtained by the first described protocol (Reaction Scheme 2.1), these dendrimers possess a very loose molecular structure characterized by five siloxane units between the neighboring branch junctures. However, the lack of their authors' interest in this approach was apparent since they reported even less experimental details than we did, and mentioned practically nothing about the properties of their products. Estimating the prospects of this synthetic strategy, we now note the complexities involved with the control over silanol functions both during the dendrimer synthesis and in the removal of the reaction water. However, a possible replacement of the hydrolysis of SiH groups by their oxidation could make this route rather promising.

The third reported protocol for the synthesis of polysiloxane dendrimers was also based on a methylsilsesquioxane branching center [16, 17], as shown in Reaction Scheme 2.3.

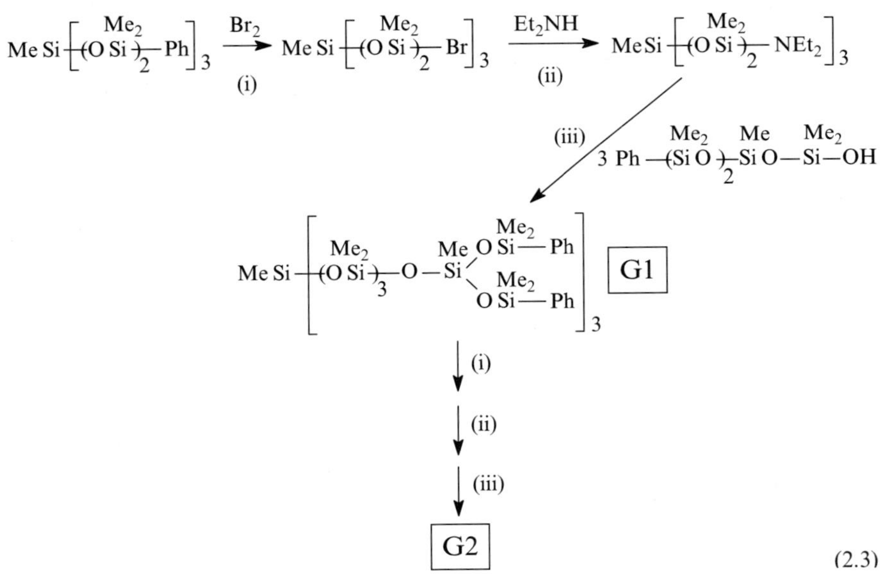

(2.3)

The use of phenyl groups as the protecting functionalities enabled isolation of stable intermediate compounds, their purification and the study of their properties. Similar to the synthetic strategy 2.1, this route provided the combinatorial possibilities of a synthetic platform, while the use of oligomeric reagents makes it similar to 2.2. Despite certain difficulties, such as those involved with silanol–amine interaction which could result in incomplete conversions, and the increasingly cumbersome removal of diethyl amine from the reaction mixtures at higher generations, this approach surely has potential for further development and improvements.

Unfortunately, in spite of a concerted start and obvious prospects for further development, none of these synthetic protocols were noticeably advanced later on. Today, one can only speculate about whether personal or objective reasons may have caused such a state of affairs. An overview of the literature, however, reveals that activities in this field of silicon-containing dendrimers did not stop but were instead shifted towards systems with mixed and carbosilane units [18–21]. There are at least two main reasons that caused such a move. These are: (a) a very convenient use of the hydrosilylation reaction for easy formation of dendrimer molecular skeletons, and (b) the chemical inertness of the carbosilane skeleton compared to that of the siloxane one. These factors are the subject of a detailed discussion in Chapter 3.

2.3 Peculiarities of Siloxane Dendrimers

It may seem strange but peculiarities of the siloxane dendrimers become more noticeable in mixed systems. Unique characteristics of the siloxane bond have been the subject of a great number of articles and a monograph entitled "*Siloxane Bond*" [22]. Therefore, we limit ourselves here to only one of this exceptional bond's main features, which it manifests in the dendrimer structures: its pronounced inherent flexibility. Qualitatively, the unique flexibility of the siloxane fragments can be easily seen from the comparison of properties of carbosilane (**1**) and carbosiloxane (**2**) dendrimers having a fluorescent probe in their cores:

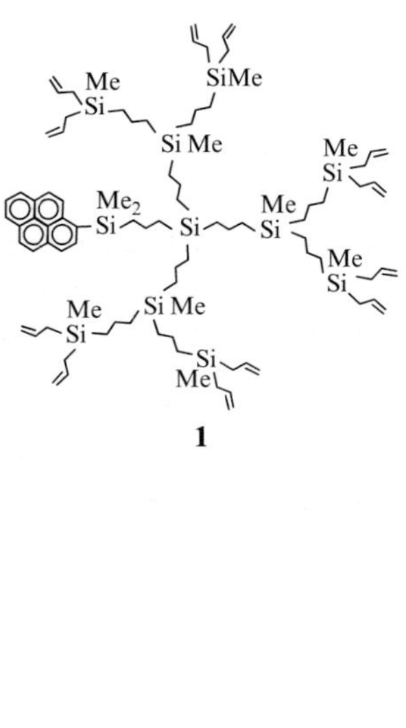

1

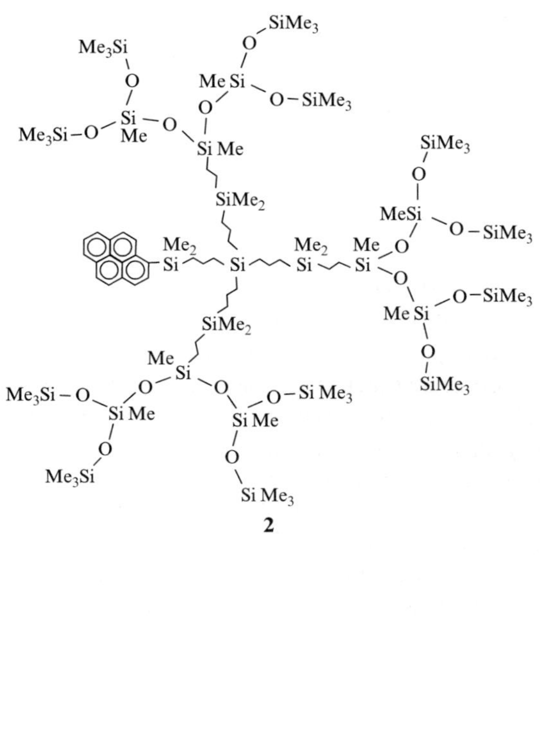

2

Thus, it was shown that in the carbosilane dendrimers **1**, even of the second generation, the accessibility of the pyrenyl fragments of the neighboring molecules is extremely limited [23]. Hence, even at high concentrations or in bulk, **1** does not form excimer complexes, while in the system with siloxane spacers, **2**, excimer formation occurs even when three times larger dendrons are grafted to a branching center (see Structure **3**) [24]:

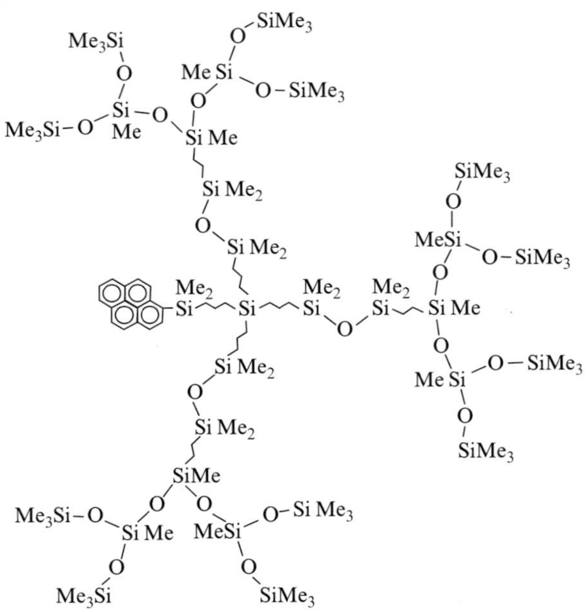

3

The usefulness of the siloxane spacers was also observed in other sterically hindered systems where their flexibility is noticeable even in comparison with pliable aliphatic spacers. For example, introduction of a siloxane bridge as a part of the spacer to mesogenic groups was successful in the synthesis of liquid crystalline dendrimers [25–27] (see also Chapter 10). In another example, the same method also enabled attachment of bulky fluorocontaining groups to the surface layer of a carbosilane dendrimer structure [28, 29].

A carbosilane–siloxane dendrimer, **4**, prepared according to the so-called universal scheme [30], by a combination of divergent and convergent approaches, is an illustration of a transitional structure from purely siloxane dendrimers to the carbosilane ones.

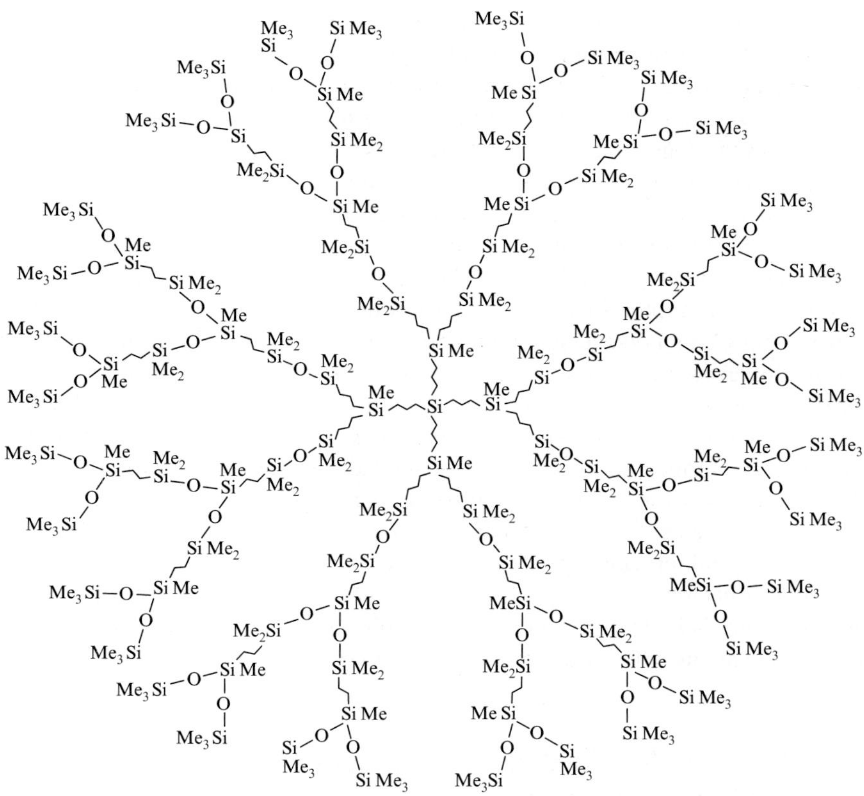

4

Although the introduction of flexible siloxane joints was shown to increase mobility of a mixed (i.e., "copolymeric") dendrimer skeleton, its thermal stability is determined by the weakest link, making the future of this direction in silicon-containing dendrimers rather uncertain. The stability of these dendrimers towards thermo-oxidative degradation will not exceed 200°C, typical for carbosilane dendritic systems, while ionic impurities will affect the stability of the siloxane fragments.

Furthermore, carbosilane–siloxane dendrimers in which the siloxane constituent contains alkoxysilyl groups [31–34] are sensitive towards moisture or alcohol. Although this sensitivity was efficiently used in analogous hyperbranched systems [35], it is unlikely to be so in the case of these dendrimers. The prospects may be quite different, however, when researchers have a clear idea of the nature of the chemical bond and apply it appropriately. The best example is poly(amidoamine–organosilicon), PAMAMOS, dendrimers where polyamidoamine interiors are encapsulated into a methylsilsesquioxane envelope [36–39]. These systems are described in Chapter 11, but it should be noted here that the authors made an excellent choice of components from a vast variety of other dendritic possibilities.

The idea of assembling copolymeric objects such as core-shell systems was also effective for other reagent pairs. For example, dendrimers with carbosilane cores and siloxane shells were realized in different ways of which the most significant one was with methylsilsesquioxane functional surface layers [40–42]. Fine control over the thermal degradation of these dendrimers allowed preparation of nano-porous polymethylsilsesquioxane film with narrow nano-pore size distribution.

A further important feature which siloxane units introduce into copolymeric systems is the low surface energy of methylsiloxanes which facilitates spreading of dendrimers containing such units over the water–air interface. This allows investigation of dendrimer monolayers and even study of the kinetics of their hydrolysis [43]. Considering the monodispersity of these polymers, this can be the source of unique information about molecular organization of various dendrimer structures.

Another important aspect of siloxane dendrimers and siloxane structural units in copolymeric systems is their versatility for further transformations. Siloxane bridges can be easily formed and destroyed under relatively mild conditions and without affecting the hydrocarbon part of the dendrimer. Experiments on the demarcation lines between the exterior and interior regions of such dendrimers, where the ease of the siloxane bond formation enables detection of interactions of dendrimers of various molecular structures, are examples of the usefulness of siloxane bonds in the carbosilane molecules [44]. The formation of cyclic units in the synthesis of dendrimers with fluorine containing exterior layers [28, 29] is also worth mentioning.

2.4 Prospects for Further Development in the Chemistry of Siloxane Dendrimers

The preceding analysis of peculiarities of the molecular structure of dendrimers containing siloxane units leads us to conclude that there is a promising future for the siloxane containing hybrids, while for dendrimers with a polysiloxane skeleton, the situation is more complex. Although there are very few data available for an in-depth analysis, there is a certain driving force for the development of this area. The high resistance towards thermal and thermo-oxidative degradation, common to siloxane dendrimers, makes their application in catalysis of oxidation and reduction processes very hopeful. Although there is no direct evidence for this, if one recalls that inorganic silicas are among the most popular catalyst supports and that polymethylsilsesquioxanes are their nearest organic analogues, this idea certainly seems reasonable. (For more on dendrimers in catalysis see Chapter 9.)

The recent report on the linear analogues of polymethylsilsesquioxane dendrimers [45] (see Reaction Scheme 2.4) offers an opportunity to compare linear and dendritic molecular architectures, similar to the earlier classic work of Hawker and Malmström [46]. A preliminary study of this indicates that in the siloxane case the difference may not be so striking, emphasizing the specificity of these systems and becoming another significant motivation for further development of the siloxane dendrimer synthesis.

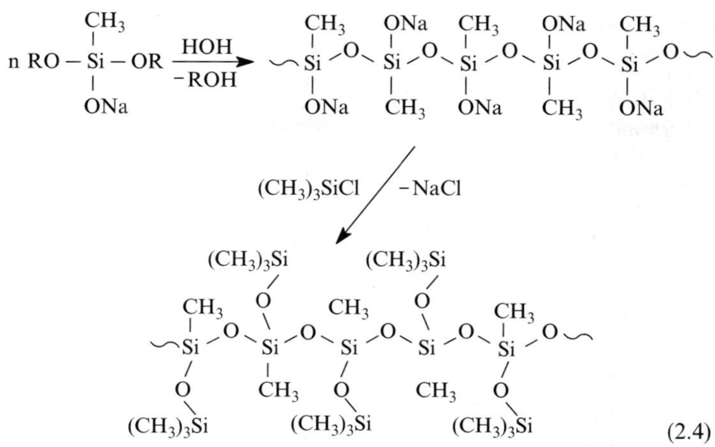

(2.4)

Reasons for optimism can also be found in the results obtained for the hyperbranched polysiloxane systems (see also Chapter 16). A variety of different dendritic structures has been obtained combining a fine understanding of the nature of the siloxanes with the benefits of the hydrosilylation reactions [47]. The unique transformation of the inorganic hyperbranched structure into a nanogel by controlled cyclization [48], and the regulated blocking of the molecular silica growth at different stages [49] are examples that favorably presuppose developments of their regular analogues.

There are still many unknown characteristics of siloxane dendrimers which in comparison to the well known classic siloxane forms may be fraught with difficulties. Rheological properties and solubility in supercritical CO_2, behavior at low and high temperatures, ability to form ordered structures, opportunity for further reduction of the organic content and ease of functionalization of the dendrimer interior, together with other, not yet studied attractive possibilities, are a mighty driving force for the development not only of new protocols for the synthesis of siloxane dendrimers, but also for the creation of new synthetic platforms with powerful combinatorial potential.

References

1. Bazant V, Chvalovsky V, et al. (1965) Organosilicon Compounds. Academic, New York.
2. Andrianov KA (1965) Metalloorganic Polymers. Wiley, New York.
3. Fleming I (1984) Organic Silicon Chemistry in Comprehensive Organic Chemistry, Part 3. Pergamon, New York.
4. Boileau S (1980) Am Chem Soc Polym Prepr 21, 1:25.
5. Rebrov EA, Muzafarov AM, et al. (1988) Doklady Chemistry USSR 302, 2:346.
6. Tomalia DA, Baker H, et al. (1986) Macromolecules 19:2466.
7. Tomalia DA, Hall M, et al. (1987) JACS 109:1601.
8. Muzafarov AM, Rebrov EA, et al. (1989) Doklady Chemistry USSR 309, 2:376.
9. Tomalia DA (1984) 1st International Polymer Conference Society of Polymer Science. Kyoto, Japan.
10. Tomalia DA, Baker H, et al. (1985) Polym J (Tokyo) 17:117.

11. Muzafarov AM, Rebrov EA, et al. (1991) Usp Khim 60:1596.
12. Rebrov EA, Muzafarov AM (2006) Heteroatom Chem 17, 6:514.
13. Uchida H, Kabe Y, et al. (1990) JACS 112, 19:7077.
14. Uchida H, Kabe Y, et al. (1991) Kao Corp JP 03263431 (Chem Abstr 116:236379).
15. Uchida H,Yoshino K, et al. (1991) Kao Corp JP 03263430 (Chem Abstr 116:130381).
16. Morikawa A, Kakimoto MA, et al. (1991) Macromolecules 24:3469.
17. Morikawa A, Kakimoto MA, et al. (1992) Polym J 24:573.
18. Roovers J (1993) Macromolecules 26:4324
19. Muzafarov AM, Gorbatsevich OB, et al. (1993) Polym Sci 35:1575.
20. Frey H, Lorenz K, et al. (1996) Macromol Symp 102:19.
21. Frey H, Mühlhaupt R, et al. (1995) Polym Mater Sci Eng 73:127.
22. Voronkov MG, Mileshkevich VP (1976) Siloxane Bond. Nauka, USSR.
23. Krasovskii VG, Sadovskii NA, et al. (1994) Polym Sci Ser A 36:589.
24. Krasovskii VG, Ignat'eva GM, et al. (1996) Polym Sci Ser A 38:1070.
25. Ponomarenko SA, Rebrov EA, et al. (1996) Liq Cryst 21:1.
26. Ponomarenko SA, Rebrov EA, et al. (1998) Polym Sci Ser A 40:763.
27. Richardson RM, Ponomarenko SA, et al. (1999) Liq Cryst 26:101.
28. Shumilkina NA, Myakushev VD, et al. (2005) Doklady Chem USSR 403:155.
29. Shumilkina NA, Myakushev VD, et al. (2006) Polym Sci Ser A 48, 12:1240.
30. Ignat'eva GM, Rebrov EA, et al. (1997) Polym Sci Ser A 39:843
31. Brüning K, Lang H (1998) J Organomet Chem 571:145.
32. Kim C, Kwon A (1998) Synthesis 105.
33. Kim C, Jeong Y, et al. (1998) J Organomet Chem 570:9.
34. Brüning K, Lang H (1999) Synthesis 1931.
35. Muzafarov AM, Golly M, et al. (1995) Macromolecules 28:8444.
36. Dvornic PR (2006) J Polym Sci Part A Polym Chem 44:2755.
37. Dvornic PR, de Leuze-Jallouli AM, et al. (2000) Macromolecules 33:5366.
38. de Leuze-Jallouli AM, Swanson DR, et al. (1997) Polym Mater Sci Eng 77:67.
39. Dvornic PR, Hu J, et al. (2002) Silicon Chem 1:177.
40. Bystrova AV, Tatarinova EA, et al. (2005) Polym Sci Ser A 47, 8:820.
41. Bystrova AV, Tatarinova EA, et al. (2007) Science and Technology of Silicones and Silicone-Modified Materials. ASC Symposium Series 964. ISBN10: 0841239436
42. Bystrova AV, Parshina EA, et al. (2007) Nanotechnol Russ 2, 1:83.
43. Tereshchenko AS, Getmanova EV, et al. (2007) Russ Chem Bull Intern Edit 56:2200.
44. Getmanova EV, Tereshchenko AS, et al. (2004) Russ Chem Bull Intern Edit 53, 1:137.
45. Obreskova MV, Rogul' GS, et al. (2008) Doklady Chem 419, 1:69.
46. Hawker CJ, Malmstrom EE (1997) JACS 119:9903.
47. Chojnowski J, Cypryk M (2003) Macromolecules 36:3890.
48. Kazakova VV, Rebrov EA, et al. (2000) Silicones and Silicone-Modified Materials. ACS Symposium series 792:503. ISBN 0-8412-3613-5.
49. Voronina NV, Meshkov IB et al. (2008) Nanotechnol Russ 3, 5-6:321.

Chapter 3*
Carbosilane Dendrimers

Jacques Roovers and Jianfu Ding

3.1 Introduction

The concept of highly symmetrical, perfectly branched macromolecules prepared in a generational fashion was introduced in 1978 [1]. The synthesis of polylysine dendrimers [2] and the seminal research by Tomalia and Newkome in the mid-1980s established that such molecules could indeed be prepared [3, 4]. Tomalia et al. used trifunctional nitrogen branch points and Newkome chose tetrafunctional carbon branch points. These dendrimers contained ether, ester, amine and amide polar bonds.

Carbosilane dendrimers with a silicon branch point in an exclusively carbon–silicon skeleton are non-polar, inert, neutral and thermally and hydrolytically stable compounds. The absence of polar bonds facilitates the use of many derivatization reactions and creates the possibility of strong physico-chemical contrast between the core and the outer corona. The synthesis of carbosilane dendrimers is almost always by a *divergent* process from the core to the interior generations and to the periphery, with the number of reactions per dendrimer increasing geometrically with each generation (see Chapter 1). The divergent synthesis of carbosilane dendrimers consists of the generational repetition of a sequence of two clean, high-yield reactions: (a) hydrosilylation and (b) nucleophilic substitution by Grignard or organolithium reagents. The hydrosilylation reaction 3.1 introduces the branch juncture and creates the next generation:

J. Roovers
Formerly of National Research Council of Canada, Canada
E-mail: jnroovers@rogers.com

J. Ding
Institute for Chemical Process and Environmental Technology, National Research Council, Ottawa, Ontario, Canada K1A 0R6
E-mail: Jianfu.Ding@nrc.ca

P.R. Dvornic and M.J. Owen (eds.), *Silicon-Containing Dendritic Polymers*,
Advances in Silicon Science 2, © Springer Science+Business Media B.V. 2009

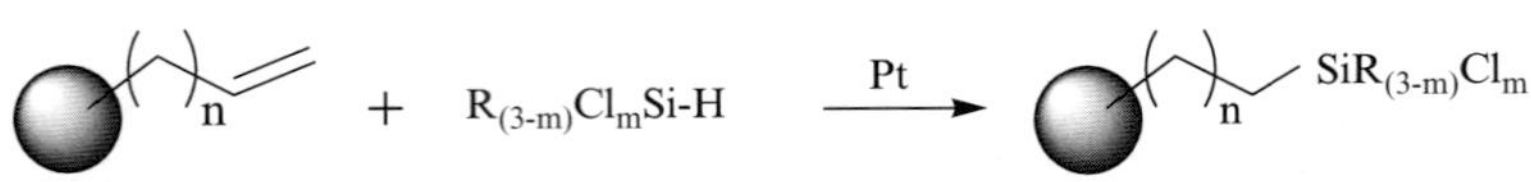

(3.1)

while the substitution reaction 3.2 introduces the branch:

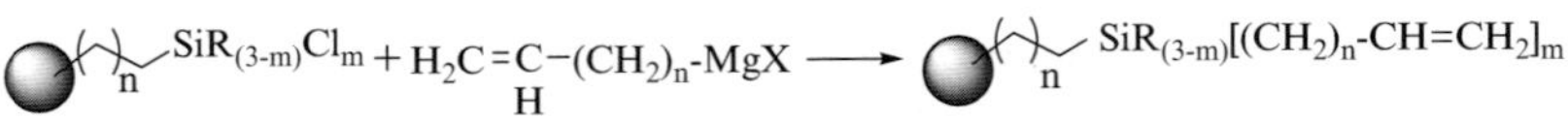

(3.2)

where n = 0 (vinyl) or 1 (allyl) defines the length (2 or 3 methylene groups) of the branch and m = 1, 2 or 3 defines the functionality (i.e., multiplicity) of the branch point (i.e., juncture). Both reactions are performed with excess of the highly reactive hydrosilane and Grignard or organolithium compounds in order to promote quantitative conversions. Due to the high sensitivity of these reagents to oxygen, moisture, carbon dioxide, etc., these reactions have to be protected by high vacuum and/or Schlenk techniques.

In a *convergent* synthesis, a dendron containing the peripheral groups is repeatedly added via its focal group to a branched molecule and in the final step to a multifunctional core molecule (see also Chapter 1). A possible scheme for a convergent synthesis of a carbosilane dendron consists of the repetition of the following two reactions:

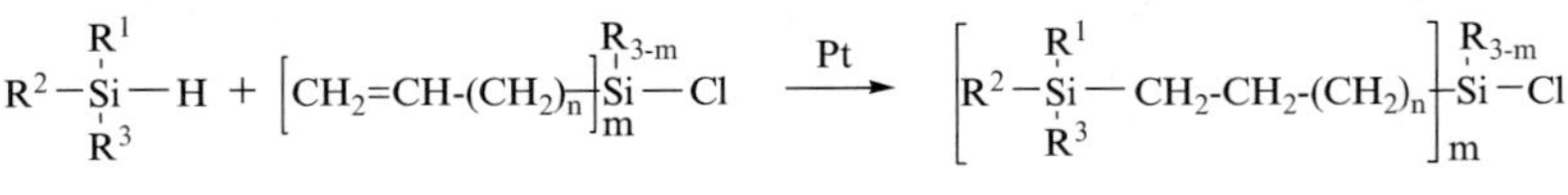
(3.3)

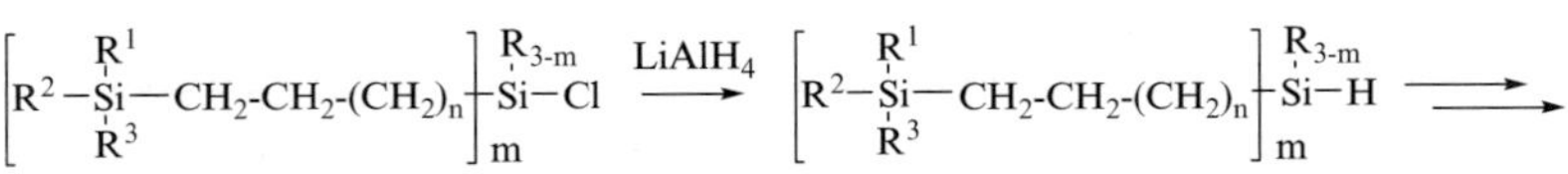
(3.4)

where the definitions of subscripts m and n are the same as for the divergent synthesis. The convergent approach has seldom been applied to carbosilane dendrimers.

Embryonic carbosilane dendrimers with 12 and 18 Si–Cl reactive peripheral (i.e., end) groups were described as early as 1978 and 1980 [5, 6]. The first synthesis of carbosilane dendrimers was reported almost simultaneously by three groups in the early 1990s [7–11], and several reviews of the field have been published previously [12–17].

3.2 Synthesis of Carbosilane Dendrimers

Dendrimers consist of a core, interior generational layers and the exterior or corona (see Chapter 1). These three elements of carbosilane dendrimers will be discussed separately in the following subsections. The core molecule is considered the zeroth generation (0G). A 1G dendrimer has but a core and a corona. In this chapter, dendrimers are named by their last generation and the multiplicity of each generation specifying the branching architecture is given in brackets. For example, 4G(4,3,2,2,2) is a fourth generation dendrimer with a tetrafunctional core and multiplicities 3,2,2 and 2 in the four consecutive generations. The number of end-groups is then given

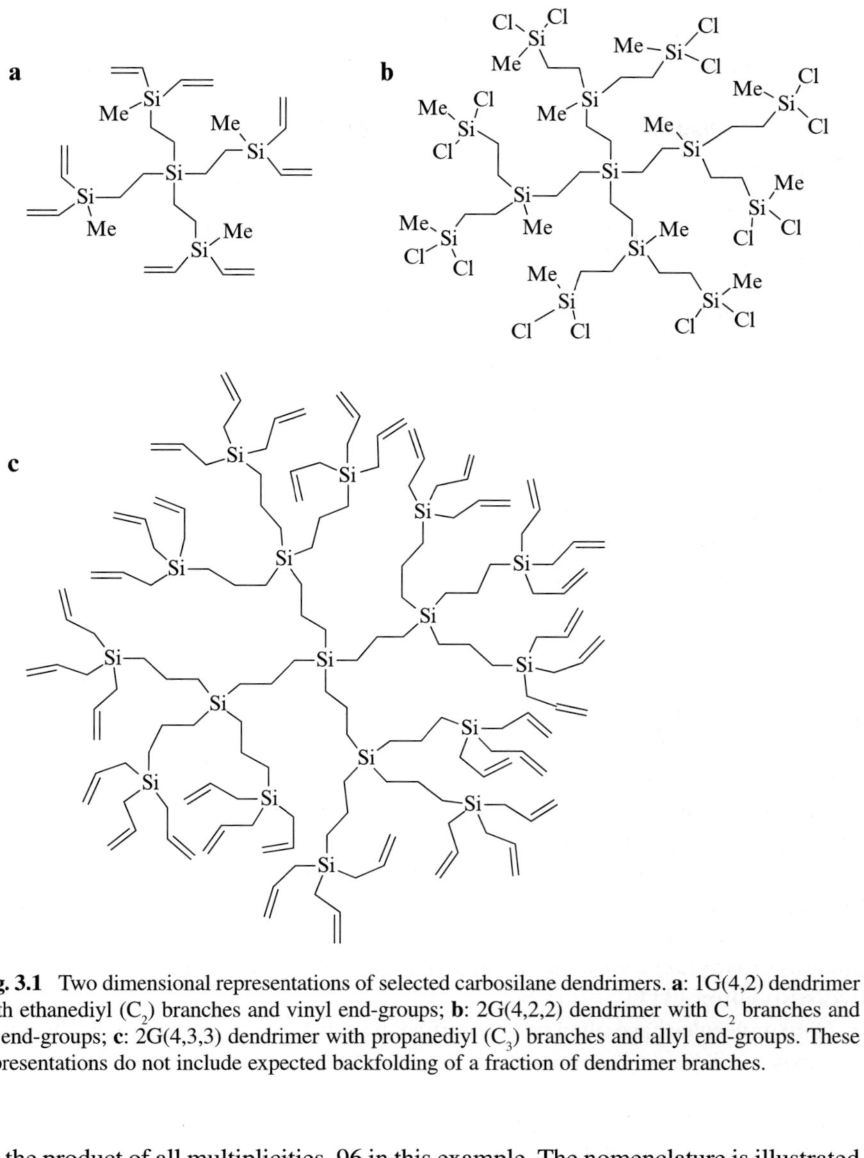

Fig. 3.1 Two dimensional representations of selected carbosilane dendrimers. **a**: 1G(4,2) dendrimer with ethanediyl (C_2) branches and vinyl end-groups; **b**: 2G(4,2,2) dendrimer with C_2 branches and Cl end-groups; **c**: 2G(4,3,3) dendrimer with propanediyl (C_3) branches and allyl end-groups. These representations do not include expected backfolding of a fraction of dendrimer branches.

by the product of all multiplicities, 96 in this example. The nomenclature is illustrated by three examples of dendrimers in Fig. 3.1.

3.2.1 Core Molecules

The inert core molecules that have been used in the divergent synthesis of classical carbosilane dendrimers are listed in Table 3.1.

Table 3.1 Classical carbosilane dendrimers

	Core	Branch point	Branch	Generation	References
1	$Si(vinyl)_4$	$HSiMeCl_2$	Vinyl	1G–5G (4,2,2,2,2,2)	[7, 18–20]
2	$Si(allyl)_4$	$HSiCl_3$	Allyl	1G–5G* (4,3,3,3,3,3)	[8, 9]
			Vinyl	1G–3G* (4,3,3,3)	
3	$MeSi[OSiMe(allyl)_2]_3$	$HSiMeCl_2$	Allyl	1G–7G (3,2,2,2,2,2,2,2)	[10, 11]
4	$Si(vinyl)_4$	$HSiCl_3$	Vinyl	1G–4G* (4,3,3,3,3)	[21]
5	$Si(allyl)_4$	$HSiCl_3$	Allyl	1G–3G* (4,3,3,3)	[22]
6	$MeSi(allyl)_3$	$HSiMeCl_2$	Allyl	1G–5G (3,2,2,2,2,2)	[23]
7	$[–CH_2Si(allyl)_3]_2$	$HSiMeCl_2$	Allyl	1G–3G* (6,2,2,2)	[24, 25]
8	c-$[–Si(Me)(vinyl)O]_4$	$HSiCl_3$ + $HSiMeCl_2$	Allyl	1G–4G (4,3,2,2,2)	[26–29]
		$HSiMeCl_2$	Allyl	1G–3G (4,2,2,2)	
9	$Si_8O_{12}(vinyl)_8$	$HSiCl_3$	Allyl	1G–2G (8,3,3)	[30]
		$HSiCl_3$ + $HSiMeCl_2$	Allyl	1G–2G (8,3,2)	
		$HSiCl_3$ + $HSiMe_2Cl$	Allyl	1G–2G (8,3,1)	
10	$Si(C{\equiv}CPh)_4$	$HSiMeCl_2$	Ph-ethenyl	1G–3G* (4,2,2,2)	[31, 32]
	$MeSi(C{\equiv}CPh)_3$	$HSiMeCl_2$	Ph-ethenyl	1G–4G* (3,2,2,2,2)	
	$Me_2Si(C{\equiv}CPh)_2$	$HSiMeCl_2$	Ph-ethenyl	1G–5G* (2,2,2,2,2,2)	

* The dendrimer is sterically saturated

The first three rows of this table have historical relevance that already cover the common variation in chemistry and multiplicity of carbosilane dendrimers. The core molecules for most carbosilane dendrimers are vinyl or allylic compounds with functionality (i.e., multiplicity) from 2 to 8. A large core functionality reduces the number of steps required to produce a dendrimer with the desired number of peripheral groups but early steric saturation may set limits to this approach. The dendrimers marked with an asterisk in Table 3.1 have been confirmed to be sterically saturated. The synthesis of the next higher generation dendrimer was recorded to have failed catastrophically. Line ten in Table 3.1 should be particularly noted: the ethynyl triple bond can be selectively monohydrosilylated with the resulting double bond being resistant to further hydrosilylation [31].

Core molecules have often been modified in such a way that a functional "handle" becomes available at the root of the dendron, as shown in Reaction Scheme 3.5. These core molecules are all of the $(allyl)_{4-m}$ Si-X_m type, where m is almost always equal to one and X is compatible with the divergent hydrosilylation-Grignard substitution sequence of the dendron synthesis.

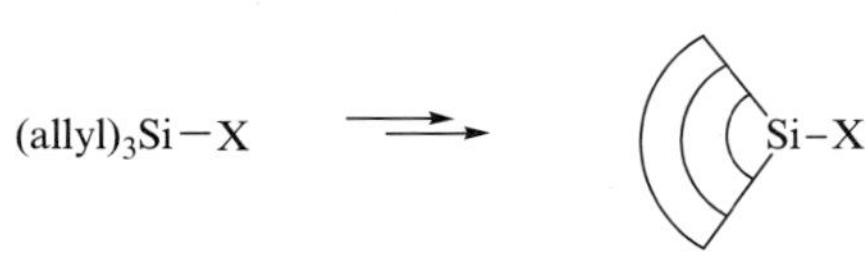

(3.5)

These core-functionalized dendritic molecules have also been called "wedges" [33, 34]. For example, Russian workers prepared carbosilane dendrons from a perylene root starting from 3-bromoperylene [35] (see also Chapter 2). The subsequent dendron synthesis was extended to 6G(1,2,2,2,2,2,2):

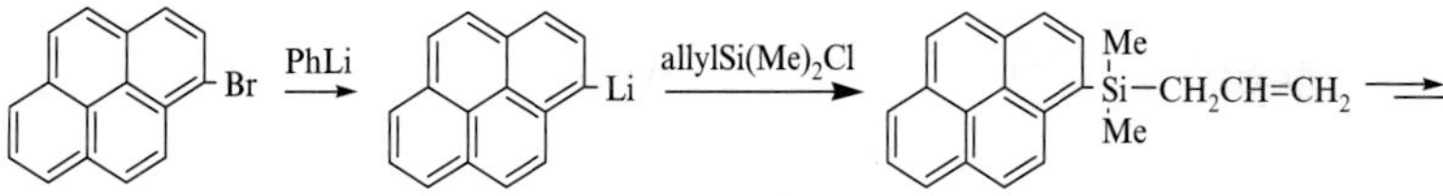

(3.6)

with 64 peripheral allyl groups. The effect of the dendron crowding around the perylene has been investigated by fluorescence studies [35, 36]. Kim et al. prepared dendrons starting from diallylmethylphenylsilane to 4G(2,3,3,3,3) and 5G(2,2,2,2,2,2), respectively [37]. In these prototypes the phenyl group can be displaced by triflic acid. The triflate is then a good leaving group for further modification [38]. The detailed reaction conditions for displacement of the phenyl ring with triflate have been described [39]. This procedure was followed by Andrés et al. for the direct attachment of a cyclopentadienyl ring onto the silicon focal atom of 1G, 2G and 3G dendrimers with fully saturated peripheral shells consisting of $SiMe_2Bz$ groups [40]:

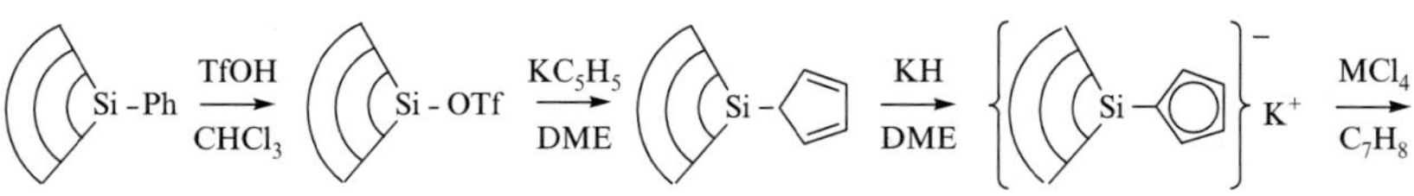

(3.7)

Coupling of 1G via its potassium salt with $TiCl_4$ and $ZrCl_4$ yields metal complexes that have catalytic activity. The coupling of 2G and 3G dendrimers, however, failed. An almost identical approach was also followed by Meder et al. [41].

The first core modified dendrons used in the *convergent* synthesis of the next generation dendrimer were described in 1998 [42, 43]:

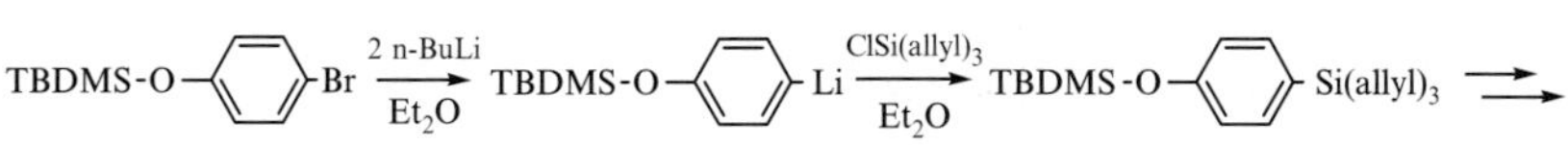

(3.8)

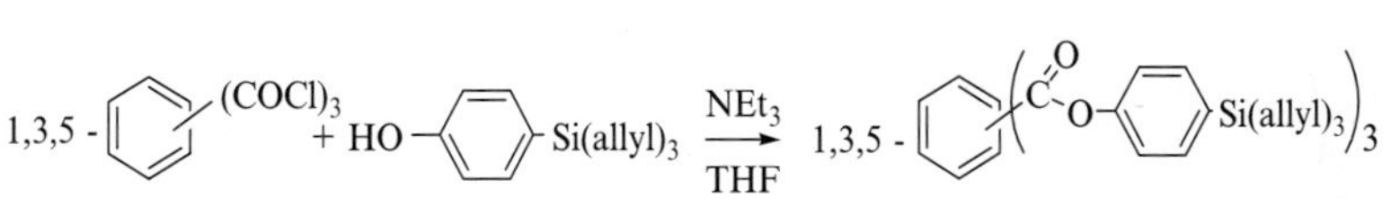

(3.9)

The convergent step is illustrated here with the 0G but has also been realized with 1G and 2G dendrons. The *tert*-butyldimethylsilyl protected phenol, which can be carried through the dendron synthesis, is then deprotected with tetrabutylammonium fluoride and linked with 1,3,5-benzenetricarbonyl trichloride to form mixed ester carbosilane dendrimers. A similar strategy has also been followed for a mixed amide carbosilane dendrimer [34]. Starting from the 3-bromopropyltrichlorosilane core the desired dendrimer is constructed as shown in Reaction Scheme 3.10. The bromide is converted to the primary amine in liquid NH_3 (15 bar) at 70°C and then reacted with a multifunctional carbonyl chloride compound, as shown in Reaction Scheme 3.11:

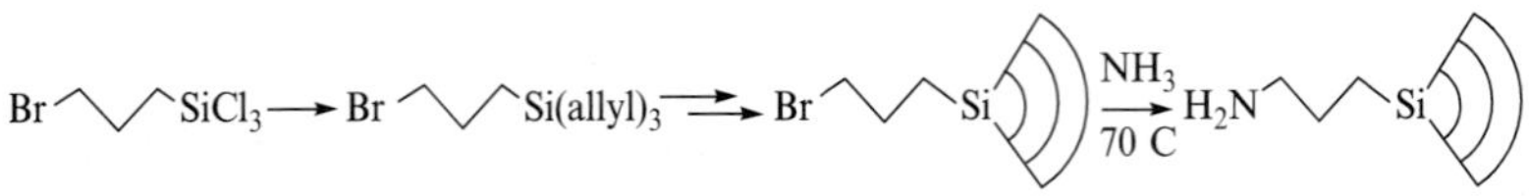

(3.10)

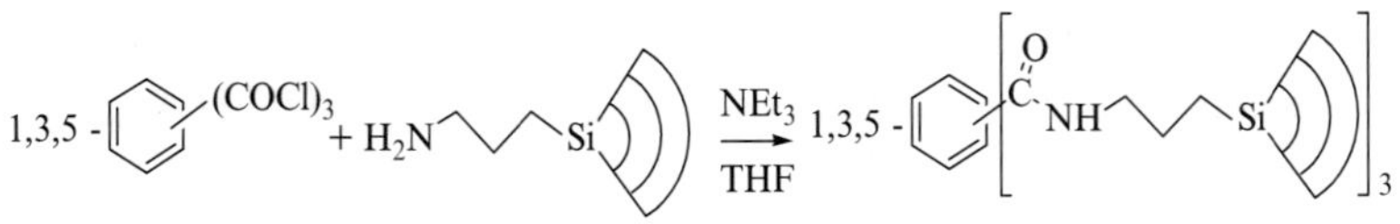

(3.11)

This amine functionalized 2G(3,3,3) dendron has also been used in a formation of highly efficient enantiopure intramolecular bidentate P,N-ligand on a 1,1′-binaphthyl [44]. Further modification of the functionalized core of a dendron prior to coupling into a dendrimer leads to materials with an active function in the direct vicinity of the core and surrounded by a hydrophobic shell. The incorporation of an alanyl group is illustrated in Reaction Scheme 3.13 [45]:

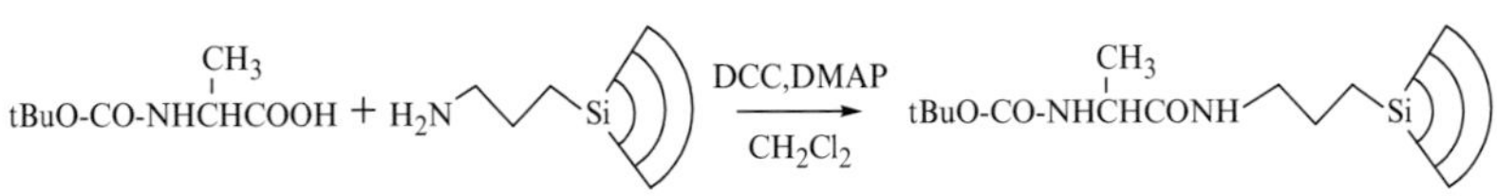

(3.12)

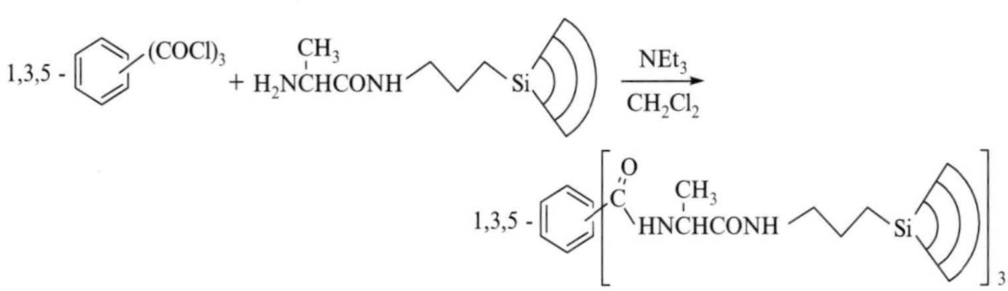

(3.13)

The influence of the dendrimer environment on the binding constants of acids with the amide bonds has been investigated.

The first catalytically active carbosilane center was obtained by reacting a dendron containing a phenyllithium focal point with a ferrocenyl compound [46]:

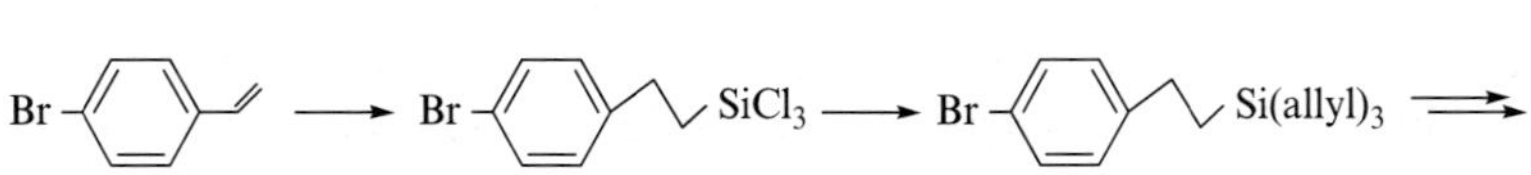

(3.14)

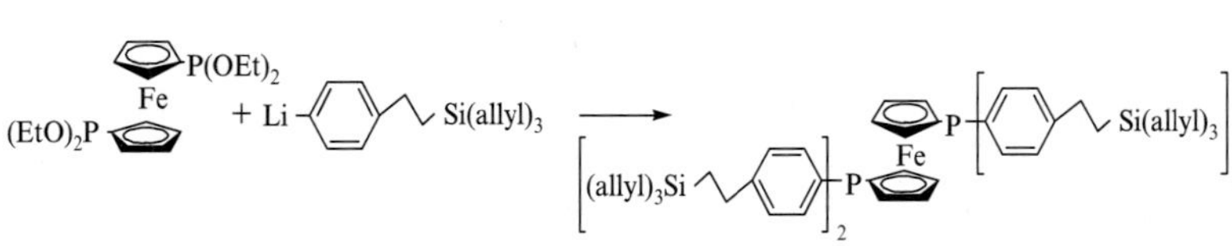

(3.15)

The example is given here for the 0G dendron. The coupling reaction with the ferrocenyl compound produces a 1G(4,3) dendrimer with 12 allyl groups. The 0G triallyl dendron (Reaction Scheme 3.14) was also expanded to 1G(3,3) and 2G(3,3,3) dendrons so that the final ferrocenyl-centered products are 2G(4,3,3) and 3G(4,3,3,3) dendrimers, respectively. The catalytic properties of carbosilane dendrimer-metal complexes are reviewed in Chapter 9. By the same procedure, P,O-ligands for Ni catalysts as well as a tridendrophosphine have been obtained [47, 48]:

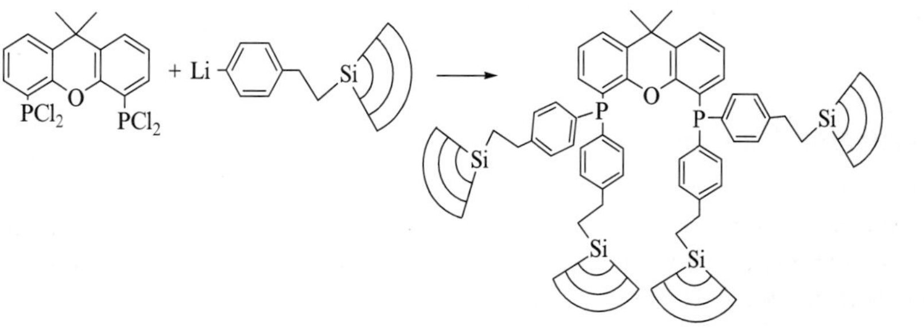

(3.16)

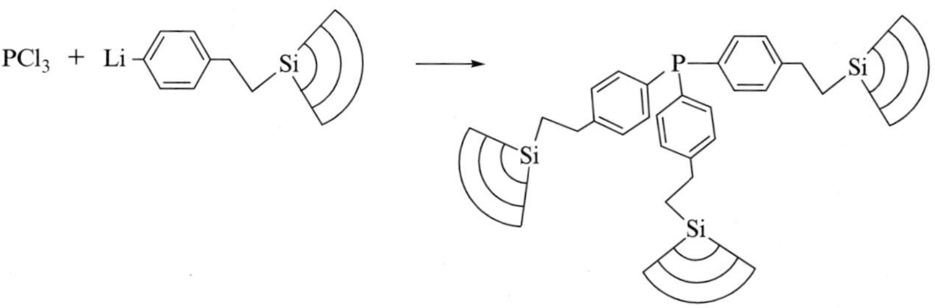

(3.17)

A most detailed report concerning the synthesis of carbosilane carbodiimides and their use in the cyclization of dipeptides for the formation of seven-membered rings was published recently [49]. The reactions needed to convert the focal functional group into a carbodiimide require progressively more stringent conditions with increase of the dendrimer size. The carbodiimide is prepared convergently via the urea followed by dehydration with triphenylphosphine-bromine complex. For the 3G(2,3,3,3) dendron a direct coupling between the isocyanate and an activated amine gave a better (72%) yield (Reaction Scheme 3.20). The intramolecular cyclization of two different dipeptides with the dendritic carbodiimides doubled the yield when compared to the same cyclization with dicyclohexylcarbodiimide.

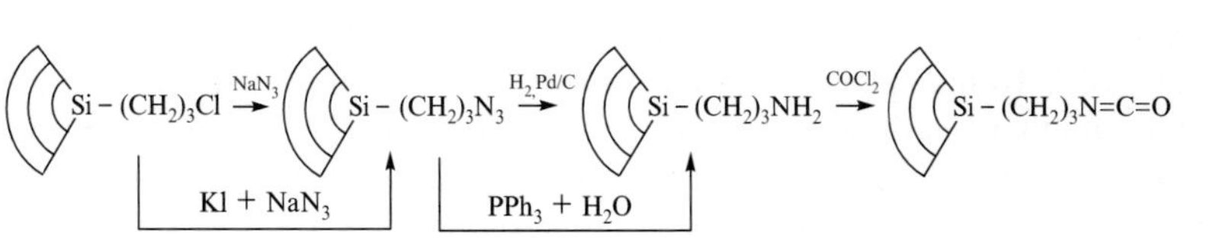

(3.18)

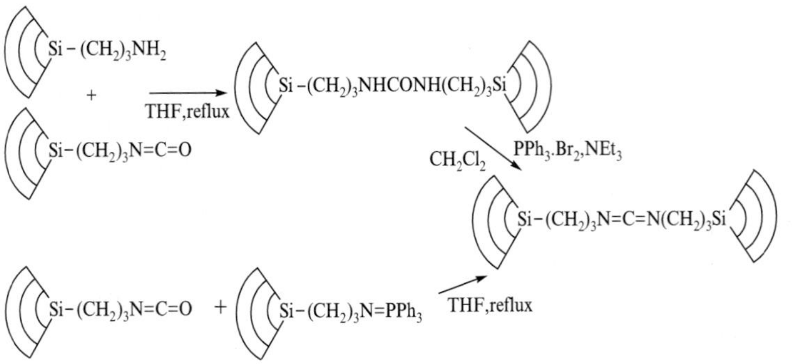

(3.19/3.20)

3.2.2 Interior Generations

Each interior generation (i.e., tier or layer) of a carbosilane dendrimer is made up of branch points introduced via a hydrosilylation reaction and branches introduced via an alkenylation step. These two reactions as applied to the synthesis of these dendrimers are next briefly reviewed.

3.2.2.1 Hydrosilylation

The homogeneous hydrosilylation reaction (see also Chapters 7 and 13 as well as the first book of this series: B. Marciniec, "*Hydrosilylation – A Comprehensive review on recent Advances*" is catalyzed by platinum compounds and is considered the more finicky step. General reviews are available [50, 51]. Either soluble Pt(IV) compounds such as Speier's catalyst ($H_2PtCl_6.2H_2O$ in isopropanol) [52] or Lukevics' catalyst $[(n\text{-}Bu)_4N]_2PtCl_6$ [45, 48, 49, 53] or Pt(II) $[(C_8H_{17})_3BzN]_2Pt(NO_2)_4$ [54] are used. Alternatively, Pt(0) complexes such as Karstedt's catalyst [Pt-(1,3-divinyltetramethyldisiloxane)$_{1.5}$] (PC072) [55] and the Silopren™ (Platinum siloxane complex) [56] are used in a molar ratio of 10^{-5} to 10^{-3} relative to olefin. Exceptionally, Kim et al. used a heterogeneous Pt/C catalyst throughout most of their work [37].

It is generally believed that Pt(IV) and Pt(II) catalysts are first reduced to a more active Pt(0) complex by hydrosilane and that a common hydrosilylation mechanism operates [57]. Lewis has devoted considerable research to elucidate several details of the mechanism of Pt(0) catalyzed hydrosilylation, pointing out an initial induction period and the need for traces of oxygen; first, to create small active colloids, and later, to avoid formation of inactive Pt–Pt agglomerates [58]. Present understanding of the catalytic hydrosilylation mechanism, including its homogeneous/heterogeneous aspect, has been succinctly reviewed [59]. Hydrosilylation tolerates a wide variety of solvents such as benzene, hexane, diethylether, tetrahydrofuran (THF), dichloromethane and their mixtures with alcohols. Initial warming may be required in order to reduce the induction period but temperature control of the exothermic reaction is desirable in order to minimize side reactions. Detailed descriptions of the best procedures for carbosilane dendrimer production have been given [7, 20, 21, 45, 60].

Hydrosilylation reactions are usually performed with excess (1.5-fold, in difficult cases up to 2.5-fold [60]) hydrosilane in order to drive the reaction to completion. The presence of electron withdrawing groups on silicon facilitates the reaction so that the rate order $HSiCl_3 \approx HSi(Me)Cl_2 > HSi(Me)_2Cl > HSiEt_3 \approx HSiMe_3$ is observed. The order $HSiMeCl_2 > HSiMePh_2$ has also been reported [61]. The general rule is that electron donating groups on the olefin and electron withdrawing groups on the hydrosilane increase the hydrosilylation rate [57]. Methylchlorosilanes are the most common branching agents for carbosilane dendrimers but $HSiPhCl_2$ has also been used [39]. The addition to the double bond is almost exclusively β or anti-Markovnikov which creates a linear branch. Occasionally, traces of α-addition are observed [21, 62]. As clearly indicated by the work of Iovel et al., α-addition occurs when less reactive hydrosilanes, e.g. $HSi(aliphatic)_3$ compounds, are reacted at high temperature or without solvent [53]. Although α-addition changes the structure of the dendrimer, often little attention is paid to its complete elimination because α-placement does not immediately limit the growth. The number of Si–Cl bonds in the hydrosilane determines the multiplicity of the branch point. This multiplicity is often kept constant throughout the dendrimer synthesis but can also be changed at each generation (see also Chapter 1, Section 1.4). Kriesel and Tilley described a 4G(4,3,3,2) dendrimer with C_3 branches with 72 triethoxysilyl end-groups [63]. It was pointed out that branch point variation helps in the interpretation of NMR spectra of carbosilane dendrimers [64].

Hydrosilylation with $HSi(Me)_2Cl$ serves one of two purposes. When used in an interior generation as a branch point with multiplicity of 1 it forms a $-Si(CH_3)_2-$ linker between consecutive linear branches and helps to decrease steric crowding [65, 66]. Alternatively, it is very often used in the last generation where it allows each chain end to be individually functionalized. Several examples of this use are described in Section 3.2.3.2 on the peripheral modification of dendrimers.

3.2.2.2 Substitution with Organometallics

The nucleophilic substitution on Si–Cl with an excess of either vinylmagnesiumbromide or allylmagnesiumchloride or -bromide (Grignard reagents) is an alkenylation reaction. In conjunction with a further hydrosilylation step the former introduces C_2 (ethanediyl), the latter C_3 (propanediyl) branches in the reacting dendrimer generation. Introduction of longer ω-alkenyl chains with primary Grignard reagents has been proposed but no details have been reported [9]. It is suspected that in that case incomplete alkenylation due to steric hindrance and side reactions prevents efficient dendrimer formation [12, 67, 68] but the reactions of multiple Si–Cl groups on large dendrimers with *n*-BuLi and methylmagnesiumbromide appear to proceed satisfactory [69].

Carbosilane dendrimers with phenylethenyl branches have also been synthesized. The prototype reaction is illustrated by the formation of tetra(phenylethynyl) silane from $SiCl_4$ and lithium phenylacetylene [31]. Hydrosilylation of tetra(phenylethynyl)silane with $HSiMeCl_2$ followed by substitution with lithium phenylacetylene provides the first generation [31]. It is noted that the phenylethenyl double bond is not further hydrosilylated:

$$\mathrm{Si}\left[\mathrm{C{\equiv}C\text{-}Ph}\right]_4 \rightarrow \mathrm{Si}\left[\overset{\mathrm{H}}{\mathrm{C}}\mathrm{{=}C(Ph)\text{-}Si(Me)Cl_2}\right]_4 \rightarrow \mathrm{Si}\left[\overset{\mathrm{H}}{\mathrm{C}}\mathrm{{=}C(Ph)\text{-}Si(Me)}\left[\mathrm{C{\equiv}C\text{-}Ph}\right]_2\right]_4 \quad (3.21)$$

The sequence had been repeated to 3G(4,2,2,2) but further hydrosilylation with $HSiMeCl_2$ failed. On the other hand, hydrosilylation with $HSiMe_2Cl$ was successful. The limit of growth of pure phenylethenyl branched dendrimers was investigated starting from the following three core molecules with different multiplicities [32]:

$$\mathrm{Me_2Si}\left(\mathrm{C{\equiv}C\text{-}Ph}\right)_2 \qquad \mathrm{Me{-}Si}\left(\mathrm{C{\equiv}C\text{-}Ph}\right)_3 \qquad \mathrm{Si}\left(\mathrm{C{\equiv}C\text{-}Ph}\right)_4$$

By generational reiteration from the di- tri- and tetrafunctional cores perfect 5G(2,2,2,2,2,2), 4G(3,2,2,2,2) and 3G(4,2,2,2) dendrimers with 64, 48 and 32 peripheral phenylethynyl groups, respectively, can be obtained and it was well established that further hydrosilylation was only partly successful. A larger trifunctional core, $C_6H_3[Si(Me)_2vinyl]_3$ branched with $HSiCl_3$ leads only to 2G(3,3,3) [70]. However, when mixed multiplicity is introduced higher generations become obtainable. Repetitive hydrosilylation-alkynylation of the previous 1G(3,3) dendrimer with $HSiMeCl_2$ affords a 5G(3,3,2,2,2,2) with 144 phenylethynyl end-groups.

Probably the most efficient method to introduce peripheral phenylethynyl groups consists of preparing a standard carbosilane dendrimer with trimethylene (C_3) branches and decorating the final generation with phenylethynyl groups. In this way, 3G(6,2,2,2) and 4G(6,2,2,2,2) layered dendrimers with 48 and 96 phenylethynyls, respectively, were obtained [71]. In another variation, a double layer of phenylethynyl groups was added so that a 3G(4,3,2,2) dendrimers was formed [72]. This last dendrimer is interesting because not only is the chemistry changed in different generations but also the multiplicity is lowered. The phenylethenyl layer has also been placed exclusively in an interior generation [73]. A standard 3G(4,2,2,2) dendrimer with 32 Si–Cl bonds was first reacted with lithium phenylacetylene. The fourth generation was continued with a sequence of hydrosilylation with $HSiMeCl_2$ and alkenylation with allylmagnesiumchloride and the fifth generation was finally capped with 128 peripheral butyl groups, to yield the following structure:

$$\mathrm{Si}\left\{\mathrm{C_3H_6}\left(\mathrm{C_3H_6}\left(\mathrm{C_3H_6}\left(\mathrm{CH{=}CPh}\left(\mathrm{C_3H_6}\left(\mathrm{C_4H_9}\right)_2\right)_2\right)_2\right)_2\right)_2\right\}_4$$

Very recently, a method was described for the synthesis of segmented carbosilane dendrimers that have two different pairs of dendrons emanating from the central silicon atom [60]:

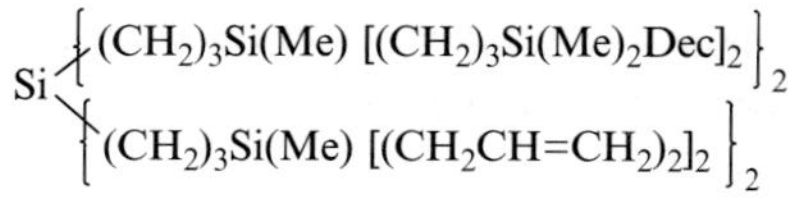

Aromatic groups have been introduced into the branches of carbosilane dendrimers in order to expand the dimensions between the branch points, reduce the internal density, create larger cavities and relieve surface crowding [74]. The process was semi-convergent as the introduction of the phenyl or biphenyl (via Suzuki coupling) was always the last step in the synthesis. The construction of the dendrimers started from 1,4-dibromobenzene, and the building blocks were obtained via an interesting set of hydrosilylation/substitution reactions as shown in Reaction Scheme 3.22:

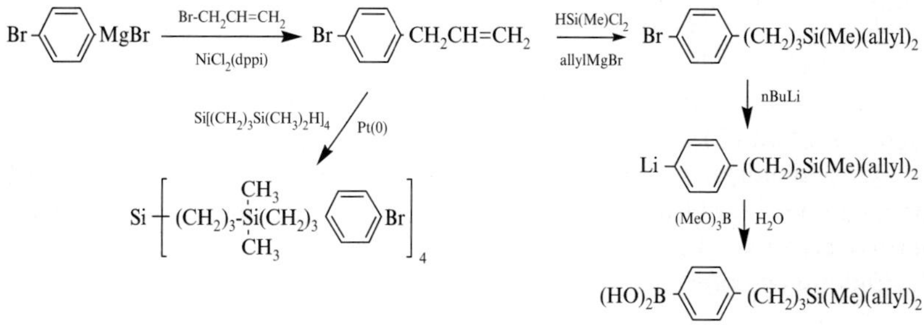

(3.22)

The final convergent formations of the phenyl and biphenyl branch are depicted for the 1G(4,2) dendrimer in the following reactions:

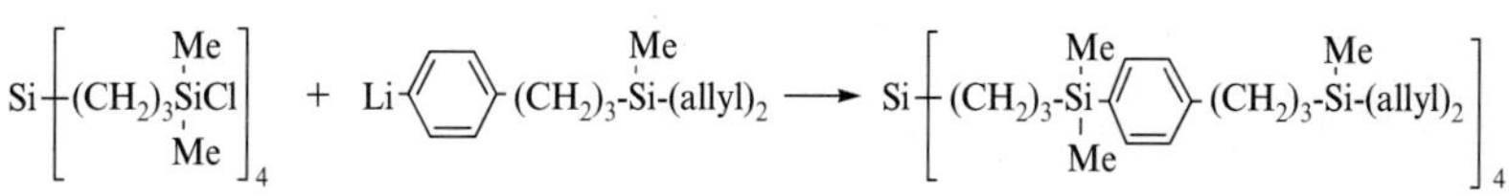

(3.23)

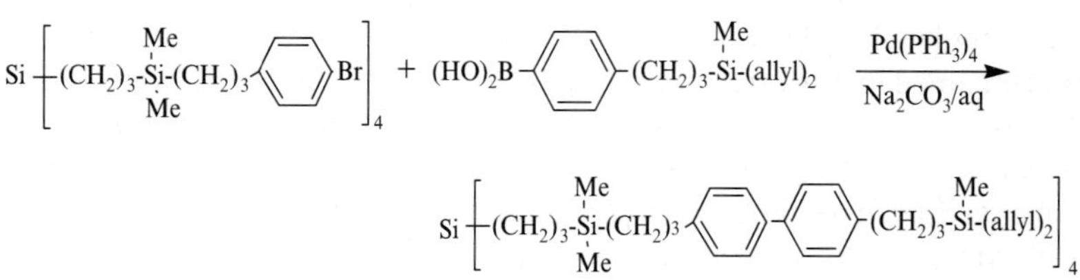

(3.24)

For the 2G(4,2,2) biphenyl analogue the same coupling reaction was employed starting with $(HO)_2B–C_6H_4–(CH_2)_3–Si(Me)[(CH_2)_3–Si(Me)(allyl_2]_2$.

To conclude this section, we mention a few extraordinary carbosilane dendrimers whose chemistry and structure are not directly related to the classical ones. Interestingly, these examples are often based on silicon halide and organolithium coupling reactions. A carbosilane dendrimer with 16 thiophene rings, four functioning as branches and 12 as peripheral groups was prepared from tetra(2-thienyl) silane and tri(2-thienyl)methoxysilane with BuLi in 19% yield [75]. The basic reaction was the controlled coupling of tetramethoxysilane with four and three 2-thienyllithium molecules, respectively:

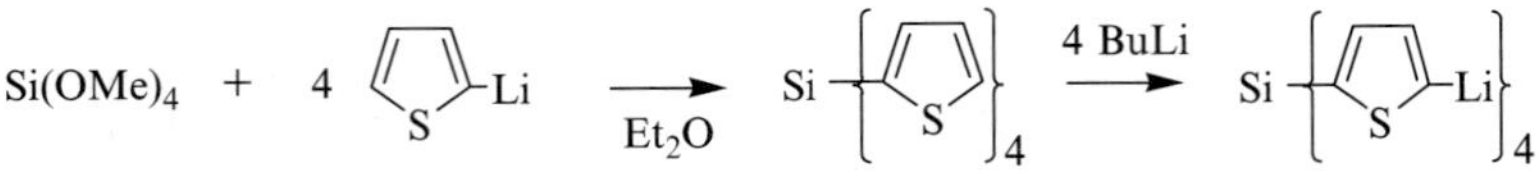
(3.25)

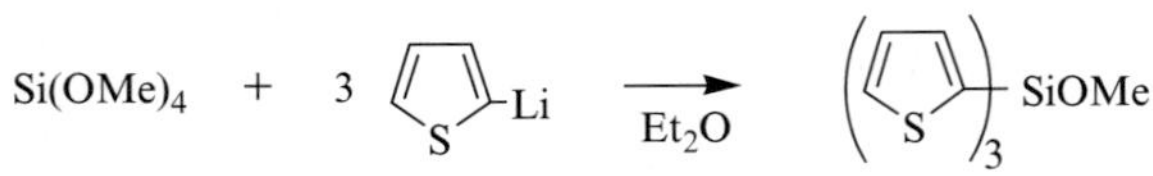
(3.26)

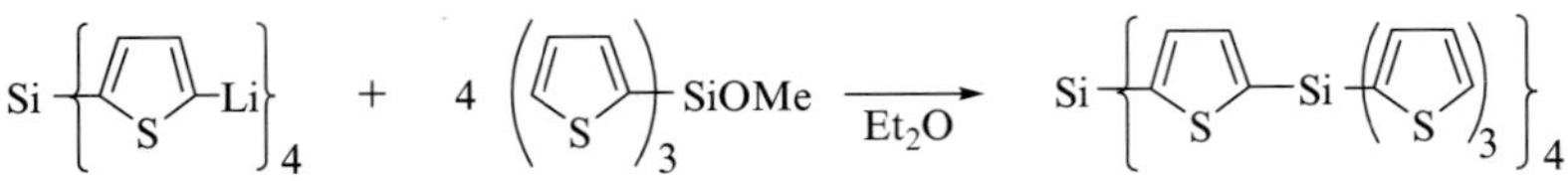
(3.27)

Incomplete lithiation of tetrathienylsilane leads to partial dendrimers that have been isolated and identified [75].

Silylacetylene dendrimers with Si–C≡C repeat units were prepared convergently starting from trimethylsilyl acetylene Grignard reagent and methyldichlorosilane [76]. Conversion of the silicon hydride to silicon chloride allowed for coupling to a central trifunctional core:

$$Me_3Si\text{-}C{\equiv}C\ MgBr + HSi(Me)Cl_2 \longrightarrow (Me_3Si\text{-}C{\equiv}C)_2Si(Me)H \xrightarrow[CCl_4]{PdCl_2} (Me_3Si\text{-}C{\equiv}C)_2Si(Me)Cl \qquad (3.28)$$

$$3\,(Me_3Si\text{-}C{\equiv}C)_2Si(Me)Cl \;+\; MeSi(C{\equiv}C\text{-}MgBr)_3 \longrightarrow MeSi\{C{\equiv}C\text{-}Si(Me)\text{-}(C{\equiv}C\text{-}SiMe_3)_2\}_3 \qquad (3.29)$$

The 1G dendrimer with matrix assisted laser desorption ionization-time of flight (MALDI-TOF) molecular weight M = 850.8 (M + Na^+) and four Si and nine ethynyl branches was obtained in 48% yield. For the 2G dendrimer the $(Me_3SiC{\equiv}C)_2SiMeCl$ product of Reaction Scheme 3.28 was converted in four steps to $[(Me_3SiC{\equiv}C)_2\text{-}SiMeC{\equiv}C]_2SiMeCl$ and then coupled to the same trifunctional Grignard core of Reaction Scheme 3.29. The 2G dendrimer had ten Si branch points and 21 ethynyl branches with MALDI-TOF molecular weight of 1836 (M + Na^+) [76].

A completely different dense carbosilane dendrimer was prepared through a Si–C bond formation process. It consisted of trifunctional carbon branch points and $Si(Me)_2$ branches [77]:

$$HC(SiMe_2Br)_3 \;+\; 3\,LiC(SiMe_2CH{=}CH_2)_3 \longrightarrow HC\{SiMe_2\text{-}C(SiMe_2CH{=}CH_2)_3\}_3 \qquad (3.30)$$

The first generation compound had four carbon branch points and nine peripheral vinyl groups.

3.2.3 Peripheral (Corona; End-Groups) Modification

In this section we describe modification of the peripheral groups of carbosilane dendrimers. Such dendrimers modified with multiple functional groups are proposed for applications in catalysis (see Chapter 9), sol-gel processes (see Chapter 11), liquid crystalline materials (see Chapter 10), star polymers (Section *Star Polymers with Carbosilane Dendrimer Core*, pp. 54–57; see also Chapter 10 and Section 11.3) and biorecognition [78, 79].

3.2.3.1 Modification of Si–Cl End-Groups

The very reactive Si–Cl bonds at the intermediate stage of the carbosilane dendrimer synthesis are ideally suited for introduction of multiple, compactly-placed functional groups. A large number of such transformations have been described. In this chapter, only Si–C compounds are described, while the reactions leading to carbosiloxane and carbosilazane compounds are described in Chapters 2 and 5.

The first modification to be considered is the reduction of Si–Cl bonds to Si–H bonds with $LiAlH_4$ in diethylether. The reaction was pioneered on $SiCl_3$ terminal groups in dendrimers up to 4G(4,3,3,3,3) affording a dendrimer with 324 Si–H bonds [21]:

(3.31)

The conversion is easily recognized by the ^{1}H NMR resonance of Si–H at 3.5 ppm and the ^{29}Si resonance at −53 ppm. The same reaction has also been applied to $Si(Me)Cl_2$ [80] and $Si(Me)_2Cl$ end-groups [16, 80–82]. Dendrimers with $Si(Me)_2H$ end-groups are reagents for "inverse hydrosilylation" reactions wherein the dendrimer carries multiple Si–H bonds. These applications are described in Section 3.2.3.3.

In order to obtain inert dendrimers, it is desirable to create a saturated aliphatic or aromatic external corona. This has been realized in a number of ways starting from the Si–Cl intermediates. For example, Tatarinova et al. reacted $Si(Me)Cl_2$ groups with n-BuLi in hexane [69]:

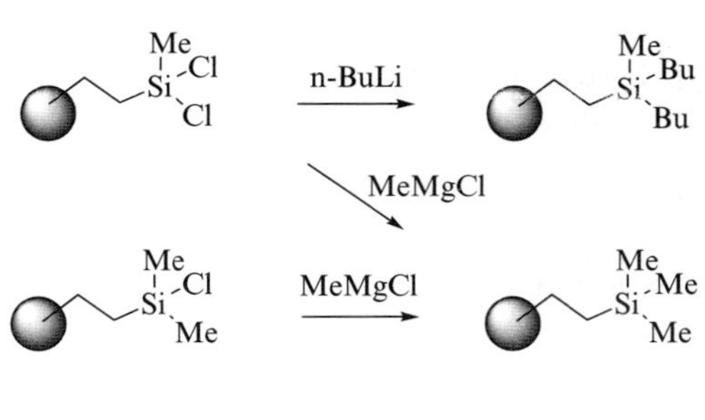

(3.32)

to obtain the product that required chromatographic purification. Others chose reactions of methylmagnesiumchloride with $Si(Me)Cl_2$ [39] and $Si(Me)_2Cl$ end-groups [48]. It is worth noting that the latter approach is preferred over the difficult direct hydrosilylation of allylic double bonds with gaseous trimethylsilane. Similarly, $Si(Ph)Cl_2$ end-groups were converted to $Si(Ph)(Me)_2$ [39]. Terminal $SiCl_3$ and $Si(Me)Cl_2$ were also treated with benzylmagnesiumchloride [83].

Functional groups linked to silicon via a stable Si–C bond are usually introduced by means of organometal, more specifically, organolithium chemistry. The use of organolithium compounds combined with a protection/deprotection strategy is practically inexhaustible. Seyferth et al. produced peripheral ethynyl groups that form $Co(CO)_4$ complexes when treated with $Co_2(CO)_8$ [84]:

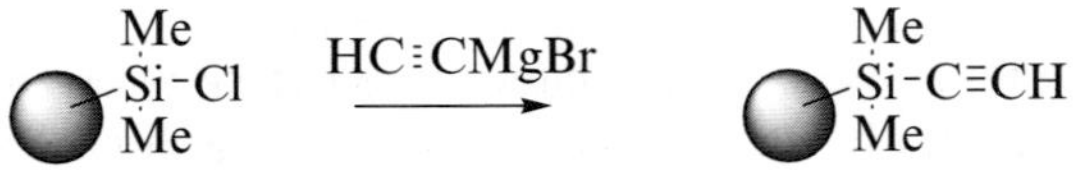

(3.33)

The same reaction was also applied when $Si(Me)Cl_2$ was treated with (Ph)C≡CLi [85] and $Si(Me)_2Cl$ with $[(Ph)_2PCH_2]_2C(CH_3)CH_2C{\equiv}CLi$ or $RO(CH_2)_2C{\equiv}C{-}Li$ [41, 86].

The ferrocenyl group was introduced by reaction of mono- and disubstituted silicon with ferrocenyllithium (see also Chapter 8) [87–89]:

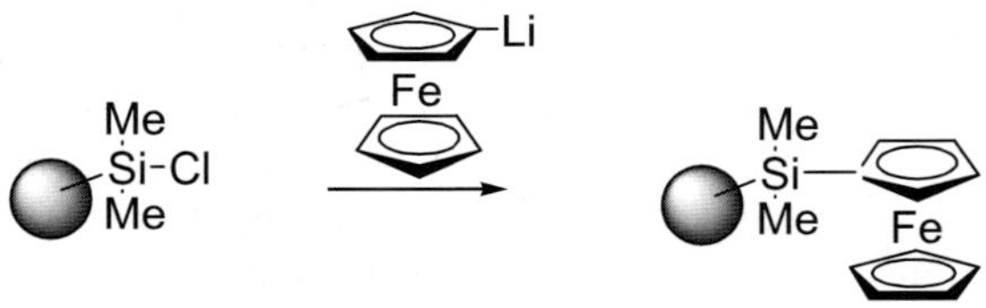

(3.34)

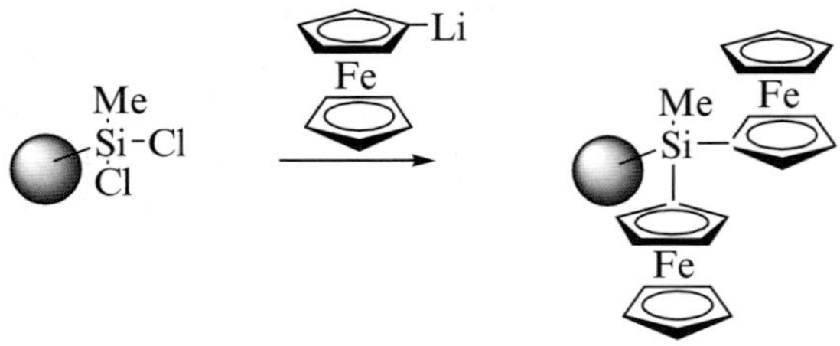

(3.35)

In the monosubstituted dendrimer all ferrocenyl groups have identical redox properties but electronic coupling occurs in the disubstituted dendrimers [89]. A similar method for preparing dendrimers peripherally substituted with carboranes has also been described [90]. 1-Methyl-1,2-*closo*-$C_2B_{10}H_{11}$ carborane was easily lithiated with *n*-BuLi and then reacted with the 1G (4,1) dendrimer:

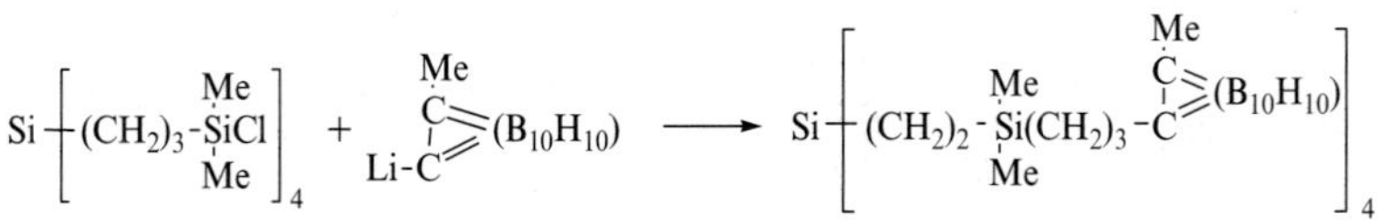

(3.36)

Another example involves the attachment of diphenylmethylphosphine groups via the easily formed lithiummethyl reagent in the presence of tetramethylethylenediamine [91–97]:

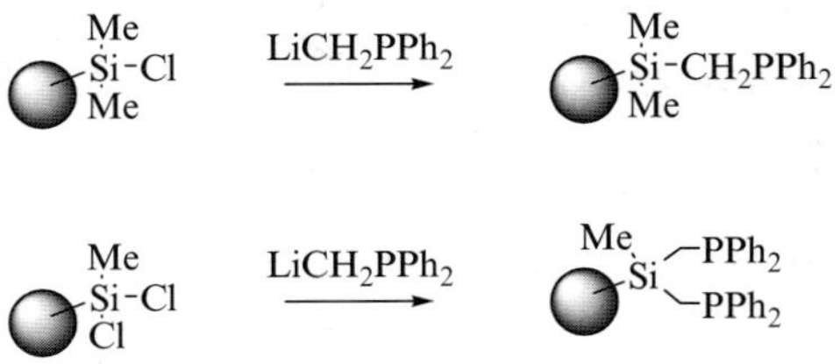

(3.37/3.38)

The phosphines are ligands for organometallic compounds and many of the resulting metallodendrimers have been assessed for their catalytic properties (see Chapter 9) [92, 97–99]. An optically active phosphorus center has also been introduced [65].

A wide variety of functionalized phenyl rings has been placed on the periphery of carbosilane dendrimers. The phenyl ring creates a rigid 5 Å extension of the dendrimer branch. This work has been driven by the search for effective complexing agents for homogeneneous multifunctional organometallic catalysts and is often limited to 2G(4,3,3) dendrimers. In almost all cases, branch extension is on a terminal –Si(Me)$_2$Cl group and the required lithium (occasionally, Grignard reagent) is prepared *in situ* from the phenylbromide or phenyliodide precursor by lithium–halogen exchange with one equivalent of *n*-butyllithium or *t*-butyllithium.

Monolithiation of *para*-dibromobenzene and substitution of Si–Cl leads to carbosilane dendrimers with 4-bromophenyl groups [100]. The 4-bromo substituent itself can be converted quantitatively to the 4-lithiophenyl, a versatile reagent for reactions with many electrophiles, including Cl–P(C$_6$H$_5$)$_2$, Cl–P(C$_6$H$_{11}$)$_2$ [100, 101], R–CO–R and R–COH [100]:

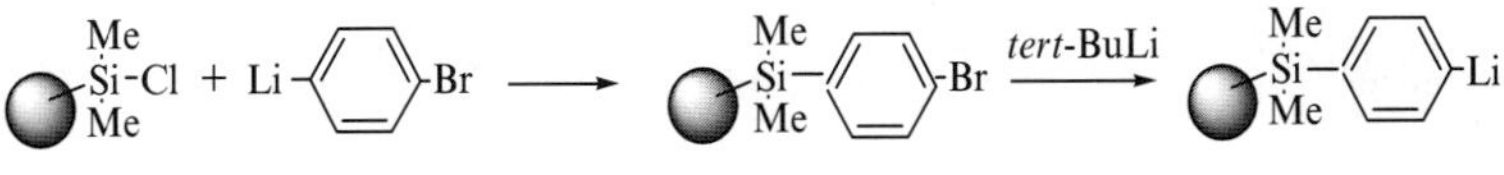

(3.39)

Phenol groups are introduced after protection with *tert*-butyldimethylsilyl groups which can be cleanly cleaved with Et$_3$N HF [100]:

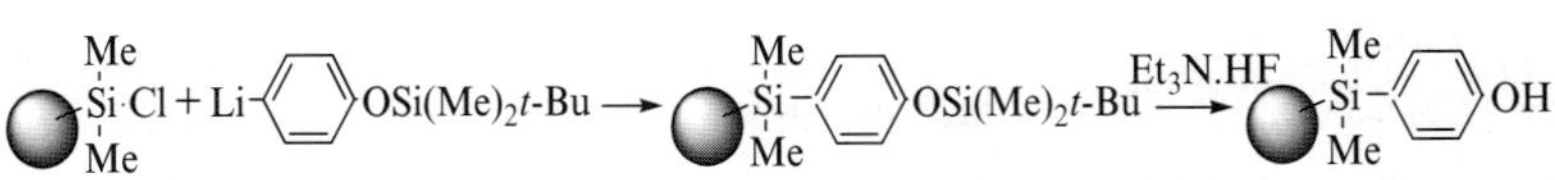

(3.40)

Dimethoxy substituted phenyl rings are introduced via reaction with 2,4-dimethoxyphenylmagnesiumbromide [102]:

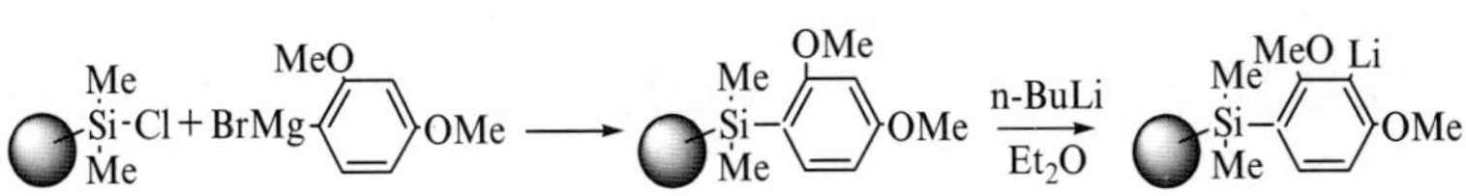

(3.41)

The resulting carbosilanes can be directly lithiated at the 3-(*meta*) position where the lithium atom is strongly stabilized by the two neighboring methoxy groups. However, the related 3,5-dimethoxytolyl group, when attached to carbosilane dendrimers, is simultaneously lithiated at the benzyl position.

Introduction of hydroxyl groups requires prior protection with *tert*-butyldimethylsilyl groups. The general scheme for R = CH_3 or H is [103–105]:

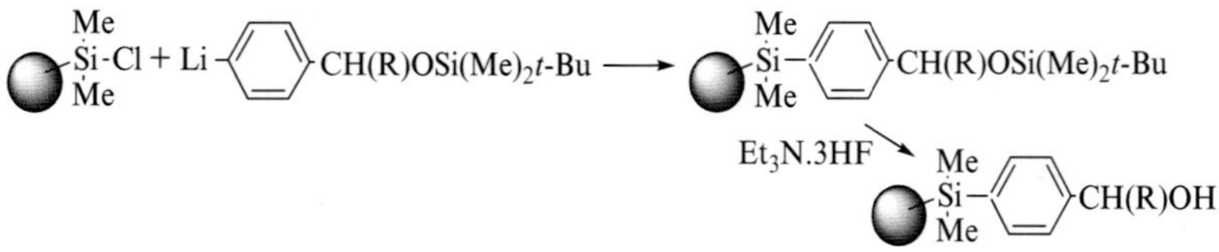

(3.42)

The hydroxyl group is another versatile starting point for further modification, such as esterification or conversion of the hydroxymethyl group to a bromomethyl group [106].

A special case involves the modification of $(Ph)_2SiCl_2$ at the 0G(2) level with 3-hydroxymethylphenyl groups and conversion to the benzyl bromide groups [107]:

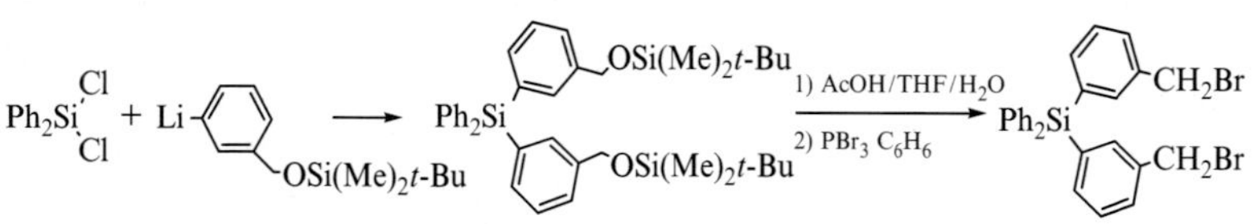

(3.43)

3,5-Di(methoxymethyl)phenyliodide is the starting point for introduction of 3,5-di(methoxymethyl)phenyl groups that can be converted cleanly to benzyl bromide groups [108]:

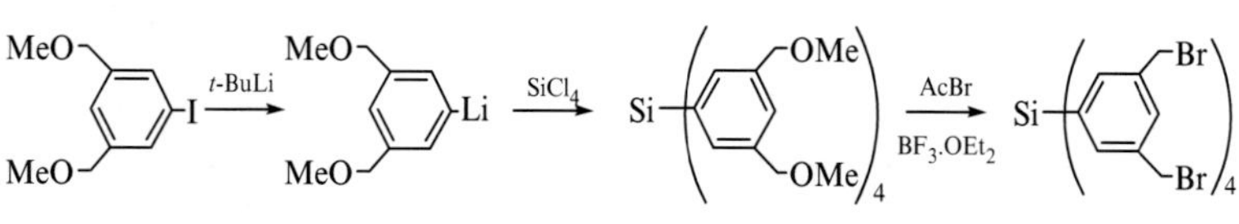

(3.44)

Anilino substitution of peripheral Si–Cl groups proceeds by a similar strategy after primary and secondary amines are protected [100, 109, 110]:

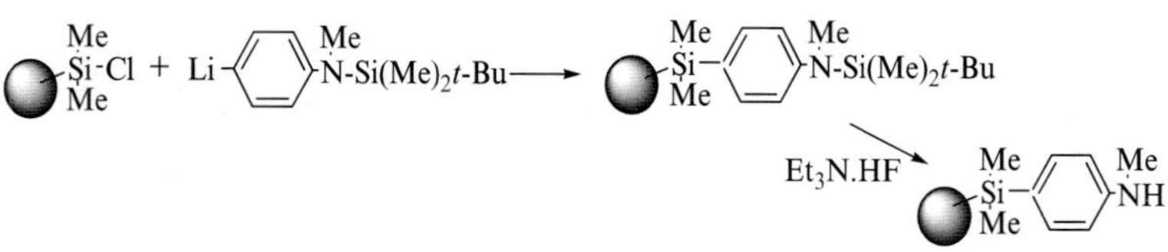

(3.45)

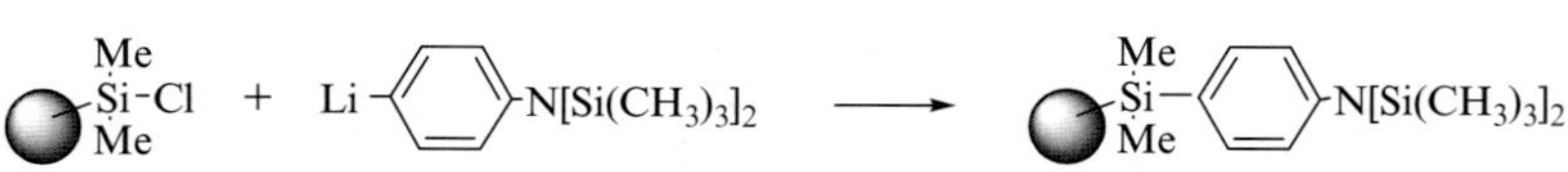

(3.46)

Tertiary amines can be introduced directly, monoamines starting from 4-[(dimethyl-amino)methyl]phenylbromide and diamines starting from 3,5-[bis(dimethylamino) methyl]phenylbromide [111–114]:

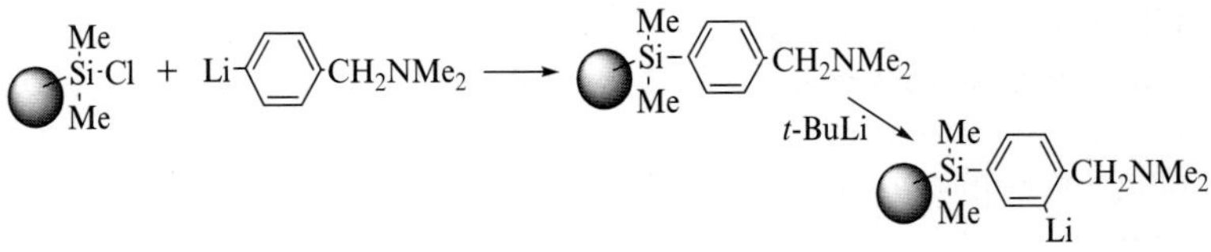

(3.47)

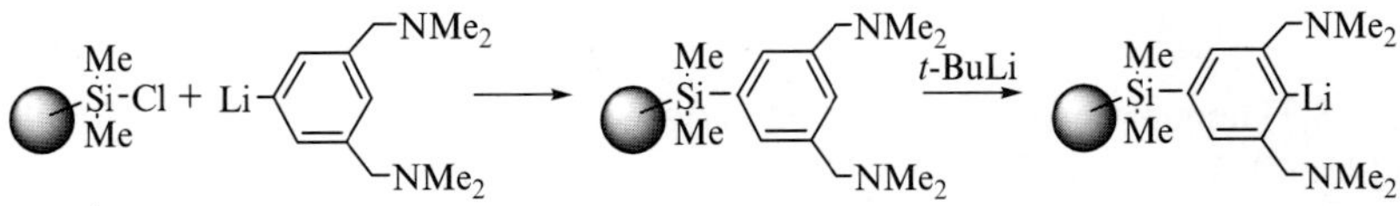

(3.48)

In both cases, lithiation of the position ortho to the tertiary amine functions has been studied extensively for further attachment of catalytic metal centers.

A special case is the introduction of chiral 1R,2S–N-(4-bromobenzyl)–O-(*tert*-butyldimethylsilyl)ephedrine [115, 116] for the construction of carbosilane dendrimers with chiral ligands that are used in the enantioselective addition of dialkylzinc to phosphonylimines:

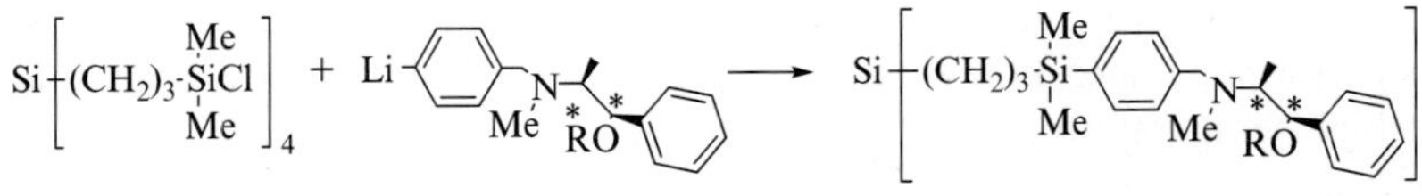

(3.49)

Post modification steps on dendrimers are preferably kept to a minimum in order to optimize the yield of the final product. This is illustrated with the placement of a complex peripheral group on a 2G(4,3,3) dendrimer [117]:

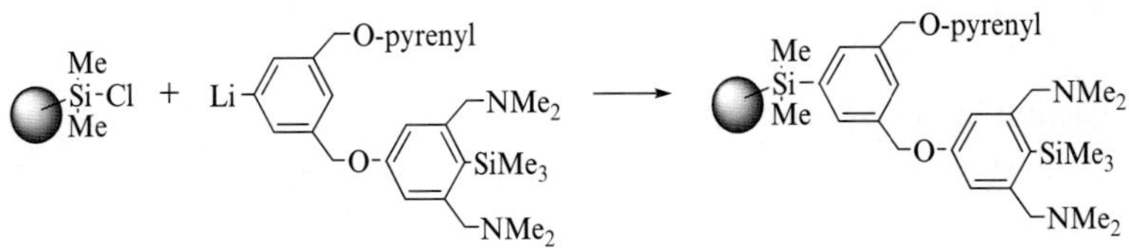

(3.50)

A 1G(4,3) dendrimer was decorated with *meta*-bromopyridine [118], and the resulting compound was used as a reagent for multiple Suzuki coupling reactions with *para*-tolylboronic acid:

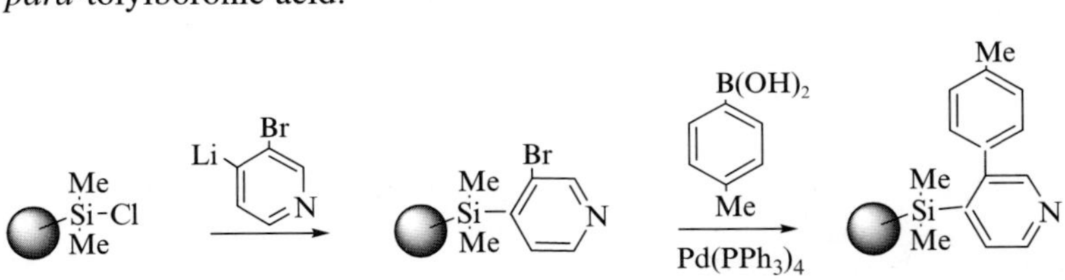

(3.51)

3.2.3.2 Modification of Si–$(CH_2)_n$–CH=CH_2 End-Groups

Carbosilane dendrimers with multiple allyl (n = 1) or vinyl (n = 0) double bonds (see Reaction Scheme 3.1) are sensitive to slow oxidation and polymerization in air and must be stored at –20°C in the absence of oxygen [69]. As a consequence, in order to work with core modified dendrimers or to study physicochemical properties of carbosilane dendrimers it is desirable to decorate them with stable alkyl or aryl terminal groups. For example, peripheral vinyl and allyl double bonds have been hydrogenated directly with H_2 under mild conditions to provide ethyl and *n*-propyl end groups:

(3.52)

0G to 2G(3,3,3) dendrons were hydrogenated over 10% Pd/C in EtOAc/MeOH with 1 atm H_2 [45, 49, 119]. The 2G(3,3,3) dendron was also treated over 10% PtO_2 [49]. An alterative approach is hydrosilylation with R_3SiH compounds which adds one generation to the dendrimer. A variety of alkyl and aryl substituents have been used:

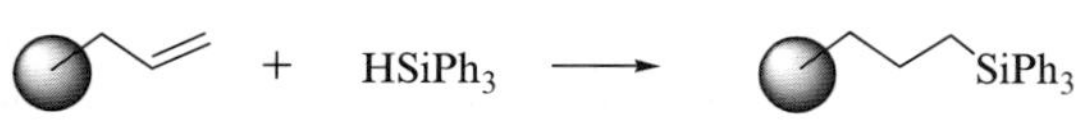

(3.53)

The phenyl substituted reagents are probably more reactive due to the electron withdrawing property of the phenyl ring [119–121]. Triphenylsilane end-groups impart crystallinity to dendrimers [121] so that dimethylphenylsilane [120] and diphenylmethylsilane [40, 122] are alternatives. With all-aliphatic substituents the reaction is performed under more forcing conditions with excess silane and without solvent at high temperature [40, 122]. Triethylsilane (b.p. = 107°C) and benzyldimethylsilane (b.p. = 71°C) have been used for this purpose [40, 119, 122].

Hydrosilylation of the allylic double bonds allows introduction of a variety of functional groups. For example, perfluoroalkanes are peripherally placed on dendrimers in this way, without the need for –S– or –O– bridging [123]:

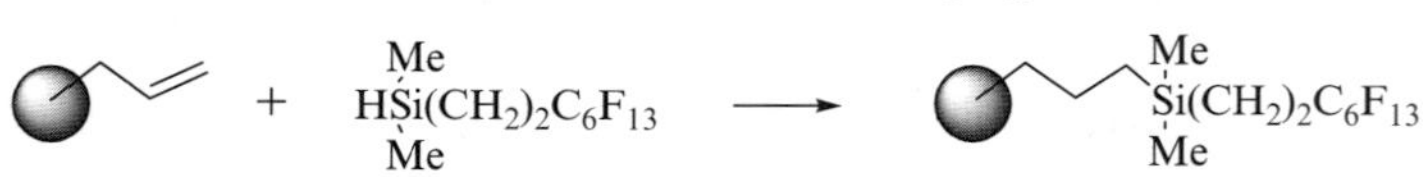

(3.54)

This type of reaction has been extensively used by the Russian school. It was originally used to produce carbosilane-based multi-mesogen liquid crystalline materials with a long aliphatic spacer and a flexible dimethylsiloxane bridge [54]:

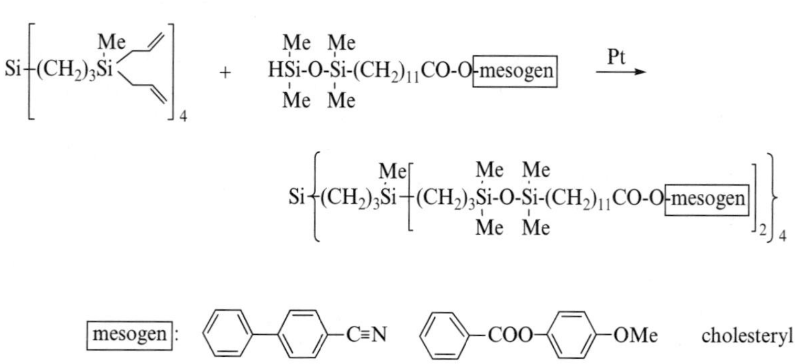

(3.55)

The procedure was successful for the preparation of a 5G(4,2,2,2,2,2) dendrimer with 128 *p*-cyanodiphenyl groups [124]. A variety of other substituents, including methoxyundecylate groups [125], chiral mesogens [126–128], bent mesogens [129], photosensitive cinnamoyl [130] and azobenzene [131, 132] groups have also been introduced by this technique. The synthesis and properties of such dendrimers are reviewed in detail in Chapter 10.

Muzafarov and co-workers performed hydrosilylations with trimethylsilyl protected mono- and trihydroxy compounds [133, 134]:

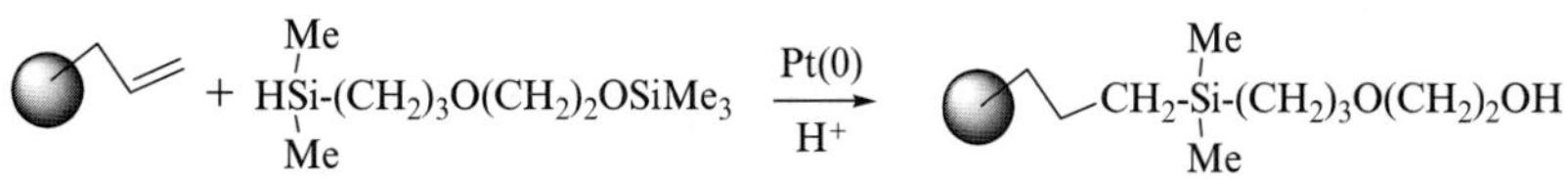

(3.56)

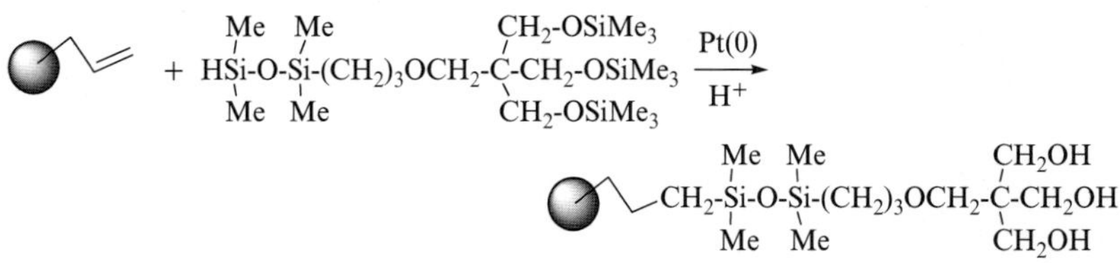

(3.57)

The trimethylsilyl protecting groups were removed by transetherification in MeOH with 20% acetic acid. The resulting triol substituted dendrimer can be considered a hybrid with a carbosilane core and a Newkome-type polyol periphery [135]. The fifth generation dendrimers 5G(3,2,2,2,2,2) and 5G(4,2,2,2,2,2) had 288 and 384 HO groups, respectively. One particularly striking extension is the construction of a mixed dendrimer with carbosilane core and methylsiloxane corona [136]:

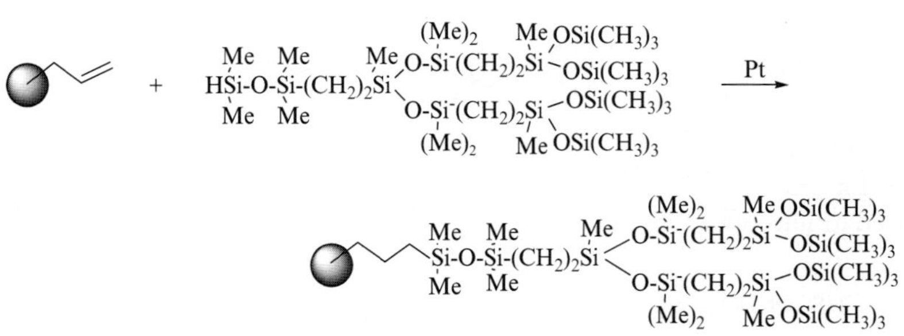

(3.58)

Hydrosilylation of 1G and 2G(4,2,2) vinyl terminated dendrimers with di(ω-hydroxyhexyl)methylsilane proved to be more satisfactory than direct substitution of Si–Cl terminated dendrimers with tetrahydropyranyl (THP)-protected ω-hydroxyhexylmagnesiumbromide [67]:

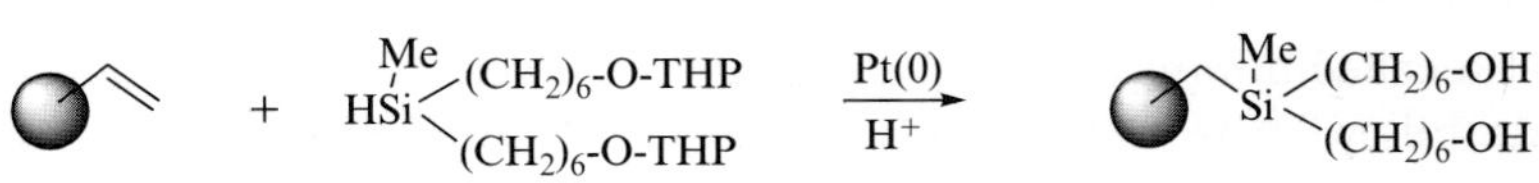

(3.59)

Moreover, one extra generation is added to the dendrimer in the process. After deprotection, the polyhydroxy dendrimers were good initiators for anionic polymerization of ethylene oxide [137].

In another approach, vinyl terminated dendrimers were first converted to alkylchloride derivatives by the hydrosilylation with $HSiMe_2CH_2Cl$ [62]. The CH_2Cl group was then reacted under basic conditions with numerous thiol compounds to introduce a variety of functional groups:

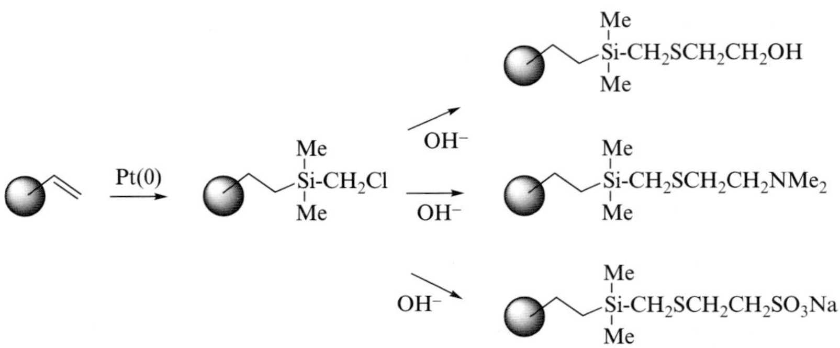

(3.60)

The same procedure was also used to introduce thioglycolic acid, its ethylester and various perfluoroalkyl groups [138]. Hydroxyl terminated carbosilane dendrimers are soluble in methanol and ethanol but not in water [62, 133]. The charged dendrimers with $-SO_3^-Na^+$ and $-NHMe_2^+\ Cl^-$ groups are water soluble [62].

The radical addition-hydrogen abstraction reaction was performed with a large excess of thio-compound and used to form a perfluoroalkyl corona around a carbosilane core [139]:

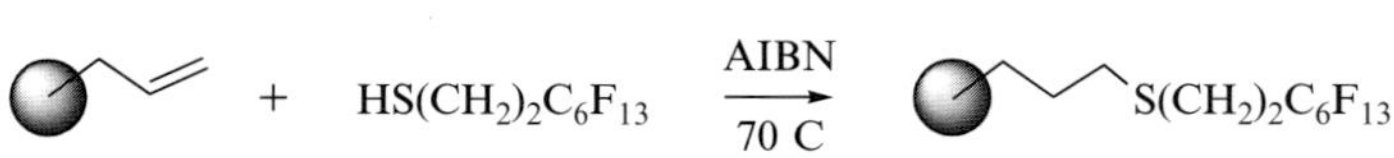

(3.61)

The contrast between carbosilane and fluoroalkyl groups leads to liquid crystalline mesophases for 1G to 3G(4,3,3,3) materials. The same reaction with thiolacetic acid was used to introduce terminal thiol groups [140]:

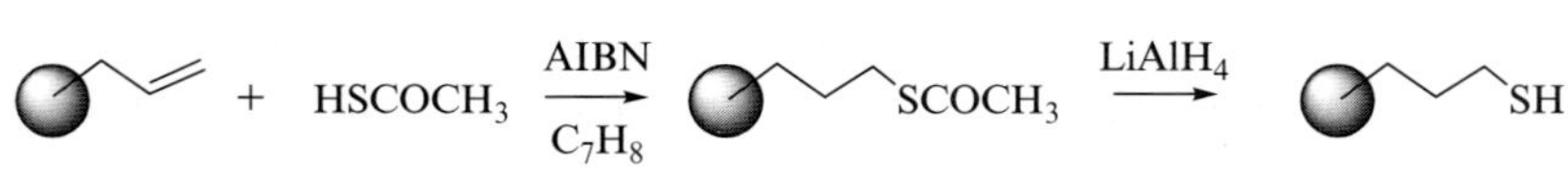

(3.62)

The hydrogen abstraction-addition reaction, initiated with azobisisobutyronitrile (AIBN) as the initiator at 50°C, has also been used to attach phosphines HPR_2 (R = Et, Ph) to vinylsilane double bonds [141, 142].

Allyl terminated dendrimers are good substrates for direct introduction of terminal hydroxyl groups to form a hydrophilic shell around the hydrophobic skeleton and for use in further modifications. The first carbosilane-based dendritic polyols, sometimes also called *arborols* (see [21] of Chapter 1), were obtained via direct hydroboration followed by alkaline H_2O_2 oxidation of peripheral allyl groups

[22, 143, 144]. The reactions must be carried out carefully in order to avoid potential crosslinking [145]:

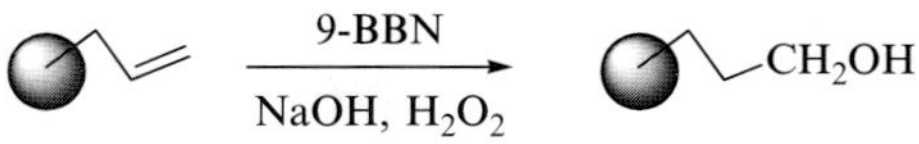

(3.63)

The MALDI-TOF analysis of 3G(4,3,3,3) dendrimer with a polyol corona revealed dendrimers with 94 to 108 hydroxy groups, in agreement with the analysis of the multifunctional allyl parent. The same method was followed by others to prepare 4G(4,2,2,2,2) and 2G(8,3,3) polyols [30, 146]. Treatment of the 2G polyol with carbonyl chloride was used to introduce mesogenic groups [147, 148]. Subsequent modification of hydroxyl groups to bromopropyl, iodopropyl, azidopropyl, and aminopropyl has also been described [144, 149]. These dendrimers offer gateways to further derivatization; for example, with saccharides [150, 151].

3.2.3.3 Modification of Si–H End-Groups

The Si–H peripheral groups are introduced into carbosilane dendrimers by reaction of Si–Cl functionalized precursors with $LiAlH_4$ in diethylether. The Si–H bond is relatively stable when care is exercised and it is often used as a starting function for further dendrimer functionalization via hydrosilylation when functional molecules with vinyl or allyl double bonds are available. We call this approach "*inverse hydrosilylation*" in contrast to the classical divergent hydrosilylation with $HSiR_nX_{3\text{-}n}$ compounds. Preliminary studies of inverse hydrosilylation were described in 1993 and 1995 for the addition of vinylferrocene to octakis(dimethylsiloxy)octasilsesquioxane [152] and to tetramethylcyclotetrasiloxane [153], respectively (see also Chapter 8). The approach was first systematically applied to the construction of bis(ferrocenyl)silyl substituted carbosilane dendrimers [89]:

$$\mathrm{Si}\left[(\mathrm{CH_2})_3\text{-}\underset{\mathrm{Me}}{\overset{\mathrm{Me}}{\mathrm{SiH}}}\right]_4 + \mathrm{CH_2{=}CH\text{-}Si(Me)(ferrocenyl)_2} \xrightarrow{\mathrm{Pt}} \mathrm{Si}\left[(\mathrm{CH_2})_3\text{—}\underset{\mathrm{Me}}{\overset{\mathrm{Me}}{\mathrm{Si}}}\text{—}\mathrm{CH_2\text{-}CH_2\text{-}}\overset{\mathrm{Me}}{\mathrm{Si}}\mathrm{(ferrocenyl)_2}\right]_4 \qquad (3.64)$$

and a larger 1G tetraferrocenyl compound was also used. Such inverse hydrosilylation can be considered the final step in a convergent dendrimer synthesis.

The same reaction has been applied to modify the carbosilane dendrimer periphery when vinyl or allyl substituted functional compounds are easily available. In most cases substitutions are limited to the 0G(4) compound and product purification is often indicated. Arévalo et al. carefully studied the hydrosilylation of eugenol, $CH_2{=}CHCH_2C_6H_3(OCH_3)OH$, with Et_3SiH and established that the phenolic groups require trimethylsilyl protection in order to prevent side reactions including O-silylation and simultaneous isomerization of allyl to propenyl [154–156]. Reaction in a minimum amount of THF at 70°C for 9 h provided a 90–95% yield and could then be extended to 3G(4,2,2,2) dendrimers [154]:

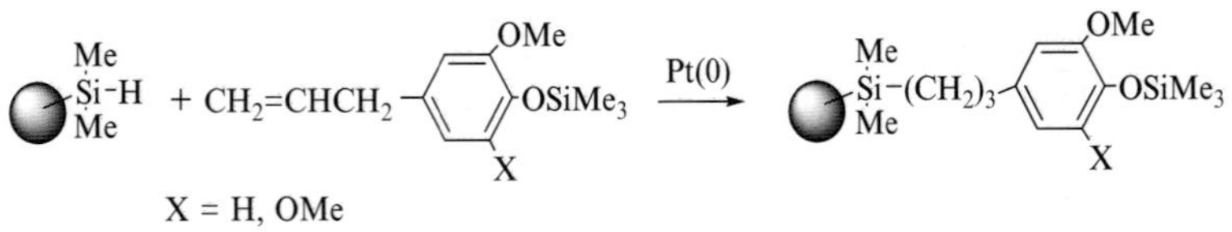
(3.65)

Primary amines were directly introduced via hydrosilylation with excess allylamine in a minimum amount of THF at 120°C giving moderate 80% yield for 0G(4) and 55% yield in the case of the 1G(4,2) dendrimer. N-silylation is a possible side reaction here. However, it was not studied in detail [157]:

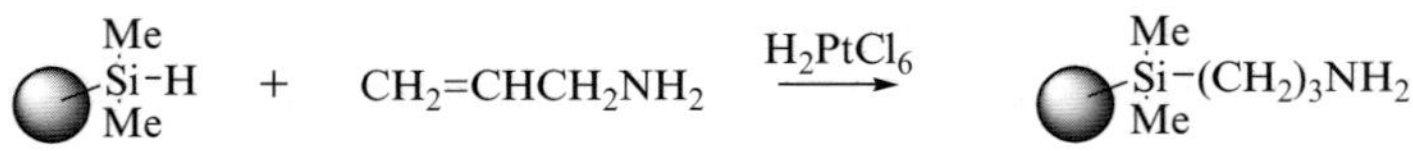
(3.66)

Findeis and Gade used inverse hydrosilylation as an alternative to modification of Si–Cl with Li–C≡C–R so that no triple bonds appear in the dendrimer. However, the reaction was performed only on the 0G(4) model [86]:

$$\mathrm{Si}\left[(CH_2)_3\text{-}\underset{Me}{\overset{Me}{Si}}H\right]_4 + CH_2{=}CHCH_2\text{-}\overset{Me}{C}(CH_2Cl)_2 \xrightarrow{Pt(0)} \mathrm{Si}\left[(CH_2)_3\text{-}\underset{Me}{\overset{Me}{Si}}(CH_2)_3\text{-}\overset{Me}{C}(CH_2Cl)_2\right]_4 \quad (3.67)$$

Allyloxy substituted reagents seem particularly suited for inverse hydrosilylation, as shown by a comparison of their effectiveness with that of α-alkenyl substituted reagents [158]:

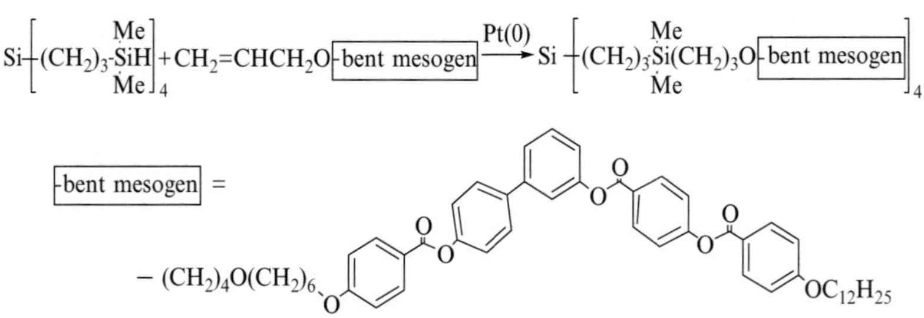
(3.68)

The yields for introduction of four bent-core mesogens (see also Chapter 10) into 0G(4) and 16 bent-core mesogens into 2G(4,2,2) dendrimers were 32% and 39%, respectively. Similarly, ionophores, either triethyleneglycol or crown ethers (15-crown-5, 18-crown-6), with allyloxy handles were introduced into 0G(4) or 1G(4,2) dendrimers [159–161]. The same reaction was used to attach the monoallylether of pentaerythrytol suitably protected with tosylates [162]:

$$\mathrm{Si}\left[(CH_2)_3\text{-}\underset{Me}{\overset{Me}{Si}}H\right]_4 + CH_2{=}CHCH_2O\text{-}C\text{-}(CH_2OTs)_3 \xrightarrow{Pt(0)} \mathrm{Si}\left[(CH_2)_3\text{-}\underset{Me}{\overset{Me}{Si}}(CH_2)_3O\text{-}C(CH_2OTs)_3\right]_4 \quad (3.69)$$

The complete reaction creates a mixed carbosilane-Newkome-type dendrimer [135] and the protecting tosylate groups are versatile leaving groups that facilitate further derivatization; for example, by thiocyanate to form thiol end-groups [162].

Allyl diisopropyloxyboron reacted quantitatively with 0G(4) dendrimer, as shown in Reaction Scheme 3.70 [163]:

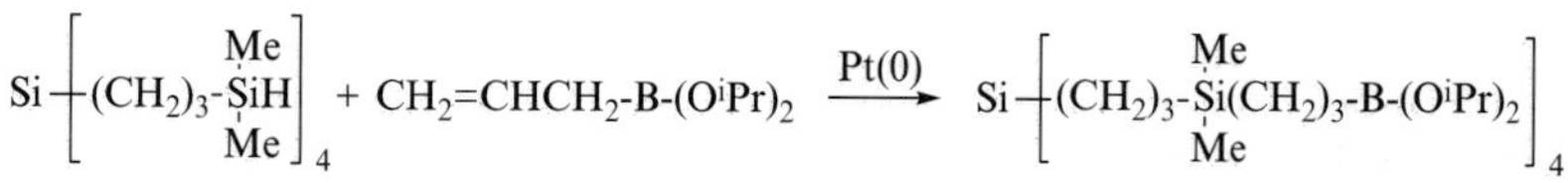

(3.70)

In a recent study of end-functionalization of carbosilane dendrimers with carboranes, hydrosilylation and inverse hydrosilylation have been closely compared:

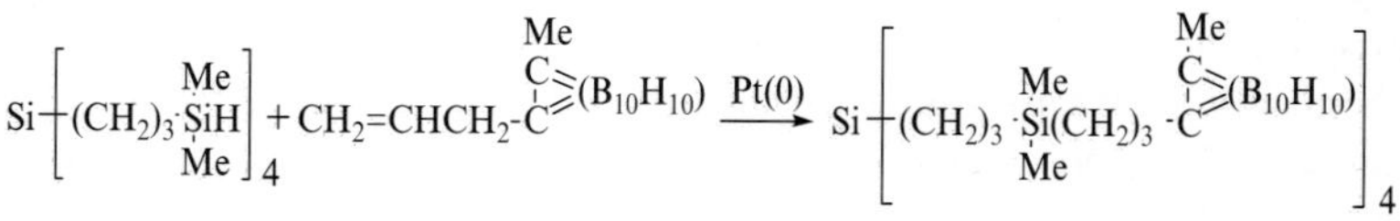

(3.71)

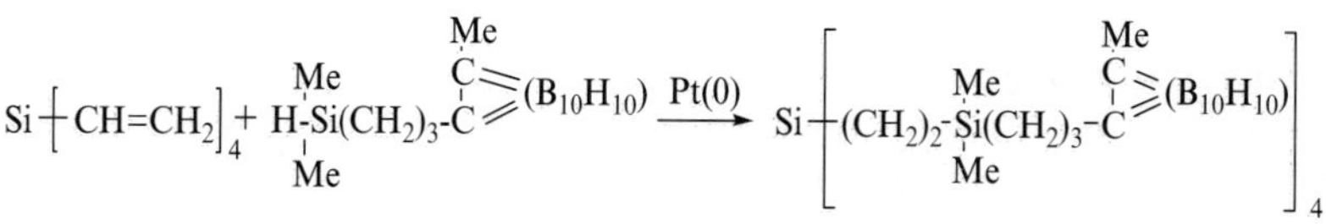

(3.72)

When 1-phenyl or 1-methyl-2-allyl-1,2-*closo*-carborane was added to a 0G(4) carbosilane hydride, as shown in Reaction Scheme 3.71, the yield was 36%. On the other hand, when the same carboranes were first hydrosilylated with $HSi(Me)_2Cl$ and then reduced with $LiAlH_4$ [164], hydrosilylation of tetraallylsilane failed but hydrosilylation of tetravinylsilane gave 99% yield [165] (see Reaction 3.72). Direct reaction of the Si–Cl terminated 0G dendrimer with 1-methyl-2-lithio-1,2-*closo*-carborane has also been performed [90], as shown in Reaction Scheme 3.36.

Overall, the available evidence suggests that inverse hydrosilylation is a reaction with limited yield that can only be applied effectively to low generation multifunctional hydrides after carefully establishing reaction conditions. There are several possible reasons for this, including the bulky size of the hydride which may limit the formation of the active Pt intermediate, and the impossibility of using excess hydride which prevents promoting the yield in the usual fashion.

Finally, it should be mentioned that the Si–H bond can be easily converted to Si–Metal as shown in the following example [166]:

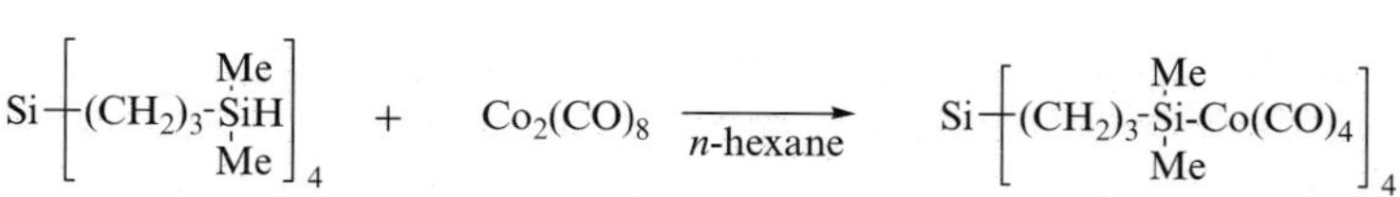

(3.73)

3.2.3.4 Carbosilane Dendrimer–Polymer Hybrids

In this section we deal with two types of dendrimer-linear polymer hybrids (sometimes also referred to as architectural copolymers), including: (a) carbosilane dendrimers to which linear polymers chains are attached, usually referred to as *star polymers* (see also Section 11.3), and (b) linear polymers decorated with pendant carbosilane dendrons, the so-called *dendronized polymers* (see also Section 8.2.2.5). Three general reviews of such dendrimer-polymer hybrids can be found in the literature [167–169].

Star Polymers with Carbosilane Dendrimer Core

Derivatization of dendrimers with linear polymers is an efficient use of expensive dendrimers because of the large molecular mass increase that is realized in the process. Moreover, the resulting star polymers have unusual colloidal properties [170]. The chemically inert carbosilane dendrimers are uniquely suited for modification with anionic living polymers. The basic reaction involved in this process is exemplified with polybutadienyllithium:

$$SiCl_4 + Bu\text{-}(CH_2CH{=}CHCH_2)_n\text{-}Li \longrightarrow Si\left[(CH_2CH{=}CHCH_2)_n\text{-}Bu\right]_4 \quad (3.74)$$

It has been shown that this reaction is essentially free of side reactions when performed in hydrocarbon solvents [171, 172]. Earlier work led to two crucial observations: (1) Si–Cl bonds on a single Si atom are successively substituted, and (2) in the substitution, the least sterically hindered anion is favored, i.e. polybutadienyllithium over polyisoprenyllithium over polystyryllithium, independent of the molecular weight [173, 174]. Even in the least favorable case, coupling of polystyryllithium with dendrimers containing multiple $-Si(Me)Cl_2$ end-groups is successful [173]:

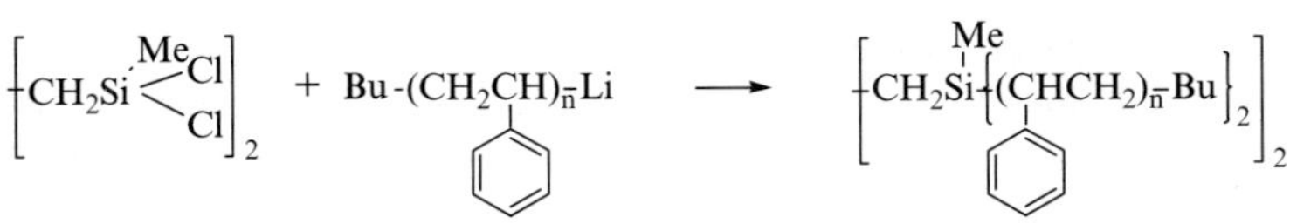

(3.75)

3G(4,2,2,2), 4G(4,2,2,2,2) and 5G(4,2,2,2,2,2) C_2 carbosilane dendrimers with $-Si(CH_3)Cl_2$ end-groups have been reacted with excess polybutadienyllithium [19, 20, 175, 176]:

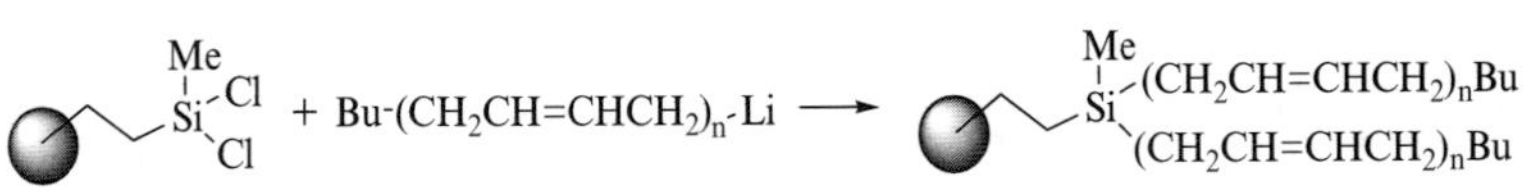

(3.76)

In this reaction, narrow molecular weight distribution (MWD) linear polymers are transformed into narrow MWD star polymers with 32, 64, and 128 arms, respectively.

Application of successive substitution of multiple Si–Cl bonds on one silicon atom has lead to miktoarm star polymers (ABC, A_nB_n, etc.) [174, 177]:

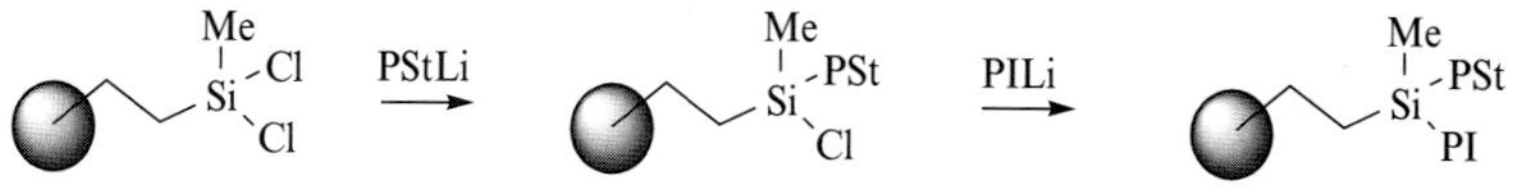

(3.77)

A 2G(4,2,2) dendrimer with 16 Si–Cl bonds is first reacted with 0.5 equivalents of the sterically demanding polystyryllithium followed by the addition of a 30% excess polyisoprenyllithium. The intermediate eight-arm star polymer and the final star blockcopolymer were fully characterized. The same scheme was also followed with a 2G(3,2,2) dendrimer [178]. The special placement of the blocks influences the microphase behavior of the miktoarm stars.

In addition to the "*grafting to*" reactions described above, a "*grafting from*" approach using carbosilane dendrimers with multiple lithiated sites has also been proposed. However, multifunctional lithium compounds are notoriously insoluble in hydrocarbon solvents due to their great tendency to associate intermolecularly. Nevertheless, an ingenious method has been devised to decorate the interior of carbosilane dendrimers with an exact number of C–Li sites for initiation of the polymerization of styrene and polar monomers [179, 180]:

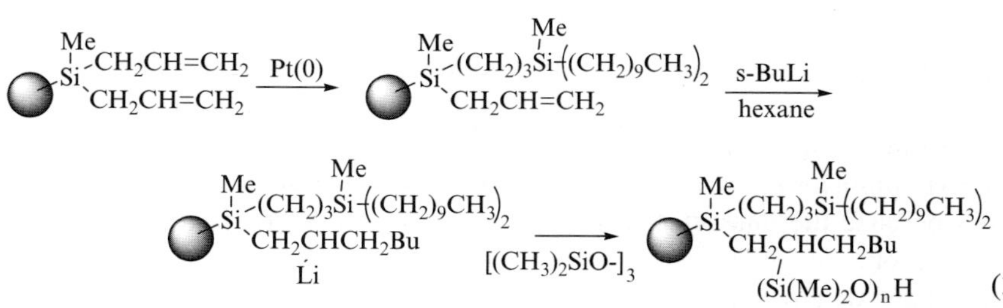

(3.78)

The allylic end-groups of a 2G(4,2,2) and 3G(4,2,2,2) dendrimer were first hydrosilylated with 0.5 equivalents of didecylmethylsilane and then the remaining allyl groups were reacted with *sec*-BuLi in the presence of a stoichiometric amount of tetramethylethylenediamine (TMEDA). The solubility of the resulting octa- and hexadecalithium initiators was explained by a reduction of the lithium association and/or by a preference for intra-dendrimer association. However, on polymerization of hexamethylcyclotrisiloxane the living polydimethylsiloxane solutions formed physical gels indicating that the $–Si(Me)_2OLi$ end-groups associate outside the dendrimer cores. Note that this type of initiator leads to somewhat broader molecular weight distribution polymers (M_w/M_n ranging between 1.2 and 1.3). The method was expanded to 3G(3,2,2,2) and 5G(3,2,2,2,2,2) dendrimers for the preparation of 12 and 48-arm polydimethylsiloxane stars [181]. Intrinsic viscosities of the stars clearly revealed their size contraction relative to linear polymers but the contraction factors showed unusual molecular weight dependence [181].

This grafting-from method was also used to prepare 8, 16 and 32-arm poly(ethylene oxide) star polymers [137, 182]:

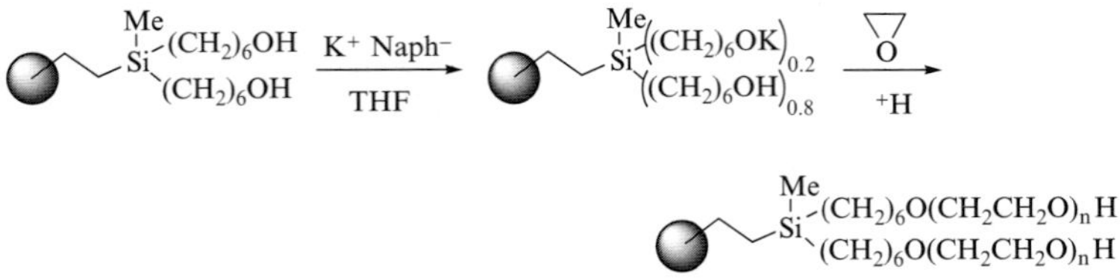

(3.79)

In this case, a fraction of the peripheral hydroxyl groups was converted to potassium alkoxide with potassium naphthalene in THF. This prevented ionic association and consequent insolubility of the initiator. The rapid exchange of potassium ions between CH_2OK and CH_2OH groups relative to the propagation step guaranteed narrow molecular weight distribution star polymers.

The 0G and 1G dendrimers with 2-bromoisobutyrate end-groups were used for the initiation of the pseudo-living radical polymerization of methyl methacrylate in the presence of Cu(I) complex [183]:

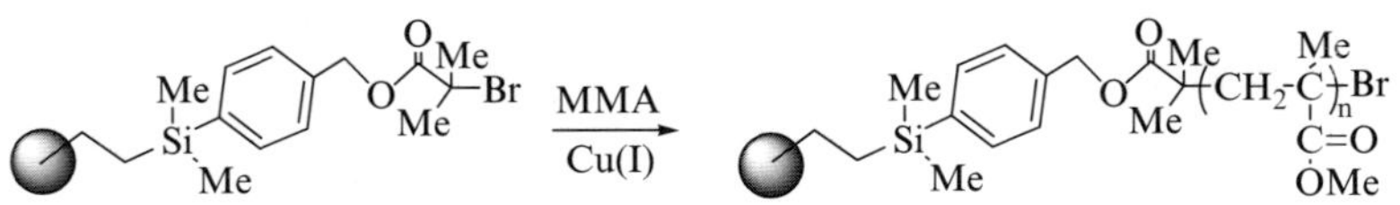

(3.80)

This process works well and yields narrow MWD star polymers provided the initiator concentration is relatively high and monomer conversion is kept below 30%. At higher conversion, coupled star polymers appear due to chain termination by recombination. Incompletely substituted dendrimers are poor initiators, possibly because the unreacted allyl groups on the dendrimer participate in the polymerization.

Carbosilane dendrimers have also been modified with ring opening metathesis (ROMP) initiator sites. In a prototype reaction, tetraallylsilane was treated with a tungsten carbene in an olefin metathesis reaction:

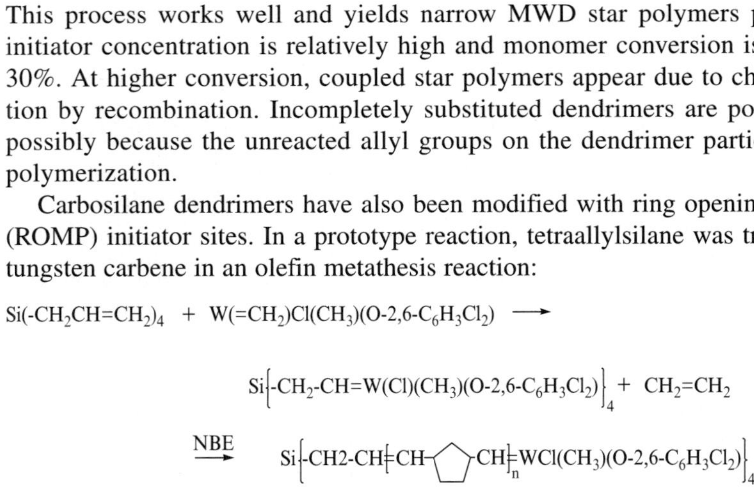

(3.81)

and successfully used to initiate the polymerization of norbornene (NBE) [184]. The molecular weight and molecular weight distribution increased rapidly after the propagation step due to dismutation of the living ends. The arm coupling of stars results in high molecular weight and highly branched materials. Multiple Ru-carbenes on 0G (tetraallylsilane) and 1G(4,2) dendrimer were excellent initiators for norbornene [185, 186]. A combination of end-group analysis and size exclusion chromatography (SEC) indicated that narrow molecular weight distribution star-branched polymers were obtained [186].

In an exceptional case, the amine-functionalized focal point of a dendron was reacted with a nickel complex to form a carbene polymerization initiator for L-alanine-L-alanine isocyanide [187]:

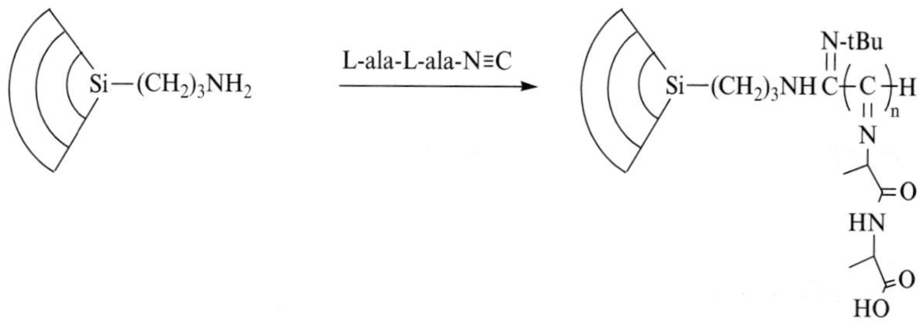

(3.82)

Isocyanide polymers are characterized by a substituent on every chain carbon. The resulting dense packing and the interaction of the functional groups in the side chains lead to stiff, rod-like, helical polymers. In the presence of Ag^+ ions the block copolymers composed of flexible spherical carbosilane dendrimers and rigid polyisocyanides self-assemble into discrete nanoarrays. On irradiation of some third generation arrays the Ag^+ ions were reduced to metallic silver that was predominantly oriented as [111] silver crystals.

Carbosilane Dendronized Polymers

Most carbosilane dendronized polymers have been obtained by growing the dendrons from every repeat unit or from a fraction of repeat units of a linear polymer chain. In a sense, the linear polymer chain serves as a multifunctional core for these dendrons. Dendronized monomers have also been polymerized into dendronized polymers to ensure that an identical dendron is attached to every repeat unit of the resulting polymer.

Kim and collaborators developed a method to grow carbosilane dendrons from a linear carbosilane polymer [38, 188]. This is a unique design as both polymer-core and dendrons have similar chemical compositions. The linear carbosilane polymer ($n \geq 10$) was obtained by the self-hydrosilylation of diphenylallylsilane, following which one or both phenyl rings were displaced by triflic acid in toluene at −78°C:

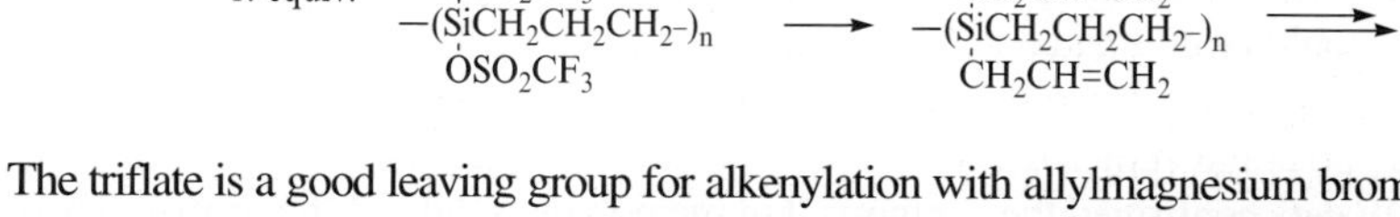

(3.83)

The triflate is a good leaving group for alkenylation with allylmagnesium bromide. Further construction of the dendronized polymer through reiteration of hydrosilylation and alkenylation reactions led to a linear singly or doubly dendronized polymer with 3G(2,3,3,3) dendrons, each carrying 27 peripheral allyl groups. Analysis of such polymers, however, is difficult as the dendrons and the polymer backbone have the same chemical composition.

Dendronized polysiloxanes with carbosilane dendrons to the second generation (four allyl groups) were prepared from poly(methylhydrosiloxane) [189]:

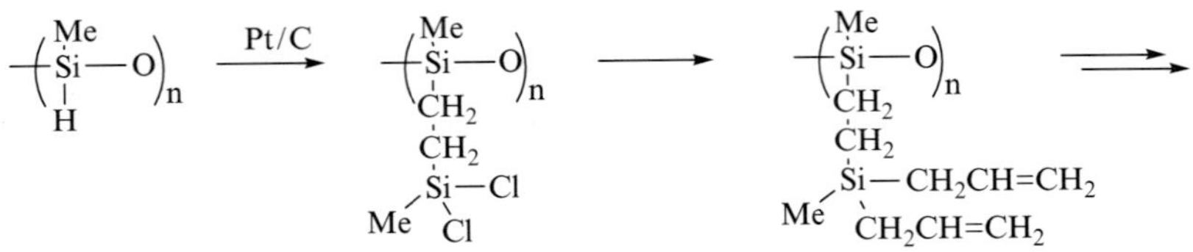

(3.84)

The presence of methyl groups on the polymer backbone is a convenient marker for quantitative ^1H NMR and SEC analyses which can prove the polymer molecular weight integrity. The same poly(methylhydrosiloxane) linear polymer was also dendronized with trichlorosilane and alkenylated with allylmagnesium bromide [190]. The 1G(1,3) dendronization yielded 100% conversion while the 2G(1,3,3) conversion was limited to 84%. Molecular weight and radii of gyration from small angle neutron scattering (SANS) in THF-d_8 were obtained for narrow molecular weight distribution fractions of the linear precursor and the 1G and 2G dendronized polymer. Since dendronization is known to stiffen the linear polymer backbone, the exponent ν in

$$<R_g^2> = K \cdot M_w^{2\nu} \qquad \text{(E.3.1)}$$

increased from 0.53 (typical for linear polymers in good solvent) to 0.84 and 0.93 for the 1G and 2G dendronized polymers, respectively. Note that the last value was near unity which is characteristic of rod-like polymers. Similarly, the persistence length, a good measure of backbone stiffness, increased from 14 to 44 Å for the 1G and to 200 Å for the 2G dendronized polymer.

Chang and Kim synthesized and characterized a triblock copolymer in which carbosilane dendrons were constructed on the chain-ends of a narrow molecular weight distribution poly(ethylene oxide) (PEO) (M_w = 1,000, degree of polymerization (DP) = 22) [191]. The PEO was diallylated and cycles of hydrosilylation with $HSiCl_3$ and alkenylation with allylmagnesiumbromide were repeated on each end to build 3G(1,3,3,3) dendrons with 27 allyl groups:

$$\left[CH_2O(CH_2CH_2O)_{10.5}H\right]_2 \longrightarrow \left[CH_2O(CH_2CH_2O)_{10.5}\text{-}CH_2CH{=}CH_2\right]_2 \Longrightarrow$$

$$\left[CH_2O(CH_2CH_2O)_{10.5}\text{-}(CH_2)_3Si(CH_2CH{=}CH_2)_3\right]_2 \qquad (3.85)$$

This synthesis offered an excellent opportunity to assess the perfection of the reactions since the 12, 36 and 108 CH_2= hydrogens of the terminal allyl groups in the 1G, 2G and 3G dendrons could be compared to 88 hydrogens of the central polymer chain in the ^1H NMR spectrum. The use of monomolecular PEO would avoid possible changes in these ratios due to potential molecular weight fractionation during the dendron synthesis. MALDI-TOF data confirmed the synthesis, but the polydispersity of the primary chain (m_0 = 44.05) made assignments of defects (e.g. missing allyl group, m_0 = 42.08) difficult. The absence of coupled products was verified by SEC. These amphiphilic materials – the 3G product had 80% carbosilane content – formed micelles in water. The crystallization/melting behavior of the PEO chains was strongly affected by the increase in dendrimer generation, due to increasing isolation of the PEO chains.

Dendronized polymers can also be obtained by the polymerization of dendronized monomers. Méry et al. described anionic polymerization of 1G carbosilane dendronized styrenics with different multiplicities [192]:

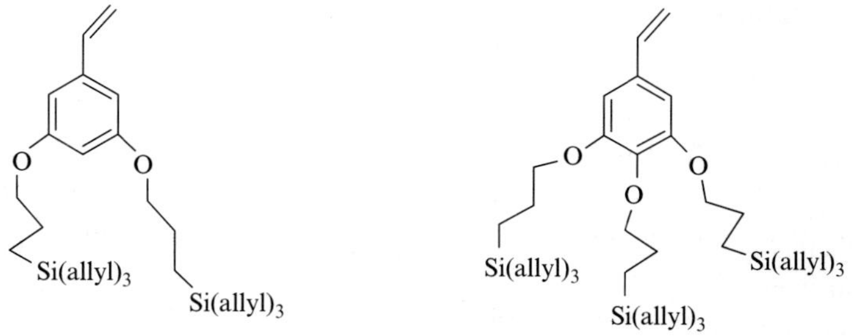

In principle, this method assures structurally perfect polymers with potentially narrow MWD. Best polymerization results have been obtained with potassium naphthalene initiation in THF at −78°C. The propagation step was limited to DP ≤ 30 for the bis-substituted monomer, and to DP ≤10 for the tris-substituted monomer. Steric crowding was presumably responsible for this limitation.

Diederich and his collaborators have decorated a tri(acetylene) monomer (C≡C–CR = CR–C≡C) with Fréchet-type polybenzylether dendrimers [193] and with a variety of carbosilane dendrimers. Polymerization of the latter under end-capping conditions with phenylacetylene produced a series of oligomers [119]:

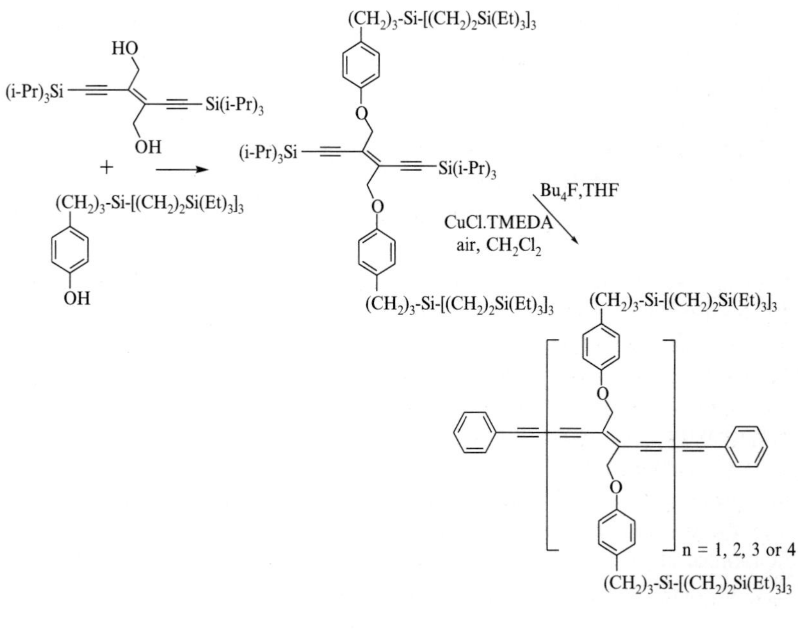

(3.86)

The UV-Vis spectra indicated that insulating dendritic wrapping did not alter the electronic properties of these polyacetylene molecular wires.

3.3 Carbosilane Dendrimer Characterization

The sequence of reactions used in dendrimer synthesis suggests a particular chemical structure that requires verification by analytical methods commonly used in organic chemistry. However, the repeated catenation of "identical" building blocks lends dendrimers a polymeric character, and creates analytical difficulties not otherwise encountered in macromolecules. As a consequence, analytical methods of polymer science with their statistical approach, must often compliment more rigorous methods, especially for high generation dendrimers. The products of the hydrosilylation reaction stage, with their multiple reactive Si–Cl bonds, are not usually analyzed beyond qualitative observation by ^{1}H NMR that all olefinic double bonds have reacted [21, 24]. Quantitative analysis is, therefore, limited to the carbosilane dendrimer generations with vinyl, allyl or other stable end-groups.

Organic chemistry methods, including elemental analysis and quantitative absorption spectroscopies (UV–Vis, IR), are limited in their capability to confirm dendrimer structures and detect defects because dendrimer compositions tend towards asymptotes with increasing generation. Generally, a property, X_i, associated with one, some or all dendrimer subunits with mass m_i, drifts to a limiting value with increasing generation:

$$\lim_{G \to \infty} X = (\Sigma\ X_i\ m_i)/(\Sigma\ m_i). \qquad \text{(E.3.2)}$$

For example, elemental composition is often inconclusive since it rapidly approaches its asymptote at generation 2 or 3 as the composition of the core, internal subunits and end-units of carbosilane dendrimers are very similar [24, 28].

Similarly, in quantitative ^{1}H NMR intensity ratios also tend to asymptotes. A signal associated with the core will tend to zero with increasing generation. In some cases, a marker, such as phenyl groups [39, 60], perylene [35] or a polymer chain attached to the core [191], allows useful comparison of the core and periphery proton signals to higher generations. In all cases, attention to quantitative spectrum acquisition (pulse power, relaxation times and digitalization of signals) is required. Broadening of resonances associated with atoms at or near the core is commonly observed and makes their detection and quantification more difficult. A complete structural analysis of 1G(4,1) and 2G(4,2,1) C_2 carbosilane dendrimers with $-Si(CH_3)_2H$ end-groups was accomplished with a high resolution 600 MHz spectrometer based on the intensity of the Si resonances as the primary assignment and ^{1}H/^{13}C/^{29}Si three-dimensional analysis for full identification of all resonances [81]. A similar study was also made with 1G(4,1) and 2G(4,2,1) C_3 dendrimers with phenyl end-groups [194]. Note that such analysis remains limited to 2G dendrimers in which the core, first and second generation Si atoms have different substituents. Complete resolution of higher generation dendrimer spectra requires introduction of chemical variation, such as $–Si(C_6H_5) =$ branch points [39] or $–CH = CH(C_6H_5)–$ branches [73].

It has been remarked that ^{1}H and ^{13}C spectra of carbosilane dendrimers should be clean and not contain *traces* of spurious resonances, the presence of which indicates structural defects. For example, proton resonances at 0.1–0.2 ppm from tetramethylsilane indicate the presence of methoxysilane groups (CH_3–O–Si–) originating from the methanolysis of Si–Cl bonds [7, 28]. α-Addition in the hydrosilylation step is most easily recognized by spurious resonances in the ^{13}C spectrum and can be quantified by the doublet at 1.00 ppm assigned to CH_3 protons in –SiCH(CH_3)Si– units [21, 62, 152]. Isomerization of the allylic double bond has been observed under some hydrosilylation conditions. The resulting propenyl group is then immune to further hydrosilylation [119].

Fast atom bombardment mass spectroscopy (FAB-MS) is the most commonly used method for identification of low molecular mass (0G–2G) carbosilane dendrimers [49, 166, 195]. A careful study of the fragmentation products is required in order to identify possible incomplete dendritic compounds [48]. Electron spray ionization (ESI) mass spectroscopy [83] and atmospheric pressure chemical ionization (APCI) mass spectroscopy [80] have also been used.

Matrix assisted laser desorption ionization-time of flight (MALDI-TOF) mass spectroscopy has proven ideally suited for the analysis of high molecular mass carbosilane dendrimers and their derivatives because of its mass range and resolution. Under ideal conditions, the mass spectra consist exclusively of positively ionized parent species: [M^+], $[M + Na]^+$, $[M + Ag]^+$ etc. dependent on the added salt. The first published MALDI-TOF data probed the perfection of 2G(4,3,3) and 3G(4,3,3,3) carbosilane dendrimers with allyl terminal groups. In the 2G dendrimer, two species were observed, one with the expected 36 allyl groups and the other with 34 allyl groups [M-152], which corresponded to one defect in the hydrosilylation step. In the 3G dendrimer about ten species were observed. They were spaced by 152 mass units down from the perfect dendrimer with 108 allyl groups. From random probability considerations [62, 196] it was concluded that the hydrosilylation reactions were 97.3% and 96.6% complete for the 2G and 3G dendrimers, respectively [22, 143]. The defective species had therefore a kinetic rather than thermodynamic origin. These results were further confirmed by the MALDI-TOF spectra of the hydroxypropyl derivatives. Beside the perfect structure, species from one and two missing hydrosilylation reactions (out of a possible 32) were also observed at the 3G level in dendrimers with $Si(Me)(Bz)_2$ end-groups [83].

Allgaier et al., on the other hand, concluded that the pairwise reduction in the number of end-groups was due to the incorporation of ethyl end-groups into the dendrimer, as a result of the use of vinylbromide contaminated with 1% ethylbromide in the alkenylation step [19]. Mass spectra with identified masses of dendrimers with 14, 15 and the correct 16 PPh_2 groups were also observed [197]. In another example, the product of the hydrosilylation reaction had 15 out of 16 bent mesogens, the remaining Si–H of the dendrimer being converted with methanol into SiOMe during chromatographic purification [158]. In principle, lower mass ions that are due to defects caused during the synthesis should differ from the parent ion by multiples of the monomer mass. Their relative intensity should follow statistical rules [196].

Instructions for obtaining MALDI-TOF spectra of carbosilane dendrimers with multiple polar end-groups were given [62, 198]. Fragmentation processes are rare but tend to become more frequent in the spectra of high generation dendrimers and dendrimers with large polar substituents [62, 97, 114, 199]. In most cases, fragmentation occurs in the substituents and, therefore, does not disprove the integrity of the carbosilane dendrimer skeleton. Occasionally, ions are observed that seem to arise from the loss of $(Me)_2Si = CHCl$ and $CH_3CH_2Si(Me)_2CH_2Cl$ which suggests true fragmentation at the C–Si bond of the dendrimer but may also be evidence for incomplete dendrimer construction [62]. Missing allyl groups have also been reported [45]. In such cases recourse to evidence from ^1H NMR or other analytical method may be required to decide the case. Compared to synthetic defects, fragmentation products are expected to exhibit a more random relative intensity distribution that can be manipulated by changing parameters of the spectroscopy.

Size exclusion chromatography (SEC) fractionates samples according to their hydrodynamic volume but does not provide the high resolution of the MALDI-TOF technique. The elution trace of a high quality carbosilane dendrimer should be narrow and result in a small apparent polydispersity index that essentially measures the axial dispersion of the chromatographic system at the size of the dendrimer. In all cases this polydispersity will be narrower than that of the best narrow-distribution polymer from a living polymerization. SEC is commonly used to qualitatively assess the monodispersity of carbosilane dendrimers. It is most useful for quick detection of higher and much lower molecular mass species [60, 73]. The presence of higher molecular mass species (dendrimer dimers, trimers and/or multiplets) is most likely due to trace hydrolysis (with formation of intermolecular Si–O–Si bonds) at the intermediate Si–Cl rich stage of the synthesis [19, 97]. This result also suggests simultaneous intramolecular hydrolysis that is nearly impossible to detect at low levels. Flash chromatography on silica gel is the standard purification method [7, 20] but preparative SEC has also been used to purify carbosilane dendrimers [60]. Ultimately, combined NMR and MALDI-TOF analyses are the best way to prove the structure of the purified dendrimer.

In principle, the universal calibration of SEC applies to carbosilane dendrimers, i.e. the elution volume depends on the product M [η] where the standard calibrant is polystyrene [7]. The only exception to this appears to be dendrimers with perfluoro end-groups where clear deviations from universal calibration were observed [123].

3.4 Properties of Carbosilane Dendrimers

3.4.1 Molecular Dimensions of Carbosilane Dendrimers

The molecular size of dendrimers, including carbosilanes, places them at the lower end of the nanoscopic size domain (see Chapter 1). While much research has been

Table 3.2 Core-to-periphery dimensions of carbosilane dendrimers (nm)[a]

Generation	C_3 – allyl end-groups	C_2 – vinyl end-groups
0G Tetraallylsilane	0.395	
Tetravinylsilane		0.276
1G	0.95	0.70
2G	1.50	1.13
3G	2.06	1.56
4G	2.61	1.99
5G	3.17	2.41
6G	3.72	2.84
7G	4.27	3.27

[a]Values are calculated assuming fully stretched all-trans configurations.

devoted to the synthesis and functionalization of carbosilane dendrimers, the study of their physical properties has considerably lagged behind. However, the general knowledge (although still somewhat controversial) about size, conformation, density profiles and dynamics of other types of dendrimers should, properly adjusted, also be applicable to carbosilanes. Table 3.2 lists the core-to-periphery distances of the fully stretched all-trans configurations of single arms of the two common types of carbosilane dendrimers [9, 190]. These calculated lengths are based on tetrahedral (109.5°) bond angles and on 0.154 nm C–C and 0.185 nm C–Si bond lengths.

The obtained dimensions are equivalent to the contour lengths of linear polymer chains and can be compared to the crystallite lengths per backbone carbon (0.126 nm) in aliphatic chains. Note that the core-to-periphery is the largest dimension in the dendrimer that is independent of the core and branch functionalities. Clearly, most real core-to-periphery distances in dendrimers are reduced by one or more ± gauche configurations at the branch points and in the branches. Experimentally measured dimensions, like the radius of gyration, R_g, from small angle X-ray scattering (SAXS) or small angle neutron scattering (SANS), or the hydrodynamic radius, R_H, from self-diffusion, or R_v from intrinsic viscosity measurements, are therefore considerably smaller.

Radii of gyration of a number of carbosilane dendrimers in solution have been determined by SAXS or SANS [200–202]. The scattering profiles (scattering intensities as a function of scattering vector q or angular dependence) were matched to different particle models. It was concluded that some anisotropy was present, as well as some local density variations [202]. No observable solvent effects (in benzene, chloroform) were detected. Experimental R_g data are collected in Table 3.3.

Based on $R_H = (5/3)^{1/2} R_g$, the internal mass density of carbosilane dendrimers in solution was determined to be of the order of 0.6–0.7 g/cm^3 [200], a value later corrected to 0.7–0.8 g/cm^3 [202]. In the case of a peripherally substituted trifluoropropyl dendrimer, indirect evidence indicates that peripheral Si–O–Si($CH_2CH_2CF_3$)$_3$ groups reside mostly in the corona [200]. This may be due to the chemical contrast

Table 3.3 Experimental dimensions of C_3 carbosilane dendrimers in solution (nm)

Sample	MW[a]	R_g [b]	D_0 [c]	R_H	$[\eta]$ [d]	R_v
5G(3,2...)-allyl	11,971	1.4	3.04	1.43		
6G(3,2...)-allyl	24,093	1.9	2.28	1.89		
7G(3,2...)-allyl	48,337	2.3	1.74	2.46		
1G(4,2...)-allyl	696		4.5[e]	0.89	–	–
2G(4,2...)-allyl	1,704		3.6[e]	1.12	3.0	0.93
3G(4,2...)-allyl	3,720		2.9[e]	1.38	3.7	1.30
4G(4,2...)-allyl	7,752		2.2[e]	1.89	3.5	1.63
3G(3,2...)-butyl	3,315		2.18[f]	0.99		
4G(3,2...)-butyl	6,705		1.82[f]	1.19		
5G(3,2...)-butyl	13,539		1.49[f]	1.45		
3G(4,2...)-butyl	4,299		4.3	1.0	3.6[g]	1.36
4G(4,2...)-butyl	8,946		3.3	1.3	4.0[g]	1.78
5G(4,2...)-butyl	17,906	2.0	2.55	1.7	4.2[g]	2.28
6G(4,2...)-butyl	36,122	2.5	2.07	2.1	3.6[g]	2.74
7G(4,2...)-butyl	72,554	3.0	1.70	2.5	4.6[g]	3.75
8G(4,2...)-butyl	145,421				3.8[g]	4.44
9G(4,2...)-butyl	291,155[h]				4.3[g]	5.83

[a] Calculated molecular weight.
[b] In C_6D_6 and $CDCl_3$ from Figure 5 of [200], Table 1 and Figure 5 of [201] and table of [202].
[c] 10^6 cm^2/s in $CDCl_3$ at 303 K from Figure 7 of [208] and Figure 18 of [207].
[d] cm^3/g in $CHCl_3$ at 294 K from table of [210].
[e] In $CHCl_3$ at 298 K from table of [210].
[f] In CCl_4 at 293 K from Figure 2 of [209].
[g] In toluene [69].
[h] MW = 295,600 by light scattering (LS) in toluene [69].

between the peripheral groups and the dendrimer core which may have led to an enthalpically driven internal microphase separation. A similar conclusion was also drawn from SANS data [203].

A molecular dynamics simulation did not directly assess the dimensions of carbosilane dendrimers but probed the size of the internal cavity as a function of the branch length of the 0G shell and its accessibility for different size molecules through the corona [204]. For a 5G(3,2,2,2,2,3) dendrimer with C_3 branches and terminal methyl groups (M_w = 9,373) in CCl_4 this yielded R_g = 1.65 nm [205, 206]. The radial density had a maximum at the core, a plateau of 0.6 g/cm^3 between r = 0.5 and 1.3 nm and a tail reaching to r = 2.6 nm which was slightly less than the core-to-periphery distance of 2.9 nm estimated for a fifth generation dendrimer with methyl end-groups. Although the methyl groups were found mostly in the peripheral zone (1.6–2.6 nm), some were found in all regions of the structure as a result of backfolding of the branches. Zhang et al. performed molecular dynamics simulations on 2G(8,3,n) dendrimers with n = 1, 2 or 3 hydroxypropyl terminal groups [30]. They found that the large octafunctional core promoted a small asymmetry (less than 1.15), and noted that hydroxyl groups were found in all parts of the dendrimer with R_g = 0.97, 1.02 and 1.08 nm for n = 1, 2 and 3, in agreement with increasing number of hydroxypropyl groups.

Self-diffusion and intrinsic viscosity measurements allow determination of an equivalent-sphere hydrodynamic radius via the Stokes-Einstein and Einstein relations, respectively:

$$R_H = kT/6\pi\eta_s D_0 \tag{E.3.3}$$

and

$$R_V = \{3[\eta]M/10\pi N_A\}^{1/3} \tag{E.3.4}$$

For equal density spheres, which is a good approximation for high generation dendrimers, $R_H \approx R_v = (5/3)^{1/2} R_g$ is expected. The self-diffusion coefficient, D_0, was measured by Pulsed-Field Gradient NMR in either chloroform [207, 208] or carbon tetrachloride [209]. The data and corresponding R_H values are collected in Table 3.3. The R_H data are internally consistent; for example, 5G(3,2,2,2,2,2)-allyl and 5G(3,2,2,2,2,2)-butyl have approximately the same dimensions, while all butyl dendrimers with a trifunctional core are slightly smaller than the same generation butyl dendrimers with a tetrafunctional core. Unfortunately, comparison of R_H with R_g, where possible, is less satisfactory. Hydrodynamic radii (R_v), based on intrinsic viscosity measurements of the first four generations of dendrimers with tetrafunctional cores, C_3 branches and allyl end-groups are also given in Table 3.3 [210]. R_v values of the corresponding dendrimers with tetrafunctional cores, C_2 branches and vinyl end-groups are 1G (M_w = 528) 0.48, 2G (M_w = 1,312) 0.74, 3G (M_w = 2,880) 1.06 and 4G (M_w = 6,016) 1.48 nm, respectively [7]. Therefore, the C_2 dendrimers are smaller than C_3 dendrimers and all radii are less than the calculated core-to-periphery lengths of Table 3.2. It should be noted that there is no evidence that generational evolution of intrinsic viscosities of carbosilane dendrimers shows a maximum as reported for polyamidoamine, PAMAM, and benzylether dendrimer series [211, 212]. Indeed, where data are available, the experimental radii scale approximately as $M^{1/3}$ indicating that internal density of the dendrimers is little changed with generation number. The spherical nature and relatively constant density of carbosilane dendrimers, even at low generations, have also been deduced from molecular mechanics simulations [213].

Other studies of dimensions were performed on highly functionalized carbosilane dendrimers. In these cases, the properties tend to be dominated by the large mass of the peripheral groups and their individual and collective properties. In many cases, microphase separation is observed [214]. In one interesting analysis, the thickness of the corona of mesogenic groups was estimated by subtracting the experimental radius of the carbosilane core from the overall dendrimer radius [210, 215]. On the assumption that the presence of the corona had no effect on the dendritic core size it was concluded that the corona density was about one half that of the dendrimer proper.

3.4.2 Dynamics of Carbosilane Dendrimers

Carbosilane dendrimers are very flexible molecules as indicated by their low glass transition temperatures (T_g). The most extensive measurements were reported by Lebedev et al. on C_3 dendrimers by means of adiabatic vacuum calorimetry [216–219]. The data obtained are collected in Table 3.4.

The glass transition region was about 10 K wide and the ΔCp (JK^{-1} mol^{-1}) was proportional to the molecular weight. Previous differential scanning calorimetry (DSC) determinations of T_g gave similar results for 3G to 7G dendrimers [11]. A T_g of 175 K was originally obtained for a 3G(1,2,2,2)-allyl dendrimer with a perylene core [35]. The 1G dendrimer of each series had a somewhat lower T_g, but the T_g of the higher generation dendrimers became rapidly constant. The dendrimers with multiplicity 3 [22, 139] had a 10 K higher T_g than the corresponding dendrimers with multiplicity 2. Most importantly, T_g was strongly affected by the end-groups of the dendrimer. As seen from Table 3.4, the T_g of the dendrimers with butyl end-groups [219] was about 10 K higher than the T_g of dendrimers with allyl end-groups, despite their lower core multiplicity [216–218]. Larger differences were observed when the end-groups were changed to hydroxyl groups. Polyols 1G(4,3), 2G(4,3,3) and 3G(4,3,3,3) had T_g of 233, 239 and 241 K, respectively, 50–60 K higher than the parent dendrimers with allyl end-groups [22, 139]. This large difference was ascribed to hydrogen bonding of the hydroxyl groups. All carbosilane dendrimers functionalized by multiple large substituents had T_gs that were mostly determined by the thermal properties of their substituents. In no case have two glass transitions been observed in calorimetric measurements.

A single T_g, related to both components, was also observed in carbosilane–dendrimer hybrids. For example, the T_g of poly(methylsiloxane) is 129 K, but it increased to 179 K when dendronized by 0G(3)-allyl groups and to 205 K when

Table 3.4 Glass transition temperatures of C_3 carbosilane dendrimers

Dendrimer	Tg/K	Reference
1G(4,2)-allyl	154	[216, 217]
2G(4,2,2)-allyl	172	
3G(4,2,2,2)-allyl	173	
4G(4,2,2,2,2)-allyl	172	
5G(4,2,2,2,2,2)-allyl	162	
6G(4,2,2,2,2,2,2)-allyl	180	[218]
7G(4,2,2,2,2,2,2,2)allyl	181	
1G(4,3)-allyl	170	[139]
2G(4,3,3)-allyl	184	
3G(4,3,3,3)-allyl	188	[22, 139]
3G(3,2,2,2)-butyl	179.8	[219]
4G(3,2,2,2,2)-butyl	186	
5G(3,2,2,2,2,2)-butyl	186	
6G(3,2,2,2,2,2,2)-butyl	186	

dendronized by 1G(3,3)-allyl [190]. When dendronizing polystyrene oligomers, carbosilane dendrons had the opposite effect and lowered the T_g [192]. Polystyrene ($T_g \approx 343$ K) dendronized with a 0G3,5-di-(γ-triallylsilyl)propyloxy groups had a T_g of 243 K and when dendronized with 0G3,4,5-tri-(γ-triallylsilyl)propyloxy groups T_g dropped to 223 K. Such changes were qualitatively explained by the weighted molecular mixing of components with low (carbosilane) and high (polystyrene) T_gs. Clearly, the available data were insufficient to establish a mixing rule [220] and to recognize more specific effects due to dendron structure.

The melting temperature, T_m, of poly(ethylene oxide) dendronized at both ends with carbosilane dendrons decreased from 317 K to 307, 273 and 260 K for generation OG, 1G and 2G, respectively [191]. This change was not correlated with lamellar thicknesses of the crystallites.

Dynamic mechanical spectra of carbosilane dendrimers (20% in an immiscible polystyrene matrix) revealed a strong loss peak at about 210 K (at 150 Hz) with a large Vogel-Fulcher type activation energy assigned to the α (T_g) transition in agreement with a T_g of 175 K by DSC. A weaker relaxation process was observed at 165 K (at 150 Hz) with an Arrhenius-type activation energy dependence. No specific molecular process was assigned to this β relaxation, but it was observed that incorporation of *n*-decyl end-groups shifted the relaxation to lower temperatures [221]. Both signals had a dielectric counterpart. Merger of the dielectric α- and β-processes has also been suggested [222].

High resolution NMR spectra obtained on 600 MHz instruments provided sufficient resolution to resolve all resonances in 1G(4,1) and 2G(4,2,1) C_3 dendrimers with phenolic end-groups [194] and 1G(4,1) and 2G(4,2,1) C_2 dendrimers with $Si(CH_3)_2H$ end-groups [81]. All resonances were assigned by means of their generational intensities and via two-dimensional $^1H/^{13}C$ analysis [194] or three-dimensional $^1H/^{13}C/^{29}Si$ analysis [81]. An investigation of the T_1 spin lattice relaxation times in toluene-d_8 showed that T_1 increased with the radial distance of the atom from the core. The slower average motions experienced by the atoms near the core were possibly due to the higher density in the core [194]. It is not established that this conclusion is general or that it depends on the quality of the solvent. Nevertheless, a similar conclusion was also reached in a T_1 study of poly(propylene imine) dendrimers [223].

In favorable cases, the spin echo technique of quasi-elastic neutron scattering (NSE) has allowed probing internal (segmental) dynamics of polymers and this method was, therefore, also applied to some carbosilane dendrimers with perfluoro $-SCH_2CH_2C_6F_{13}$ end-groups [214]. In the case of 1G(4,3), 2G(4,3,3) and 3G(4,3,3,3) dendrimers, the available q and time ranges were such that only center of mass diffusion was observed. This is in agreement with observations on bare PAMAM dendrimers in which no internal motions were detected by NSE [224]. In one particularly large 4G dendrimer ($R_g = 2.7$ nm, $R_H = 4.9$ nm) with an extended core and 324 (theoretical) perfluoro end-groups, additional relaxation processes were also observed. Since the maximum of the effective diffusion coefficient $D_{eff} = (\tau' q^2)^{-1}$ coincided with the minimum in the structure factor in the Kratky representation ($S(q) = I' q^2$) it was argued that the additional processes involved shape changes on the surface of the dendrimer [214], but it could not be excluded that internal density fluctuations were responsible [225]. Introduction of long chains between branch points was required to make internal motions observable [225].

Molecular dynamics simulations are a natural tool to probe intramolecular mobility of carbosilane dendrimers. Mazo et al. have shown that a one-barrier mechanism with an activation energy that does not exceed that of the isolated local environment governs the conformational transition between *trans* and *gauche* conformers [205]. It was shown that, in good solvents, the frequency of the transitions was only mildly dependent on the radial position of the bond [206]. In poor solvents, where the internal density approached the bulk density, the end-groups were clearly more mobile than internal ones. The effect of temperature decrease was most prominent in poor solvents as all transitions were frozen out and fluctuations became local – a signature of the liquid-to-glass conversion.

References

1. Buhleier E, Wehner W, Vögtle F (1978) Synthesis 155–158
2. Denkewalter RG, Kolc J, Lukasavage WJ (1981) US Patent 4,289,872
3. Tomalia DA, Baker H, Dewald J, Hall M, Kallos G, Martin S, Roeck J, Ryder J, Smith P (1985) Polym J (Tokyo) 17: 117–132
4. Newkome GR, Yao ZQ, Baker GR, Gupta VK (1985) J Org Chem 50: 2003–2004
5. Hadjichristidis N, Guyot A, Fetters LJ (1978) Macromolecules 11: 668–672
6. Hadjichristidis N, Fetters LJ (1980) Macromolecules 13: 191–193
7. Zhou LL, Roovers J (1993) Macromolecules 26: 963–968
8. van der Made AW, van Leeuwen PWNM (1992) J Chem Soc Chem Commun 1400–1401
9. van der Made AW, van Leeuwen PWNM, de Wilde JC, Brandes RAC (1993) Adv Mat 5: 466–468
10. Muzafarov AM, Gorbatsevich OB, Rebrov EA, Ignat'eva GM, Chenskaya TB, Myakushev VD, Bulkin AF, Papkov VS (1993) Polym Sci (USSR) 35: 1575–1580
11. Ignat'eva GM, Rebrov EA, Myakushev VD, Muzafarov AM, Il'ina MN, Dubovik II, Papkov VS (1997) Polym Sci Ser A 39: 874–881
12. Schlenk C, Frey H (1999) Monatsh Chem 130: 3–14
13. Frey H, Schlenk C (2000) Top Curr Chem 210: 70–123
14. Majoral JP, Caminade AM (1999) Chem Rev 99: 845–880
15. Lukevics E, Arsenyan P, Pudova O (2002) Main Group Metal Chem 25: 135–154
16. Cuadrado I, Morán M, Losada J, Casado CM, Pascual C, Alonso B, Lobete F (1996) In: Newkome GR (ed) Advances in Dendritic Macromolecules, vol. 3, JAI Press, Greenwich, CT, pp. 151–195
17. Krska SW, Son DY, Seyferth D (2000) In: Jones RG, Ando W, Chojnowski J (eds) Silicon-Containing Polymers, Kluwer, Dordrecht, pp. 615–641
18. Roovers J, Toporowski PM, Zhou LL (1992) Polym Preprints 33(1): 182–183
19. Allgaier J, Martin K, Räder HJ, Müllen K (1999) Macromolecules 32: 3190–3194
20. Munam A (2007) Ph.D. thesis, University of Waterloo, Waterloo Ontario, Canada
21. Seyferth D, Son DY, Rheingold AL, Ostrander RL (1994) Organometallics 13: 2682–2690
22. Lorenz K, Mülhaupt R, Frey H, Rapp U, Mayer-Posner FJ (1995) Macromolecules 28: 6657–6661
23. Kim C, Park E, Kang E (1995) J Korean Chem Soc 39: 799–805
24. Kim C, Park E, Kang E (1996) Bull Korean Chem Soc 17: 592–595

25. Kim C, Jeong Y (1998) Main Group Met Chem 21: 593–599
26. Kim C, An K (1997) Bull Korean Chem Soc 18: 164–170
27. Kim C, Choi SK (1997) Main Group Met Chem 20: 143–150
28. Kim C, An K (1997) J Organomet Chem 547: 55–63
29. Kim C, Park E (1998) J Korean Chem Soc 42: 277–284
30. Zhang X, Haxton KJ, Ropartz L, Cole-Hamilton DJ, Morris RE (2001) Dalton Trans 3261–3268
31. Kim C, Kim M (1998) J Organomet Chem 563: 43–51
32. Kim C, Jeong K, Jung I (2000) J Polym Sci Part A: Polym Chem 38: 2749–2759
33. Hawker CJ, Fréchet JMJ (1990) J Am Chem Soc 112: 7638–7647
34. van Heerbeek R, Reek JNH, Kamer PCJ, van Leeuwen PWNM (1999) Tetrahedron Lett 40: 7127–7130
35. Krasovskii VG, Sadovskii NA, Gorbatsevich OB, Muzafarov AM, Myakushev VD, Il'ina MN, Dubovik II, Strelkova TV, Papkov VS (1994) Polym Sci (USSR) 36: 589–594
36. Sluch MI, Scheblykin IG, Varnavsky OP, Vitukhnovsky AG, Krasovskii VG, Gorbatsevich OB, Muzafarov AM (1998) J Lumin 76–77: 246–251
37. Kim C, Sung DD, Chung DI, Park E, Kang E (1995) J Korean Chem Soc 39: 789–798
38. Kim C, Kwon A (1998) Main Group Met Chem 21: 9–12
39. Tuchbreiter A, Werner H, Gade LH (2005) Dalton Trans 1394–1402
40. Andrés R, de Jesús E, de la Mata FJ, Flores JC, Gómez R (2005) Eur J Inorg Chem 3742–3749
41. Meder MB, Haller I, Gade LH (2005) Dalton Trans 1403–1415
42. Gossage RA, Muñoz-Martínez E, van Koten G (1998) Tetrahedron Lett 39: 2397–2400
43. Gossage RA, Muñoz-Martínez E, Frey H, Burgath A, Lutz M, Spek AL, van Koten G (1999) Chem Eur J 5: 2191–2197
44. Botman PNM, David O, Amore A, Dinkelaar J, Vlaar MT, Goubitz JF, Schenk H, Hiemstra H, van Maarseveen JH (2004) Angew Chem Int Ed 43: 3471–3473
45. van Heerbeek R, Kamer PCJ, van Leeuwen PWNM, Reek JNH (2006) Org Biomol Chem 4: 211–223
46. Oosterom GE, van Haaren RJ, Reek JNH, Kamer PCJ, van Leeuwen PWNM (1999) Chem Commun 1119–1120
47. Müller C, Ackerman LJ, Reek JNH, Kamer PCJ, van Leeuwen PWNM (2004) J Am Chem Soc 126: 14960–14963
48. Oosterom GE, Steffens S, Reek JNH, Kamer PCJ, van Leeuwen PWNM (2002) Top Catal 19: 61–73
49. Amore A, van Heerbeek R, Zeep N, van Esch J, Reek JNH, Hiemstra H, van Maarseveen JH (2006) J Org Chem 71: 1851–1860
50. Marciniec B, Gulinski J et al. (1992) In: Marciniec B (ed) Comprehensive Handbook on Hydrosilylation, Pergamon, Oxford
51. Ojima I (1989) In: Patai S, Rappoport Z (eds) The Chemistry of Organic Silicon Compounds, vol. 2, Wiley Interscience, New York, pp. 1479–1526
52. Musolf MC, Speier JL (1964) J Org Chem 29: 2519–2524
53. Iovel IG, Goldberg YS, Shymanska MV, Lukevics E (1987) Organometallics 6: 1410–1413
54. Ponomarenko SA, Rebrov EA, Bobrovsky AY Boiko NI, Muzafarov AM, Shibaev VP (1996) Liq Crystals 21: 1–12
55. Karstedt BD (1973) US Patent 3,775,452
56. Boysen MMK, Lindhorst TK (2003) Tetrahedron 59: 3985–3988
57. Lewis LN (1990) J Am Chem Soc 112: 5998–6004
58. Stein J, Lewis LN, Gao Y, Scott RA (1999) J Am Chem Soc 121: 3693–3703
59. Faglioni F, Blanco M, Goddard III WA, Saunders D (2002) J Phys Chem B 106: 1714–1721
60. Rebrov EA, Ignat'eva GM, Lysachkov AI, Demchenko NV, Muzafarov AM (2007) Polym Sci Ser A 49: 483–495
61. Garcia B, Casado CM, Cuadrado I, Alonso B, Morán M, Losada J (1999) Organometallics 18: 2349–2356
62. Krska SW, Seyferth D (1998) J Am Chem Soc 120: 3604–3612

63. Kriesel JW, Tilley TD (2001) Adv Mater 13: 1645–1648
64. Gossage RA, Hooper R, Stobart SR (1999) Unpublished results
65. Rodríguez LI, Rossell O, Seco M, Grabulosa A, Muller G, Rocamora M (2006) Organometallics 25: 1368–1376
66. Rodríguez LI, Rossell O, Seco M, Muller G (2007) J Organomet Chem 692: 851–858
67. Comanita B, Roovers J (1999) Des Monom Polym 2: 111–124
68. Wakefield B (1995) Organomagnesium Methods in Organic Chemistry, Academic Press, London, pp. 219–222
69. Tatarinova EA, Rebrov EA, Myakushev VD, Meshkov IB, Demchenko NV, Bystrova AV, Lebedeva OV, Muzafarov AM (2004) Russ Chem Bull Int Ed 53: 2591–2600
70. Kim C, Jung I (2000) J Organomet Chem 599: 208–215
71. Kim C, Choi SK, Kim B (2000) Polyhedron 19: 1031–1036
72. Kim C, Son S (2000) J Organomet Chem 599: 123–127
73. Vodop'yanov EA, Tatarinova EA, Rebrov EA, Muzafarov AM (2004) Russ Chem Bull Int Ed 53: 358–363
74. Casado MA, Stobart SR (2000) Org Lett 2: 1549–1552
75. Nakayama J, Lin JS (1997) Tetrahedron Lett 38: 6043–6046
76. Matsuo T, Uchida K, Sekiguchi A (1999) Chem Commun 1799–1800
77. Kowaleska A, Stańczyk WA (2006) ARKIVOC 110–115
78. Zhou M, Roovers J, Robertson GP, Grover CP (2001) Anal Chem 75: 6708–6717
79. Yamada A, Hatano K, Matsuoka K, Koyama T, Esumi Y, Koshino H, Hino K, Nishikawa K, Natori Y, Terunuma D (2006) Tetrahedron 62: 5074–5083
80. Omotowa BA, Keefer KD, Kirchmeier RL, Schreeve JM (1999) J Am Chem Soc 121: 11130–11138
81. Chai M, Pi Z, Tessier C, Rinaldi PL (1999) J Am Chem Soc 121: 273–279
82. Ramírez-Oliva E, Cuadrado I, Casado CM, Losada J, Alonso B (2006) J Organomet Chem 691: 1131–1137
83. Kriesel JW, König S, Freitas MA, Marchall AG, Leary JA, Tilley TD (1998) J Am Chem Soc 120: 12207–12215
84. Seyferth D, Kugita T, Rheingold AL, Yap GPA (1995) Organometallics 14: 5362–5366
85. Kim C, Jung I (1999) J Organomet Chem 588: 9–19
86. Findeis RA, Gade LH (2002) Dalton Trans 3952–3960
87. Alonso B, Cuadrado I, Morán M, Losada J (1994) J. Chem Soc Chem Commun 2575–2576
88. Alonso B, Morán M, Casado M, Lobete F, Losada J, Cuadrado I (1995) Chem Mater 7: 1440–1442
89. Cuadrado I, Casado CM, Alonso B, Morán M, Losada J, Belsky V (1997) J Am Chem Soc 119: 7613–7614
90. Núñez R, González-Campo A, Viñas C, Teixidor F, Sillanpää R, Kivekäs R (2005) Organometallics 24: 6351–6357
91. de Groot D, Eggeling EB, de Wilde JC, Kooijman H, van Haaren RJ, van der Made AW, Spek AL, Vogt D, Reek JNH, Kamer PCJ, van Leeuwen PWNM (1999) Chem Commun 1623–1624
92. de Groot D, Emmererink PG, Coucke C, Reek JNH, Kamer PCJ, van Leeuwen PWNM (2000) Inorg Chem Commun 3: 711–713
93. Benito M, Rossell O, Seco M, Segalés G (1999) Inorg Chim Acta 291: 247–251
94. Benito M, Rossell O, Seco M, Segalés G (1999) Organometallics 18: 5191–5193
95. Benito M, Rossell O, Seco M, Segalés G (2001) J Organomet Chem 619: 245–251
96. Angurell I, Lima JC, Rodríguez LI, Rodríguez L, Rossell O, Seco M (2006) New J Chem 30: 1004–1008
97. de Groot D, Reek JNH, Kamer PCJ, van Leeuwen PWNM (2002) Eur J Org Chem 1085–1095
98. Angurell I, Muller G, Rocamora M, Rossell O, Seco M (2003) Dalton Trans 1194–2000
99. Angurell I, Muller G, Rocamora M, Rossell O, Seco M (2004) Dalton Trans 2450–2457

100. Wijkens P, Jastrzebski JTBH, van der Schaaf PA, Kolly R, Hafner A, van Koten G (2000) Org Lett 2: 1621–1624 [1093]
101. Beerens HI, Wijkens P, Jastrzebski JTBH, Verpoort F, Verdonck L, van Koten G (2000) J Organomet Chem 603: 244–248
102. Harder S, Meijboom R, Moss JR (2004) J Organomet Chem 689: 1095–1101
103. Hovestad NJ, Eggeling EB, Heidbüchel HJ, Jastrzebski JTBH, Kragl U, Keim W, Vogt D, van Koten G (1999) Angew Chem Int Ed 38: 1655–1658
104. Hovestad NJ, Ford A, Jastrzebski JTBH, van Koten G (2000) J Org Chem 65: 6338–6344
105. Eggeling EB, Hovestad NJ, Jastrzebski JTBH, Vogt D, van Koten G (2000) J Org Chem 65: 8857–8865
106. Hovestad NJ, Hoare JL, Jastrzebski JTBH, Canty AJ, Smeets WJJ, Spek AL, van Koten G (1999) Organometallics 18: 2970–2980
107. Rodríguez G, Lutz M, Spek AL, van Koten G (2002) Chem Eur J 8: 46–57
108. Dijkstra HP, Kruithof CA, Ronde N, van de Coevering R, Ramón DJ, Vogt D, van Klink GPM, van Koten G (2003) J Org Chem 66: 675–685
109. Benito JM, de Jesús E, de la Mata FJ, Flores JC, Gómez R, Gómez-Sal P (2002) J Organomet Chem 664: 258–267
110. Benito JM, de Jesús E, de la Mata FJ, Flores JC, Gómez R, Gómez-Sal P (2006) J Organomet Chem 691: 3602–3608
111. Kleij AW, Kleijn H, Jastrzebski JTBH, Smeets WJJ, Spek AL, van Koten G (1999) Organometallics 18: 268–276
112. Kleij AW, Gossage RA, Klein Gebbink RJM, Brinkmann N, Reijerse EJ, Kragl U, Lutz M, Spek AL, van Koten G (2000) J Am Chem Soc 122: 12112–12124
113. Kleij AW, Klein Gebbink RJM, van den Nieuwenhuijzen PAJ, Kooijman H, Lutz M, Spek AL, van Koten G (2001) Organometallics 20: 634–637
114. Kleij AW, Klein Gebbink RJM, Lutz M, Spek AL, van Koten G (2001) J Organomet Chem 621: 190–196
115. Sato I, Hosoi K, Kodaka R, Soai K (2002) Eur J Org Chem 3115–3118
116. Sato I, Kodaka R, Hosoi K, Soai K (2002) Tetrahedron: Asymmetry 13: 805–808
117. Slagt MQ, Jastrzebski JTBH, Klein Gebbink RJM, van Ramesdonk HJ, Verhoeven JW, Ellis DD, Spek AL, van Koten G (2003) Eur J Org Chem 1692–1703
118. Le Nôtre J, Firet JJ, Sliedregt LAJM, van Steen BJ, van Koten G, Klein Gebbink RJM (2005) Org Lett 7: 363–366
119. Schenning APHJ, Arndt JD, Ito M, Stoddart A, Schreiber M, Siemsen P, Martin RE, Boudon C, Gisselbrecht JP, Gross M, Gramlich V, Diederich F (2001) Helv Chim Acta 84: 296–334
120. Lobete F, Cuadrado I, Casado CM, Alonso B, Morán M, Losada J (1996) J Organomet Chem 509: 109–113
121. Friedmann G, Guilbert Y, Wittmann JC (1997) Eur Polym J 33: 419–426
122. Andrés R, de Jesús E, de la Mata FJ, Flores JC, Gómez R (2002) Eur J Inorg Chem 2281–2286
123. Casado MA, Roovers J, Stobart SR (2001) Chem Commun 313–314
124. Ponomarenko SA, Rebrov EA, Boiko NI, Muzafarov AM, Shibaev VP (1998) Polym Sci Ser A 40: 763–768
125. Lebedev BV, Kulagina TG, Ryabkov MV, Ponomarenko SA, Makeev EA, Boiko NI, Shibaev VP, Rebrov EA, Muzafarov AM (2003) J Therm Anal Calorimetry 71: 481–492
126. Zhu XM, Vinokur RA, Ponomarenko SA, Rebrov EA, Muzafarov AM, Boiko NI, Shibaev VP (2000) Polym Sci Ser A 42: 1263–1271
127. Boiko N, Zhu X, Vinokur R, Rebrov E, Muzafarov A, Shibaev V (2000) Ferroelectrics 243: 59–66
128. Boiko NI, Lysachkov AI, Ponomarenko SA, Shibaev SA, Richardson RM (2005) Coll Polym Sci 283: 1155–1162
129. Kosata B, Tamba GM, Baumeister U, Pelz K, Diele S, Pelzl G, Galli G, Samaritani S, Agina EV, Boiko NI, Shibaev VP, Weissflog W (2006) Chem Mater 18: 691–701

130. Boiko N, Zhu X, Bobrovsky A, Shibaev V (2001) Chem Mater 13: 1447–1452
131. Bobrovsky A, Ponomarenko S, Boiko N, Shibaev E, Rebrov E, Muzafarov A, Stumpe J (2002) Macromol Chem Phys 203: 1539–1546
132. Bobrovsky AY, Pakhomov AA, Zhu XM, Boiko NI, Shibaev VP (2001) Polym Sci Ser A 43: 431–437
133. Getmanova EV, Chenskaya TB, Gorbatsevich OB, Rebrov EA, Vasilenko NG, Muzafarov AM (1997) React Funct Polym 33: 289–297
134. Getmanova EV, Tereshchenko AS, Ignat'eva EA, Tatarinova EA, Myakushev VD, Muzafarov AM (2004) Russ Chem Bull Int Ed 53: 137–143
135. Newkome et al. reported on the divergent synthesis of dendrimers characterized by the $-C(CH_2-)_3$ branch points [4]. For a review of the many variants of such dendrimers: Newkome GR, Moorefield CN, Vögtle F (1996) Dendritic Molecules. Concept-Syntheses-Perspectives. VCH, Weinheim
136. Ignat'eva GM, Rebrov EA, Myakushev VD, Chenskaya TB, Muzafarov MA (1997) Polym Sci Ser A 39: 843–852
137. Comanita B, Noren B, Roovers J (1999) Macromolecules 32: 1069–1072
138. Omotowa BA, Shreeve JM (2003) Macromolecules 36: 8336–8345
139. Lorenz K, Frey H, Stühn B, Mülhaupt R (1997) Macromolecules 30: 6860–6868
140. Camerano JA, Casado MA, Ciriano MA, Tejel C, Oro LA (2005) Dalton Trans 3092–3100
141. Miedaner A, Curtis CJ, Barkley RM, Dubois DL (1994) Inorg Chem 33: 5482–5490
142. Ropartz L, Morris RE, Foster DF, Cole-Hamilton DJ (2002) J Mol Catal A: Chem 182–183: 99–105
143. Frey H, Lorenz K, Mülhaupt R, Rapp U, Mayer-Posner FJ (1996) Macromol Symp 102: 19–26
144. Matsuoka K, Terabatake M, Esumi Y, Hatano K, Terunuma D, Kuzuhara H (2006) Biomacromolecules 7: 2284–2290 [928]
145. Chung TC, Raate M, Berluche E, Schulz DN (1988) Macromolecules 21: 1903–1907
146. Kim C, Son S, Kim B (1999) J Organomet Chem 588: 1–8
147. Terunuma D, Kato T, Nishio R, Matsuoka K, Kuzuhara H, Aoki Y, Nohira H (1998) Chem Lett 59–60
148. Terunuma D, Nishio R, Aoki Y, Nohira H, Matsuoka K, Kuzuhara H (1999) Chem Lett 565–566
149. Matsuoka K, Terabatake M, Saito Y, Hagihara C, Esumi Y, Terunuma D, Kuzuhara H (1998) Bull Chem Soc Jpn 71: 2709–2713
150. Matsuoka K, Kurosawa H, Esumi Y, Terunuma D, Kuzuhara H (2000) Carbohyd Res 329: 765–772
151. Yamada A, Hatano K, Koyama T, Matsuoka K, Esumi Y, Terunuma D (2006) Carbohyd Res 341: 467–473
152. Morán M, Casado CM, Cuadrado I, Losada J (1993) Organometallics 12: 4327–4333
153. Casado CM, Cuadrado I, Morán M, Alonso B, Lobete F, Losada J (1995) Organometallics 14: 2618–2620
154. Arévalo S, de Jesús E, de la Mata FJ, Flores JC, Gómez R, Gómez-Sal MP, Ortega P, Vigo S (2003) Organometallics 22: 5109–5113
155. Arévalo S, Benito JM, de Jesús E, de la Mata FJ, Flores JC, Gómez R (2000) J Organomet Chem 602: 208–210
156. Arévalo S, de Jesús E, de la Mata FJ, Flores JC, Gómez R (2001) Organometallics 20: 2583–2592
157. Bermejo JF, Ortega P, Chonco L, Eritja R, Samaniego R, Müllner M, de Jesús E, de la Mata FJ, Flores JC, Gómez R, Muñoz-Fernandez A (2007) Chem Eur J 13: 483–495
158. Hahn H, Keith C, Lang H, Amaranatha Reddy R, Tschierke C (2006) Adv Mater 18: 2629–2633
159. Chandra S, Buschbeck R, Lang H (2006) Anal Sci 22: 1327–1332
160. Chandra S, Buschbeck R, Lang H (2006) Talanta 70: 1087–1093
161. Buschbeck R, Lang H (2005) J Organomet Chem 690: 696–703

162. Camerano JA, Casado MA, Ciriano MA, Lahoz FJ, Oro LA (2005) Organometallics 24: 5147–5156
163. Casado MA, Hack V, Camerano JA, Ciriano MA, Tejel C, Oro LA (2005) Inorg Chem 44: 9122–9124
164. Núñez R, González A, Viñas C, Teixidor F, Sillanpää R, Kivekäs R (2005) Org Lett 7: 231–233
165. González-Campo A, Viñas C, Teixidor F, Núñez R, Sillanpää R, Kivekäs R (2007) Macromolecules 40: 5644–5652
166. Cuadrado I, Morán M, Moya A, Casado CM, Barranco M, Alonso B (1996) Inorg Chim Acta 251: 5–7
167. Roovers J, Comanita B (1999) Adv Polym Sci 142: 179–228
168. Malenfant PRL, Fréchet JMJ (2001) In: Fréchet JMJ, Tomalia DA (eds) Dendrimers and Other Dendritic Polymers, Wiley, Chichester, pp. 171–193
169. Frauenrath H (2005) Prog Polym Sci 30: 325–384
170. Vlassopoulos D (2004) J Polym Sci Part B: Polym Phys 42: 2931–2941
171. Morton M, Helminiak TE, Gadkary SD, Bueche F (1962) J Polym Sci 57: 471–482
172. Zelinsky RP, Wofford CF (1965) J Polym Sci Part A 3: 93–103
173. Roovers J, Bywater S (1972) Macromolecules 5: 384–388
174. Iatrou H, Hadjichristidis N (1992) Macromolecules 25: 4649–4651
175. Zhou LL, Hadjichristidis N, Toporowski P, Roovers J (1992) Rubber Chem Technol 65: 303–314
176. Roovers J, Zhou LL, Toporowski P, van der Zwam M, Iatrou H, Hadjichristidis N (1993) Macromolecules 26: 4324–4331
177. Avgeropoulos A, Poulos Y, Hadjichristidis N, Roovers J (1996) Macromolecules 29: 6076–6078
178. Polyakov DK, Ignat'eva GM, Rebrov EA, Vasilenko NG, Sheiko SS, Möller M, Muzafarov AM (1998) Polym Sci Ser A 40: 876–883
179. Vasilenko NG, Getmanova EV, Myakushev VD, Rebrov EA, Möller M, Muzafarov AM (1997) Polym Sci Ser A 39: 977–983
180. Vasilenko NG, Rebrov EA, Muzafarov AM, Esswein B, Striegel B, Möller M (1998) Macromol Chem Phys 199: 889–895
181. Vasilenko NG, Ignat'eva GM, Myakushev VD, Rebrov EA, Möller M, Muzafarov AM (2001) Dokl Chem 377: 84–88
182. Comanita B, Roovers, J (1999) Unpublished results
183. Hovestad NJ, van Koten G, Bon SAF, Haddleton DM (2000) Macromolecules 33: 4048–4052
184. Beerens H, Verpoort F, Verdonck L (2000) J Mol Catal A: Chem 159: 197–201
185. Beerens H, Verpoort F, Verdonck L (2000) J Mol Catal A: Chem 151: 279–282
186. Beerens H, Wang W, Verdonck L, Verpoort F (2002) J Mol Catal A: Chem 190: 1–7
187. Cornelissen JJLM, van Heerbeek R, Kamer PCJ, Reek JNH, Sommerdijk NAJM, Nolte RJM (2002) Adv Mater 14: 489–492
188. Kim C, Park E, Jung I (1996) J Korean Chem Soc 40: 347–356
189. Kim C, Kang S (2000) J Polym Sci Part A: Polym Chem 38: 724–729
190. Ouali N, Méry S, Skoulios A, Noirez L (2000) Macromolecules 33: 6185–6193
191. Chang Y, Kim C (2001) J Polym Sci Part A: Polym Chem 39: 918–926
192. Moingeon F, Masson P, Méry S (2007) Macromolecules 40: 55–64
193. Hawker and Fréchet reported on the convergent synthesis of dendrimers with 3,5-dioxo–C_6H_3–CH_2– repeat units [33]. For a review see also [168]
194. Welch KT, Arévalo S, Turner JFC, Gómez R (2005) Chem Eur J 11: 1217–1227
195. Benito M, Rossell O, Seco M, Muller G, Ordinas JI, Font-Bardia M, Solans X (2002) Eur J Inorg Chem 2477–2487
196. Krupková A, Čermák J, Walterová Z, Horský J (2007) Anal Chem 79: 1639–1645
197. Ropartz L, Morris RE, Foster DF, Cole-Hamilton DJ (2001) Chem Commun 361–362
198. Wu Z, Biemann K (1997) Int J Mass Spectrom Ion Processes 165/166: 349–361

199. Zhou M, Roovers J (2001) Macromolecules 34: 244–252
200. Kuklin AI, Ignat'eva GM, Ozerina LA, Islamov AK, Mukhamedzyanov RI, Shumilkina NA, Myakushev VD, Sharipov EY, Gordelii VI, Muzafarov AM, Ozerin AN, (2002) Polym Sci Ser A 44: 1273–1280
201. Kuklin AI, Ozerin AN, Islamov AK, Muzafarov AM, Gordeliy VI, Rebrov EA, Ignat'eva GM, Tatarinova EA, Mukhamedzyanov RI, Ozerina LA, Shapirov EY (2003) J Appl Cryst 36: 679–683
202. Ozerin AN, Muzafarov AM, Kuklin AI, Islamov AK, Gordelyi VI, Ignat'eva GM, Myakushev VD, Ozerina LA, Tatarinova EA (2004) Doklady Chem 395: 59–62
203. Lyulin A, Davies G, Adolf D (2000) Macromolecules 33: 6899–6900
204. Lach C, Brizzolara D, Frey H (1997) Macromol Theor Simul 6: 371–380
205. Mazo MA, Shamaev MY, Balabaev NK, Darinskii AA, Neelov IM (2004) Phys Chem Chem Phys 6: 1285–1289
206. Mazo MA, Zhilin PA, Gusarova EB, Sheiko SS, Balabaev NK (1999) J Mol Liq 82: 105–116
207. Sagidullin A, Skirda VD, Tatarinova EA, Muzafarov AM, Krykin MA, Ozerin AN, Fritzinger B, Scheler U (2003) Appl Magn Res 25: 129–156
208. Sagidullin A, Muzafarov AM, Krykin MA, Ozerin AN, Skirda VD, Ignat'eva GM (2002) Macromolecules 35: 9472–9479
209. Krykin MA, Volkov VI, Volkov EV, Surin NM, Ozerina LA, Ignat'eva GM, Muzafarov AM, Ozerin AN (2005) Doklady Chem 403: 115–117
210. Lezov AV, Mel'nikov AB, Polushina GE, Antonov EA, Novitskaya ME, Boiko NI, Ponomarenko SA, Rebrov EA, Shibaev VP, Ryumtsev EI, Muzafarov AM (2001) Doklady Chem 381: 313–316
211. Tomalia DA, Hedstrand DM, Wilson LR (1990) Encyclopedia of Polymer Science and Engineering, 2nd edn., Index Volume, Wiley, New York, pp. 46–92
212. Mourey TH, Turner SR, Rubinstein M, Fréchet JMJ, Hawker CJ, Wooley KL (1992) Macromolecules 25: 2401–2406
213. Elshahre M, Atallah AS, Santos S, Grigoras S (2000) Compt Theor Polym Sci 10: 21–28
214. Stark B, Lach C, Farago B, Frey H, Schlenk C, Stühn B (2003) Coll Polym Sci 281: 593–600
215. Lezov AV, Mel'nikov AB, Polushina GE, Ponomarenko SA, Boiko NI, Kossmehl E, Ryumtsev EI, Shibaev VP (1998) Doklady Phys Chem 362: 338–342
216. Ryabkov MV, Kulagina TG, Lebedev BV (2001) Russ J Phys Chem 75: 1988–1996
217. Lebedev BV, Ryabkov MV, Tatarinova EA, Rebrov EA, Muzafarov AM (2003) Russ Chem Bull Int Ed 52: 545–551
218. Smirnova NN, Lebedev BV, Khramova NM, Tsvetkova LY, Tatarinova EA, Myakushev VD, Muzafarov AM (2004) Russ J Phys Chem 78: 1196–1201
219. Smirnova NN, Stepanova OV, Bykova TA, Markin AV, Muzafarov AM, Tatarinova EA, Myakushev VD (2006) Thermochim Acta 440: 188–194
220. Fox TG (1956) Bull Am Phys Soc 1: 123–128
221. Perov NS, Martirisov VA, Gritsenko OT, Aulov VA, Nikol'skii OG, Ozerin AN (2000) Doklady Chem 372: 77–80
222. Trahasch B, Stühn B, Frey H, Lorenz K (1999) Macromolecules 32: 1962–1966
223. Chai M, Niu Y, Youngs WJ, Rinaldi PL (2001) J Am Chem Soc 123: 4670–4678
224. Rathgeber S, Monkenbusch M, Kreitschmann M, Urban V, Brulet A (2002) J Chem Phys 117: 4047–4062
225. Rathgeber S, Monkenbusch M, Hedrick JL, Trollsås M, Gast AP (2006) J Chem Phys 125: 204908

Chapter 4
Polysilane Dendrimers

Masato Nanjo and Akira Sekiguchi

4.1 Introduction

Polysilanes, $-(Si)_n-$, are polymers that contain catenated silicon atoms, and their chemistry has attracted considerable interest during the last 30 years because of their electronic, optical, structural, and chemical properties [1]. In particular, the σ-conjugation of the –Si–Si– backbone has attracted much attention compared with analogous carbon polymer systems. Although, in contrast to the numerous reports on polysilanes with linear main chains, little attention has been devoted to their branched counterparts; hyperbranched polysilanes [2], ladder polysilanes [3], and organosilicon nanoclusters [4] have nevertheless been described. However, with the exception of ladder polysilanes, the precise structures of these branched polymers have not been sufficiently elucidated. This chapter deals with polysilane dendrimers from the initial [6] to the most recent report [20]. These dendrimers, which contain silicon atoms attached to three or four other silicon atoms, exhibit some interesting properties compared with their linear homologues [5]. As in other chapters of this book, the "1G(4,^{0}3) end-group" nomenclature system (see Fig. 4.1) is used. In this system, 1G denotes generation 1, the first number in parentheses denotes the branching functionality of the core (in this example a tetradendron dendrimer with first digit 4); the following superscript represents the number of spacer silicon atoms between the core and the next branching point; the following numbers represent functionalities of the silicon atoms in each subsequent generational layer; and finally the end-groups are specified by their chemical formulas.

M. Nanjo
Department of Chemistry and Biotechnology, Graduate School of Engineering,
Tottori University,
Koyamacho-minami Tottori 680-8552 Japan
E-mail: nanjo@chem.tottori-u.ac.jp

A. Sekiguchi
Department of Chemistry, Graduate School of Pure and Applied Sciences,
University of Tsukuba,
Tsukuba, Ibaraki 305-8571, Japan
E-mail: sekiguch@chem.tsukuba.ac.jp

P.R. Dvornic and M.J. Owen (eds.), *Silicon-Containing Dendritic Polymers*,
Advances in Silicon Science 2, © Springer Science+Business Media B.V. 2009

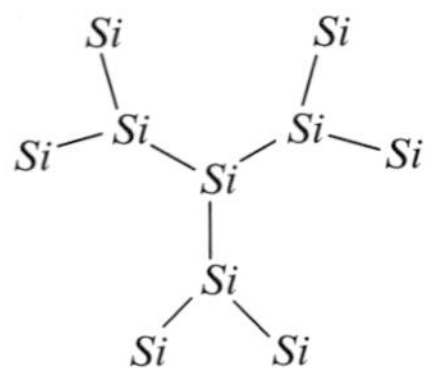

$1G(3,^{0}2)$

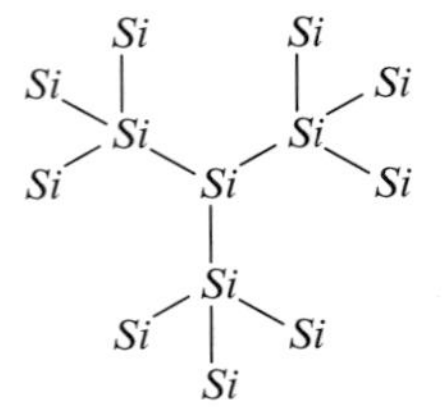

$1G(3,^{0}3)$

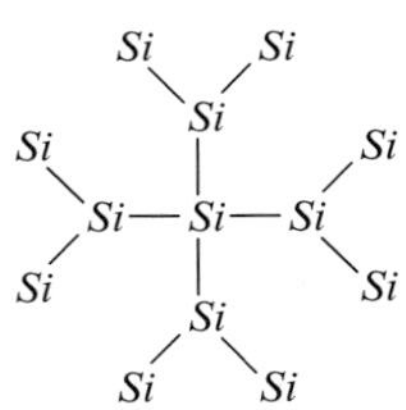

$1G(4,^{0}2)$

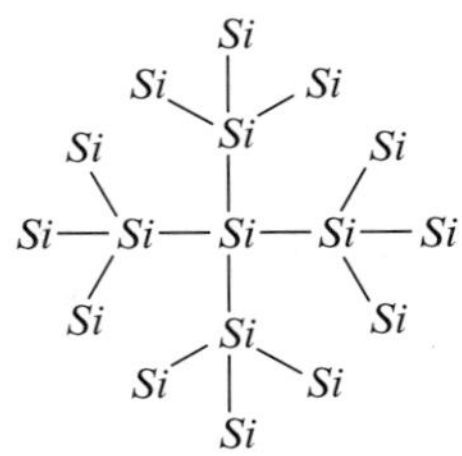

$1G(4,^{0}3)$

Fig. 4.1 Polysilane dendrimers without spacer groups. ***Si*** indicates $SiMe_n$ (n = 3, 2, 1, 0).

4.2 Synthetic Approaches to Polysilane Dendrimers

Polysilane dendrimers represent one of the very few types of dendrimer compositions whose preparation has almost equally utilized both convergent and divergent synthetic methods (see also Chapter 8).

4.2.1 Convergent Methods

As described in Chapter 1, the convergent dendrimer synthesis involves preparation of individual dendrons starting from what will later in the synthesis become the dendrimer surface and proceeding inwards, followed by the coupling of the resulting presynthesized dendrons to a multifunctional molecule (or anchor) that will become the dendrimer core. A Si–Si bond is generally formed by a metal–halogen exchange reaction between a metallosilane (Si–M) and a halosilane (Si–X). Using this chemistry, the first syntheses of polysilane dendrimers by a convergent method were reported independently by Lambert et al. and Suzuki et al. in 1995 [6, 7]. Three equivalents of hypersilyllithium, $(Me_3Si)_3SiLi$ [8], were reacted with tris(chlorodimethylsilyl)methylsilane, $MeSi(SiMe_2Cl)_3$, to give the first generation polysilane dendrimer $1G(3,^{1}3)$, **1**, as shown in Reaction Scheme 4.1. Dendrimer **1** was fully characterized by X-ray diffraction, as well as by ^{1}H, ^{13}C, and ^{29}Si NMR spectroscopy. It was a single core, tridendron dendrimer with three dimethylsilylene ($–Me_2Si–$) moieties as spacers between the branching points.

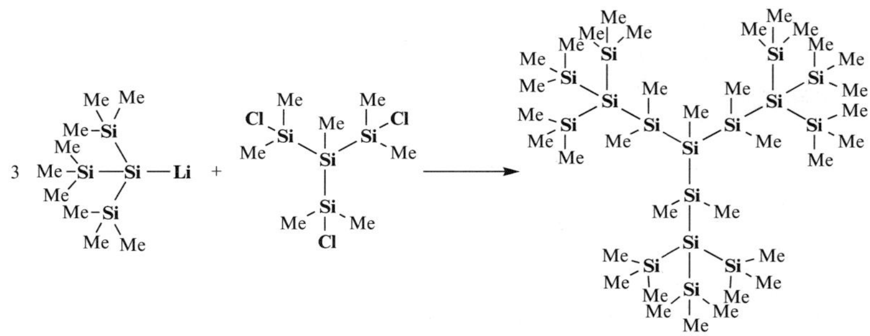

1

1G(3,¹3) (4.1)

In principle, the smallest permethylated polysilane dendrimers of the first generation, without $-SiMe_2-$ spacers, can assume one of the four structures shown in Fig. 4.1. However, instead of the desired product, an attempted synthesis of the tetradendron dendrimer 1G(4,⁰3), shown bottom right in Fig. 4.1, by the reaction of hypersilyllithium, $(Me_3Si)_3SiLi$, with silicon tetrachloride, $SiCl_4$, gave only tris(trimethylsilyl)silyltrichlorosilane, $(Me_3Si)_3SiSiCl_3$, in which only one of the $(Me_3Si)_3Si$ groups replaced the Cl atom in the silicon tetrachloride [9]. It was concluded that further reaction of this bulky branch with the core was difficult because of steric hindrance.

The only polysilane dendrimer without any spacer units that has been successfully isolated so far is 1G(3,⁰2) (top left in Fig. 4.1) [10], obtained by the reaction of bis(trimethylsilyl)methylsilyllithium, $(Me_3Si)_2MeSiLi$ [11], with methyltrichlorosilane, $MeSiCl_3$. This silyllithium branch reagent is smaller than $(Me_3Si)_3SiLi$, and the anchoring core is three-functional, which decreases the crowding around the dendrimer center. However, in the course of the reaction the cyclotetrasilane derivative was also formed as a byproduct in a 1:1 mole ratio with 1G(3,⁰2). The reaction mechanism is still unclear, but it seems that dimerization occurred between the very active disilene intermediates $Me_3SiMeSi{=}SiMe[SiMe(SiMe_3)_2]$ generated by a fragmentation reaction of the bulky polysilane dendrimer 1G(3,⁰2).

To reduce the steric hindrances around the central core, the synthesis of 1G(4,¹2), **2**, was attempted using a tetrafunctional core with dimethylsilylene ($-Me_2Si-$) spacer groups (see Reaction Scheme 4.2) [12]. The tetrakis(chlorodimethylsilyl)silane, $Si(SiMe_2Cl)_4$, was reacted with the less hindered silyllithium, $(Me_3Si)_2MeSiLi$, as a branch reagent to give the expected tetradendron polysilane dendrimer **2**. However, the generated dendrimer **2** was unstable, and a fragmentation reaction took place to form hydrosilanes **3** and **4** in 91% and 4% yield, respectively [12]. In a detailed investigation of the mechanism of this reaction, it was found that in the first step the tetradendron dendrimer **2** is formed, and then the central (or next to the central) Si–Si bond is cleaved homolytically to generate the corresponding silyl radicals. These silyl radicals then react with a hydrogen source (probably the solvent) to give the corresponding hydrosilanes **3** or **4**, respectively. The generation of these transient silyl radicals was confirmed by a trapping experiment using phenylacetylene, while with methyl iodide no methylated compound was formed, suggesting that the corresponding anionic intermediates are not involved. Hence, although the targeted tetradendron dendrimer **2** with bulky dendrons would be initially generated, its subsequent fragmentation resulted in the formation of the tridendron dendrimers in order to reduce the steric bulkiness.

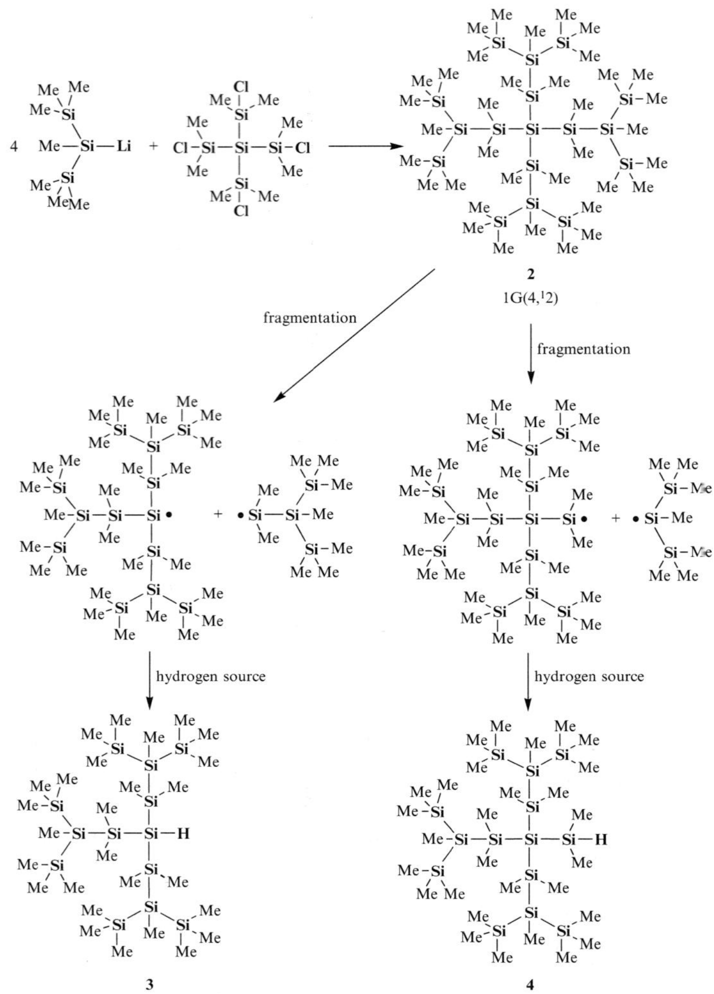

(4.2)

The convergent method has the advantage of being able to form large polysilane dendrimers in a single reaction step, but the key factor for its success is incorporation of a branch extender into the large branch reagents or dendrons. For this purpose, 1,2-dichlorotetramethyldisilane, $ClMe_2SiSiMe_2Cl$, was reacted with two equivalents of $(Me_3Si)_3SiLi$ to give 2,2,5,5-tetrakis(trimethylsilyl)decamethylhexasilane, in which two hypersilyl groups were introduced (see Reaction Scheme 4.3) [13]. Next, methyllithium was added to the resulting hexasilane to obtain pentasilanyllithium together with the inert tetramethylsilane as a byproduct. The resulting pentasilanyllithium, which is a very large branch extender, was then reacted with tris(chlorodimethylsilyl)methylsilane, $MeSi(SiMe_2Cl)_3$, to yield a tridendron polysilane dendrimer 1G(3,[4]3), **5**. This dendrimer contained 31 silicon atoms, and its longest silicon chain consisted of 13 silicon atoms, with a distance between the antipodal terminal methyl groups of 1.7 nm [13].

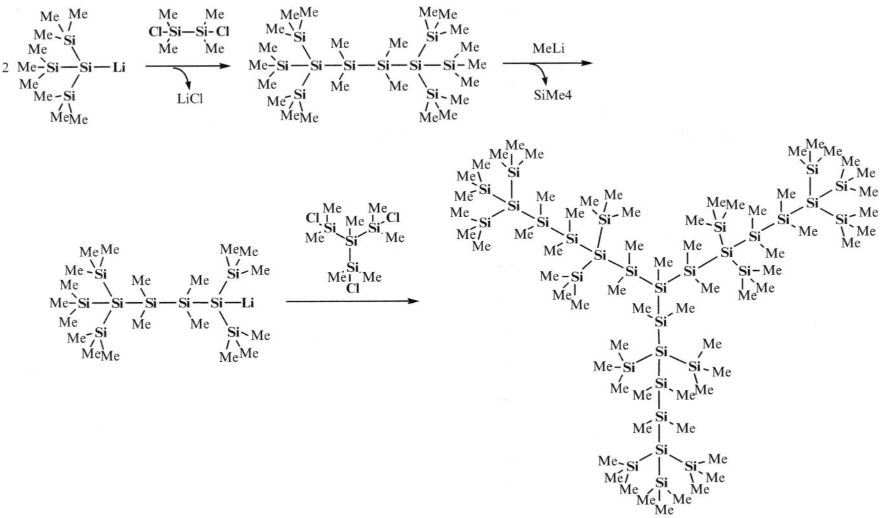

5
$1G(3,^{4}3)$ (4.3)

4.2.2 Divergent Methods

The synthesis of polysilane dendrimers by a divergent method (see Chapter 1) was reported by Sekiguchi et al. for the first time in 1995 (see Reaction Scheme 4.4) [14], the same year that the first synthesis by a convergent method was also reported. The key to a successful polysilane dendrimer synthesis by this approach is the rational design of suitable building blocks. For this, the authors used silyllithium with two peripheral phenyl groups where the Si–Ph bonds were inert under the reaction conditions. However, silylbenzene undergoes a proto–desilylation reaction with superacids or hydrogen halides, HX, in the presence of Lewis acids to give the corresponding functionalized silane derivatives (see Reaction Scheme 4.5), so that the phenyl group actually worked as a protecting group. Tris(dimethylphenylsilyl)silane, $(PhMe_2Si)_3SiMe$, was treated with three equivalents of trifluoromethanesulfonic acid, TfOH, to give the corresponding silyl triflate, $(TfOMe_2Si)_3SiMe$, which was then reacted with three equivalents of the branching reagent (i.e., building block), $Me(PhMe_2Si)_2SiLi$, to obtain the first generation polysilane dendrimer $1G(3,^{1}2)Ph_6$, **6**. Using permethyl-substituted silyllithium, $Me(Me_3Si)_2SiLi$, instead of $Me(PhMe_2Si)_2SiLi$, led to the permethylated polysilane dendrimer $1G(3,^{1}2)$, **7**. Alternatively, dendrimer **7** can also be synthesized by replacement of the six phenyl groups in dendrimer **6** with methyl groups. For this, dendrimer **6** was first allowed to sequentially react with six equivalents of TfOH and an excess of ammonium chloride, NH_4Cl, to give the hexachloro dendrimer $1G(3,^{1}2)Cl_6$. Subsequent treatment of this hexachloro intermediate with methylmagnesium iodide led to the permethylated dendrimer **7**.

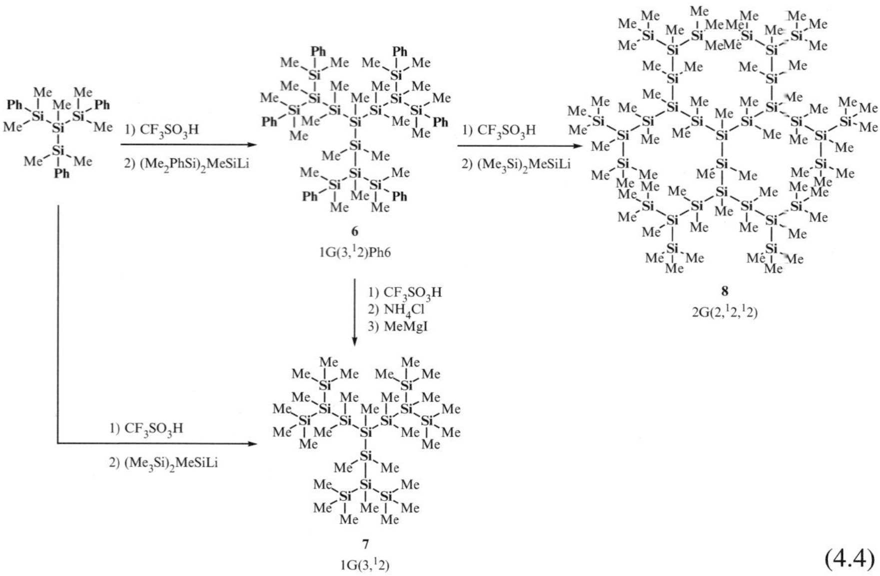

(4.4)

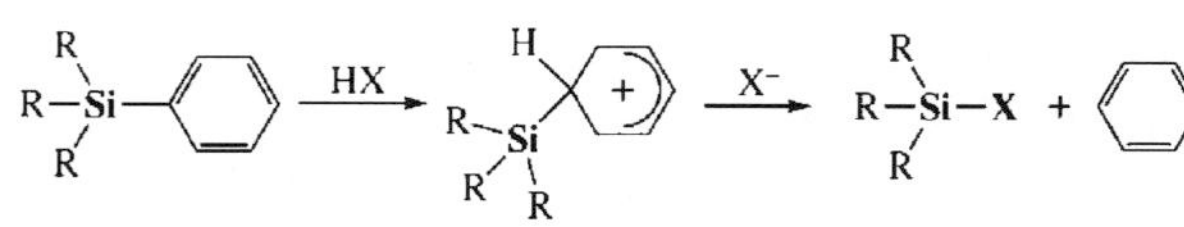

(4.5)

The reaction of the protected polysilane dendrimer $1G(3,{}^{1}2)Ph_6$, **6**, with six equivalents of TfOH affords the functional dendrimer $1G(3,{}^{1}2)OTf_6$, which has six triflate groups on its periphery. When this deprotected polysilane dendrimer was then reacted with permethylated silyllithium, $Me(Me_3Si)_2SiLi$, the second generation permethyl-substituted polysilane dendrimer $2G(3,{}^{1}2,{}^{1}2)$, **8**, was obtained as colorless crystals. This second generation dendrimer **8** has a total of 31 silicon atoms and molecular weight of 1,833. It also contains ten three-functional branch points ($\geq$SiMe), nine dimethylsilylene spacers ($>SiMe_2$), 12 terminal trimethylsilyl groups ($-SiMe_3$), and its longest –Si–Si– chain consists of 11 silicon atoms. This $2G(3,{}^{1}2,{}^{1}2)$ dendrimer still remains the only example of a regular polysilane dendrimer of the second generation.

When using the divergent growth method, the backbone architecture of dendrimers can be readily varied. Thus, dendrimers with alternating Si and Ge atoms in the main branches have been reported (see Reaction Scheme 4.6) [15]. A tris(dimethylphenylgermyl)methylsilane, $(Me_2PhGe)_3SiMe$, core reagent was first synthesized by the reaction of dimethylphenylchlorogermane, $Me_2PhGeCl$, with dimethylphenylgermyl-substituted silyllithium, $Me(Me_2PhGe)_2SiLi$. After deprotection of this core reagent, three equivalents of $Me(Me_2PhGe)_2SiLi$ were added to give the coupling product of the first generation Si/Ge hybrid dendrimer **9**. The six phenyl groups in **9** are readily replaced by methyl groups by the same procedure as that used for polysilane dendrimer **7**. The molecular structure of the resulting permethylated Si/Ge hybrid dendrimer **10** was confirmed by X-ray diffraction.

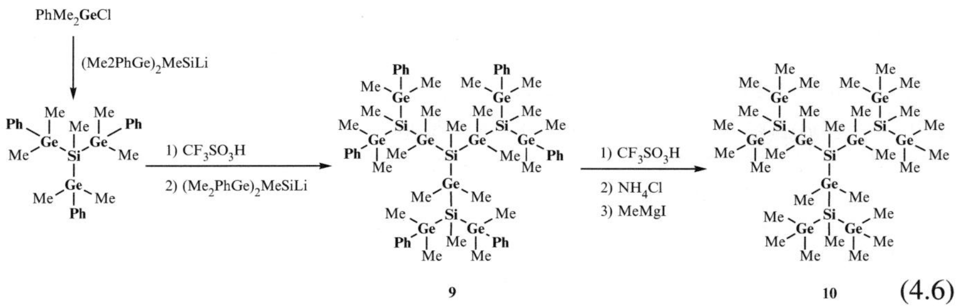

(4.6)

4.2.3 Double-Cored Polysilane Dendrimers

The first attempts to synthesize oligosilanes with branched structures and two silicon atoms in their cores were reported by Gilman and Harrell in 1965 [16]. The reaction of hypersilyllithium, $(Me_3Si)_3SiLi$, with hypersilylchloride, $(Me_3Si)_3SiCl$, gave hexakis(trimethylsilyl)disilane, $(Me_3Si)_3SiSi(SiMe_3)_3$, as a coupling product, although in low yield. This suggested that the synthetic approach by nucleophilic substitution is not suitable for branched oligosilanes because of steric hindrance. Following this, Kumada and Ishikawa successfully synthesized the same compound in high yield (92%) by a radical reaction of tris(trimethylsilyl)silane, $(Me_3Si)_3SiH$, in the presence of di-*tert*-butyl peroxide by heating in a sealed tube [17]. This radical reaction is particularly convenient when reagents with very bulky substituents are involved.

An oligosilane with a more branched structure was formed by a convergent method [13]. Lambert et al. tried to synthesize a large branched silyllithium extender reagent by a mercury–lithium exchange reaction (see Reaction Scheme 4.7). Bis{methylbis(trimethylsilyl)silyl}methylsilane, $\{(Me_3Si)_2MeSi\}_2MeSiH$ was reacted with di-*tert*-butylmercury, $t\text{-}Bu_2Hg$, to give the corresponding bis(silyl)mercury compound **11**. Although the resulting bis(silyl)mercury does not react with lithium metal, its heating led to compound **12** accompanied by mercury elimination. Compound **12** contains 14 silicon atoms, its longest silicon chain consists of six silicon atoms, and it can be considered a double-cored polysilane dendrimer.

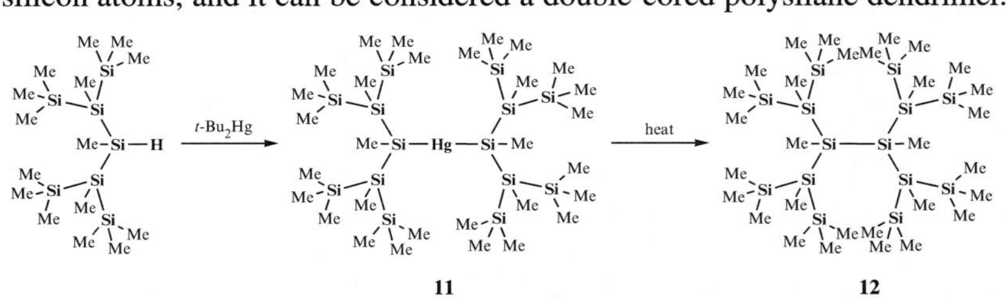

(4.7)

A series of double-cored polysilane dendrimers was prepared by Krempner et al. (Reaction Scheme 4.8) [18]. These authors succeeded in synthesizing the large dendron reagent **13** by a lithium–halogen exchange reaction. Treatment of **13** with 1,2-dichlorotetramethyldisilane, $ClMe_2SiSiMe_2Cl$, gave the double-cored polysilane dendrimer **14**, which has the structure of two dendrons joined by a tetramethyldisilane bridge. This double-cored dendrimer has 16 silicon atoms, with the longest silicon chain containing eight atoms. The authors also synthesized a double-cored polysilane dendrimer with dimethylsilylene spacers between the branching points (Reaction Scheme 4.9). For this, bis(trimethylsilyl)(methyl)silylpotassium, $Me(Me_3Si)_2SiK$, was reacted with tetrakis(chlorodimethylsilyl)disilane as a double core to give the didendron polysilane dendrimer **15**, while the double-cored dendrimer **16** was synthesized by the Würtz-type coupling of the corresponding dendritic chlorosilane using an Na–K alloy [19]. Dendrimer **16** has four dimethylsilylene spacers, six three-functional silicon branching junctures, eight trimethylsilyl terminal groups, and a tetramethyldisilane core bridge. The longest silicon chain consists of ten silicon atoms.

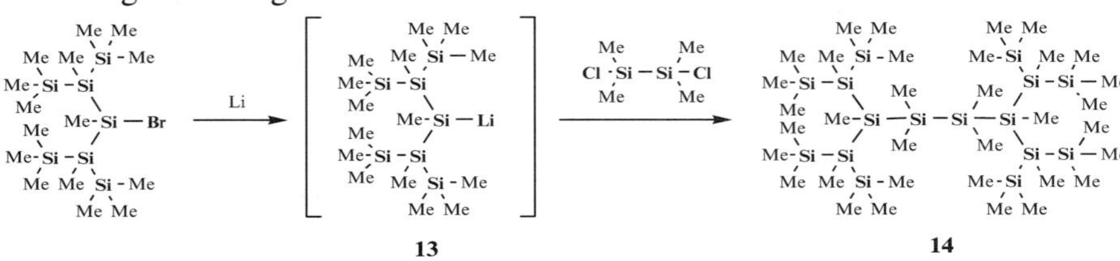

(4.8)

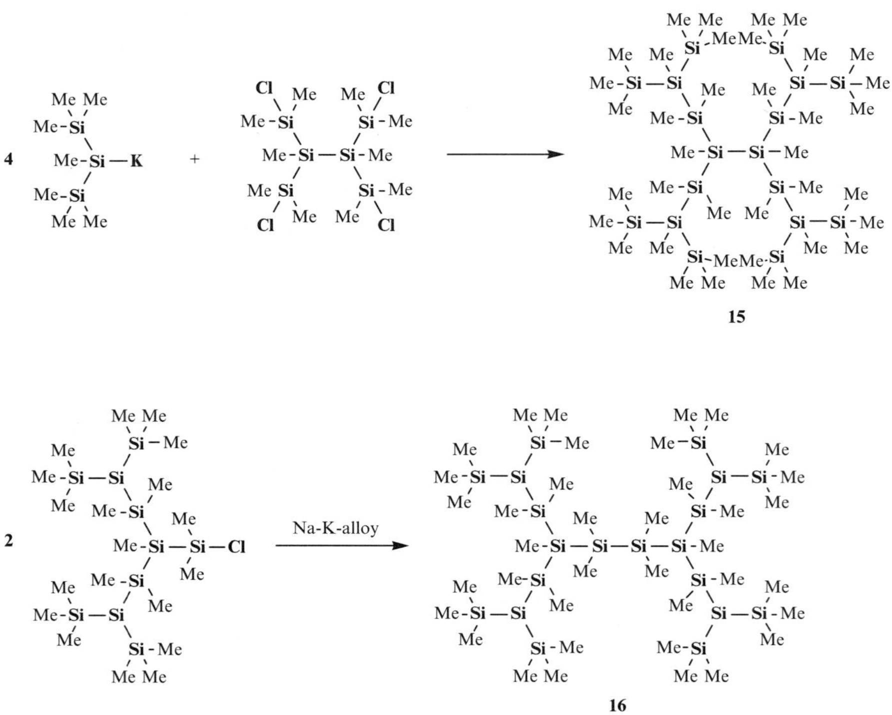

(4.9)

Very recently, Krempner and Köckerling developed twin-type dendrimers with the longest silicon chain bearing 14 silicon atoms as shown in Reaction Scheme 4.10 [20]. 1,4-Dipotassiumtetrasilane **17** was reacted with two equivalents of dendron **18** to give double-cored dendrimer **19** in high yield. The resulting dendrimer **19** was fully characterized by NMR, UV, elemental analysis, and X-ray diffraction. It should be noted that dendrimer **19a** consisting of 32 silicon atoms is the largest well-defined polysilane dendrimer reported to date. Its molecular size reaches 2.5 nm.

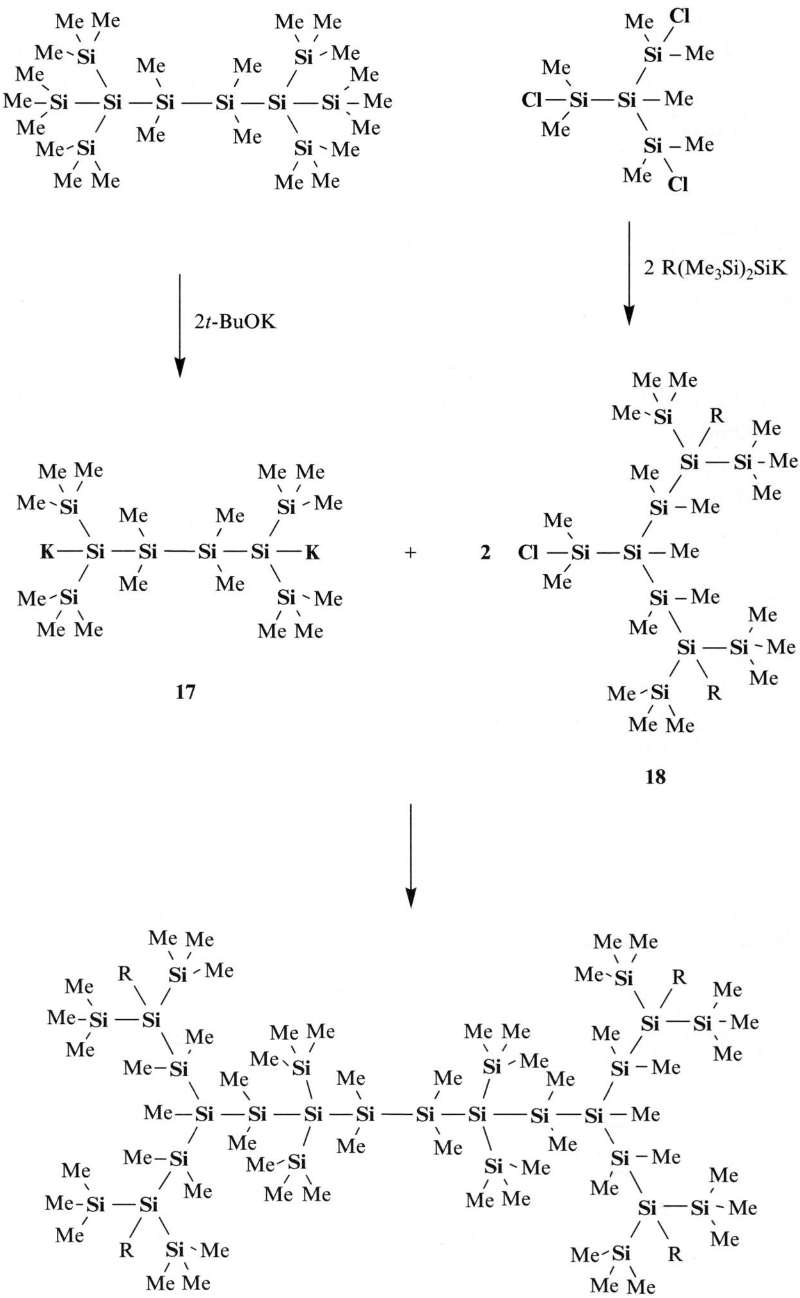

19 (4.10)

(a: R = $SiMe_3$, b: R = Me)

It should also be noted that "double-cored" polysilane dendrimers are unique among silicon-containing dendrimers, which generally contain either a single silicon atom or a silicon-containing group (e.g., cyclosiloxanes or POSS, see Chapters 6 and 7, respectively) in their cores.

4.2.4 Functionalized Polysilane Dendrimers

Alkyl and aryl groups are the most common substituents of silicon atoms in polysilane dendrimers. The methyl group on silicon, Si–CH_3, is inert under the usual reaction conditions, and it represents a key substituent of polysilane dendrimers. The phenyl group on silicon, Si–C_6H_5, is also inert, except under very strong acidic conditions, such as superacids or hydrogen halides in the presence of a Lewis acid, when it can split off in a proto–desilylation reaction. On the other hand, a hydrogen atom directly bound to silicon, Si–H, can be readily replaced by halogen, or react with olefins in the presence of a platinum catalyst to produce the corresponding addition products via the hydrosilylation reaction. Lambert et al. reported a hydrogen-substituted polysilane dendrimer, in which the hydrogen was attached to the three-functional central core [10, 12]. Interestingly, this dendrimer was formed accidentally during the intended synthesis of a polysilane dendrimer with a tetrafunctional core. However, transformation of the central Si–H group of this dendrimer to another functional group has not yet been reported.

The chlorinated polysilane dendrimer **20** was reported by Krempner et al. in 2006 [21]. The tetrachlorinated core reagent, $(ClMe_2Si)_3SiCl$, was reacted with hypersilylpotassium, $(Me_3Si)_3SiK$, or methylbis(trimethylsilyl)silylpotassium, $Me(Me_3Si)_2SiK$, to give the tridendron dendrimers **20a** and **20b**, respectively (see Reaction Scheme 4.11), where both **20a** and **20b** possessed a chlorine atom attached to the central silicon core atom. This Si–Cl bond was found to be very stable, because it survived hydrolysis for 3 days.

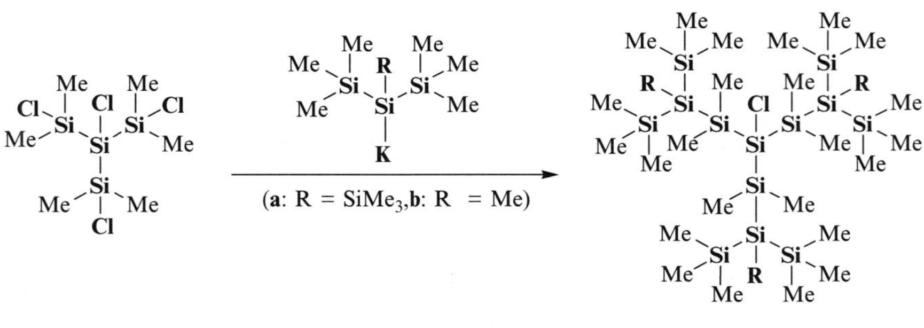

20

(4.11)

Krempner et al. were also successful in the introduction of a hydroxyl group on the silicon spacer atoms (see Reaction Scheme 4.12) [22]. For this purpose, three equivalents of $(Me_3Si)_2MeSiK$ were reacted with tris(dichloromethylsilyl)methylsilane, $(MeCl_2Si)_3SiMe$, to give the corresponding chlorine-substituted dendrimer **21**, where only one $(Me_3Si)_2MeSi$ group could be introduced into each dendron. The resulting dendrimer **21** was reacted with water to obtain the trihydroxy-substituted dendrimer **22** in the form of two diastereomers, *l,l*-**22** and *l,u*-**22**, in 21% and 64% yields, respectively.

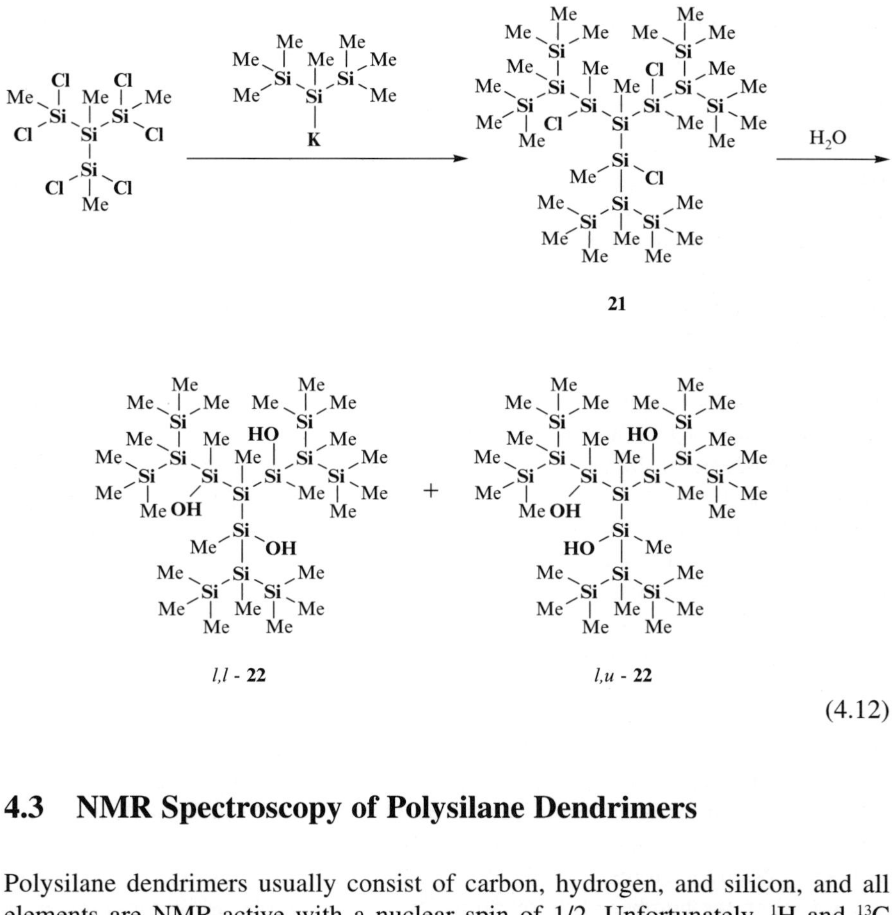

(4.12)

4.3 NMR Spectroscopy of Polysilane Dendrimers

Polysilane dendrimers usually consist of carbon, hydrogen, and silicon, and all elements are NMR-active with a nuclear spin of 1/2. Unfortunately, ^1H and ^{13}C NMR signals of polysilane dendrimers are observed in a narrow region at high magnetic field around $\delta = 0$, and consequently ^1H and ^{13}C NMR spectroscopy are not very useful techniques for structural identification of these products. However, ^{29}Si NMR gives very useful information for analyzing these structures because different silicon atoms constitute the dendrimer skeleton.

The silicon atoms of polysilane dendrimers can be primary, secondary, tertiary, and quaternary, denoted *P*, *S*, *T*, and *Q*, respectively. Of these, primary silicons correspond

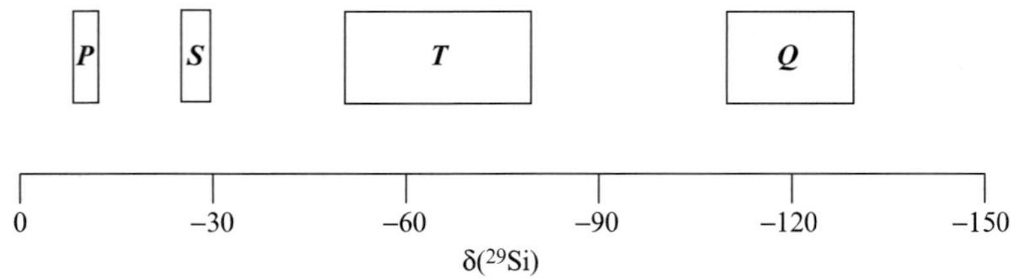

Fig. 4.2 Various ^{29}Si NMR chemical shifts in polysilane dendrimers. *P*: primary, *S*: secondary, *T*: tertiary, *Q*: quarternary.

to the end-groups, whereas secondary, tertiary, and/or quaternary silicons are spacers and/or branching points, respectively. A schematic representation of possible ^{29}Si NMR chemical shifts from polysilane dendrimers is shown in Fig. 4.2. Primary silicons, *P*, are observed in the range from −9 to −11 ppm, secondary silicons, *S*, from −25 to −30 ppm, tertiary silicons, *T*, from −44 to −80 ppm, and quaternary silicons, *Q*, from −112 to −128 ppm, respectively. These signals and their shifts are of critical importance in determining the structure of polysilane dendrimers.

Sometimes, identification of ^{29}Si NMR signals of a polysilane dendrimer is clear and easy from comparisons of their chemical shifts and intensities. However, this becomes much more difficult for more complicated dendrimer structures where traditional one-dimensional ^{29}Si NMR spectroscopy becomes insufficient and more complex methods must be used. For example, for the polysilane dendrimer **5**, 1G(3,^{4}3), the ^{29}Si–^{29}Si 2-D INADEQUATE NMR technique proved to be an important tool in determining its structure [23].

The ^{29}Si NMR chemical shift values of several permethyl-substituted polysilane dendrimers are listed in Table 4.1. It can be seen from these data that as the number of internal silicon atoms with the same environment increases, their ^{29}Si NMR signals shift to lower magnetic field. For example, the ^{29}Si NMR chemical shifts of the three branch points in dendrimer 2G(3,^{1}2,^{1}2), **8**, are observed at −80.1 (outermost), −64.8 (middle), and −62.3 ppm (core), respectively, while the signals of $-SiMe_2-$ spacers are found at −30.6 (outer) and −26.7 ppm (inner), respectively. This tendency is common to all polysilane dendrimers, and appears to be related to the number of β-silicon atoms, such that the ^{29}Si NMR signals shift to lower field with an increase in the number of these atoms [5c]. For example, while the number of β-silicons in the outermost branch SiMe in **8** is 1, their number in the inner branching points is 3. Although the difference in the chemical shifts between the outermost branches and the inner parts is very large (~15 ppm), there is little difference between the two inner chemical shifts at the branching points. In addition, the ^{29}Si chemical shift of the core in dendrimer 1G(3,^{0}2) is −44 ppm, which is the lowest value for the central core of a tridendron polysilane dendrimer resulting when the number of β-silicon atoms is six.

Table 4.1 ^{29}Si NMR chemical shifts of permethylated polysilane dendrimers

Compound		Q	T	S	P
	1G(3, ^{0}2)	–	−76, −44	–	−11
1	1G(3, ^{1}3)	−124	−69.6	−26.7	−9.4
5	1G(3, ^{4}3)	−127.3, −112.5	−66.1	−29.1, −27.7, −25.5	−9.3, −9.1
7	1G(3, ^{1}2)	–	−80.4, −66.3	−30.0	−11.4
8	2G(3, ^{1}2, ^{1}2)	–	−80.1, −64.8, −62.3	−30.0, −26.7	−11.3, −11.4
14		–	−76.8, −54.4	−31.4	−11.0, −11.3

4.4 Crystallography and Conformation of Polysilane Dendrimers

Although dendrimers are macromolecular compounds, they are also highly monodisperse because of the high level of control of their synthesis (see Chapter 1). As a consequence, molecular structure determination techniques can be used to evaluate their conformations, and more than ten examples of X-ray crystallography of polysilane dendrimers have been reported. For example, Fig. 4.3 shows the molecular structure of the second generation polysilane dendrimer 2G(3,^{1}2,^{1}2), **8** [14], which is the only example of a crystallographically defined second generation dendrimer reported to date. The structure includes three clathrate benzene molecules, which were used as the crystallization solvent, and these benzene molecules are arranged in such a fashion that they occupy the entire space inside the dendrimer molecule.

In general, the average Si–Si bond length of all examined polysilane dendrimers is approximately 2.34 Å, except for the Si–Si bonds near the core, which are usually slightly longer than those outside the core. This is clearly seen from the Si_{core}–Si_{core} bond lengths of double-cored polysilane dendrimers, where in dendrimer **15**, for example, the central Si_{core}–Si_{core} distance is 2.405 Å, which is about 2.3% longer than the average Si–Si bond length [21]. The Si_{core}–C bond lengths in tridendron polysilane dendrimers range between 1.90 and 2.13 Å, and are longer than normal Si–C single bonds of 1.87 Å, which seems to be a result of steric repulsion between the dendrons. All silicon atoms in polysilane dendrimers feature sp^3-geometry, but many Si–Si_{spacer}–Si bond angles around the spacer silicons expand to 110–120°. These values are larger than those usually found in normal *sp^3* hybrids of about 109.5°, and clearly show how spacer moieties reduce steric hindrance in these dendrimers.

Theoretical and experimental studies were performed in order to evaluate σ-conjugation and Si–Si chain conformation in linear polysilanes [1], where the smallest unit that shows σ-conjugation consists of four silicon atoms, Si–Si–Si–Si. Conformations of such chains can be described by torsional angles, as shown in Fig. 4.4, and can be classified as *syn* (*S*, ~0°), *gauche* (*G*, ~60°), *ortho* (*O*, ~90°), *eclipsed* (*E*, ~120°), *deviant* (*D*, ~150°), and *anti* (*A*, ~180°) [24]. It can be expected that σ-conjugation reaches its minimum in the *O*-conformation because the two Si–Si σ-orbitals do not overlap with each other, although, according to a recent

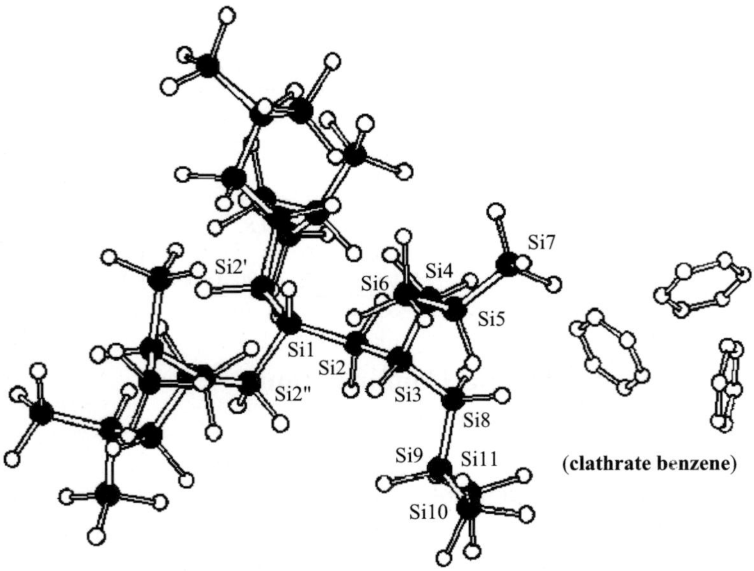

Fig. 4.3 Molecular structure of 2G(3,[1]2,[1]2) (**8**). Hydrogen atoms are omitted for clarity.

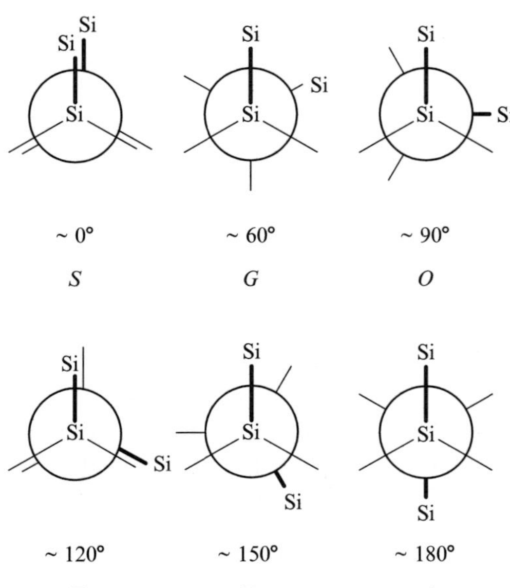

Fig. 4.4 Schematic representation of typical conformations for Si–Si–Si–Si units.

study, even *G*-conformations do not contribute significantly to the elongation of σ-conjugation [25].

In the polysilane dendrimer 1G(3,[1]2), **7**, the longest Si–Si chain consists of a heptasilane unit with a total of 12 possible pathways, listed in Table 4.2 [26]. It can be seen from this table that all Si–Si backbones consist of *O*- and *D*-conformations, and that four conformational patterns: *DDOD*, *DDOO*, *DODO*, and *ODOO* are observed. In fact, every route includes at least one *O*-conformation and there is neither an all-*anti* heptasilane nor an all-*gauche* arrangement in any of these units.

Table 4.3 shows all conformations of the second generation polysilane dendrimer 2G(3,[1]2,[1]2), **8**, [14]. The longest silicon chain is an undecasilane unit, and there is a total of 48 pathways through the core silicon atom (Si1). However, the conformation of this silicon chain should involve 16 pathways of undecasilane arrangements, because the molecular structure of **8** has a three-fold axis along the core silicon center and its methyl group. As in the case of the first generation dendrimer **7**, all conformations of silicon chains consist exclusively of *O*- and *D*-contributions, and the longest *D*-arrangement also consists of only a pentasilane unit. This feature is also confirmed by the UV absorption spectrum, which is discussed in the next section.

The longest silicon chain in 1G(3,[1]3), **1**, is a heptasilane unit, and a total of 27 pathways exist. However, in its solid state this dendrimer has a hexagonal crystal system, and the molecule has a three-fold axis of symmetry through the core silicon and methyl carbon atoms, so that the conformation of the silicon backbone should

Table 4.2 Conformations of 1G(3, [1]2) (**7**)

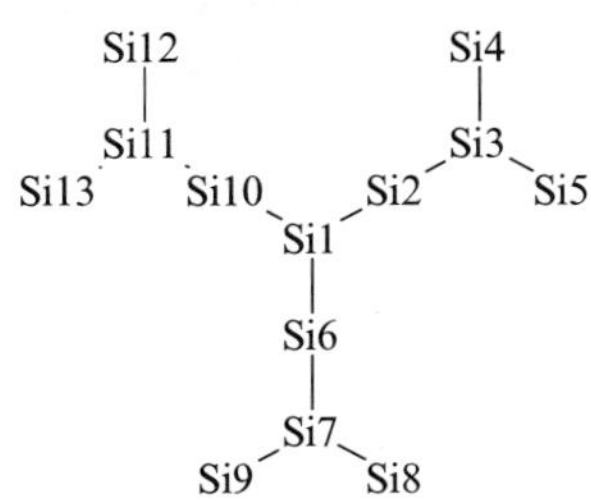

Si chain	Conformation
Si4–Si3–Si2–Si1–Si6–Si7–Si8	O O D O
Si4–Si3–Si2–Si1–Si6–Si7–Si9	O O D D
Si4–Si3–Si2–Si1–Si10–Si11–Si12	O D O O
Si4–Si3–Si2–Si1–Si10–Si11–Si13	O D O D
Si5–Si3–Si2–Si1–Si6–Si7–Si8	D O D O
Si5–Si3–Si2–Si1–Si6–Si7–Si9	D O D D
Si5–Si3–Si2–Si1–Si10–Si11–Si12	D D O O
Si5–Si3–Si2–Si1–Si10–Si11–Si13	D D O D
Si8–Si7–Si6–Si1–Si10–Si11–Si12	O O D O
Si8–Si7–Si6–Si1–Si10–Si11–Si13	O O D D
Si9–Si7–Si6–Si1–Si10–Si11–Si12	D O D O
Si9–Si7–Si6–Si1–Si10–Si11–Si13	D O D D

Table 4.3 Conformations of 2G(3, [1]2, [1]2) (**8**)

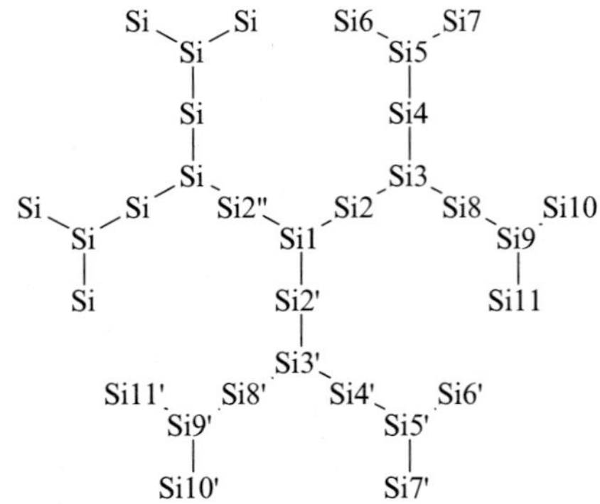

Si chain	Conformation
Si6–Si5–Si4–Si3–Si2–Si1–Si2′–Si3′–Si4′–Si5′–Si6′	*D D O D O O D D*
Si6–Si5–Si4–Si3–Si2–Si1–Si2′–Si3′–Si4′–Si5′–Si7′	*D D O D O O D O*
Si6–Si5–Si4–Si3–Si2–Si1–Si2′–Si3′–Si4′–Si5′–Si10′	*D D O D O D O O*
Si6–Si5–Si4–Si3–Si2–Si1–Si2′–Si3′–Si4′–Si5′–Si11′	*D D O D O D O D*
Si7–Si5–Si4–Si3–Si2–Si1–Si2′–Si3′–Si4′–Si5′–Si6′	*O D O O D O D D*
Si7–Si5–Si4–Si3–Si2–Si1–Si2′–Si3′–Si4′–Si5′–Si7′	*O D O O D O D O*
Si7–Si5–Si4–Si3–Si2–Si1–Si2′–Si3′–Si4′–Si5′–Si10	*O D O O D D O O*
Si7–Si5–Si4–Si3–Si2–Si1–Si2′–Si3′–Si4′–Si5′–Si11′	*O D O O D D O D*
Si10–Si9–Si8–Si3–Si2–Si1–Si2′–Si3′–Si4′–Si5′–Si6′	*O O D O D O D D*
Si10–Si9–Si8–Si3–Si2–Si1–Si2′–Si3′–Si4′–Si5′–Si7′	*O O D O D O D O*
Si10–Si9–Si8–Si3–Si2–Si1–Si2′–Si3′–Si4′–Si5′–Si10′	*O O D O D D O O*
Si10–Si9–Si8–Si3–Si2–Si1–Si2′–Si3′–Si4′–Si5′–Si11′	*O O D O D D O D*
Si11–Si9–Si8–Si3–Si2–Si1–Si2′–Si3′–Si4′–Si5′–Si6′	*D O D O D O D D*
Si11–Si9–Si8–Si3–Si2–Si1–Si2′–Si3′–Si4′–Si5′–Si7′	*D O D O D O D O*
Si11–Si9–Si8–Si3–Si2–Si1–Si2′–Si3′–Si4′–Si5′–Si10′	*D O D O D D O O*
Si11–Si9–Si8–Si3–Si2–Si1–Si2′–Si3′–Si4′–Si5′–Si11′	*D O D O D D O D*

involve nine heptasilane pathways. Observed conformations are *OAGG*, *OAGA*, *OAGO*, *AAGO*, *AAGA*, *AAGG*, *GAGO*, *GAGA*, and *GAGG*, so that in contrast to dendrimers **7** and **8** *G*- and *A*-conformations are present and a heptasilane unit without an *O*-conformation exists. However, all-*anti AAAA* or all-*gauche GGGG* conformations are absent.

The molecular structure of the phenyl-substituted polysilane dendrimer 1G(3,[1]2)Ph_6, **6**, which was obtained by a divergent synthesis, has also been established by X-ray crystallographic analysis [26]. Although, as in the permethylated polysilane dendrimer **7**, there are 12 pathways of heptasilane units in **6** as well, in this dendrimer an *E*-conformation is also observed. As a consequence, the conformations of silicon chains involve a total of four different conformational patterns as follows: three *ODOE*, three *ODOO*, three *OODE*, and three *EDOE*. More than one *O*-conformation is certainly included in all heptasilane units, but no *G*-conformation can be observed.

The molecular structures of the two permethylated double-cored polysilane dendrimers **14** and **15** have also been established by X-ray diffraction. Dendrimer **14** has no dimethylsilylene spacer group but possesses a tetramethyldisilane bridge group. Because the spacer group of **14** is P1-(No. 2) [18], there is an inversion center of symmetry at the center of the bridge. Therefore, the torsional angle of Si_{core}–Si_{bridge}–Si_{bridge}–Si_{core} should ideally be 180°. The longest silicon chain in dendrimer **14** consists of eight silicon atoms, and there is a total of 16 pathways of this octasilane unit with seven different conformational patterns: one *OEAEO*, two *DEAEO*, two *GOAEO*, one *DEAED*, two *GOAED*, four *OEAOG*, and four *GOAOG*. It should be noted that dendrimer **14** has the pattern of a *DEAED* arrangement, which does not include *O*- or *G*-conformations, showing that the σ-conjugation is efficiently extended along the longest octasilane chain. In the case of dendrimer **15** [19], the longest silicon chain also corresponds to an octasilane unit, but 11 conformational patterns are observed from the crystal structure: one *DDGDD*, two *DDGDO*, one *DDOOD*, one *DDOOO*, one *ODGDO*, one *ODOOD*, one *ODOOO*, two *DDODD*, two *DDODO*, two *DDDOD*, and two *DDDOO*. Neither *E*- nor *A*-conformations are observed, but because dendrimer **15** has *O*- or *G*-conformations in any octasilane chain, this suggests that its degree of σ-conjugation is smaller than that in dendrimer **14**.

The molecular structures of hydroxy-substituted polysilane dendrimers *l,l*-**22** and *l,u*-**22** have also been determined by X-ray crystallography [22]. Interestingly, these two diastereomers form two different dimers by hydrogen bonding of the hydroxy groups. In *l,l*-**22**, all hydroxy groups are involved in intermolecular hydrogen bonding to form a dimeric structure. Conformational patterns of the heptasilane chain of this diastereomer are similar to those found in 1G(3,^{1}2), **7**, and the following four patterns: *DDOD*, *DDOO*, *DODO*, and *ODOO* are observed. In contrast to this, in the *l,u*-**22** dimer, only two of the three hydroxy groups interact intermolecularly to form four hydrogen bonds, while one hydroxy group creates an intramolecular hydrogen bond. As a consequence, dendrimer *l,u*-**22** has an *E*-conformation and the *DDED* arrangement exists in a heptasilane chain. These conformational differences between *l,l*-**22** and *l,u*-**22** also appear in the UV absorption spectra, which derive from the Si–Si backbone.

4.5 Electronic Spectra

UV absorption spectra are very important for analysis of σ-conjugation systems along the Si–Si backbones, and relationships between these spectra and linear polysilane and oligosilane structures have been investigated both theoretically and experimentally [1, 27]. It was found that the UV absorption band of the σ–σ^* transition red-shift increases with the length of the linear permethylated Si–Si chains both in polysilanes and in their shorter oligomers. This shift, however, has a limit at about 296 nm, which corresponds to that of the octadecasilane, $Si_{18}Me_{36}$, unit [28] and if the number of silicon atoms is increased above 18, no significant further red shift is observed. Also, the UV absorption spectrum strongly depends on the conformation of the Si–Si skeleton, where the σ–σ^* excitation band of oligosilanes depends on the Si–Si backbone conformation. The first absorption band shows a blue shift as the dihedral angle of an Si–Si–Si–Si segment decreases from 180°.

Unfortunately, the studies of electronic spectra to date have focused mainly on linear chain polysilanes, while the σ-conjugation properties of the branched structures have received little attention. For the latter, however, polysilane dendrimers are clearly ideal molecules because they possess an ultimately branched architecture, pronounced monodispersity (i.e., ideally a single molecular weight), and unequivocally determined molecular structures. Some available electronic absorption data for permethylated polysilane dendrimers are listed in Table 4.4.

The UV absorption spectra of 1G(3,^{1}3), **1**, and 1G(3,^{1}2), **7**, in solution show distinct absorption maxima at 265 nm ($\varepsilon = 5.2 \times 10^4$ M^{-1} cm^{-1}) [20] and 269 nm ($\varepsilon = 4.9 \times 10^4$ M^{-1} cm^{-1}) [14], respectively. These values are close to that found for linear heptasilane, Si_7Me_{16}, λ_{max} = 269 nm, because the longest silicon chains in both dendrimers **1** and **7** correspond to seven silicon atoms. The UV absorption maximum of 2G(3,^{1}2,^{1}2), **8**, is observed at 279 nm ($\varepsilon = 9.6 \times 10^4$ M^{-1} cm^{-1}), which is red shifted by 10 nm compared with the first generation dendrimer **7** [14]. However, although the longest silicon chain in **8** consists of 11 silicon atoms, the wavelength of its UV absorption maximum is less than the 284 nm found for linear permethylated decasilane, $Si_{10}Me_{22}$, indicating that this behavior must be related to the conformation of the polysilane dendrimer segments. Thus, the polysilane dendrimer **7** possesses more than one *O*-conformation in its Si–Si backbone and because the overlap of Si–Si σ-bond orbitals becomes smallest in an *O*-conformation, σ-conjugation weakens at this point and, as a result, the longest silicon chain is not directly reflected in the UV absorption spectrum but instead, a blue shift is usually observed. For example, dendrimer 1G(3,^{4}3), **5**, has a total of 31 silicon atoms, which is the same as for **8**, but the longest silicon chain consists of 13 silicon atoms. Hence, its UV spectrum shows two absorption maxima at 283 nm ($\varepsilon = 1.2 \times 10^5$ M^{-1} cm^{-1}) and 260 nm ($\varepsilon = 1.8 \times 10^5$ M^{-1} cm^{-1}) [13], which are only slightly smaller than

Table 4.4 UV absorption data of polysilane dendrimers

Compound		λ_{max}/nm	ε/M^{-1} cm^{-1}	Longest Si chain
	0G^a	242		5
	1G (3, ^{0}2)	240	25,000	5
1	1G (3, ^{1}3)	265	52,000	7
5	1G (3, ^{4}2)	260	180,000	13
		283	116,000	
7	1G (3, ^{1}2)	269	49,000	7
8	2G (3, ^{1}2, ^{1}2)	279	96,400	11
14		285	56,000	8
15		271	82,000	8
16		270	72,000	10
		292	55,000	

those found in linear permethyl-substituted dodecasilane, $Si_{12}Me_{26}$, at 285 and 264 nm, respectively.

The absorption spectra of double-cored polysilane dendrimers **14** [18] and **15** [19], both of which possess octasilane units as the longest silicon chains, show absorptions at 285 nm ($\varepsilon = 5.6 \times 10^4$ M^{-1} cm^{-1}) and 271 nm ($\varepsilon = 8.2 \times 10^4$ M^{-1} cm^{-1}), respectively. The red shift of 14 nm reflects the fact that dendrimer **14** possesses a *DEAED* arrangement in the octasilane chain without *O*- or *G*-conformations. In other words, the overlap of the Si–Si σ-bonds is larger than that in dendrimer **15**. Dendrimer **16**, in which the decasilane unit represents the longest Si–Si chain, shows a maximum absorption wavelength at 292 nm ($\varepsilon = 5.5 \times 10^4$ M^{-1} cm^{-1}) together with another absorption at 270 nm ($\varepsilon = 7.2 \times 10^4$ M^{-1} cm^{-1}) [21]. The longest absorption is close to the 296 nm value found in linear polysilanes. Unfortunately, the molecular structure of dendrimer **16** has not been determined yet; however, one can expect the longest silicon backbone to adopt an all-*anti* conformation suitable for σ-conjugation.

Recently, Krempner synthesized an unusual tethered polysilane dendrimer **23**, in which two of the three dendrons were connected by a tetramethylene bridge, and a "normal" (free) dendrimer **24** without any bridging unit [29]. According to the crystallographic data, the *O*-conformation was present in all silicon pathways of **24** (similar to other previously reported dendrimers), while an *ADAD* arrangement was found in the tethered dendrimer **23**. These structural results are also reflected in the UV absorption spectra where a conspicuous red shift was observed in tethered dendrimer **23** at λ_{max} = 277 nm ($\varepsilon = 1.2 \times 10^5$ M^{-1} cm^{-1}), compared with the "normal" dendrimer **24** which has λ_{max} = 269 nm, $\varepsilon = 6.0 \times 10^4$ M^{-1} cm^{-1}. It is very likely that this behavior stems from the fact that the Si–Si chain conformations in polysilane dendrimers are strongly related to the σ-conjugation property.

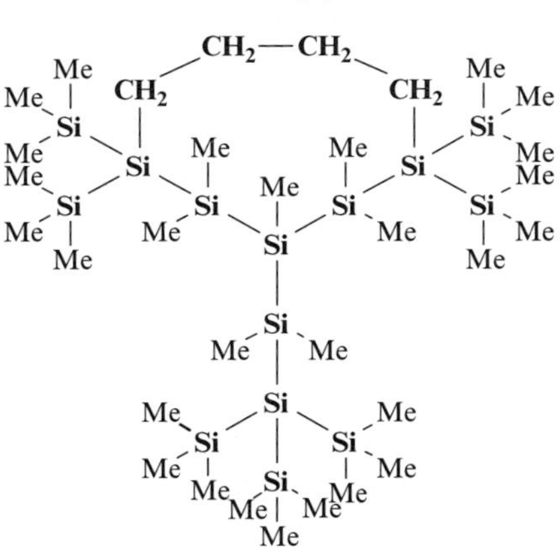

23

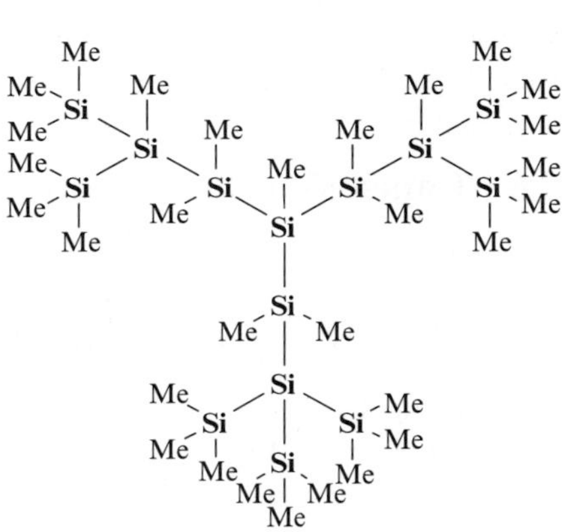

24

The emission spectra suggest a different nature for linear polysilanes and polysilane dendrimers. For example, Watanabe et al. reported the excited-state dynamics of $(Me_3SiMe_2Si)_3SiMe$, 0G, 1G(3,¹2), **7**, and 2G(3,¹2,¹2), **8**, studied using time-resolved

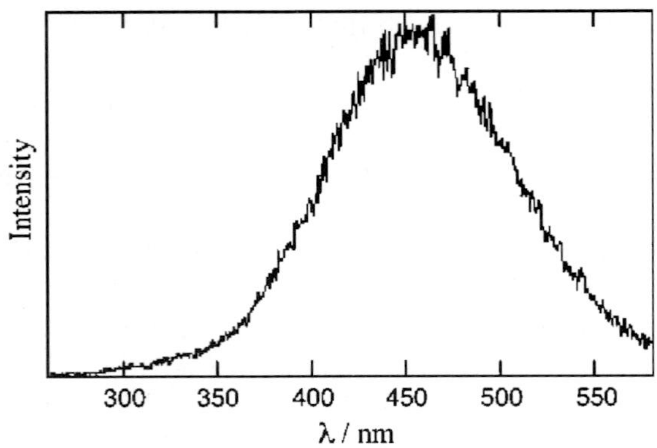

Fig. 4.5 Emission spectrum of 2G(3,^{1}2,^{1}2) (**8**) in hexane. Excitation wavelength: 279 nm.

emission spectroscopy [30]. The spectra obtained under 279 nm Ti:sapphire laser irradiation showed two bands around 350 and 460 nm, respectively. The UV emission around 350 nm was observed early (~20 ns) after the laser excitation, but with time it reduced in intensity and a new very broad emission signal appeared in the visible region at 460 nm (see Fig. 4.5). In contrast to this behavior of polysilane dendrimers, linear polysilanes show a sharp emission in the UV region, a mirror image of their absorption spectra, while network polysilanes emit in the visible range with a broad signal and a large Stokes shift. Thus, emission spectra of polysilane dendrimers show properties of both linear and branched polysilanes: emission in the UV fluorescence region resulting from excitation along the linear Si–Si chains, while the broad band with large Stokes shift in the visible region can be attributed to emission from branching points. These dual emission properties are also observed in 1G(3,^{1}3), **1**, at 77 K [7, 10].

4.6 Conclusions and Future Outlook

Since the first syntheses of polysilane dendrimers reported in 1995, various members of this family have been prepared by both convergent and divergent methods. In addition to this, structural and spectroscopic studies have also focused on the σ-conjugation along the Si–Si chains of both polysilane dendrimers and their corresponding linear homologues. According to the molecular structural analysis, it appears that most polysilane dendrimers possess *ortho*-conformations in their longest Si–Si chains, which is unfavorable to the extension of the σ-conjugation of catenated silicon atoms. As a result, the UV absorption bands for the σ–σ^* transition are usually underrepresented compared with those found in the corresponding linear polysilanes of the same Si–Si chain length. Nevertheless, the double-cored polysilane dendrimer **16** exhibits an absorption maximum at 292 nm which is the highest absorption observed for a polysilane dendrimer.

Emission spectra of polysilane dendrimers show two bands: one in the UV and one in the visible region. The fluorescence in the UV region is assigned to emission from the exited state in the linear Si–Si segments of the polysilane dendrimers, while the visible emission is attributed to the branching points. As a consequence, polysilane dendrimers may be viewed as intermediates between linear polysilanes and a silicon nanocrystal.

Although it is only a first generation dendrimer, dendrimer **5** is the largest polysilane dendrimer reported to date with respect to the length of the longest Si–Si segment. In fact, only one example of a second generation polysilane dendrimer, **8**, has been described, clearly showing how progress in the development of these dendrimers lags behind that in other dendrimer families. One of the reasons for this may be attributed to synthetic difficulties. Most polysilane dendrimers are prepared by either anionic or radical reactions that not only form the Si–Si bonds but can also break them. Hence, a new synthetic route to Si–Si bonds is clearly needed in order to further develop this area. From the application point of view, although not enough has been reported on the conducting, photoactive, third-order nonlinear, or ceramics-forming properties of polysilane dendrimers, they certainly open up some interesting possibilities and reports in these directions should be expected before long.

References

1. Michl J, Miller RD (1989) Chem. Rev. 89: 1359–1410.
2. Bianconi PA, Weidam TW (1988) J. Am. Chem. Soc. 110: 2342–2344.
3. Matsumoto H, Miyamoto H, Kojima N, Nagai Y (1987) J. Chem. Soc. Chem. Commun. 1316–1317.
4. Watanabe A, Fijitsuka M, Ito O, Miwa T (1997) Jpn. J. Appl. Phys. 36: L1265–L1267.
5. (a) Richter R, Roewer G, Böhme U, Busch K, Babonneau F, Martin HP, Müller E (1997) Appl. Organomet. Chem. 11: 71–106. (b) Sekiguchi A, Lee VYa, Nanjo M (2000) Coordin. Chem. Rev. 210: 11–45. (c) Lambert JB, Pflug JL, Wu H, Liu X (2003) J. Organomet. Chem. 685: 113–121. (d) Watanabe A (2003) J. Organomet. Chem. 685: 122–133.
6. Lambert JB, Pflug JL, Stern CL (1995) Angew. Chem. Int. Ed. Engl. 34: 98–99.
7. Suzuki H, Kimata Y, Satoh S, Kuriyama A (1995) Chem. Lett. 293–294.
8. Gutekunst G, Brook AG (1982) J. Organomet. Chem. 225: 1–3.
9. Derouiche Y, Lickiss PD (1991) J. Organomet. Chem. 407: 41–49.
10. Lambert JB, Pflug JL, Denari JM (1996) Organometallics 15: 615–625.
11. Bains K, Brook AG, Ford RR, Lickiss PD, Saxena AK, Chatterton WJ, Sawyer JF, Behnam BA (1989) Organometallics 8: 693–709.
12. Lambert JB, Liu X, Wu H, Pflug JL (1999) J. Chem. Soc., Perkin Trans. 2: 2747–2749.
13. Lambert JB, Wu H (1998) Organometallics 17: 4904–4909.
14. Sekiguchi A, Nanjo M, Kabuto C, Sakurai H (1995) J. Am. Chem. Soc. 117: 4195–4196.
15. Nanjo M, Sekiguchi A (1998) Organometallics 17: 492–494.
16. Gilman H, Harrell RL (1965) J. Organomet. Chem. 5: 199–200.
17. Ishikawa M, Nakamura A, Kumada M (1973) J. Organomet. Chem. 59: C11–C12.
18. Reinke H, Krempner C (2003) J. Organomet. Chem. 685: 134–137.
19. Krempner C, Köckerling M, Mamat C (2006) Chem. Commun. 720–722.
20. Krempner C, Köckerling M (2008) Organometallics 27: 346–352.
21. Krempner C, Reinke H (2006) Inorg. Chem. Commun. 9: 259–262.

22. Jäger-Fiedler U, Köckerling M, Ludwig R, Wulf A, Krempner C (2006) Angew. Chem. Int. Ed. 45: 6755–6759.
23. Lambert JB, Wu H (2000) Magn. Reson. Chem. 38: 388–389.
24. Michl J, West R (2000) Acc. Chem. Res. 33: 821.
25. Tamao K, Tsuji H, Terada M, Asahara M, Yamaguchi S, Toshimitsu A (2000) Angew. Chem. Int. Ed. 39: 3287–3290.
26. Nanjo M, Sunaga T, Sekiguchi A, Horn E (1999) Inorg. Chem. Commun. 2: 203–206.
27. Obata K, Kira M (1999) Organometallics 18: 2216–2222.
28. Boberski WG, Allred AL (1975) J. Organomet. Chem. 88: 65–72
29. Krempner C, Reinke H (2007) Organometallics 26: 2053–2057.
30. Watanabe A, Nanjo M, Sunaga T, Sekiguchi A (2001) J. Phys. Chem. A 105: 6436–6442.

Chapter 5
Polycarbosilazane and Related Dendrimers and Hyperbranched Polymers

David Y. Son

5.1 Introduction

Linear polysilazanes and polycarbosilazanes are well-known members of the organosilicon polymer family and can be prepared by a variety of methods [1, 2]. These polymers are characterized by having either a –Si–N– backbone (polysilazanes) or a –R–Si–N– backbone (polycarbosilazanes). On the other hand, dendritic analogs are relatively rare. Polysilazane dendrimers are essentially nonexistent, and the syntheses of polycarbosilazane dendrimers have been reported by only two groups. A primary reason for the paucity of examples is the relative reactivity of the Si–N bond. Since many reagents involved in conventional organosilicon dendrimer synthesis would react with Si–N bonds, syntheses of these systems are difficult to design. This problem has been partially circumvented in polycarbosilazane dendrimers in which the nitrogen atoms are bonded to three silicon atoms, a bonding situation that is considerably less reactive than nitrogen bonded to only one or two silicon atoms.

Despite the synthetic difficulties, organosilicon dendrimers with nitrogen in the structure are of interest from a fundamental perspective. First, the nitrogen atoms represent potential binding sites, as in the poly(amidoamine), PAMAM, poly(amidoamine-organosilicon), PAMAMOS (see Chapter 11), or poly(ethylene imine), PEI, dendrimer systems. Secondly, the lability of Si–N bonds raises the possibility of controlled degradation of the dendrimers. Furthermore, the presence of planar trisilyl-substituted amine groups [3–5] throughout the structure would impose some rigidity and interesting configurational constraints on the dendrimer. As a consequence, when combining these factors with the underlying synthetic challenges involved, one can see that this field should hold much interest for the synthetic dendrimer chemist.

This chapter summarizes the developments in the synthesis of polysilazane and polycarbosilazane dendrimers, and also describes an example of a related silatrane system. Finally, the issue of synthesizing hyperbranched derivatives is also addressed.

D.Y. Son
Department of Chemistry, Southern Methodist University,
Dallas, TX 75275-0314, USA
E-mail: dson@smu.edu

P.R. Dvornic and M.J. Owen (eds.), *Silicon-Containing Dendritic Polymers*,
Advances in Silicon Science 2, © Springer Science+Business Media B.V. 2009

5.2 Polycarbosilazane Dendrimers

As stated above, polycarbosilazane dendrimers are known but have been reported by only two groups, those of Son [6, 7] and Veith [8]. In both cases, essentially the same divergent synthetic route was utilized. Starting from tris(vinyldimethylsilyl)amine as a core, layers were added by hydrosilylation with chlorodimethylsilane (using Karstedt's catalyst) followed by reaction with either lithium or potassium bis(vinyldimethylsilyl)amide in THF (see Reaction Scheme 5.1). In the initial report by Son and Hu [6, 7], the first and the second generation dendrimers were prepared in excellent yields (88% and 87%, respectively). Although complete hydrosilylation of the second generation dendrimer could be achieved, subsequent reaction with lithium or sodium bis(vinyldimethylsilyl)amide to make the third generation dendrimer was incomplete. However, in the subsequent report by Veith and coworkers [8], dendrimers up to the fourth generation could be synthesized in good yield by using potassium bis(vinyldimethylsilyl)amide as the nucleophilic reagent.

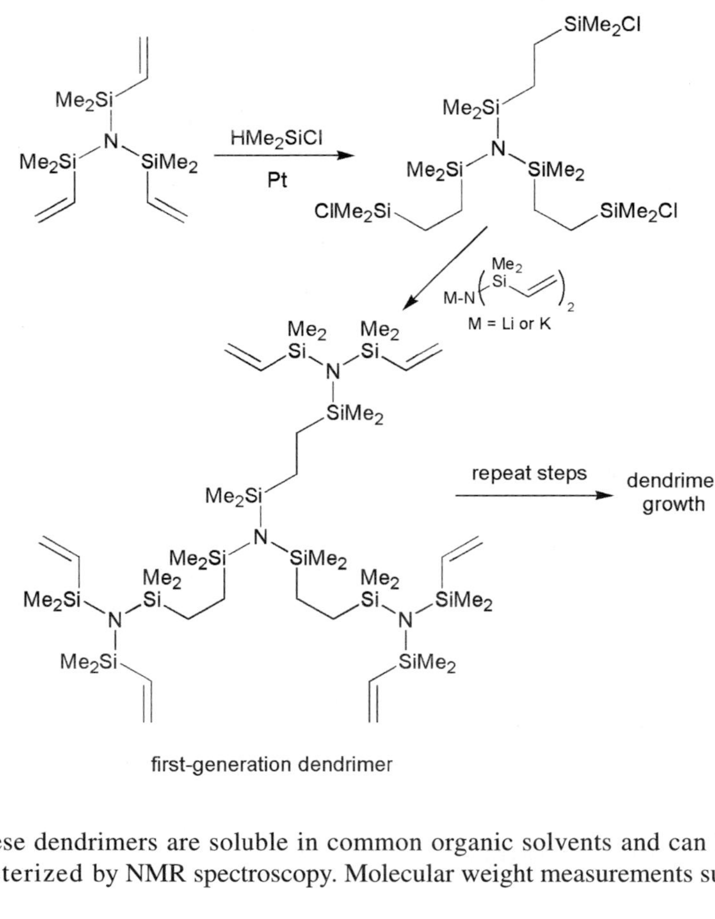

(5.1)

These dendrimers are soluble in common organic solvents and can be easily characterized by NMR spectroscopy. Molecular weight measurements supporting the dendrimer structures were obtained using either vapor pressure osmometry (VPO) [7] or matrix assisted laser desorption ionization-time of flight (MALDI-TOF) mass spectrometry [8]. Certain derivatives could be grown as single crystals; however, X-ray structural determinations were unsuccessful [8].

These dendrimers are stable to water, reflecting the relative inertness of trisilyl-substituted amines. They are also stable to anhydrous solutions of HCl in ether, although binding of HCl to the nitrogen atoms does take place as suggested by IR measurements. However, although stable to anhydrous HCl, these dendrimers are not stable to *aqueous* HCl, as decomposition occurs quickly and exothermically, suggesting the possibility of controlled degradation [7].

Since the initial synthetic reports, several papers have described additional characterization and modifications of these polycarbosilazane dendrimers. For example, freezing and glass transition behaviors have been investigated using temperature modulated differential scanning calorimetry (DSC) and Brillouin spectroscopy [9]. Among the reported observations was the expected increase in glass transition temperature (T_g) with increasing generation number. Another series of reports described the potential utility of these dendrimers as liquid crystalline materials [10–14]. For example, Mehl and Veith attached various mesogens (an example is shown in the following structure) to the periphery of the vinyl-terminated polycarbosilazane dendrimers via hydrosilylation reactions. The result was a series of dendritic liquid crystal systems that displayed enantiotropic nematic phase behavior.

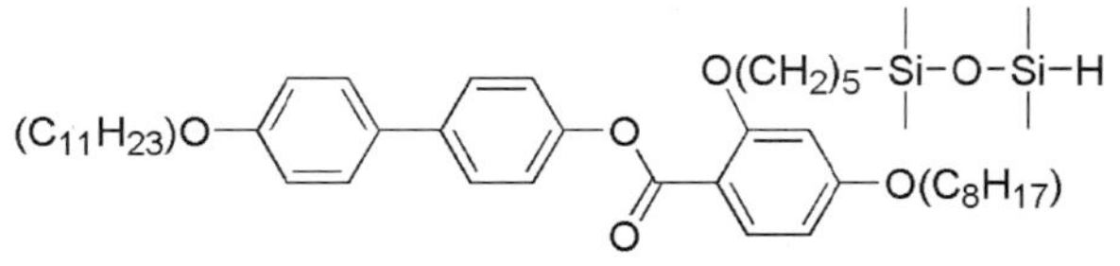

5.3 Polysilazane Dendrimers

Polysilazane dendrimers are currently unknown. As part of an effort to synthesize these novel materials, Son and Xiao investigated the synthesis of small branched silazanes that could be used as core molecules [15]. They successfully synthesized branched silazanes **1** and **2** in 83% and 72% yields, respectively, by the pathway shown in Reaction Scheme 5.2. After some experimentation, it was found that treating **2** with boron trichloride gave the chloro-substituted product **3** in quantitative yield (see Reaction Scheme 5.2). Compound **3** seems ideally suited for divergent dendrimer growth due to the presence of multiple reactive Si–Cl bonds. However, further reactions with various silylamines were unsuccessful due to product decomposition and the formation of byproducts believed to arise from intramolecular cyclization reactions.

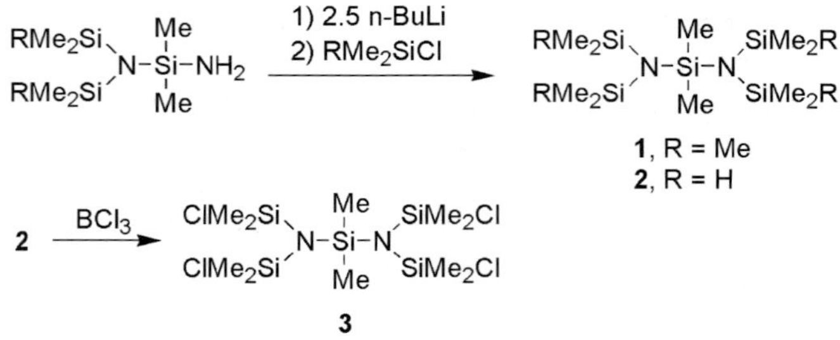

(5.2)

5.4 Related Dendrimers

In 1997, Kemmitt and Henderson reported the synthesis of two dendritic silatrane wedges containing pentacoordinate silicon atoms bonded to one nitrogen atom (see also Chapter 11) [16]. An example of this dendrimer growth is shown in Reaction Scheme 5.3. Analogous reactions of this type led to the formation of the silatrane dendrimers, which are of interest from a biological perspective. Yields were generally good, and byproducts could be easily separated by chromatography. The dendrimers and intermediates were characterized with ^{29}Si NMR spectroscopy and electrospray mass spectrometry.

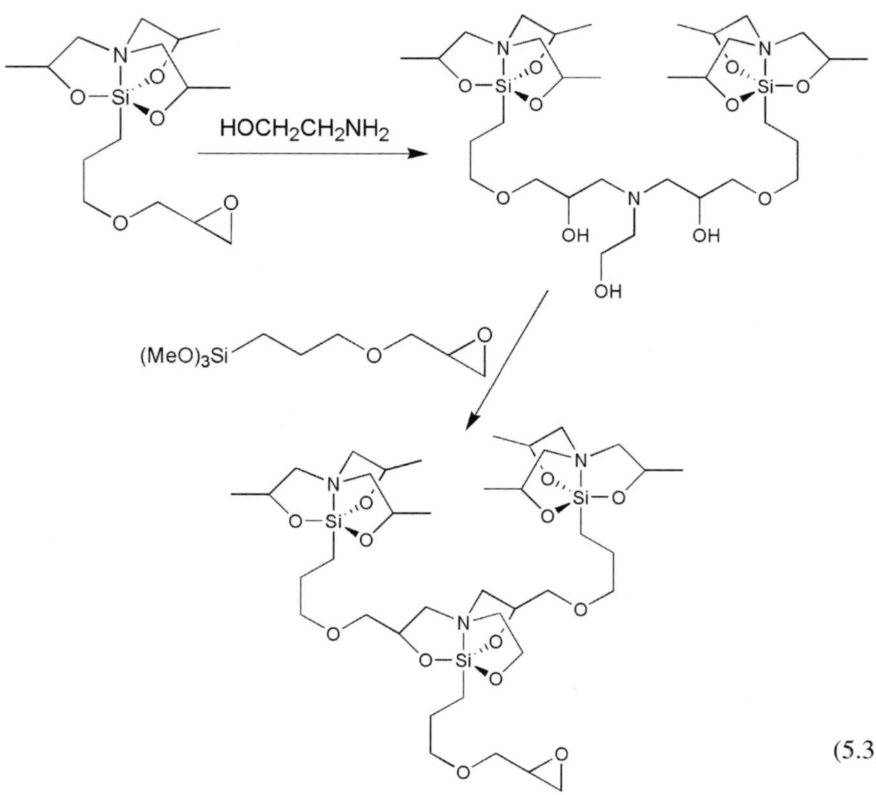

(5.3)

5.5 Hyperbranched Polycarbosilazanes

As a general rule, in order to obtain reasonable yields of hyperbranched polymers from AB_n types of monomers, intramolecular cyclization during the early stages of polymerization should be minimized (for more details on cyclization reactions in such systems see Chapter 15). In particular, if intramolecular cyclization is the preferred reaction of the *monomer*, then polymerization will clearly proceed to only a small extent. Thus, a starting point for hyperbranched polymer synthesis from AB_n monomers is to design a monomer that does not have a tendency to cyclize.

With this in mind, Son and Yoon synthesized compounds **4** and **5** as AB_2 monomers for hyperbranched polymerization (see Reaction Scheme 5.4) [17]. However, treating either of these compounds with Karstedt's catalyst gave no traces of polymer, but rather the intramolecularly cyclized four-membered ring compounds. Switching to a Rh catalyst gave a mixture of both the four- and five-membered ring products, but again no polymer was obtained.

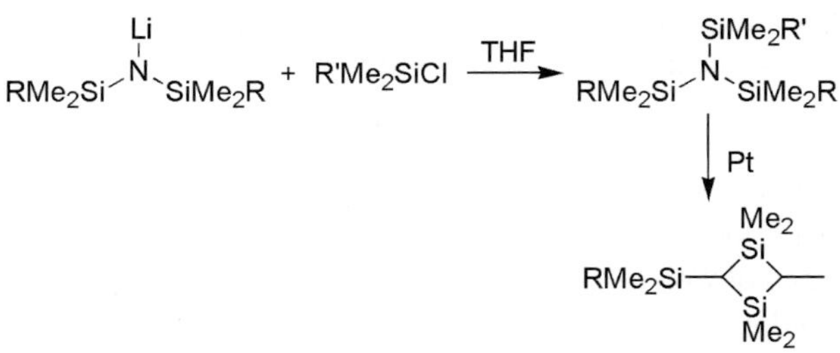

4, R = vinyl, R' = H
5, R = H, R' = vinyl (5.4)

In an effort to prevent intramolecular cyclization, Son and Hu synthesized an AB_2 monomer in which the reactive moieties were separated by a benzene ring (see Reaction Scheme 5.5) [18]. This strategy had worked previously for the synthesis of hyperbranched polycarbosilarylenes [19]. Treatment of the monomer with a Pt catalyst in refluxing THF resulted in total loss of Si–H and gave a quantitative recovery of product. Number-average molecular weight of the product was determined to be 5,700 by VPO. Additional characterization was carried out using NMR spectroscopy.

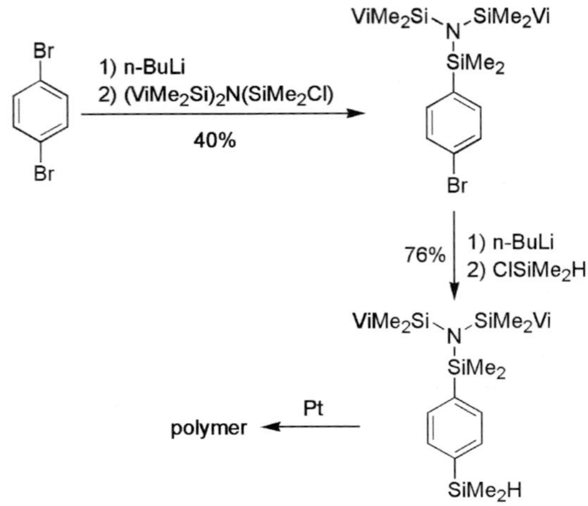

(5.5)

In a recent paper [20], Fan and coworkers reported the synthesis of hyperbranched polycarbosilazanes by the hydrosilylation polymerization of the monomers **6–8** shown below. On treatment with H_2PtCl_6 as the catalyst and *p*-quinone as an inhibitor, the authors obtained hyperbranched polymer from all three monomers. The polymers were isolated from low molecular weight species by sequential dissolution and precipitation from solvents of varying polarity. Characterization was carried out using ^{1}H, ^{13}C, and ^{29}Si NMR spectroscopy, and IR spectroscopy. M_w values were obtained using multi-angle laser light scattering (MALLS) and ranged from 3,900 to 13,500. These results are surprising in light of the results described above as obtained by Son and the work of Tamao and coworkers [21, 22] who found that (N-allyl)silylamines cyclize intramolecularly in very high yields. Unfortunately, yields were not reported for any of the obtained hyperbranched polymers; thus, it is impossible to determine the extent of any monomer cyclization.

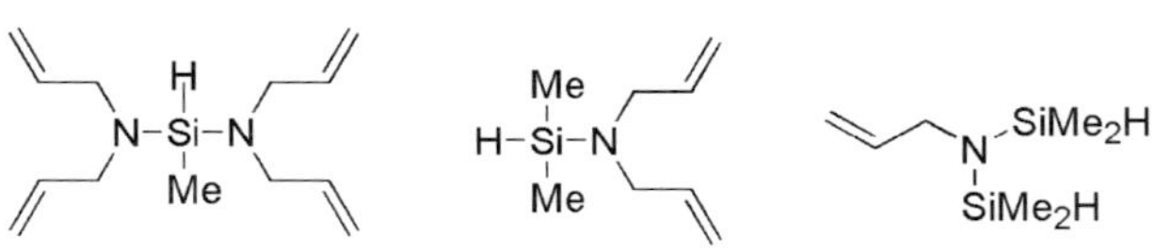

6 7 8

5.6 Concluding Remarks

Clearly, the field of silazane and carbosilazane dendrimers remains open and underdeveloped. The reactivity of the Si-N bond will continue to pose a challenge for synthetic chemists. However, the unique structural and chemical characteristics of these materials justify continuing research efforts in this fertile area of investigation.

Acknowledgements The author thanks the Robert A. Welch Foundation (grant no. N-1375) and the National Science Foundation (grant no. 0092495) for support in this area of research.

References

1. Aylett BJ (1968) Organomet Chem Rev 3:151
2. Bouquey M, Brochon C, Bruzaud S, Mingotaud A-F, Schappacher M, and Soum A (1996) J Organomet Chem 521:21
3. Ebsworth EAV (1987) Accounts Chem Res 20:295
4. Mitzel N, Schier A, and Schmidbaur H (1992) Chem Ber 125:2711
5. Mitzel NW, Riede J, Schifer A, and Schmidbauer H (1995) Acta Crystallogr, Sect C: Cryst Struct Commun C51:756
6. Hu J, and Son DY (1998) Polym Prepr 39(1):410
7. Hu J, and Son DY (1998) Macromolecules 31:8644
8. Veith M, Elsässer R, and Krüger R-P (1999) Organometallics 18:656
9. Krüger JK, Veith M, Elsässer R, Manglkammer W, Le Coutre A, Baller J, and Henkel M (2001) Ferroelectrics 259:27
10. Mehl GH, Elsässer R, Goodby JW, and Veith M (2001) Mol Cryst Liq Cryst 364:219
11. Tajber L, Kocot A, Vij JK, Merkel K, Zalewska-Rejdak J, Mehl GH, Elsässer R, Goodby JW, and Veith M (2002) Macromolecules 35:8601
12. Elsässer R, Goodby JW, Mehl GH, Rodriguez-Martin D, Richardson RM, Photinos DJ, and Veith M (2003) Mol Cryst Liq Cryst 402:237
13. Elsässer R, Mehl GH, Goodby JW, and Veith M (2001) Phosphorus, Sulfur, Silicon 169:17
14. Elsässer R, Mehl GH, Goodby JW, and Veith M (2001) Angew Chem Int Ed Engl 40:2688
15. Xiao Y, and Son DY (2004) Organometallics 23:4438
16. Kemmitt T, and Henderson W (1997) J Chem Soc, Perkin Trans 1:729
17. Yoon K, and Son DY (1999) Org Lett 1:423
18. Hu J, and Son DY:unpublished results
19. Yoon K, and Son DY (1999) Macromolecules 32:5210
20. Zhang G-B, Fan X-D, Kong J, Liu Y-Y, Wang M-C, and Qi Z-C (2007) Macromol Chem Phys 208:541
21. Tamao K, Nakagawa Y, and Ito Y (1990) J Org Chem 55:3438
22. Tamao K, Nakagawa Y, and Ito Y (1993) Organometallics 12:2297

Chapter 6
Silyl Ether Containing Dendrimers with Cyclic Siloxane Cores

Chungkyun Kim

6.1 Introduction

About three decades ago a remarkable cascade-type molecule was reported by Vögtle and his co-workers [1, 2]. This development set the stage for new types of polymers with a high degree of isomolecularity that are now widely known as dendrimers (see Chapter 1). In the years that followed, a number of different compositions of dendrimers, including amidoamine-, ether-, amine-, and ester-type dendrimers, etc. as well as their hetero-atom homologues has been prepared by many organic and inorganic chemists [3–6]. Among these, the introduction of silicone and organosilicon moieties into dendrimer structures has resulted in very unique silicon-containing dendrimers with considerable structural versatility [7–9].

One of the most versatile groups of organosilicon dendrimers is the carbosilanes (see Chapter 3), which often have four allylic branches emanating from a single central silicon atom (core) (0G(4-n); where n = 1–2). Physical properties of such dendrimers are gradually altered with increasing number of silicon-based moieties, branch units, and by addition of other functional groups to the silicon atoms in the peripheral region. The siloxane dendrimers with Si–O bonds in their main skeleton were reported earlier than the carbosilane dendrimers with Si–C bonds by Muzafarov et al. (see Chapter 2). The synthesis of these dendrimers, where silicon atoms form branch junctures with three Si–O bonds, was performed starting from trichloromethylsilane as a core, 0G(3)-Cl, using repetitive substitution of the chlorosilyl bond with ethoxy groups and their subsequent conversion back to chlorosilanes by thionylchloride (see Chapter 2).

This chapter describes the preparation, characterization and potential uses of the Si–O–C containing dendrimers by a reiterative sequence of hydrosilylation and alcoholysis in a basic medium. Other preparative methods for such dendrimers are summarized in other reviews [10, 12]. Of several different types of siloxane cores (0G(n)) available, such as carbosiloxane monomers [11], cyclic siloxane rings [12], solid

C. Kim
Dong-A University, Busan 604-714, Korea
E-mail: ckkim@dau.ac.kr

P.R. Dvornic and M.J. Owen (eds.), *Silicon-Containing Dendritic Polymers*,
Advances in Silicon Science 2,

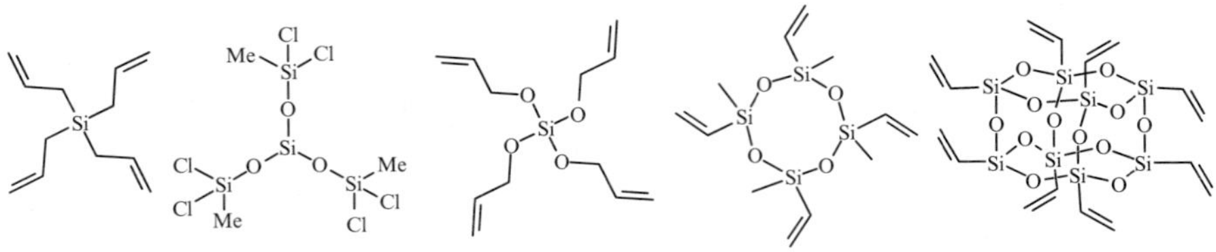

Fig. 6.1 Selected cores of carbosilane and siloxane dendrimers.

cubic polyhedral oligosilsesquioxane (POSS) moieties [13], linear polysiloxane chains (for dendronized polysiloxanes; see also Section *Carbosilane Dendronized Polymers* of Chapter 3) [14] and hyperbranched siloxane polymers (see also Chapter 16) [9, 15], this chapter focuses on tetravinyl cyclic siloxane (0G(4)-vinyl) and octavinyl POSS (0G(8)-vinyl) (see also Chapter 7) cores (see Fig. 6.1).

6.2 Siloxane Dendrimers with Cyclic Siloxane Core (0G(4)-Vinyl)

The synthesis of this type of dendrimers is carried out by a divergent growth method starting from a siloxane tetramer core (1,3,5,7-tetravinyl-1,3,5,7-tetramethylcyclotetrasiloxane, $D_{4,}^{Vi}$ (0G(4)-vinyl), shown fourth from the left in Fig. 6.1, utilizing hydrosilylation with chlorosilanes (Me_nSiHCl_{3-n}; n = 1–3) and allyl alcohol addition in a nonpolar solvent such as toluene. For example, the first generation 1G(4,3)-Cl dendrimer with 12 Si–Cl end-groups (see Reaction Scheme 6.1) has been prepared by hydrosilylation of 0G(4)-vinyl with an excess of trichlorosilane either neat or under reflux in toluene [12]. Higher generations are then obtained by repetitive application of the same reaction sequence and the structures of the products of both hydrosilylation and branching with allyloxy groups are confirmed by NMR spectroscopy. The characteristic signals are from the C=C double bonds and allyloxysilyl moieties which form β-silylated branches ($–OCH_2(CH_2)_2SiCl_3$) in an anti-Markovnikov manner. In nonpolar toluene, hydrosilylation of unsaturated groups of the core moiety is achieved in high yields, but in the polar solvent tetrahydrofuran (THF) undesired ring opening results in the polymerization of THF itself [16].

Allyloxy groups are introduced onto the terminal Si atoms of dendrimer precursors, as shown in Reaction Scheme 6.1, by reacting their Si–Cl end-groups with allyl alcohol in toluene in the presence of tetramethylethylenediamine (TMEDA) as an acid acceptor [12]. All products of both hydrosilylation and alcoholysis are generally obtained in high yields due to the simplicity and cleanness of these reactions. However, purification at each generational stage is very critical if monodisperse products with no structural defects are to be obtained. A common method for effective purification is simple column chromatography in mixed solvents which usually gives highly purified products in good yields [16]. The very low and consistent polydispersity index (PDI) values obtained for these dendrimers by gel permeation chromatography (GPC) indicate their high purity, while regular shift patterns of their retention times reflect generational growth without significant structural defects [17].

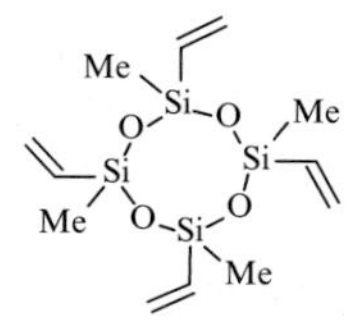

0G(4) vinyl $\xrightarrow{(a)}$ $(Cl_3SiCH_2CH_2MeSiO)_4$ **1G(4,3) Cl**

1G(4,3) Cl $\xrightarrow{(b)}$ $((CH_2{=}CHCH_2O)_3SiCH_2CH_2MeSiO)_4$ **1G(4,3) allyloxy**

1G(4,3) allyloxy $\xrightarrow{(c)}$ $((Cl_2MeSiCH_2CH_2CH_2O)_3SiCH_2CH_2MeSiO)_4$ **2G(4, 3,2) Cl**

2G(4, 3,2) Cl $\xrightarrow{(d)}$ $(((CH_2{=}CHCH_2O)_2MeSiCH_2CH_2CH_2O)_3SiCH_2CH_2MeSiO)_4$ **2G(4,3,2) allyloxy**

2G(4,3,2) allyloxy $\xrightarrow{(e)}$ $(((Cl_2MeSiCH_2CH_2CH_2O)_2MeSiCH_2CH_2CH_2O)_3SiCH_2CH_2MeSiO)_4$ **3G(4,3,2,2) Cl**

3G(4,3,2,2) Cl $\xrightarrow{(f)}$ $((((CH_2{=}CHCH_2O)_2MeSiCH_2CH_2CH_2O)_2MeSiCH_2CH_2CH_2O)_2MeSiCH_2CH_2MeSiO)_4$ **3G(4,3,2,2) allyloxy**

3G(4,3,2,2) allyloxy $\xrightarrow{(g)}$ $((((ClMeSiCH_2CH_2CH_2O)_2MeSiCH_2CH_2CH_2O)_2MeSiCH_2CH_2CH_2O)_2MeSiCH_2CH_2MeSiO)_4$ **4G(4,3,2,2,1) Cl**

3G(4,3,2,2) allyloxy $\xrightarrow{(h)}$ $((((Cl_2MeSiCH_2CH_2CH_2O)_2MeSiCH_2CH_2CH_2O)_2MeSiCH_2CH_2CH_2O)_3SiCH_2CH_2MeSiO)_4$ **4G(4,3,2,2,2) Cl**

4G(4,3,2,2,2) Cl $\xrightarrow{(i)}$ $(((((CH_2{=}CHCH_2O)_2MeSiCH_2CH_2CH_2O)_2MeSiCH_2CH_2CH_2O)_2MeSiCH_2CH_2CH_2O)_3SiCH_2CH_2MeSiO)_4$ **4G(4,3,2,2,2) allyloxy**

4G(4,3,2,2,2) allyloxy $\xrightarrow{(j)}$ $(((((ClMe_2SiCH_2CH_2CH_2O)_2MeSiCH_2CH_2CH_2O)_2MeSiCH_2CH_2CH_2O)_2MeSiCH_2CH_2CH_2O)_3SiCH_2CH_2MeSiO)_4$ **5G(4,3,2,2,2,1) Cl**

(a) + 12 $HSiCl_3$, Pt/C, toluene, reflux, (b) + 12 $CH_2{=}CHCH_2OH$, TMEDA, RT ~ 50°C
(c) + 24 $HSiMeCl_2$, Pt/C, toluene, reflux, (d) + 24 $CH_2{=}CHCH_2OH$, TMEDA, RT ~ 50°C
(e) + 48 $HSiMeCl_2$, Pt/C, toluene, reflux, (f) + 48 $CH_2{=}CHCH_2OH$, TMEDA, RT ~ 50°C
(g) + 48 $HSiMe_2Cl$, Pt/C, toluene, reflux, (h) + 96 $HSiMe_2Cl$, Pt/C, toluene, reflux
(i) + 96 $CH_2{=}CHCH_2OH$, TMEDA, RT ~ 50°C , (j) + 96 $HSiMe_2Cl$, Pt/C, toluene, reflux (6.1)

6.3 Synthesis of Dendrimers with Si–O–C Units from Cyclic Siloxane Cores

6.3.1 General Synthetic Strategy from 0G(4)-Vinyl Core

By changing the branching groups, various dendrimers of this type can be prepared as shown in Reaction Scheme 6.1. For example, 5G(4,3,2,2,2,1)-Cl with 96 Si–Cl end-groups is obtained by hydrosilylation of 4G(4,3,2,2,2)-allyloxy with Me_2SiHCl, which, in turn, is prepared by the reaction of 4G(4,3,2,2,2)-Cl with allyl alcohol. The same simple synthetic procedure has been applied to the preparation of other dendrimers [18, 19]. Different branching groups are introduced by using different silanes $Me_{3-n}(H)SiCl_n$ (n = 1–3) to yield dendrimers with different number of Si–Cl functional groups in the periphery. Combinations of different reactants with the same siloxane tetramer core provide different types of these dendrimers whose generation numbers are defined by the number of reiterations of the same synthetic procedure (see Reaction Schemes 6.1 and 6.2).

Preparation of monodisperse products is a very challenging and critical subject, which is not easy to accomplish, particularly for higher generations [18, 19]. It seems that reaction conditions play a crucial role in accomplishing this goal. In principle, in order to obtain monodisperse products at every generational stage the number of the reactant molecules should be larger than the number of reactive end-groups in the precursor dendrimer. However, even when this condition is met, an additional difficulty, particularly at higher dendrimer generations, becomes the increase in steric congestion of surface end-groups with increase in dendrimer generation (see Chapter 1) [19].

6.3.2 Dendrimers with Organic Functional End-Groups

The reaction of Si–Cl bonds on the dendrimer periphery with ROH reagents, such as 4-hydroxypropylpyridine, 4-hydroxyoximethylpyridine, 9-anthracenemethanol, 2-hydroxymethylanthraquinone, hydroxycholesterol, 8-hydroxyquinoline, 4-hydroxyazobenzene, or 5-(2-hydroxyethyl)-4-methylthiazole, crown-4 and crown-6, in the presence of TMEDA, gives the corresponding Si–O–R containing dendrimers, as shown in Reaction Scheme 6.2 [18–21]. The characteristic features of this approach may be summarized as follows: (1) the Si–O–C bonds at the outermost dendrimer periphery are very stable in the presence of air and moisture; (2) the end-functionalized carbosilane dendrimers with OR groups are produced in very high yields; and (3) the high purity dendrimers are obtained after silica gel column chromatography with a mixture of chloroform and higher polarity solvents. Characterization of end-functionalized dendrimers by GPC gave, for example, for 4G(4,3,2,2,1)-anthracenemethyloxy and 5G(4,3,2,2,2,1)-anthracenemethyloxy derivatives PDI values as low as 1.04 and 1.03 with retention times of 15.59 and 14.20 min, respectively [18]. These low PDI values are taken as an indication that these dendrimers are monodisperse and without defects in their outermost peripheries [19].

The Si–O–C bonds in the periphery of these dendrimers are quite stable to atmospheric conditions and even to wet solvents. Similar stability is observed for the dendrimers containing allyloxy groups prepared by other preparative methods [20]. However, dendrimers containing phenylmethyloxy or pyridylmethyloxy groups on the peripheral silicon atoms are slightly unstable to moist conditions. As an example, after 10 days at room temperature in wet chloroform, a fourth generation dendrimer with 48 1-pyridylmethyloxy groups was partially (~20%) decomposed to 1-pyridylcarbinol and polymer. However, under the same conditions, other dendrimers containing more than one, i.e., 2, 3, 4, and 6 pyridylpropyloxy, allyloxy, butadienyloxy, and farnesyloxy groups, respectively, did not decompose showing considerable chemical stability [21].

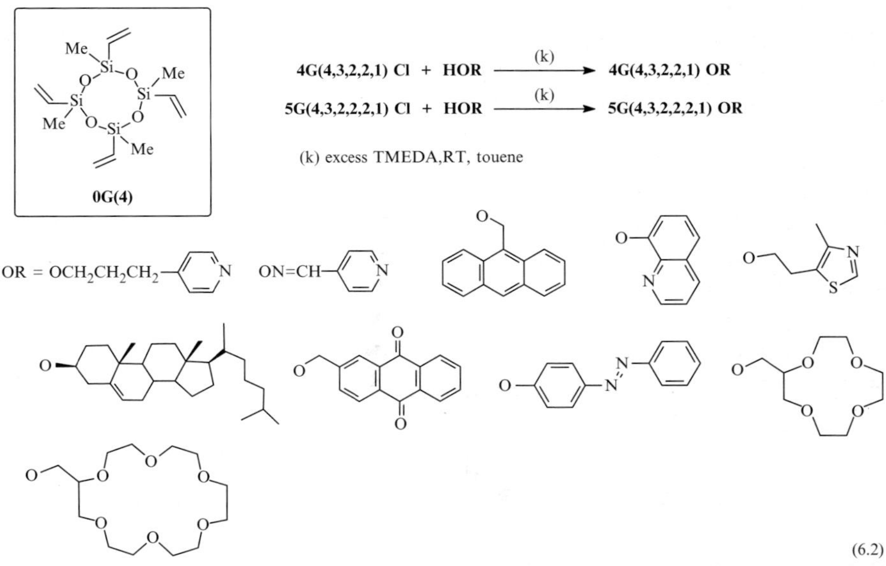

(6.2)

6.3.3 *Dendrimers with Triple Bonds*

The propargyloxy-functionalized dendrimers, **1**, were prepared by the reaction of propargyl alcohol and Si–Cl functionalized dendrimer precursors (nG(4)-Cl (n = 1–4)) [28]. Hydrosilylation of propargyloxy dendrimer end-groups with dichloromethylsilane leads to allyloxy derivatives containing dichloromethylsilyl groups, while double bonds in the dendrimer interior are left intact even after successive hydrosilylations. Purification and identification procedures were performed as described in the previous sections. All generations were fully characterized by NMR, matrix assisted laser desorption ionization-time of flight (MALDI-TOF MS), and GPC. GPC data for the generations 1 to 4, (nG(4)-Cl (n = 1–4)), give low PDI values of 1.006, 1.016, 1.022, and 1.070, respectively. The 2G(4,2,2)-propargyloxy, 3G(4,2,2,2)-propargyloxy and 4G(4,2,2,2,2)-propargyloxy dendrimers with double bonds in their inner shells and propargyloxy groups in the periphery are stable in the presence of moisture (see Structure 1) [22].

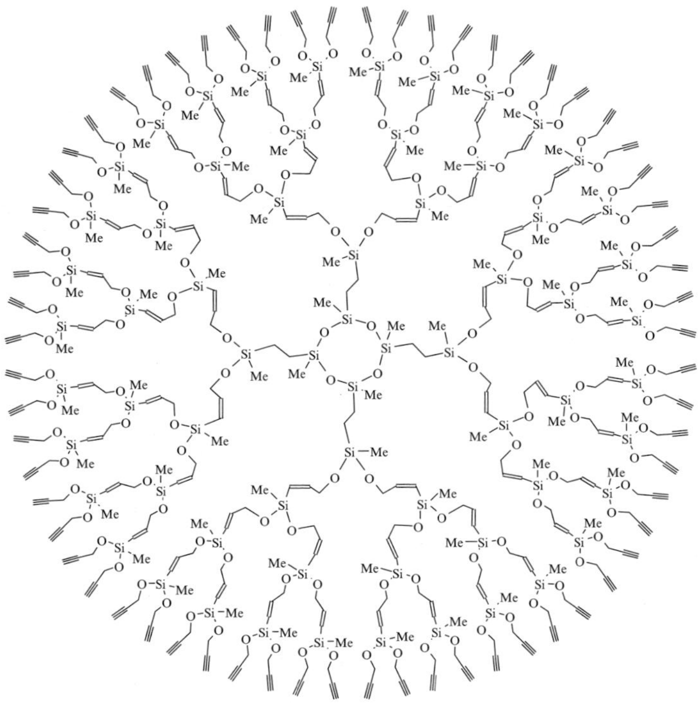

1

6.3.4 "Double-Layered" Dendrimers with Conjugated Branches

From the same tetrafunctional siloxane core, 0G(4)-vinyl, three different "double-layered" dendrimers with 64 phenylethynyl groups in the outer shells were prepared [23] (see Fig. 6.2). The first one (dendrimer on the left in Fig. 6.2) was grown to the third generation 3G(4,2,2,2)-allylsiloxy, and then 64 phenylethynyl groups were added to give the 4G(4,2,2,2,2)-C≡CPh derivative. The next dendrimer (center in Fig. 6.2) was grown to the second generation 2G(4,2,2)-allylsiloxy, then hydrosilylated with 32 phenylethynyl groups to yield a dendrimer with 32 double bonds, and then 64 phenylethynyl groups were added to its periphery. The dendrimer on the right of Fig. 6.2 was constructed from a 1G(4,2)-allyloxy core, to which two layers of ethenyl groups of 16 and 32 double bonds were added as inner shells, and 64 phenylethynyl groups were then added as the outermost shell. Thus, each of these dendrimers contained 64 phenylethynyl groups on its periphery, but with different numbers of double bonds in the interior. They were all characterized by NMR spectroscopy, and their conductivities were measured with palladium ions to give $\sigma = 5.4 \times 10^{-7}$, 1.0×10^{-6}, and 6.0×10^{-5} S/cm, respectively. From these results it clearly follows that conducting abilities of these dendrimers depend on the number of their double bonds, even if these are not conjugated [23] (see Fig. 6.2).

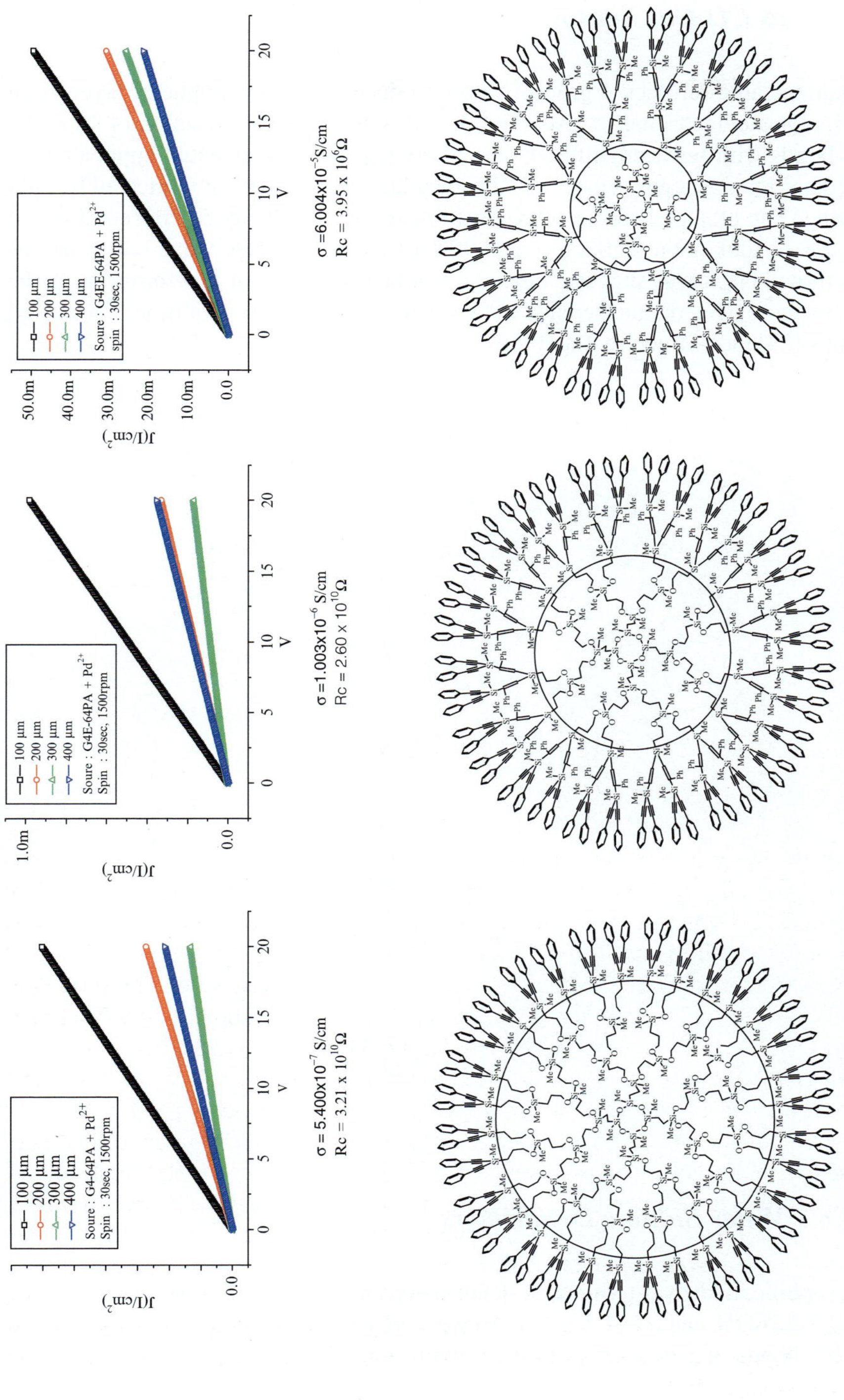

Fig. 6.2 Double layered dendrimers and their conductivity diagrams.

6.3.5 *Ferrocenyl-Functionalized Dendrimers as CO Gas Sensor*

Organometallic ferrocenyl groups were introduced into the peripheral layer of an allyl-siloxane dendrimer to give 4G(4,3,2,2,1)-ferrocene by a reaction of 48 Si–Cl end-bonds with ferrocenyllithium (see also Chapter 8). A potential application of this dendrimer as a gas sensor was demonstrated using a device fabricated by spin coating. The prepared sensor responded linearly to the change in concentration of CO gas. It took about 250 s to reach 90% of the steady-state value when the gas was turned on. It was suggested that the conduction ability of this sensor might be due to the extent of interference between the CO gas and dendrimer ferrocenyl groups [24] (see Structure **2** and Fig. 6.3).

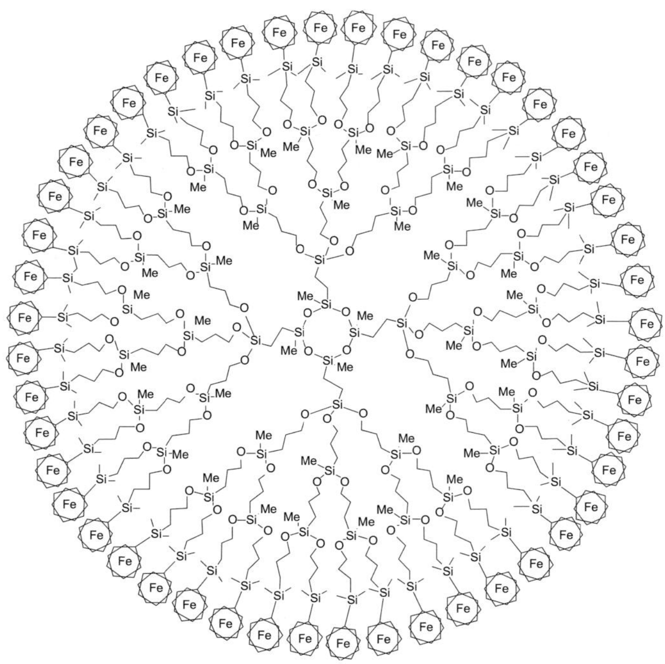

2

6.3.6 *Water Soluble Dendrimers*

Hydrophilic, hydroxyl-terminated dendrimers, such as 1G(4,2)-OH, 2G(4,2,2)OH, 3G(4,2,2,2)-OH, and, 4G(4,2,2,2,2)-OH were prepared by hydroboration of peripheral double bonds of precursor dendrimers with 9-borabicyclo[3,3,1]nonane (9-BBN)

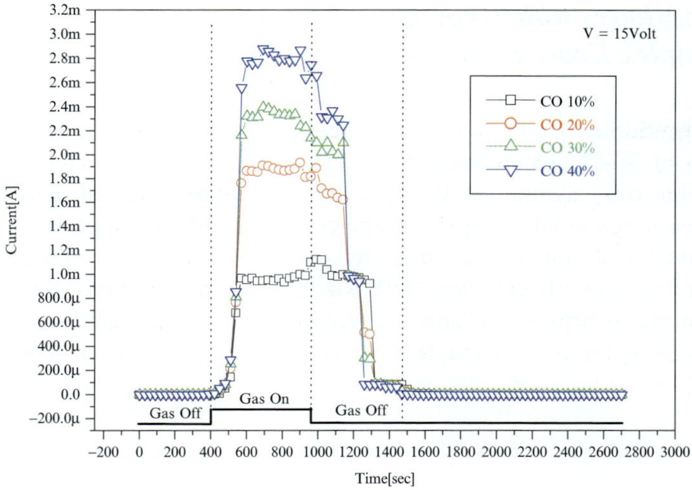

Fig. 6.3 Transient response of dendrimer CO gas sensor at room temperature.

and subsequent oxidation of the resulting hydroborated products [25]. The reaction products of both hydroboration and oxidation were characterized by ^{1}H NMR and MALDI-TOF MS (see Structure **3**).

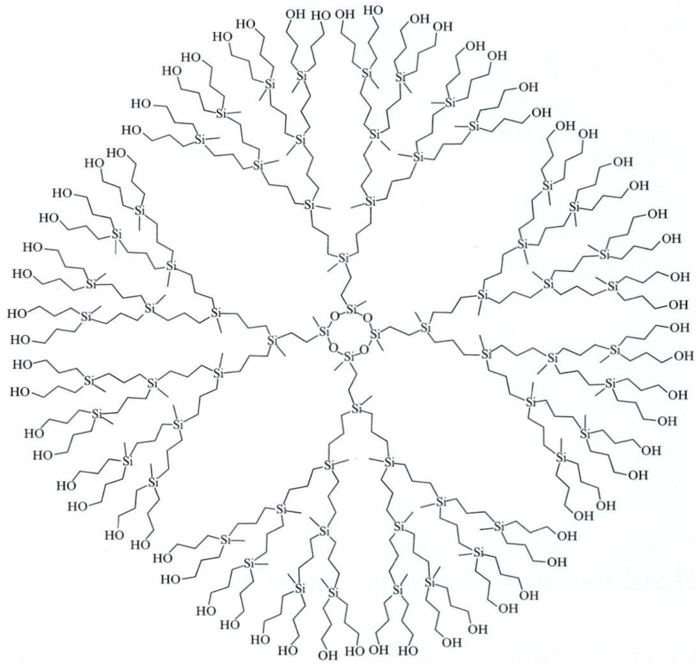

3

6.3.7 *Dendrimers with Terpyridine Ruthenium Complex End-Groups*

2,2′:6′,2″-Terpyridine ligands were introduced into the periphery of dendrimers by a reaction of Si–Cl precursors with 6-hydroxyhexa-4′-(2,2′:6′,2″-terpyridine) ether. Following this, addition of $RuCl_3$ to the terpyridine-modified dendrimers under mild reaction conditions, gave paramagnetic ruthenium complexes which, after continued addition of 2,2′:6′,2″-terpyridine, produced bis(2,2′:6′,2″-terpyridine) ruthenium (II) complex [TPY-Ru-TPY] in the outermost dendrimer shell, as illustrated by Structure **4**. Three generations of these dendrimers, of which the third generation revealed diamagnetic properties, were so prepared [26].

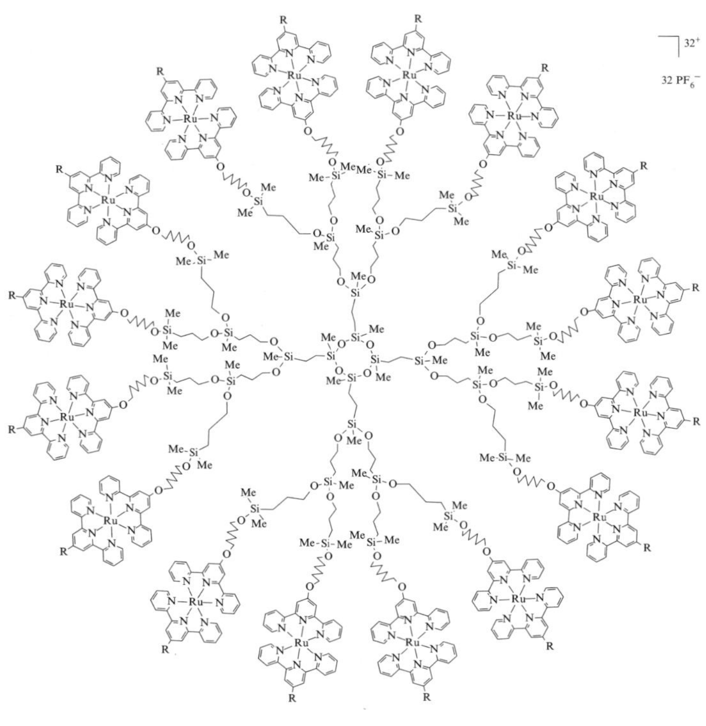

4

6.3.8 *Dendrimers with Farnesyl End-Groups*

Four generations of dendrimers with farnesyl groups on the periphery, Structure **5**, were prepared in very high yields by the reaction of farnesol with Si–Cl functionalized precursors. The obtained dendrimers were identified by NMR spectroscopy, GPC, and elemental analysis. Molar masses of the first and the second generation were also determined by MALDI-TOF mass spectroscopy.

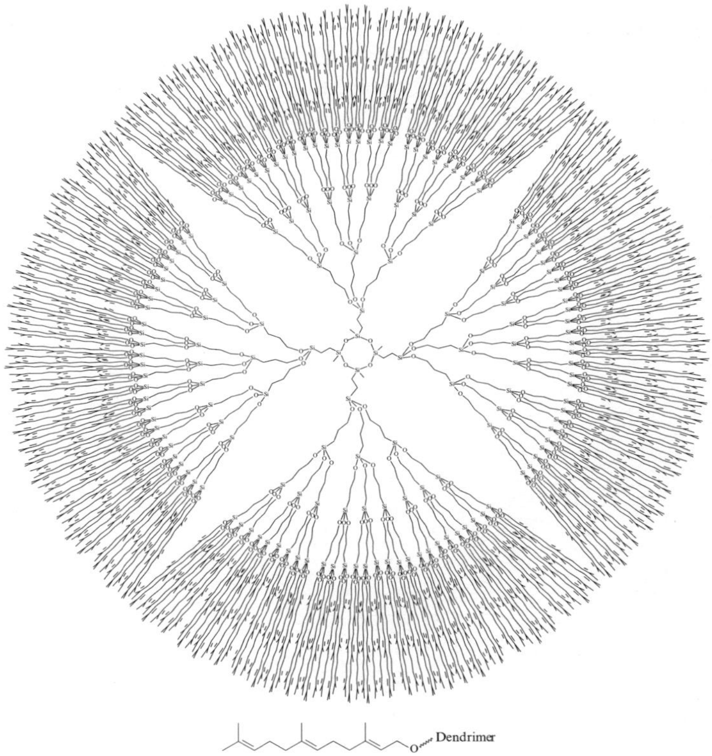

5

These stem-type, long-branched (star-type; see also Chapters 3, 10 and 11) dendrimers can possess more functionalities than other dendrimers because of their spacious structures. For example, 1G(4,3)-farnesyl contains only 12 farnesyl groups but 4G(4,3,3,3,3)-farnesyl contains a remarkable number of 324 farnesyl groups in its periphery. High purity farnesyl dendrimers were obtained by simple flash silica gel chromatography using toluene solvent, and characterized by NMR spectroscopy, GPC, and elemental analysis. The very low PDI values and regular retention times confirmed their high purity [21].

6.3.9 Diels-Alder Reaction on Dendrimer Periphery

The Diels-Alder reaction between anthracene on the dendritic periphery and maleimide reagents was also investigated [27]. Structural information of the resulting products was obtained from the hyperfine structure of their ^{1}H NMR spectra. The very low PDI values and regular retention times in GPC indicated pure dendrimer products with structural perfection. Furthermore, the Diels-Alder reaction of anthracene-modified dendrimers with 1,4-benzoquinone and 1,4-naphthoquinone provided good yields of products [27] (see Structure **6**).

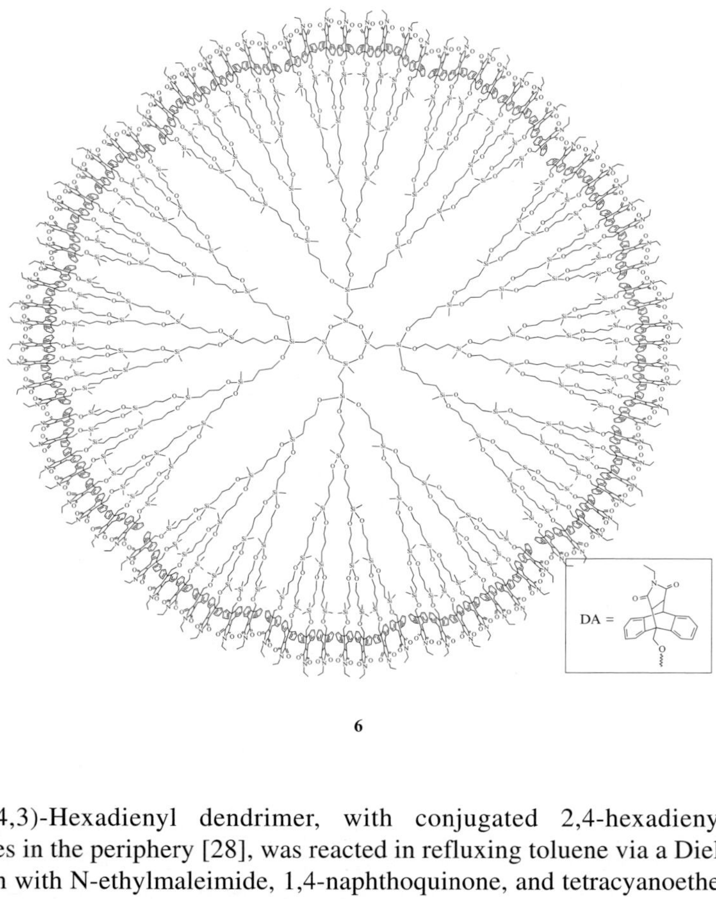

6

1G(4,3)-Hexadienyl dendrimer, with conjugated 2,4-hexadienyl-1-oxy branches in the periphery [28], was reacted in refluxing toluene via a Diels-Alder reaction with N-ethylmaleimide, 1,4-naphthoquinone, and tetracyanoethene. The first generation with 12 functional groups (see Structure **7**) was obtained as a rather unimolecular product, but the second generation, which was supposed to have 36 functional groups, was not due to the lack of enough surface area for packing of the required number of surface groups (the de Gennes dense packing stage, see Chapter 1) [28].

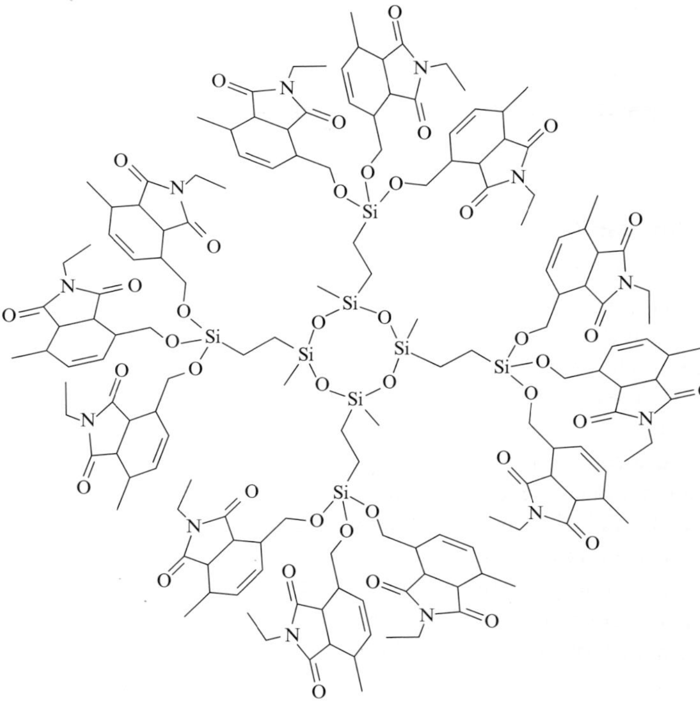

7

6.4 Dendrimers with Silsesquioxane Core

Dendrimers with POSS cores may be considered close relatives of those grown from the cyclic siloxanes discussed in the previous sections of this chapter (for more detail on the POSS-containing dendrimers see Chapter 7). Illustrative examples of dendritic growth from a silsesquioxane core with vinyl groups were reported in [29–33]. The syntheses employed divergent growth methods and formation of dendrimer architecture was accomplished by just a few reactions. As described in the previous sections of this chapter, to form dendrimer branches, hydrosilylation of double bonds with hydrosilanes was employed, and to introduce Si–O moieties, the reaction of Si–Cl bonds with alcohols was used in basic toluene medium. To ensure complete hydrosilylation and alcoholyses, all reaction steps were monitored by spectroscopic methods and the reaction conditions were controlled to obtain monodisperse dendrimers without structural defects. Isolation and purification of crude products were performed using column chromatography, and the resulting purified dendrimers were characterized by MALDI-TOF MS, with no difficulties encountered with low generations.

Silsesquioxane-core carbosilane dendrimers containing phosphine [32], ferrocenyl dendrimers containing phosphine [31] and poly-L-lysine dendrimers [33] have also been prepared. The last mentioned ones were synthesized divergently in aqueous solution with good yield and high purity. The structure was confirmed by MALDI-TOF mass spectrometry for the lower generation dendrimers and complete substitution of the surface amino groups and a nano-size globular architecture were confirmed.

6.5 Conclusion

This chapter presents divergent growth methods for the synthesis of Si–O–C containing dendrimers from cyclic siloxane cores such as siloxane tetramer D_4^{Vi} and $POSS_8^{Vi}$. The Si–O–C moieties are formed by the reaction of halosilanes with alcohols, and the dendrimers are built by repetitive sequence of hydrosilylation and alcoholysis reactions. Various organic and organometallic moieties can be introduced into these dendrimers periphery by simple coupling and complexation reactions of the parent dendrimers which can be prepared without significant defects. Using the same general approach, dendrimers with various transition metal complexes have also been synthesized and they exhibit high potential in catalysis and photochemical applications. Dendrimers with highly flexible long chain moieties such as farnesyl or hexadienyl groups in the peripheries have also been synthesized in good yields and were characterized unambiguously.

References

1. Buhleier E, Wehner W, Vögtle F (1978) Synthesis 154.
2. Fischer M, Vögtle F (1999) Angew. Chem. Int. Ed. Engl. 38: 884.
3. Tomalia DA, Naylor AM, Goddard III WA (1990) Angew. Chem. Int. Ed. Engl. 29: 138.
4. Cuadrado I, Morán M, Losada J, Casado CM, Pascual C, Alonso B, Lobete F (1996) Organometallic dendritic macromolecules. In: Newkome GR (ed) Advances in Dendritic Macromolecules, JAI Press, Greenwich, Volume 3, pp. 151–195.
5. Grayson SM, Frechét JMJ (2001) Chem. Rev. 101: 3819
6. Majoral J-P, Caminade A-M (1999) Chem. Rev. 99: 845.
7. van der Made AW, van Leeuwen PWNM, de Wilde JC, Brandes RAC (1993) Adv. Mater. 5: 466.
8. Frey H, Lach C, Lorenz K (1998) Adv. Mater. 10: 279.
9. Frey H, Schlenk C (2000) Silicon based dendrimers. In: Vogtle F (ed), Topics in Current Chemistry, Springer, Berlin, Volume 210, pp. 69–129.
10. Lang H, Lühmann B (2001) Adv. Mater. 20: 1523.
11. Morikawa A, Kakimoto M, Imai Y (1991) Macromolecules 24: 3469.
12. Kim C, Jeong Y, Jung I (1998) J. Organomet. Chem. 570: 9.
13. Murfee HJ, Thoms TPS, Greaves J, Hong B (2000) Inorg. Chem. 39: 5902.
14. Kim C, Kwark J (2002) J. Polym. Sci. A: Polym. Chem. 40: 976.
15. Kim C, Kim H (2001) J. Polym. Sci. A: Polym. Chem. 39: 3287.
16. Kim, C. Kim H, Park K (2004) J. Polym. Sci. A: Polym. Chem. 42: 2155.
17. Kim C, Seo W, Oh M-J (2007) Bull. Korean Chem. Soc. 28: 1963.
18. Kim C, Park J (2001) J. Organomet. Chem. 629: 194.
19. Kim C, Park E (2001) J. Polym. Sci. A: Polym. Chem. 39: 2308.
20. Kim C, Ryu M (2000) J. Polym. Sci. A: Polym. Chem. 38: 764.

21. Kim C, Kim H (2005) Synthesis 381.
22. Kim C, Park J (1999) Synthesis 1804.
23. Koo BW, Song CK, Kim C (2001) Sensor. Actuator. B 77: 432.
24. Kim C, Park E, Song CK, Koo BW (2001) Syn. Met. 123: 493.
25. Kim C, Son S, Kim B (1999) J. Organomet. Chem. 588: 1.
26. Kim C, Kim H (2003) J. Organomet. Chem. 673: 77.
27. Kim C, Kim H, Park K (2003) J. Organomet. Chem. 667: 96.
28. Kim C, Kim H, Park K (2005) J. Organomet. Chem. 690: 4794.
29. Saez IM, Goodly JW, Rehardson RM (2001) Chem. Eur. J. 7: 2758.
30. Lo MY, Ueno K, Tanabe H, Selinger A (2006) Chem. Rec. 6: 157.
31. Casado CM, Cuadrado I, Morán M, Alonso B, Barranco M, Losada J (1999) Appl. Organomet. Chem.13: 245.
32. Ropartz L, Morris RE, Foster DF, Cole-Hamilton DJ (2002) J. Mol. Catal. A. Chem. 182: 99.
33. Kaneshiro TL, Wang X, Lu Z-R (2007) Mol Pharm. 4: 759.

Chapter 7
Polyhedral Oligomeric Silsesquioxane Dendrimers

Katherine J. Haxton and Russell E. Morris

7.1 Introduction

Dendrimers have been prepared with a wide variety of core molecules since the first patents and publications in the early 1980s (see Chapter 1) [1–3]. The most common core molecules (e.g. ammonia, ethylenediamine, pentaerythritol) permit 2–4 branches although some molecules may give greater branch multiplicity. Polyhedral oligomeric silsesquioxanes (POSS) allow eight branches to radiate from a silicon-oxygen core. Dendrimers based on POSS were first reported in 1993 and have resulted in many publications to date [4].

Siloxanes are molecules with the general formula $[RSiO_{x/2}]$ where R is an organic group or silicon species. Siloxanes may be discrete molecules, two-dimensional ladders or networks, or three-dimensional cages or polymers. The siloxane linkage, Si–O–Si is formed when different units join to form larger molecules. Siloxy groups $[R_3SiO_{1/2}]$ are good terminal groups because they halt formation of larger siloxane networks, siloxane $[R_2SiO_{2/2}]$ groups are ideal candidates for forming long chain-like molecules, while silsesquioxanes $[RSiO_{3/2}]$ and silicates $[SiO_{4/2}]$ are most commonly found in three-dimensional structures, both random polymers and oligomers, due to the number of siloxane linkages that can be created. In all cases, a large number of structures with a large number of functionalities have been reported [5].

POSS are multifunctional molecules with a cage-like (polyhedral) core of oxygen-bridged silicon atoms. These molecules have the general formula $[RSiO_{3/2}]_n$, where the value of n determines the size of the silicon-oxygen core polyhedron. For closed polyhedrons the value of n is always even, with the lowest value being 6 (see Fig. 7.1). By far the most common POSS species is the n = 8,

K.J. Haxton
School of Physical and Geographical Sciences, Keele University, Staffordshire ST5 5BG, UK

R.E. Morris
EaStChem School of Chemistry, University of St. Andrews, Purdie Building,
St. Andrews KY16 9ST, UK
E-mail: rem1@st-andrews.ac.uk

P.R. Dvornic and M.J. Owen (eds.), *Silicon-Containing Dendritic Polymers*,
Advances in Silicon Science 2,

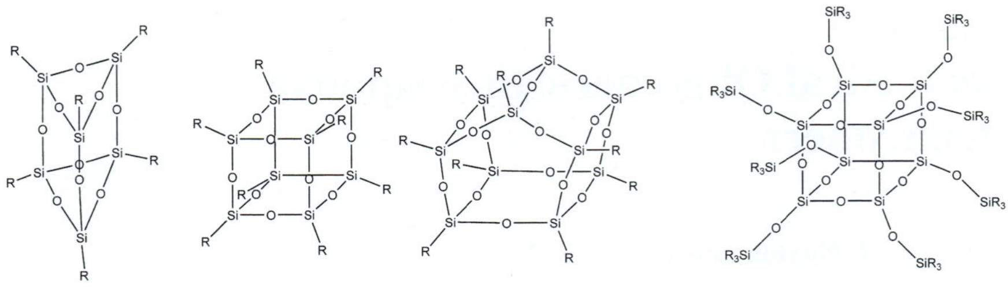

Fig. 7.1 Three different polyhedral oligomeric silsesquioxanes: R_6T_6, R_8T_8 and $R_{10}T_{10}$ containing six, eight and ten T-type silicon atoms, respectively, and the silicate molecule Q_8M_8 containing eight Q-type and eight M-type silicon atoms [11, 24]. For explanation of MDTQ nomenclature scheme, see Table 7.1.

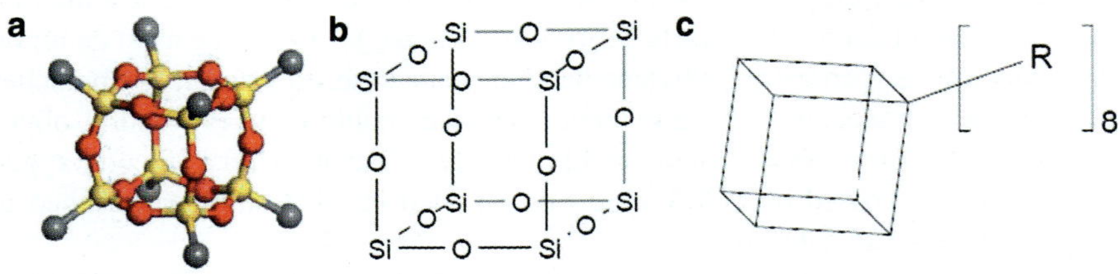

Fig. 7.2 Three representations of T_8 POSS species. (**a**) H_8T_8, Hydrido-POSS (*grey* – hydrogen, *yellow* – silicon, *red* – oxygen); (**b**) Si–O framework; (**c**) simplified representation of R_8T_8 POSS cube.

$[R_8Si_8O_{12}]$, which has an almost cubic topology with a silicon atom at the corners of the cube and bridging oxygen atoms on the cube edges (see Fig. 7.2). Organic groups decorate the corners of the cube on the outside surface, making them accessible for further functionalization. The geometry of the POSS core is often referred to as pseudospherical.

One of the interesting features of POSS is their chemical make-up which has led to them being used as molecular models for silica surfaces [6]. They are of particular interest because they have analogous topologies to the structural building units from zeolites and other important porous solids [7]. Closely related to POSS are a set of polyhedral silicate species where each silicon atom is connected to four oxygens (rather than three oxygens and an organic group as in true POSS molecules) [8, 9]. Again, the cubic topology is by far the most common, and this has sometimes been termed spherosilicate [10] or siloxysilsesquioxane [11]. The shorthand notation for silicates and polyhedral silsesquioxanes uses letters to denote the silicon type (T for silsesquioxane, Q for silicate) and subscripts and superscripts to denote the number and type of functional groups, respectively. These are summarized in Table 7.1. For example, the cubic POSS molecule with the general chemical formula $[RSiO_{3/2}]_8$ is often written as R_8T_8, where R is the organic group. This indicates that

Table 7.1 A summary of different types of silicon-oxygen species discussed in this chapter and their simplified letter nomenclature

Silicon species	General formula	Valency	Nomenclature
Siloxy	$[R_3SiO_{1/2}]$	Mono	M
Siloxane	$[R_2SiO_{2/2}]$	Di	D
Silsesquioxane	$[RSiO_{3/2}]$	Tri	T
Silicate	$[SiO_{4/2}]$	Quarternary	Q

there are eight silicon atoms with T type connectivity in the structure. The cubic silicate analogue with general chemical formula $[R_3SiOSiO_{3/2}]_8$ may be written as Q_8M_8 indicating eight silicon atoms with Q type connectivity and eight silicon atoms with M type connectivity.

POSS molecules have attracted a lot of recent interest in chemistry because they can be modified to produce a large variety of different molecules. Functionalization can take place in the siloxane core or on the organic periphery, depending on the application. Functionalizing the core of the POSS molecule has led to many different uses of these species as ligands for metals [12], and to monofunctionalized molecules that can be incorporated as pendant units in polymers to improve mechanical and thermal properties [13, 14].

Of particular interest in the field of dendrimer chemistry is the fact that POSS molecules, and the readily synthesized T_8 topologies, can be used as multifunctional cores. Each corner of the POSS polyhedron acts as a site from which dendrimer branches can be grown or to which they can be attached. Forming dendrimers from these cores leads to molecules with large numbers of terminal groups at relatively low generation numbers. In this chapter, we describe the synthesis of POSS molecules and how they can be used as highly functionalized cores in the preparation of dendrimers. We also discuss what potential applications can be envisaged for such molecules, and the specific properties imparted by the silsesquioxane core.

7.2 Synthesis of Silsesquioxanes and Silicates

7.2.1 Silsesquioxanes

Silsesquioxanes are commonly synthesized by hydrolytic condensation of trichloro or triethoxysilanes. The structure of the species formed in the reaction is highly dependent on the reaction conditions, especially the concentration of reactants (particularly the silane monomers), solvent, pH, temperature, availability of water, base or acid catalysis (or "auto-catalysis") and the solubility of the product. Sol-gels and silsesquioxane gels are often synthesized from the same monomers as POSS cages and so the reaction conditions must be carefully controlled in order to avoid the irreversible formation of such gels [10]. The fourth substituent at the silicon atom is usually a small group such as hydrogen [5, 15–17], vinyl [18, 19], allyl [20, 21], propylamine [22, 23], alkyl or aryl [24]. More complex monomers can be prepared through hydrosilylation reactions of trichloro or triethoxysilanes with unsaturated species to form larger molecules. Condensation reactions may produce a range of products including various rings and cages. T_8 cages are often formed preferentially due to lower solubility in common organic solvents than the corresponding T_{10} or T_6 analogues. Nevertheless, low yields are often obtained and so most work on POSS molecules involves the synthesis of vinyl or hydrido-POSS and subsequent reactions to attach a variety of functional groups.

POSS molecules were first isolated in 1946 by Scott [25], followed by Barry et al. [26] in 1955 and soon thereafter by the first hydridosilsesquioxanes (hydrido-POSS) in 1959 by Muller [27]. Initial yields were low but improvements made by various groups have led to reported yields to between 20% and 30% [16, 17, 28, 29]. Often a mixture of T_8 and T_{10} cages is formed that can be separated using fractional crystallization [16]. Vinyl-POSS was first synthesized by Voronkov et al. [19]. The amine-terminated POSS cube (γ-aminopropyl) silsesquioxane is synthesized by acid catalyzed hydrolysis and condensation of (γ-aminopropyl) triethoxysilane [22, 23]. The product is obtained as an octahydrochloride salt and is highly soluble in water. Aminopropyl-POSS is obtained by ion exchange but is highly unstable at room temperature, possibly due to the tendency of the POSS cube to undergo base-catalyzed rearrangements. Despite the difficulties of dealing with aminopropyl-POSS, it remains a very useful building block on the way to functionalized POSS species [22, 30, 31]. A very unusual route to a T_8 cage was reported by Richter et al. via hydrolysis of a zwitterion pentacoordinate silicate species [32]. Table 7.2 provides a summary of POSS molecules used as dendrimer cores.

7.2.2 Silicates

Silicate cages are readily formed by reaction of a silica source (silicic acid, tetraethoxysilane or rice hull ash) and a quaternary alkyl ammonium hydroxide [8, 9, 24, 33]. The size of the alkyl substituents on the organic base defines the size of the cage produced (see Table 7.3) and this has led to the synthesis of Q_6, Q_8 and Q_{10}

Table 7.2 Silsesquioxanes and silicates of relevance for dendrimer cores

R Group	Formula	Formula	Notation
H-	$[HSiO_{3/2}]_8$	$H_8Si_8O_{12}$	Hydrido-POSS
CH_2CH-	$[CH_2CHSiO_{3/2}]_8$	$C1_6H_{40}Si_8O_{12}$	Vinyl-POSS
$Ph_2PCH_2CH_2$-	$[(C_6H_5)_2PC_2H_4SiO_{3/2}]_8$	$C_{14}H_{14}Si_8O_{12}$	Diphenylphosphinoethyl-POSS
$H_2NCH_2CH_2CH_2$-	$[H_2NC_3H_6SiO_{3/2}]_8$	$C_3H_8NSi_8O_{12}$	Aminopropyl-POSS
$CH_2CH(CH_3)_2SiO$-	$[CH_2CH(CH_3)_2SiOSiO_{4/2}]_8$	$C_{32}H_{88}Si_{16}O_{28}$	VinylSilicate
$H(CH_3)_2)SiO$-	$[H(CH_3)_2SiOSiO_{4/2}]_8$	$C_{16}H_{56}Si_{16}O_{28}$	HydridoSilicate

Table 7.3 Organic bases and cages formed

$R'R_3NOH$		Cage formed
R′	R	
2-hydroxyethyl	Methyl	Q_8, $Si_8O_{20}^{8-}$
Phenyl	Methyl	Q_8
Benzyl	Methyl	Q_8
R′ = R	Methyl	Q_8
R′ = R	Ethyl	Q_6, $Si_6O_{15}^{6-}$
R′ = R	Propyl	Unknown
R′ = R	Butyl	Q_{10}, $Si_{10}O_{25}^{10-}$

molecules as well as some larger species. The silicate anion is rarely characterized as it is most often left in solution [34] and reacted further with a chlorosilane to produce a silylsilicate of the form $[R_3SiOSiO_{4/2}]_{2n}$ where $2n = 6, 8, 10$ etc. The reaction of silicate cages with a chlorosilane (R_3SiCl) produces analogues of hydrido-POSS, $Q_8M_8^H$ [11, 35] and vinyl-POSS, $Q_8M_8^{Vi}$ [36] in good yields (see Fig. 7.1).

7.2.3 *Functionalizing Silsesquioxanes and Silicates*

POSS and silicate molecules can participate in a number of chemical reactions to install new functional groups. Hydrido-POSS and $Q_8M_8^H$ mainly undergo hydrosilylation reactions with unsaturated species. Vinyl-POSS and $Q_8M_8^{Vi}$ may also undergo hydrosilylation reactions with Si–H groups. Vinyl-POSS has also been functionalized by hydrobromination [37], cross metathesis [38, 39], methylcarbonylation [37] and radical reactions [20, 21, 40–42]. Radical reactions have been used by Lucke et al. and Hong et al. to create phosphine-functionalized POSS molecules from vinyl- or allyl-POSS [20, 21, 41, 42]. Chlorination of hydrido-POSS led to the formation of Cl_8T_8 and subsequent reactions gave $(MeO)_8T_8$ [40].

Hydrosilylation (also known as hydrosilation) is the addition of Si–H to multiple bonds (see also Chapters 3 and 13), and is the most common reaction of hydrido- and vinyl-POSS. It can be catalyzed by transition metals [43, 44], Lewis acids or by radicals or nucleophilic amines. Two common catalysts for hydrosilylation are the platinum-based Speier's and Karstedt's catalysts. Speier's catalyst is a solution of hydrogen hexachloroplatinic acid first reported in 1957 and is the most widely used platinum catalyst [43], while Karstedt's catalyst is a platinum complex of divinyltetramethyldisiloxane, discovered in 1973 [44]. There have been some reports of the advantage of co-catalysts to enhance the catalytic activity and selectivity of the hydrosilylation reaction. Most commonly this is in the form of dioxygen, which is thought to act as a promoting agent in the reaction and also to prevent the deactivation of the platinum complexes [45, 46]. Electron withdrawing substituents on the Si–H bond increase the rate of hydrosilylation [47].

Addition across a double bond can occur in two ways giving two products. It is generally favorable for Si–C bond formation to occur at the terminal carbon of the alkene (see Reaction Scheme 7.1). This is usually termed α-addition as it occurs at the α-carbon, although in the case of molecules of the form $XCH_2CH = CH_2$ it has been reported as γ-addition where X is attached to the α-carbon. β-addition creates a branched product. Dittmar et al. reported that hydrosilylation reactions of small molecules with allyl groups to hydrido-POSS resulted in a mixture of isomers depending on the constitution of the small molecules [45]. It was found that α-addition was preferred when R was a weak electron withdrawing group. It is thought that lower reaction temperatures and careful choice of solvents favour α-addition, and give purer products [47].

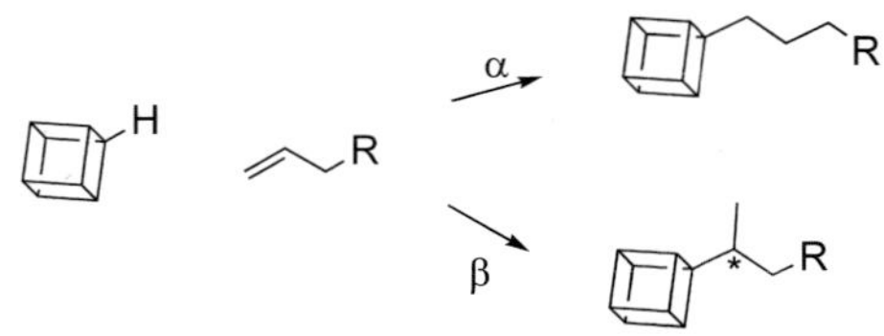

(7.1)

For all the variety of POSS molecules reported in the literature, only four have been used as dendrimer cores: hydrido-POSS, vinyl-POSS [48–56], aminopropyl-POSS [30, 57, 58] and diphenylphosphinoethyl-POSS [41, 42]. The silicate species $Q_8M_8^H$ and $Q_8M_8^{Vi}$ have been used by two groups [47, 59] (see Table 7.2).

7.3 Synthesis of POSS and Silicate Dendrimers

7.3.1 POSS Dendrimer Synthesis

There are relatively few POSS-based dendrimer synthetic strategies, and they are dominated by divergent repetitive hydrosilylation/alkenylation reactions such as those first reported by Jaffrés et al. [49]. The POSS core allows dendrimers with a large number of terminal groups to be synthesized in fewer reaction steps (compare with Chapter 3). One consequence of this is that dendrimers reach their de Gennes limit (see Chapter 1) [60], when the surface becomes too crowded for defect free growth, more rapidly and at lower generation. This creates opportunities to prepare highly functionalized dendrimer surfaces.

Divergent dendrimer synthesis requires extremely high yielding reactions with few side effects due to the eight reactive sites on the core. The divergent route has been the most common strategy of producing POSS and silicate dendrimers. Jaffrés et al. reported functionalization of vinyl-POSS with 8, 16 or 24 Si–Cl units via hydrosilylation of the corresponding chlorosilane (see Reaction Scheme 7.2). This was accomplished in near quantitative yields and without β-addition due to the bulk of the silsesquioxane core. Subsequent vinylation with vinylmagnesium bromide gave dendrimers with 8, 16 or 24 vinyl groups, respectively. A second generation dendrimer with 72 vinyl groups was also prepared. POSS cores of this type have been utilized to synthesize liquid crystal dendrimers [61–63] and other functionalized species such as those reported by Coupar [48], Ropartz [50–54] and Zhang [55]. Silanol (Si–OH) terminated dendrimers were prepared as potential mimics for silica surfaces [48]. These dendrimers were synthesized by the method of Jaffrés et al., forming the chlorosilanes which were then hydrolysed or reduced to Si–H that were subsequently exposed to catalytic hydrolysis to install the Si–OH. Chlorosilanes are, of course, subject to condensation reactions in water and so reduction followed by catalytic hydrolysis was more successful. Vinyl, chloro, hydrido and alcohol terminated POSS dendrimers are extremely useful and can be functionalized further with relative ease (see Reaction Scheme 7.2).

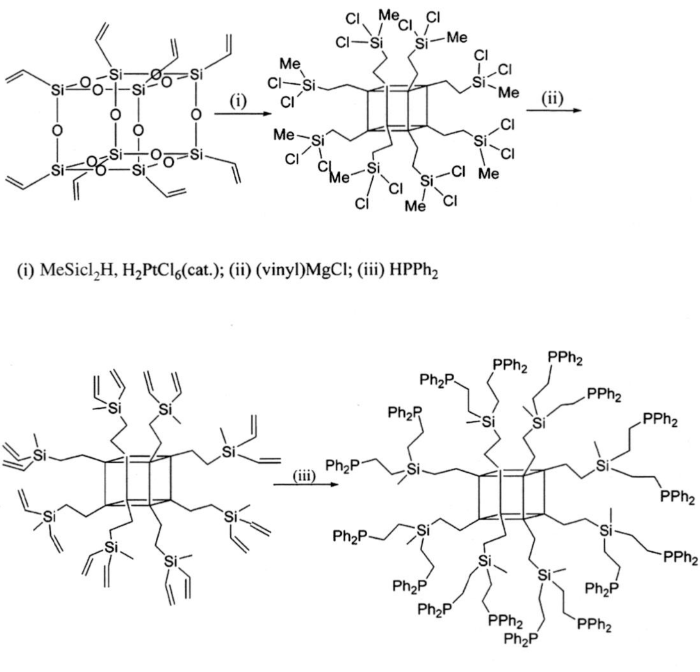

(i) $MeSicl_2H$, H_2PtCl_6(cat.); (ii) (vinyl)MgCl; (iii) $HPPh_2$

(7.2)

Poly-L-lysine dendrimers, **1**, were prepared directly from aminopropyl-POSS hydrochloride using a growth-deprotection strategy with butoxycarbonyl (BOC) protected L-lysine derivatives [58]. Poly(amidoamine) POSS dendrimers, **2**, have been prepared by standard methods from aminopropyl-POSS [30, 57]. Michael addition of methyl acrylate followed by addition of ethylene diamine leads to first generation dendrimers. Carboxylic acid-terminated dendrimers were synthesized by Naka et al. using tertiary butyl acrylate and hydrolysis [57].

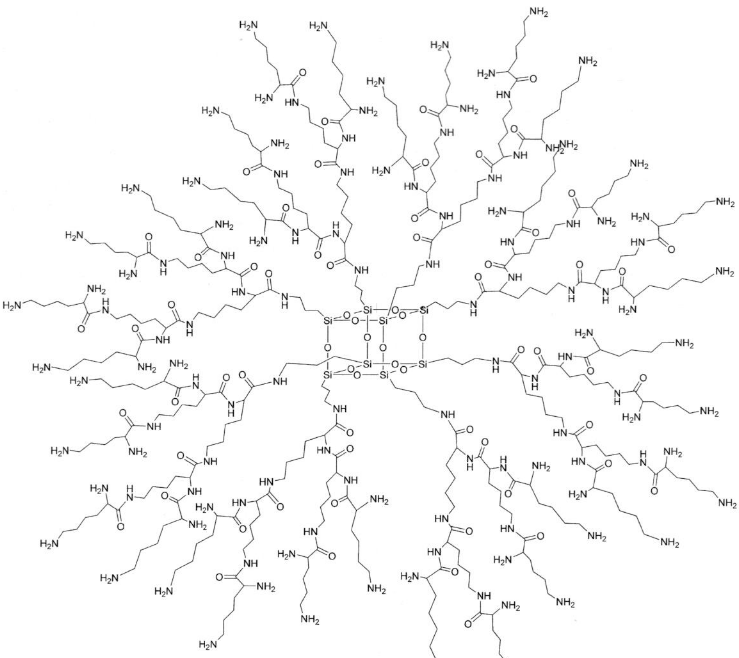

1

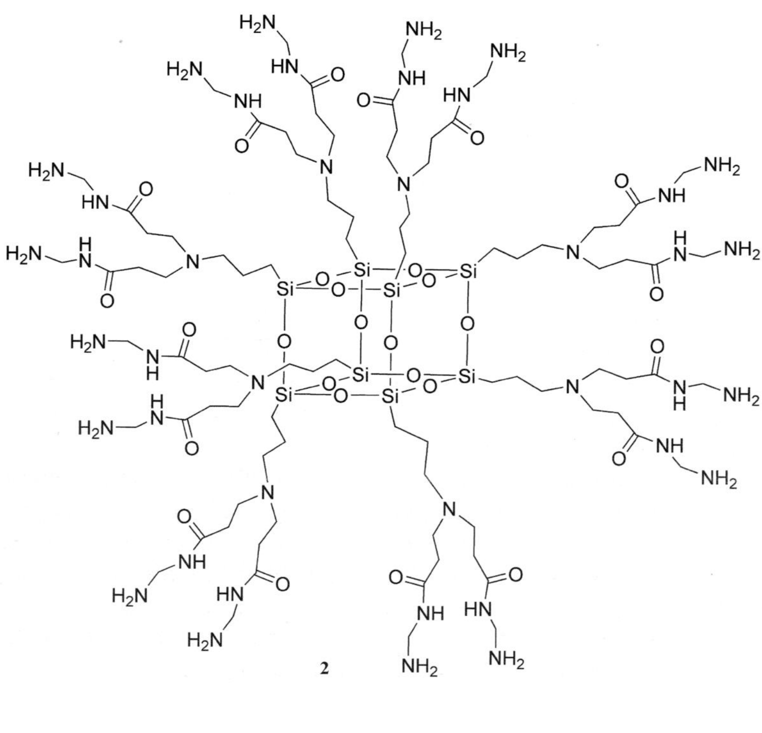

2

While the vast majority of POSS dendrimers have been synthesised divergently there has been one example of convergent synthesis. Hong et al. reported the use of a diphenylphosphinoethyl-POSS as a core for dendrimers by functionalizing the surface with small dendrons prepared from 3,5-dihydroxybenzylalcohol and terpyridine functionalized benzyl bromides as terminal units, **3** [41, 42].

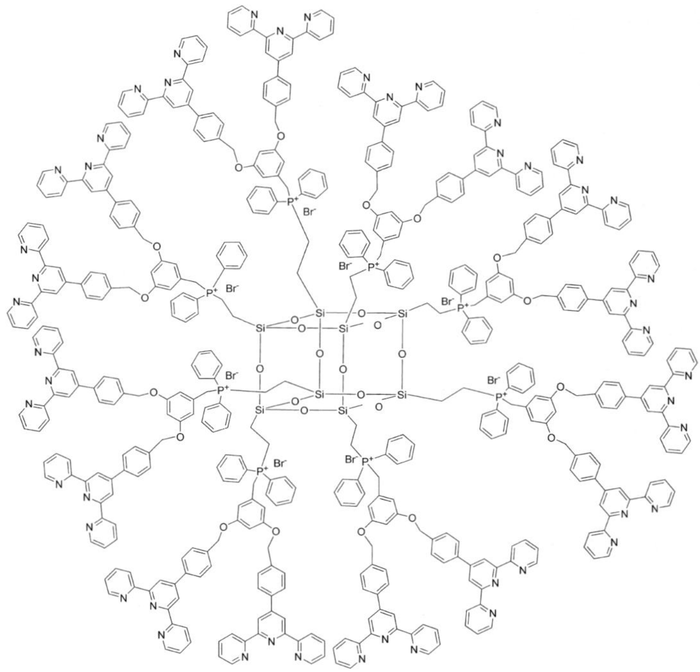

3

7.3.2 *Silicate Dendrimer Synthesis*

Muller and Edelmann reported silicate-based dendrimers with $Q_8M_8^{Vi}$ cores [64]. Hydrosilylation and vinylation, similar to those demonstrated first by Jaffrés et al., resulted in a first generation dendrimer with 24 terminal vinyl groups. $Q_8M_8^{H}$ was used by Wada et al. to synthesize giant starburst silsesquioxanes with incompletely condensed silsesquioxanes as the branching units [59]. The incompletely condensed silsesquioxanes used were essentially cyclopentane-POSS with one corner removed, creating three silanol units. The silanol groups were readily functionalized with chlorosilanes yielding vinyl or hydridosilane groups. Hydrosilylation of vinyl functionalized incompletely condensed silsesquioxanes gave the first generation dendrimer with eight peripheral silsesquioxane groups, and subsequent capping of the two remaining silanols with chlorodimethylsilane prepared the structure for further growth. The second generation with sixteen terminal silsesquioxane cages was obtained in good yield.

7.3.3 Characterization

Silsesquioxanes and silicates are generally characterized using standard analytical techniques, such as multinuclear solution NMR (^{29}Si, ^{13}C, ^{1}H), Magic Angle Spinning (MAS) solid state NMR [11, 65, 66], IR, to name but a few. The size and symmetry of POSS molecules creates many challenges for characterization by these methods, especially NMR. Matrix Assisted Laser Desorption Ionisation Time of Flight (MALDI-TOF) [42, 58, 67] is the most commonly used mass spectrometry technique generally due to the large size of POSS macromolecules. Small POSS molecules are generally solids and make ideal candidates for both crystallography and Magic Angle Spinning (MAS) NMR [11, 65, 66]. In addition to POSS molecules, silicate derivatives such as Q_8M_8 and related structures are also available. Q_8M_8 has found application as a standard for solid state NMR, giving two silicon signals enabling spectra to be referenced [11, 65, 66].

Photoluminescence, UV light induced emission; UV absorption and photoluminescence light excitation spectra of H_8T_8 and R_8T_8 molecules were studied by Azinovic et al. [68]. This systematic study of H_8T_8 and a family of R_8T_8 where R was methyl through decyl groups showed that all these molecules exhibit similar spectral structure and provided useful information to prove the formation of the T_8 cube from standard synthetic methods.

7.4 Applications of POSS and Silicate Dendrimers

7.4.1 Homogeneous Catalysis

Dendrimers have been utilized as catalytic ligands in a wide variety of reactions (see also Chapter 9). The precisely defined shape and size of a dendrimer is advantageous for separation techniques such as ultrafiltration [69], offering a means to overcome the separation and recovery of the catalyst problem that exists in homogeneous catalysis [70–72]. The minimum criterion for a useful dendrimer-based catalytic system is that it should at least equal pre-existing small (non polymeric) ligand systems in terms of selectivity and activity. There are reports that the use of dendritic systems results in some very pronounced effects ranging from total inhibition of the reaction to improved performance. Such effects are often termed the *dendritic* or *dendrimer effects* and are the subject of a review by Helms and Fréchet [73]. The authors propose that dendrimer catalysts offer enhanced stability through steric isolation of catalytic sites, positive interactions of neighboring catalytic sites and cooperative effects from the rest of the ligand.

POSS-based dendrimers are attractive molecules for use as catalytic ligands. The groups introduced in successive generations during the dendrimer synthesis (i.e. chloro- and vinyl-silane substituents) are easily functionalizable by catalytically active ligand species. One particularly fruitful area has been the addition of phosphorus-containing groups to the dendrimer's termini for use as ligands in homogenous catalysis (**4–6**). Using the strategies outlined in Sections 7.2 and 7.3 and using simple organic/inorganic reactions (nucleophilic substitution and radical addition), different phosphine substituents (PR_2, R=Cy, Et, Hex, Me, Ph) have been introduced on the periphery of these dendrimers (see Reaction Scheme 7.2). These reactions vary the size, density and number of functional groups on the exterior, allowing variations in the number and nature of the coordination sites to catalytically-active metals. To a first approximation, the resulting dendrimers are predicted to have similar chemical (and electronic properties) to small, non-dendritic phosphine complexes. Hence the dendrimer is acting merely as a means to enlarge the catalyst sufficiently so that it might be recovered by ultrafiltration, but does not prevent catalytic reaction from occurring. However, this is only a first approximation and one interesting feature of this work has been the identification of "dendrimer effects" on the catalytic activity of the molecules.

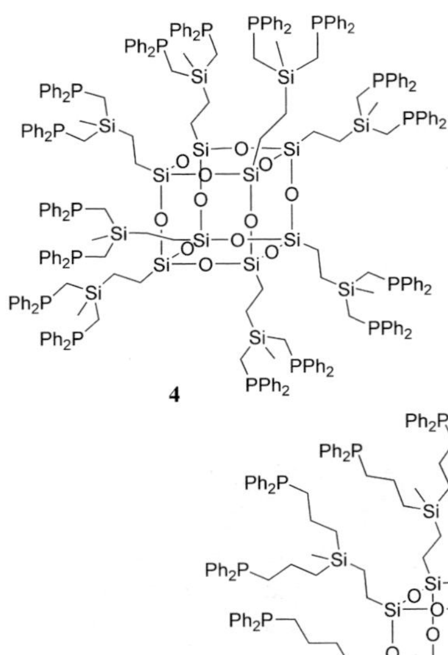

4

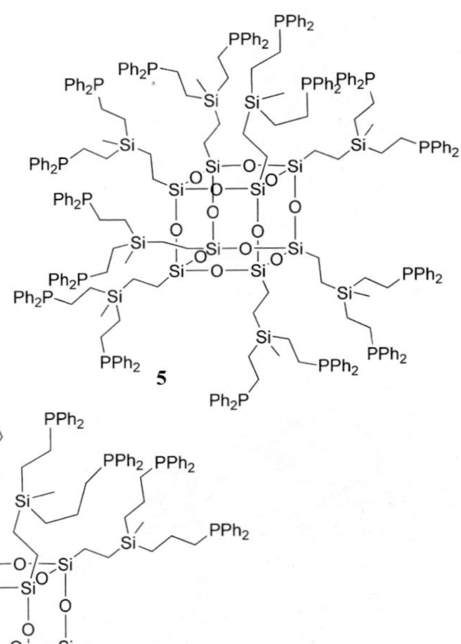

5

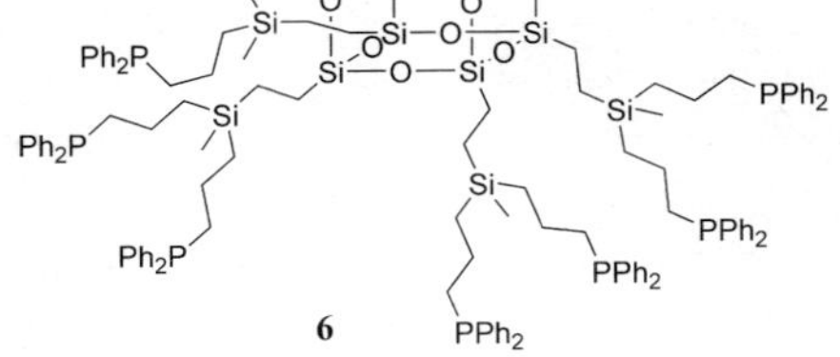

6

POSS-based dendrimers have been used in the hydroformylation of linear alkenes and certain structures show a pronounced positive dendrimer effect, leading to markedly increased selectivity towards the desired linear products [50–54]. Hydroformylation produces two main products – linear and branched aldehydes and generally the linear aldehyde is the most desired. The enhanced regioselectivity towards the linear aldehyde product [50–54] was dependent on the structure of the dendrimer. It was found that **5** has a markedly higher selectivity for the linear aldehyde than any of the other dendrimers or model compounds studied.

These dendritic ligands, in the absence of any catalytic complex, were probed by molecular dynamic techniques and several relationships between the structure and composition of the dendrimers emerged [74]. For example, lengthening the dendrimer branches created bigger dendrimers with increasing radii of gyration while substituting oxygen for a methylene group resulted in decreasing radii of gyration. The properties of dendrimers were also studied at different temperatures and under various solvent conditions. The distance between ligating groups (phosphorus–phosphorus distance) on the surface of the dendrimer was studied using molecular dynamics to give clues as to the underlying reasons for the catalytic activity. Compound **5** was found to have the narrowest distribution of phosphorus–phosphorus distances of the molecules studied but no obvious explanation for the catalytic data was obtained. This implied that phosphorus ligands were close enough to bind rhodium complexes in a bidentate fashion.

To understand more about the underlying reasons for the very interesting catalytic data reported for hydroformylation using rhodium complexes of silsesquioxane dendrimers ligands, further molecular dynamics studies were completed [75] (see Fig. 7.3). The goal of the simulations was to probe whether it is sterically possible for the dendrimer ligands to coordinate to a metal atom with bite angles of around 120° [76–78]. Such bite angles are thought to favor diequatorial binding of a transition metal species [76]. This, in turn, is believed to favor linear selectivity in catalytic

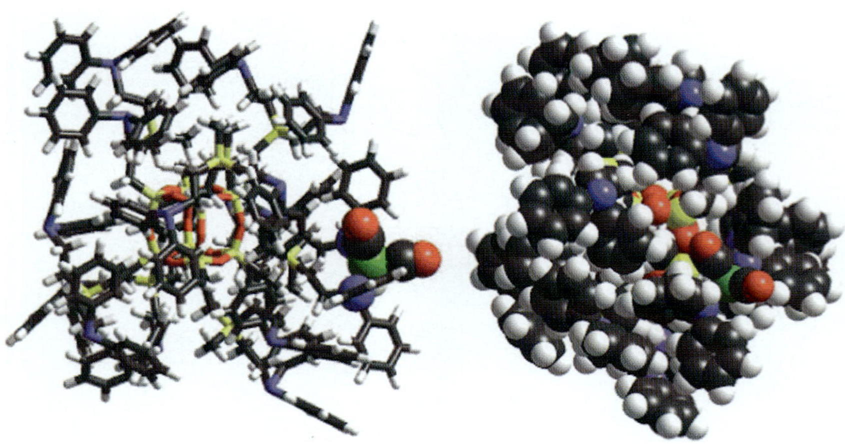

Fig. 7.3 Two views (*stick, left* and *space-filling, right*) of the molecular model of molecule **5** showing the location of a $(\text{-P})_2\text{Rh(CO)}_2\text{H}$ complex (shown as *space filling*) on the exterior surface of the dendrimer. Note that for ease of viewing the two views are slightly rotated with respect to each other. Key: Silicon = yellow, oxygen = red, carbon = black, hydrogen = grey, phosphorus = purple, rhodium = green [75].

processes by creating more steric bulk at the metal centre, making branched products less favorable. Linear products are generally more desirable. Two slightly different calculations were made, but in both cases the average bite angles calculated for structure **5** were closest to the value of 120° [75]. This seems to indicate that it is possible for **5** to coordinate to the rhodium metal complex in an equatorial-equatorial (ee) manner, and that it is possibly more likely to do so than the other two dendrimers studied. This may indeed be a clue as to why **5** gives a higher linear selectivity in hydroformylation reactions than the other dendrimers.

7.4.2 Electro- and Redox-Active Dendrimers

POSS dendrimers with organometallic terminal groups were investigated by Hong et al. [41, 42]. Three generations of terpyridine-terminated dendrimers were prepared by addition of small Frechet-style wedges with 2,2′:6′,2″-terpyridine functionalized benzylbromides to diphenylphosphinoethyl-POSS, **3**. Further reaction with ruthenium (II) species yielded spherical dendrimers with 8, 16 or 32 photo- and redox-active chromophores. The dendrimers were compared to a model compound. Photophysical analysis revealed that dendrimers had much greater extinction coefficients, indicative of the chromophore summation effect. The ligand-centered charge transfer and metal-to-ligand charge transfer processes increased proportionately to the number of Ru(II) metal centres present on the dendrimers. No drop-off in extinction coefficients was observed on moving to higher generations, a feature previously observed with PAMAM dendrimers with Ru chromophores [79]. All generations of POSS dendrimers were emissive at room temperature, confirming that the surface confined chromophores are isolated, resulting in one type of emissive excited state. The quantum yields of the third generation POSS dendrimer were lower than expected, perhaps due to nonemissive quenching as the dendrimer structure becomes more densely packed with increasing generation.

Cyclic voltametry of dendrimers demonstrated the presence of a single reversible Ru II/III redox wave, indicating simultaneous one-electron processes for the ruthenium terminal groups. Ground state interactions between the terminal groups were small or nonexistent due to a lack of peak broadening. Ligand reductions of dendrimers also indicated that there were negligible interactions between the terminal groups. Evidence of film deposition was observed during continuous potential sweeps, most notably for the third generation species.

7.4.3 Liquid Crystals

Silsesquioxanes and silicates have been investigated as cores for liquid crystalline molecules (see also Chapter10). Although silicates are pseudospherical when derivatized with mesogens such as cyanobiphenyl groups, the molecules become rodlike, in lamellar smectic A phase. Molecular modelling of these zeroth generation dendrimers indicated the mesogens could pack giving cylindrical molecules. Following this study, a first generation dendrimer with 16 mesogenic groups was synthesized using vinyl-POSS as the core [63]. A first generation 16-vinyl dendrimer was prepared by the method of Jaffrés et al. [49] followed by hydrosilylation of Si–H functionalized mesogenic groups. It was expected that such a molecule would give cubic or columnar phases and that the influence of the core to create a spherical molecule would be greater than the ability of the system to deform into rod-shaped molecules. It was found, however, that the molecule could be deformed sufficiently to support the lamellar mesophases.

Differential scanning calorimetry (DSC) and thermal polarized transmitted light microscopy indicated that enantiotropic smectic A and smectic C phases were present on heating (see Fig. 7.4). The dendritic structure was found to lower the clearing point of the smectic C to smectic A transition and the glass transition temperature (T_g) compared to the linear analogue. Later work by Saez et al. resulted in a first generation POSS dendrimer exhibiting chiral nematic, hexagonal disordered columnar and rectangular disordered columnar phases [61, 62].

7.4.4 Transition Metal Binding

Poly(amidoamine) (PAMAM) dendrimers were among the first dendrimers synthesized (see Chapter 1) [1]. The Michael addition of acrylates and amines, followed

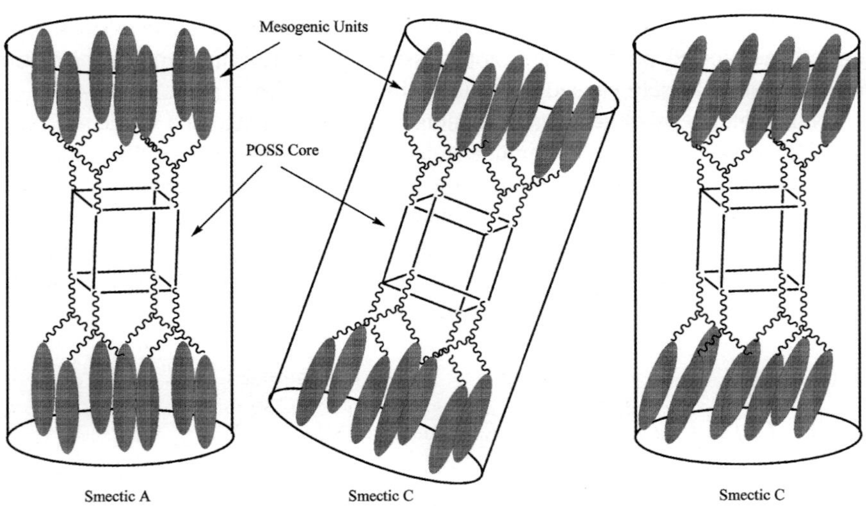

Fig. 7.4 Possible molecular topologies with silsesquioxane dendrimer liquid crystals (Reproduced by kind permission of Taylor & Francis Journals from [63]).

by amidation of the resulting esters, proceeds in high yields from amine cores, typically ammonia or ethylenediamine, resulting in many generations with a variety of terminal groups (amine, alkyl ester, carboxylic acid). Aminopropyl-POSS has been used as a core for the synthesis of PAMAM dendrimers, **2**, initially by Feher et al. [22] and subsequently by Naka et al. who reported their use as copper complexation agents (see also Chapter 11) [57].

First generation PAMAM-POSS dendrimers with carboxylic acid terminal groups were synthesized by Naka et al. [57]. POSS molecules are unstable in basic solutions, and so hydrolysis of the methyl ester was problematic. However, when first generation dendrimers were prepared with *t*-butyl ester terminal groups, subsequent acid catalysed hydrolysis yielded carboxylic acid terminated PAMAM-POSS in quantitative yields and without decomposition of the core. The sodium salt of the dendrimer was prepared through reaction with equimolar quantities of sodium hydroxide with terminal carboxylic acid groups. PAMAM dendrimers have the capacity to form complexes with copper and other metals, due to the presence of tertiary amines within the structure (see also Chapter 11). The POSS-cored first generation dendrimers were compared with first generation ethylenediamine (EDA)-cored PAMAM dendrimers [57]. Isothermal titration calorimetry was used to determine the thermodynamics of complexation, and spectrophotometric titration to determine the binding capacity of the dendrimers for copper ions (Cu^{2+}). The stoichiometry [Cu^{2+}]:[dendrimer] was found to be 4.0 for POSS dendrimers and 2.1–3.8 for the standard PAMAM equivalents.

The POSS dendrimer was found to have one complexation mode for Cu^{2+} regardless of the concentration of dendrimer, involving two carboxylate groups and two tertiary amine groups. The EDA core dendrimer demonstrated a variety of complexation modes, as evidenced by a decrease in the absorbance maximum as the concentration of the dendrimer increased. The coordination geometry of the copper ion within the EDA core dendrimer was believed to change from a system involving two carboxylate groups and two tertiary amine groups, to one involving four tertiary amine groups. The rigidity of the POSS core is the likely cause of this difference in coordination mode. The POSS core acts to decrease the mobility of the dendritic branches, restricting the conformations available and forcing the copper ions into only one coordination mode.

7.4.5 POSS-PAMAM Nanocomposites

While most POSS dendrimer research utilizes POSS molecules as the core for the molecule, Dvornic et al. reported a series of hybrid materials with PAMAM cores and POSS terminal groups [67]. The POSS molecules selected had one corner functionalized with an isocyanato group capable of reacting with terminal amines on PAMAM dendrimers, forming carbamide linkages. A range of terminal group coverage was investigated by adjusting the POSS: terminal group ratio and PAMAM dendrimers from generation 2–5 were investigated.

The properties of the resulting hybrids were found to be a mixture of POSS and PAMAM properties. Products with low degrees of substitution demonstrated PAMAM behaviour such as glass transition temperatures, albeit with POSS-like solubility in chloroform. Glass transition temperatures were observed for all four starting dendrimers by DSC, and were found to increase from generations 2–4, remaining the same for generation 5. At higher degrees of POSS substitution, a glass transition temperature was not observed. At 12.5% substitution the glass transition temperature increased with increasing dendrimer generation, and remained higher than the unsubstituted dendrimers. The lack of a clearly resolved glass transition temperature in more densely substituted hybrids may be due to the dense packing of POSS cubes on the exterior. Melting temperatures are not usually observed for dendrimers, PAMAM in particular, but were observed for all hybrids above 12.5% POSS substitution, indicating POSS-like behaviour.

Solubility depended on the degree of POSS substitution with higher substitutions demonstrating increased POSS-like behavior. Isobutyl groups on the POSS terminal groups increased the alkyl character of the surface of the dendrimer, reducing the solubility in methanol when compared with the amine-terminated PAMAM analogue. In solvents where POSS molecules show good solubility but amine-terminated PAMAM dendrimers do not, solubility was found to increase on increasing POSS substitution. This behaviour indicates the POSS cubes are located on the surface of the dendrimers, controlling the interactions with the solvent. Thermal degradation occurred through the breakdown of the less thermally stable PAMAM component, proceeding at a lower temperature than POSS molecules alone. The observed mass loss was greater than the initial organic content, implying a small quantity of silicon-oxygen volatiles were lost during heating. The properties of these hybrid materials were found to be a mixture of POSS and PAMAM properties, depending largely on the degree of substitution of the terminal groups.

7.4.6 Gene Transfection

Dendrimers based on L-lysine, **1**, and silsesquioxane core were used as gene transfection agents by Kaneshiro et al. [58]. Polycationic dendrimers have been extensively investigated as agents for gene transfection due to their varied chemistry, multiple terminal groups and biocompatibility [80]. Four generations of POSS dendrimers were synthesized divergently from aminopropyl-POSS hydrochloride with the fourth generation possessing 64 terminal lysine groups. For biomedical applications, measurement of the cytotoxicity of substances is critical and was determined for both the dendrimers and the core, and compared with the cytotoxicity of poly-L-lysine. Cytotoxicity was found to increase with increasing generation. The inhibitory concentration for 50% inhibition of cell growth (IC50) could not be determined for the core or generations one and two but was above 200 μg/mL, the highest concentration tested.

Effective gene transfection requires determining the capacity of the vector to form complexes with DNA. Generations 2–4 were capable of fully complexing DNA. The transfection efficiency of generations 1–4 was investigated through comparison with the commercially available agent SuperFect [81]. At vector to plasmid ratios of 10:1, transfection of the gene expressing luciferase was more efficient than with SuperFect. It was hypothesized that the structure of the POSS dendrimer permits the DNA to wrap itself on the surface of the dendrimer. This is different to linear polycation polymers which form an intertwined complex with DNA [81]. The more rigid, globular nature of POSS dendrimers compared to those with flexible two dimensional cores was also thought to be advantageous. POSS-cored dendrimers are denser and possess greater symmetry resulting in a more rigid and globular shape which is unlikely to change morphology when further modified. Chemical modification of the dendrimer periphery such as drug conjugation or bioconjugation of molecules to enhance targeting of the therapeutic agent normally causes morphological changes in flexible polymers. POSS dendrimers would most likely be resistant to this.

7.5 Conclusion

Polyhedral oligomeric silisesquioxanes make very attractive cores for dendrimers. There are now several different synthetic methods for functionalization of these cores in high yield and regioselectively. This has led to many different POSS-based dendrimers being synthesized, and their properties tested in applications ranging from homogeneous catalysis to gene transfection. As the syntheses of POSS species improve over the next few years we would expect to see more efficient dendrimer syntheses which will open up even more dendrimer architectures and potential applications. One particular area where we expect increased activity is in the use of POSS-based dendrimers as well-defined organic-inorganic nanostructures with interesting properties such as quantum confinement.

Acknowledgements The authors thank the EPSRC for funding. We also thank the ACS for kind permission to reproduce the chemical structures of molecules 1, 2 and 3 from references 58, 57 and 42, respectively.

References

1. Tomalia DA (1995) Sci Am 272: 62
2. Denkewalter RG, Kolc JF, Lukasavage WJ (1981) US Pat 4,289,872
3. Denkewalter RG, Kolc JF, Lukasavage WJ (1983) US Pat 4,410,688
4. Bassindale AR, Gentle TE (1993) J Mater Chem 3: 1319
5. Baney RH, Itoh M, Sakakibara A, Suzuki T (1995) Chem Rev 95: 1409
6. Feher FJ, Newman DA, Walzer JF (1989) J Am Chem Soc 111: 1741
7. Morris RE (2005) J Mater Chem 15: 931
8. Harrison PG, Hall C (1997) Main Group Met Chem 20: 515

9. Harrison PG (1997) J Organomet Chem 542: 141
10 Hoebbel D, Weber C, Schmidt H, Kruger RP (2002) J Sol-Gel Sci Technol 24: 121–129
11. Auner N, Ziemer B, Herrschaft B, Ziche W, John P, Weis J (1999) Eur J Inorg Chem 1087–1094
12. Hanssen RWJM, Van Santen RA, Abbenhuis HCL (2004) Eur J Inorg Chem 4: 675–683
13. Phillips SH, Haddad TS, Tomczak SJ (2004) Curr Opin Solid State Mater Sci 8: 21
14. Li GZ, Wang LC, Ni HL, Pittman CU (2001) J Inorg Organomet Polym 11: 123
15. Bolln C, Tsuchida A, Frey H, Mulhaupt R (1997) Chem Mater 9: 1475
16. Agaskar PA (1991) Inorg Chem 30: 2707
17. Frye CL, Collins WT (1970) J Am Chem Soc 92: 5586
18. Bonhomme C, Toledano P, Maquet J, Livage J, Bonhommecoury L (1997) J Chem Soc Dalton Trans 1617–1626
19. Voronkov MG, Lavrentyev VI (1982) Top Curr Chem 102: 199
20. Lucke S, Stoppek-Langner K (1999) Appl Surf Sci 145: 713
21. Lucke S, Stoppek-Langner K, Kuchinke J, Krebs B (1999) J Organomet Chem 584: 11
22. Feher FJ, Wyndham KD (1998) Chem Commun 323
23. Gravel MC, Zhang C, Dinderman M, Laine RM (1999) Appl Organomet Chem 13: 329
24. Laine RM (2005) J Mater Chem 15: 3725
25. Scott DW (1946) J Am Chem Soc 68: 356
26. Barry AJ, Daudt WH, Domicone JJ, Gilkey JW (1955) J Am Chem Soc 77: 4248
27. Muller R (1959) J Prakt Chem 9: 71
28. Agaskar PA, Klemperer WG (1995) Inorg Chim Acta 229: 355
29. Nyman MD, Desu SB, Peng CH (1993) Chem Mater 5: 1636
30. Feher FJ, Wyndham KD, Knauer DJ (1998) Chem Commun 2393–2394
31. Feher FJ, Wyndham KD, Soulivong D, Nguyen F (1999) J Chem Soc Dalton Trans 1491–1497
32. Richter I, Burschka C, Tacke R (2002) J Organomet Chem 646: 200
33. Hasegawa I, Sakka S, Sugahara Y, Kuroda K, Kato C (1989) American Chemical Society Symposium Series, 398: 140, American Chemical Society, Washington, DC
34. Wiebcke M, Hoebbel D (1992) J Chem Soc Dalton Trans 2451
35. Pitsch I, Hoebbel D, Jancke H, Hiller W (1991) Z Anorg Allg Chem 596: 63
36. Hoebbel D, Pitsch I, Heidemann D, Jancke H, Hiller W (1990) Z Anorg Allg Chem 583: 133
37. Drylie EA, Andrews CD, Hearshaw MA, Jimenez-Rodriguez C, Slawin A, Cole-Hamilton DJ, Morris RE (2006) Polyhedron 25: 853
38. Feher FJ, Soulivong D, Eklund AG, Wyndham KD (1997) Chem Commun 1185–1186
39. Lo MY, Ueno K, Tanabe H, Sellinger A (2006) Chem Rec 6: 157
40. Day VW, Klemperer WG, Mainz VV, Millar DM (1985) J Am Chem Soc 107: 8262
41. Hong B, Thoms TPS, Murfee HJ, Lebrun MJ (1997) Inorg Chem 36: 6146
42. Murfee HJ, Thoms TPS, Greaves J, Hong B (2000) Inorg Chem 39: 5209
43. Speier JL (1957) J Am Chem Soc 79: 974
44. Karstedt BD (1973) US Patent 3,775,452
45. Dittmar U, Hendan BJ, Florke U, Marsmann HC (1995) J Organomet Chem 489: 185
46. Lewis LN (1990) J Am Chem Soc 112: 5998
47. Casado CM, Cuadrado I, Moran W, Alonso B, Barranco M, Losada J (1999) Appl Organomet Chem 13: 245
48. Coupar PI, Jaffres PA, Morris RE (1999) J Chem Soc Dalton Trans 2183–2187
49. Jaffres PA, Morris RE (1998) J Chem Soc Dalton Trans 2767
50. Ropartz L, Foster DF, Morris RE, Slawin AMZ, Cole-Hamilton DJ (2002) J Chem Soc Dalton Trans 1997
51. Ropartz L, Haxton KJ, Foster DF, Morris RE, Slawin AMZ, Cole-Hamilton DJ (2002) J Chem Soc Dalton Trans 4323
52. Ropartz L, Morris RE, Foster DF, Cole-Hamilton DJ (2001) Chem Commun 361–362
53. Ropartz L, Morris RE, Foster DF, Cole-Hamilton DJ (2002) J Mol Catal A: Chem 182: 99
54. Ropartz L, Morris RE, Schwarz GP, Foster DF, Cole-Hamilton DJ (2000) Inorg Chem Commun 3: 714

55. Zhang XJ, Haxton KJ, Ropartz L, Cole-Hamilton DJ, Morris RE (2001) J Chem Soc Dalton Trans 3261–3268
56. Manson BW, Morrison JJ, Coupar PI, Jaffres PA, Morris RE (2001) J Chem Soc Dalton Trans 7: 1123–1127
57. Naka K, Fujita M, Tanaka K, Chujo Y (2007) Langmuir 23: 9057
58. Kaneshiro TL, Wang X, Lu Z-R (2007) Mol Pharmaceutics 4: 759
59. Wada K, Watanabe N, Yamada K, Kondo T, Mitsudo T (2005) Chem Commun 1409: 95–97
60. Gennes PGd, Hervet H (1983) J Phys Lett 44: 351
61. Saez IM, Goodby JW, Richardson RM (2001) Chem Eur J 7: 2758
62. Saez IM, Goodby JW (2001) J Mater Chem 11: 2845
63. Saez IM, Goodby JW (1999) Liq Cryst 26: 1101
64. Muller E, Edelmann FT (1999) Main Group Met Chem 22: 485
65. Auner N, Probst R, Hahn F, Herdtweck E (1993) J Organomet Chem 459: 25
67. Dvornic PR, Hartmann-Thompson C, Keinath SE, Hill EJ (2004) Macromolecules 37: 7818
68. Azinovic D, Cai J, Eggs C, Konig H, Marsmann HC, Veprek S (2002) J Lumin 97: 40
69. Freemantle M (2000) Chem Eng News 78
70. Knapen JWJ, Vandermade AW, Dewilde JC, Vanleeuwen P, Wijkens P, Grove DM, Vankoten G (1994) Nature 372: 659
71. Astruc D, Chardac F (2001) Chem Rev 101: 2991
72. Oosterom GE, Reek JNH, Kamer PCJ, van Leeuwen P (2001) Angew Chem Int Ed 40: 1828
73. Helms B, Frechet JMJ (2006) Adv Synth Catal 348: 1125
74. Haxton KJ, Cole-Hamilton DJ, Morris RE (2004) Dalton Trans 1665–1669
75. Haxton KJ, Cole-Hamilton DJ, Morris RE (2007) Dalton Trans 3415–3420
76. Dierkes P, Van Leeuwen PWNM (1999) J Chem Soc Dalton Trans 1519–1529
77. Casey CP, Whiteker GT (1990) Isr. J. Chem. 30: 299
78. Casey CP, Whiteker GT, Melville MG, Petrovich LM, Gavney JA, Powell DR (1992) J Am Chem Soc 114: 5535
79. Storrier GD, Takada K, Abruna HD (1999) Langmuir 15: 872
80. Parekh HS (2007) Curr Pharm Design 13: 2837
81. Xu M, Chen QR, Kumar D, Stass SA, Mixson AJ (1998) Mol Genet Metab 64: 193

Chapter 8
Organometallic Silicon-Containing Dendrimers and Their Electrochemical Applications*

Isabel Cuadrado

8.1 Introduction

Dendrimers constitute a unique class of macromolecular architectures that differs from all other synthetic macromolecules in its perfectly branched topology, which is constructed from a multifunctional central core and expands to the periphery that becomes denser with increasing generation number (see Chapter 1) [1–5]. Since the pioneering works published in the late 1970s and the mid-1980s [6–8], the design and synthesis of these tree-like, well-defined molecules, which exhibit a unique combination of chemical and physical properties, is a field which has sustained dramatic growth and has generated enthusiastic studies at the frontiers of organic, inorganic, supramolecular and polymer chemistry, and more recently in the fields of nanoscience, biotechnology and medicine [1–5, 9, 10]. Whereas the initial interest in dendrimers was focused on the synthetic and structural characterization challenges that pose their fractal geometries, nanometer sizes and monodisperse nature, in the last decade the emphasis has been placed mainly on modification of the properties of dendritic molecules by their functionalization.

Nowadays, one of the most active and promising research areas in dendrimer chemistry is in the integration of transition metals into dendritic structures to create metallodendrimers. Thus, the dendritic scaffold may be used for the spatial arrangement of a large number of transition metal-containing functionalities, either at the periphery or inside the dendritic skeleton (at the core or within the branches) and for the tailoring of properties through the interplay of metallic subunits. Since the first transition-metal containing dendrimers were reported in the early 1990s [11, 12], advances in the synthesis and chemistry of these molecules have not ceased to blossom. Besides

I. Cuadrado
Departamento de Química Inorgánica, Universidad Autónoma de Madrid,
Cantoblanco, 29049, Madrid, Spain
E-mail: isabel.cuadrado@uam.es

*Dedicated to the Memory of Professor Moisés Morán (1951–2002), with respect, admiration and affection

P.R. Dvornic and M.J. Owen (eds.), *Silicon-Containing Dendritic Polymers*,
Advances in Silicon Science 2, © Springer Science+Business Media B.V. 2009

the pleasant aesthetics and fundamental synthetic challenges of metallodendrimers, these molecules are also attractive because of their potential applications as functional materials in such diverse fields as catalysis, sensors, molecular electronic devices, light-harvesting antennas, nanoparticles and medical diagnostics [13–21].

Incorporation of organo-transition metal fragments into dendritic structures represents a stimulating challenging target for both organometallic and dendrimer research, because it opens the way to new nanoscaled organometallic macromolecules of the desired nuclearity, possessing branched topologies and benefits attributed to dendrimers, such as functional group multiplicity, controllable size, precise steric environments, possible cooperative effects, good solubility and recyclability [22–30]. Organo-transition metal compounds are characterized by a precise molecular geometry related to the characteristic coordination number of the metal center and also, in many cases, to the rigid structure of the organic ligand. Interestingly, the σ- or π-character of the transition metal-carbon bond in organometallics, as well as the variety of accessible stable oxidation states of transition metals, resulting from their specific electronic structure (partially occupied d-shells), has a significant influence on the reactivity of the dendrimer, providing an exceptional opportunity for tailoring organometallic dendrimers to achieve desired properties.

Synthetic organometallic chemistry, which has created many remarkable molecular compounds with fascinating structures and properties, also offers valuable synthetic routes to metallodendritic molecules with novel branched architectures and unique properties. By a suitable choice of metal and organic ligand systems, very large dendritic macromolecules having organometallic entities in precise positions and numbers can be designed and constructed for specific applications. On the other hand, several organometallic compounds have been shown to possess bioactivity and several drugs based on organometallic compounds have been developed [31]. In this context, the development of analogous organometallic-functionalized dendritic chemotherapeutic materials that would less easily diffuse through membranes is also an attractive research objective. For this, organo-transition metal-containing dendritic molecules offer attractive advantages over their more traditional linear organometallic polymeric counterparts since they possess a precisely defined three-dimensional molecular architecture, enhanced solubility and the potential to fully control their chemical constitution.

Incorporation of electroactive organometallic units into dendrimer structures is an especially attractive target area because such highly branched macromolecules are good candidates to play a key role as multielectron-transfer mediators in electrocatalytic processes of biological and industrial importance [32–36]. Understanding of electrochemical properties of organometallic units combined with synthetic control of the dendritic structures potentially allows designing of organometallic dendrimers with predetermined redox patterns and control of the number of electrons exchanged at fixed potentials.

Dendrimers also provide unique molecular scaffolding for the placement of multiple redox centres at well-defined locations within large molecular structures. From a structural point of view, electroactive organometallic moieties of the same or different types can be integrated into different topological regions of a dendritic structure, as schematically shown in Fig. 8.1. With regard to the specific envisaged redox properties,

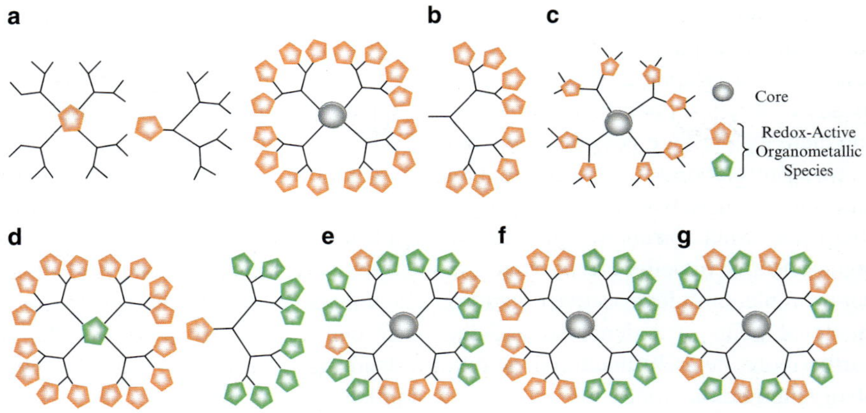

Fig. 8.1 Schematic illustration of the possible location of electroactive organometallic moieties in dendritic structures. (**a**) a single electroactive organometallic unit as a core of a dendrimer, or at the focal point of a dendron; (**b**) dendritic structures containing peripheral electroactive organometallic fragments; (**c**) organometallic units as branching centres; (**d**) heterometallic dendrimer having an electroactive organometallic located at the core (or situated at the focal point of a dendron) and also containing organometallic units of a different chemical nature at the periphery; (**e**) heterometallic dendrimer containing two dissimilar electroactive organometallic functionalities, randomly distributed on the external surface; (**f** and **g**) peripheral homogeneous bifunctionalization of a dendrimer with two different electroactive organometallic units.

the location of electroactive organometallic moieties in the dendrítìc skeleton plays a significant role in the overall electrochemical behaviour of such molecules.

When a dendritic molecule is constructed by branching out from a single electroactive organometallic unit as a core (Fig. 8.1a), the key question becomes how much the electrochemical properties (potential value, kinetic reversibility) of this encapsulated metal-based unit are modified by the shielding effect of the surrounding dendritic branches. It is, for instance, well established that encapsulation of a single redox-active unit by a dendritic scaffold leads to a decrease in electrochemical reversibility, primarily because of the branched shell hindering electron-transfer between the electrode surface and the buried redox-active moiety [32–38]. In this context, dendrimers having a single internal organometallic electroactive group can be viewed as synthetic analogues of proteins offering great potential for fundamental investigations of electron transfer reactions.

On the other hand, when many electroactive organometallic moieties constitute the peripheral units (Fig. 8.1b) or are placed into the branches of a dendrimer scaffold (Fig. 8.1c), a number of equivalent metal-based centres are present in the symmetric structure. In such metallodendrimers, multiple electroactive organometallic groups may or may not interact, depending on the distance between the neighboring metal centres, and on the chemical nature of the connectors between them. Consequently, these dendrimers can act as multielectron sources or reservoirs, where the chemically equivalent organometallic redox units may act independently in an overall multielectron process (*n* identical electroactive metal-based centres undergoing electron transfer in a single *n*-electron wave) or there may exist an electronic communication between electroactive groups in close proximity to one another, resulting in overlapping of closely spaced redox waves at different potentials [39]. In such systems, cooperative phenomena (electron delocalization, magnetism, conductivity, etc.) are possible if the radical species are sufficiently stable.

In more elaborate synthetic approaches, a controlled number of chemically different electroactive organometallic centres can be incorporated into dendritic structures. Such bifunctionalization of dendrimers or dendrons, with various organometallic moieties can be achieved in different combinations (see Fig. 8.1d–g). For example, one electroactive organometallic unit can be situated at the core (or at the focal point of a dendron) and the other one at the periphery (Fig. 8.1d). Furthermore, heterobifunctionalization of a dendritic surface, with different types of organometallic functions, can be accomplished either randomly (Fig. 8.1e), following a combinatorial approach, or by using a controlled synthetic methodology, which results in a dendritic molecule containing exactly the same number (a 1:1 ratio) of the two different organometallic moieties, precisely located at the surface (Fig. 8.1f and g). In these multi-redox heterometallic dendrimers the electrochemical behaviors are rather complicated since organometallic units are different from chemical, topological and electrochemical viewpoints, and the degree of interaction among the redox-active moieties depends on their chemical nature and distance between metal centres. Multielectron redox processes at different potentials can therefore be observed whose specific patterns are related to the extent of electronic interaction between the electroactive organometallic centres. Of course, organometallic dendrimers may belong to more than one of the aforementioned categories which in turn can be the source of even more complex electron exchange patterns.

During the last 20 years, our research group has worked on the design and construction of electroactive macromolecules, containing organo-transition metal moieties together with silicon-based linear, cyclic and cubic polymeric skeletons [40–43], as well as with silicon- and nitrogen-based dendritic structures as scaffolds [44–55]. The main goal of this chapter is to review some of the contributions in the field of organometallic dendrimers, placing significant emphasis on the work on dendritic molecules with silicon atoms as the branch junctures and having electroactive organometallic units. The first part of this chapter reviews the synthesis of such molecules and has been organized according to the synthetic approaches employed for the construction of organometallic dendrimers and chemical nature of the organometallic moieties. For this, functionalization of preformed dendritic structures containing silicon atoms as branch junctures with reactive organometallic moieties has proven to be an excellent method for preparation of well-defined peripherally functionalized organometallic molecules. Alternative routes make use of reactive functionalized organometallic entities, from which convergent and divergent growth strategies for the construction of novel silicon-based multiredox homo- and hetero-metallic dendritic structures can be developed. Throughout this chapter, special emphasis is placed on the description of the redox behavior exhibited by different families of organometallic silicon-based dendritic molecules. The second part of this chapter describes electrochemical applications of some ferrocenyl dendrimers, mainly in the field of molecular recognition of anions and in amperometric biosensors.

As a consequence of the very rapid growth of organometallic dendrimer research and particularly of the wide variety of silicon-containing organometallic dendrimers that have been reported, excellent and comprehensive reviews on these interesting molecules have been published by many of the leading authors in these fields [13–30, 34], and throughout this chapter the reader is directed to these reviews for in-depth information on selected topics. This chapter is not intended to be exhaustive in terms of listing every organometallic silicon-containing dendrimer prepared so far, but instead it aims to highlight specific aspects of organometallic dendrimers containing silicon atoms as key components of their skeletons for which well-defined redox behaviour has been reported. Particular attention is paid to the work carried out in our laboratory and to selected related examples in this field.

8.2 Synthetic Strategies and Redox Properties of Organometallic Silicon-Containing Dendritic Macromolecules

The contribution of silicon chemistry to the area of organo-transition metal dendrimers is of great interest for numerous fundamental reasons. Although a large number of metallodendrimers are built from functionalized carbon-, nitrogen- or phosphorous-based dendritic backbones, silicon-containing skeletons based on carbosilane (Si–C), siloxane (Si–O) or carbosiloxane (Si–O–C) linkages, are among the most widely used ones for the synthesis of organo-transition metal-containing dendrimers because of their kinetic and thermodynamic stability, chemical inertness, and accessibility [56–58]. Currently, a variety of key silicon-containing cores are known and they can be di-, tri-, tretra-, hexa- or octafunctionalized, allowing construction of organometallic dendrimers of various forms and sizes, possessing large numbers of organometallic moieties. In addition, the existence of several well-established, clean and high-yielding synthetic growth methodologies for silicon-based dendrimers and their general robustness and easy characterization make these skeletons attractive from the organometallic chemistry point of view. Likewise, reactive end-groups such as Si–X (X=Cl, Br), Si–Vinyl, Si–Allyl, Si–H, Si–OH, Si–NR_2, Si–Ph, Si–C≡C–, Si–C_5H_5, Si–CH_2–PPh_2, can be anchored to the surface of the silicon-containing scaffolds, remaining available for further reactions with appropriate organometallic reagents, and therefore offering the possibility of developing valuable reaction methods for organometallic chemistry. Furthermore, most of such silicon-based dendritic skeletons possess excellent solubilities in organic solvents such as THF, diethyl ether, dichloromethane and hexane, this being an important property which enables easy introduction of typical organo-transition metal starting molecules and helps purification and characterization of the resulting multi-functionalized organometallic silicon-based dendrimers. Figure 8.2 shows some

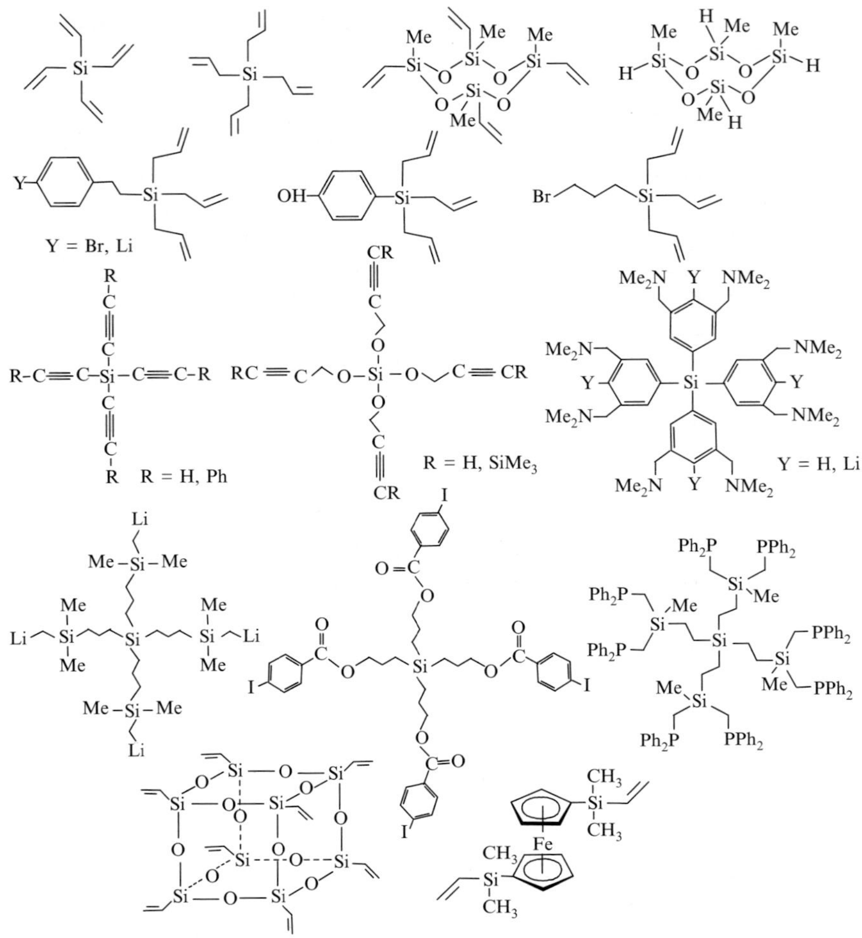

Fig. 8.2 Representative examples of polyfunctional silicon-containing cores and building blocks used in the synthesis of organo-transition metal dendrimers.

remarkable examples of silicon-based molecules functionalized with reactive groups which have been used as key starting blocks for the preparation of well-defined organo-transition metal-containing dendritic structures.

For all these reasons, it is not surprising that a broad collection of dendritic molecules with silicon skeletons and possessing organometallic units based on Ti, Zr, Nb, Ta, Cr, Mo, Mn, Re, Fe, Ru, Co, Rh, Ir, Ni, Pd, Pt and Au, as transition metals have been prepared. The state of the art in the synthesis, properties and applications of organometallic silicon-based metallodendrimers has been summarized in various comprehensive reviews [23, 27, 28, 56–58]. The following sections describe some of the synthetic approaches employed for the construction of dendritic structures with silicon atoms as the branch junctures and incorporated electroactive organometallic functions.

8.2.1 Functionalization of Silicon-Based Dendritic Scaffolds with Electroactive Organometallic Moieties

8.2.1.1 Silicon-Containing Dendrimers Functionalized with Peripheral Ferrocenyl Moieties

Our research on the chemistry of dendrimers was initiated in 1993 and we focused our early efforts mostly on the peripheral functionalization of silicon-containing dendrimers with ferrocenyl units [22, 23, 44, 59]. Our interest in such systems originated from our long-standing interest in silicon-based ferrocenyl multimetallic compounds and polymers, and from our efforts to understand the relationship between the structure and redox properties of ferrocene siloxane materials [23, 41, 42].

Due to its high stability and interesting chemical and physical properties, ferrocene has become a versatile building block for the synthesis of materials with tailor-made properties [60–63]. It is a stable organometallic 18-electron system which undergoes fast one-electron oxidation to the positively charged ferrocenium form. This electrochemical process takes place at accessible potentials and usually maintains fast kinetics (reversibility in electrochemical terminology) in mono and multiferrocenyl compounds. Thus, ferrocene-based polymers have attracted great interest for the chemical modification of electrodes, as electrochemical mediators, and as materials for the construction of electronic devices. In this context, we have prepared redox-active, ferrocenyl-containing polymers constructed from linear cyclic and cubic siloxanes [41, 42].

The same interest also provided motivation for the synthesis of ferrocene-containing dendrimers of precise sizes and perfectly branched structures. An attractive additional reason was that such macromolecules raise the possibility of combining the unique and valuable redox properties of the ferrocene nucleus with the benefits of the highly structured macromolecular geometry. This ought to provide access to materials of nanoscopic-sizes possessing unusual symmetrical architectures, as well as specific physical and chemical properties, which would be expected to differ from those of other ferrocene-based materials prepared to date.

Due to the simplicity and versatility of the synthetic methodology, functionalization of the exterior surface of preformed silicon-containing dendrimers with suitable organometallic moieties proved to be a valuable approach to gain access to well-defined, peripherally functionalized electroactive dendrimers. Initially, we focused on the construction of organometallic dendritic structures by the reactions of molecules containing Si–Cl and Si–H surface sites. The selected divergent synthesis starts from tetraallylsilane or 1,3,5,7-tetramethyl-cyclotetrasiloxane, as four-directional branching centers, and involves repetition of hydrosilylation and alkenylation reactions as growing steps in the manner developed by van der Made et al. [64], Roovers et al. [65], and Seyferth et al. [66] for the synthesis of carbosilane dendrimers (see Chapter 3). Using this synthetic approach we prepared several different families of silicon-based dendritic molecules containing 4, 8 and 16 reactive Si–H, Si–Cl or Si–Allyl functional end-groups [22, 23, 44, 50], as illustrated by selected examples **1–3**.

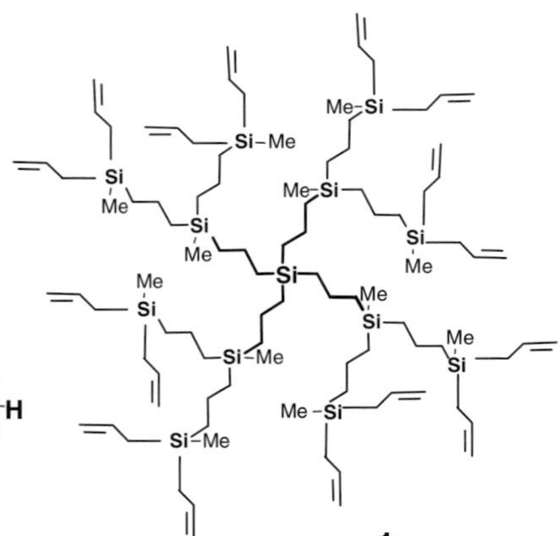

1

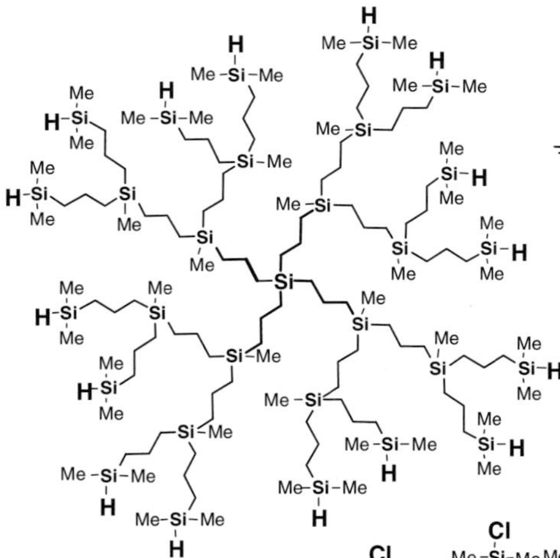

2

3

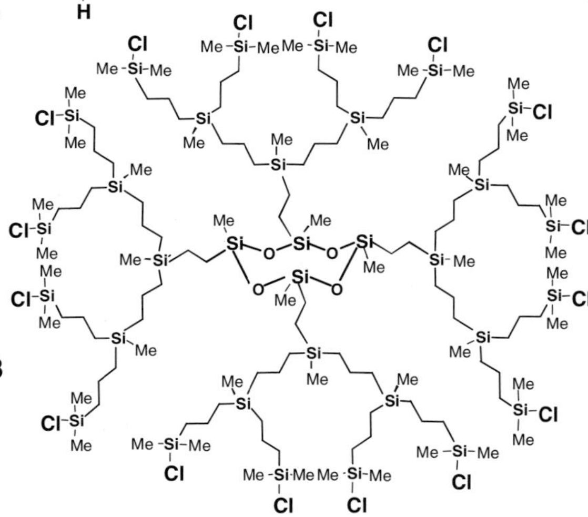

Three different synthetic routes have been developed for incorporation of ferrocenyl moieties onto the surface of such silicon-containing dendritic scaffolds, and these routes are summarized in Reaction Scheme 8.1, in which only one of the dendrimer branches, containing a peripheral functional group, is shown. The first method involved substitution of chlorine atoms of the Si–Cl end-groups by monolithioferrocene $(\eta^5\text{-}C_5H_4Li)Fe(\eta^5\text{-}C_5H_5)$ in tetrahydrofuran (THF) at low temperature [22, 23, 44, 50]. These reactions afforded the first, second and third generations of dendritic macromolecules **4–9** bearing 4, 8 and 16 peripheral ferrocenyl moieties, respectively, directly bonded to the external silicon atoms of the organosilicon dendritic scaffold.

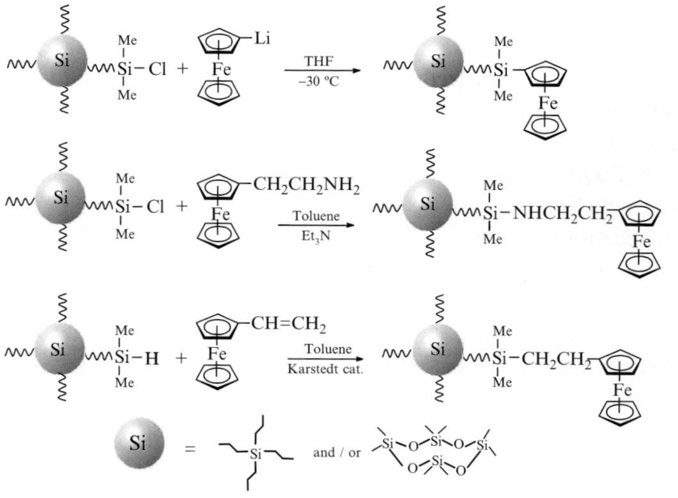

(8.1)

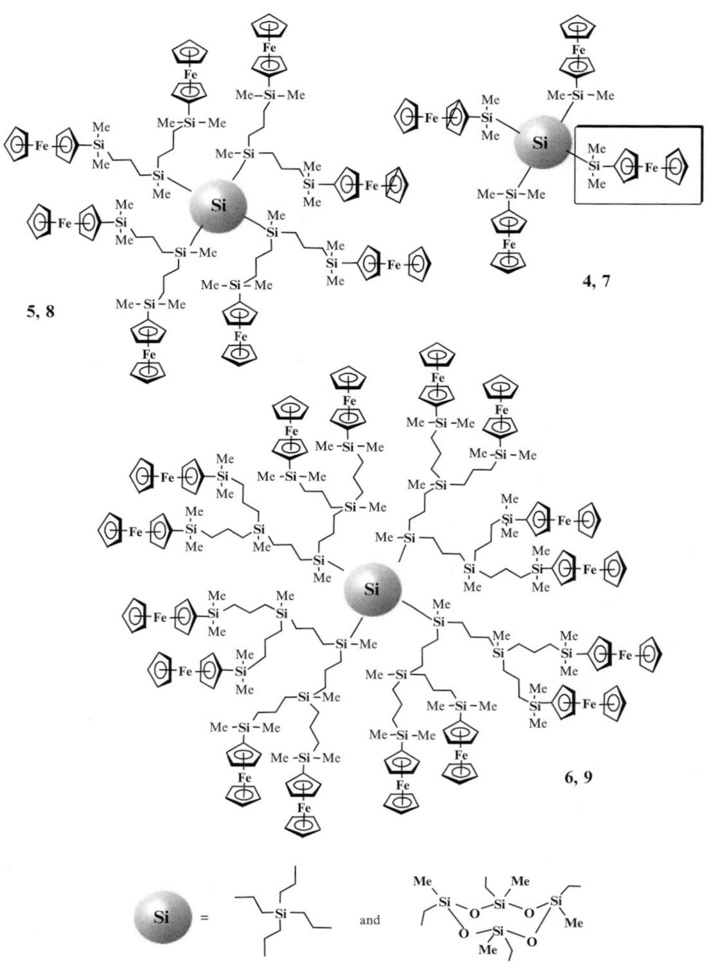

5, 8

4, 7

6, 9

4, 5, 6

7, 8, 9

The second method utilized the high reactivity of the Si–Cl bonds towards the amine groups to allow facile organometallic functionalization of the dendrimer surfaces. In this synthesis, the key organometallic monomer is (β-aminoethyl) ferrocene $(\eta^5\text{-}C_5H_4CH_2CH_2NH_2)Fe(\eta^5\text{-}C_5H_5)$ which was selected because its amino group is two methylene units removed from the ferrocene nucleus. This is of critical importance because it minimizes steric and electronic effects of the organometallic moiety, and likewise, the instability of α-functional ferrocene derivatives resulting from the α-ferrocenyl carbonium ion stability. Treatment of the Si–Cl functionalised dendrimers with appropriate amounts of (β-aminoethyl)ferrocene in toluene, at reflux temperature and in the presence of triethylamine as the acid acceptor, yielded the desired products with 4, 8, and 16 ethylferrocenyl moieties, attached to the surface of the dendritic structure through Si–NH linkages, respectively [22, 44]. The simplest dendrimers within this series were **10** and **11**.

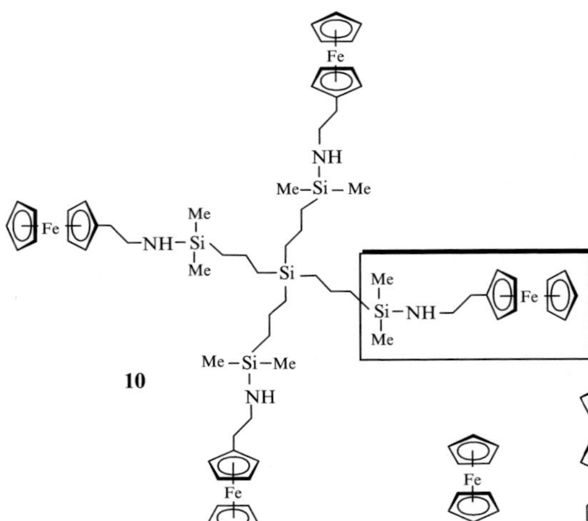

10

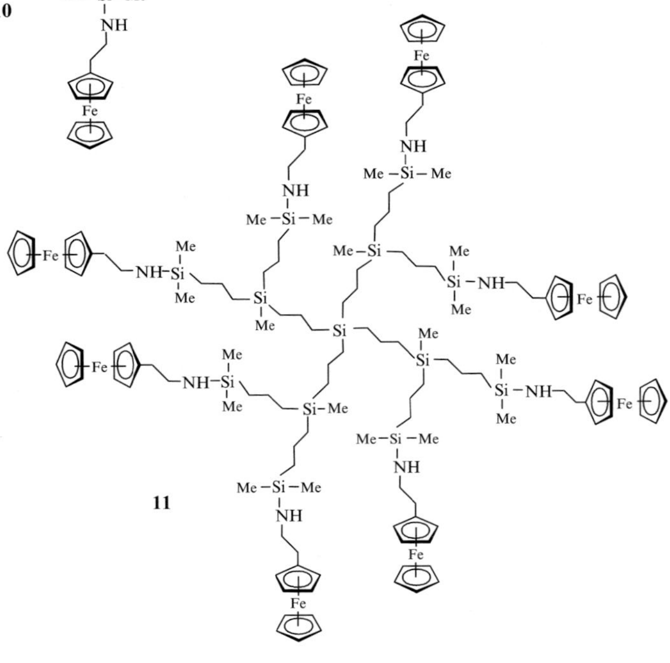

11

The third type of successful reactions on the silicon-based dendrimer surfaces, involved hydrosilylation of silicon-hydride terminated dendrimers with appropriate equivalents of vinylferrocene, in the presence of catalytic amounts of Karstedt's catalyst, bis(divinyltetramethyldisiloxane)platinum(0), in toluene as solvent [23a, 59]. This strategy resulted from our earlier work on incorporation of ferrocenyl moieties into polyfunctional siloxane frameworks via hydrosilylation reactions of vinylferrocene $(\eta^5\text{-}C_5H_4CH{=}CH_2)Fe(\eta^5\text{-}C_5H_5)$, with the Si–H containing 1,3,5,7-tetramethyl-cyclotetrasiloxane and octakis(hydrodimethylsiloxy) octasilsesquioxane, which provided the well-defined multimetallic redox-active molecules **12** and **13** shown in Reaction Scheme 8.2, as well as the ferrocenyl silicon-based structures **14–17** [42].

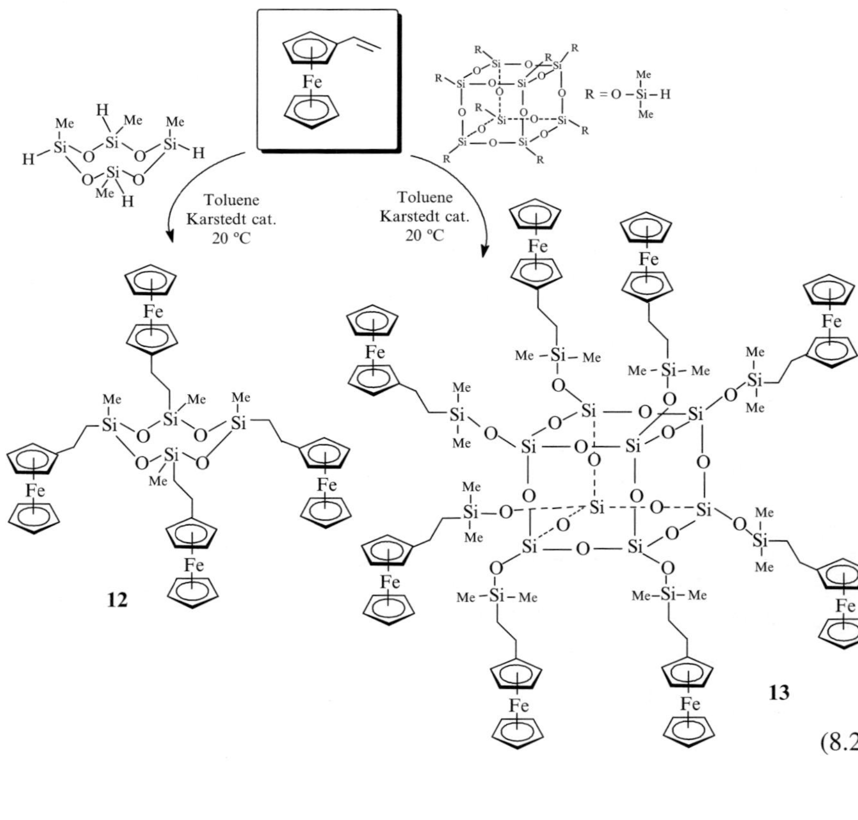

12

13

(8.2)

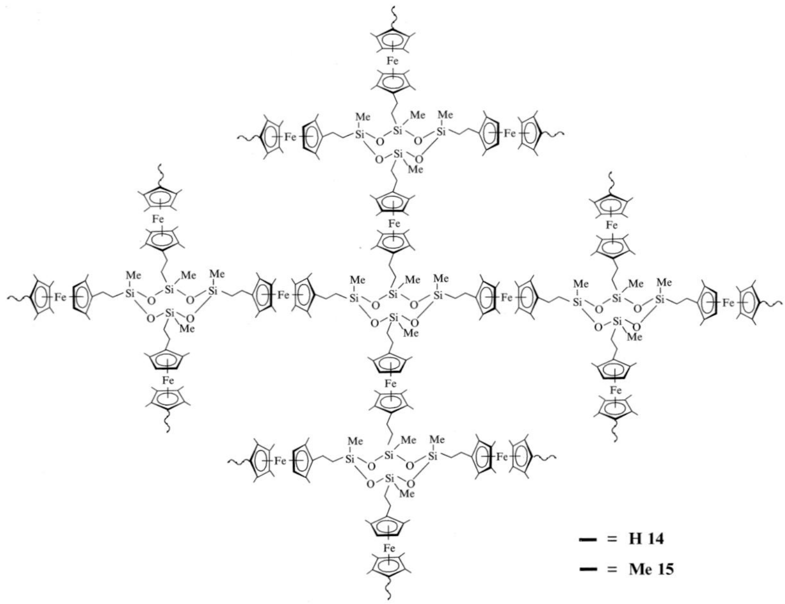

— = H 14
— = Me 15

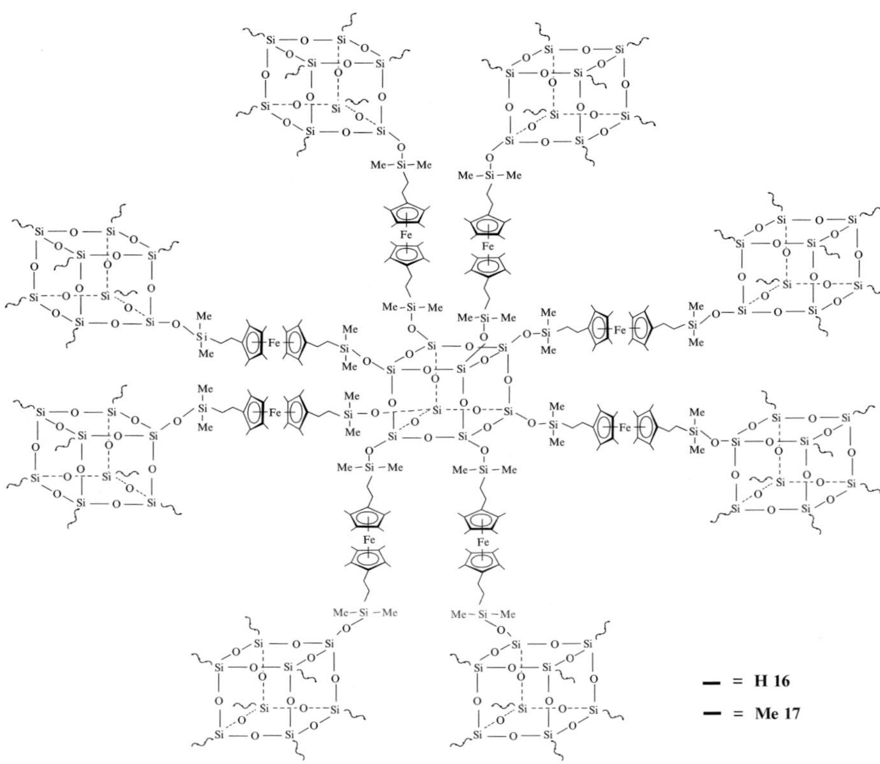

— = H 16
— = Me 17

These reactions cleanly afforded the ferrocene-containing dendrimers **18–20**, in which ferrocenyl moieties are attached to external silicon atoms of the dendritic scaffold through a two-methylene spacer.

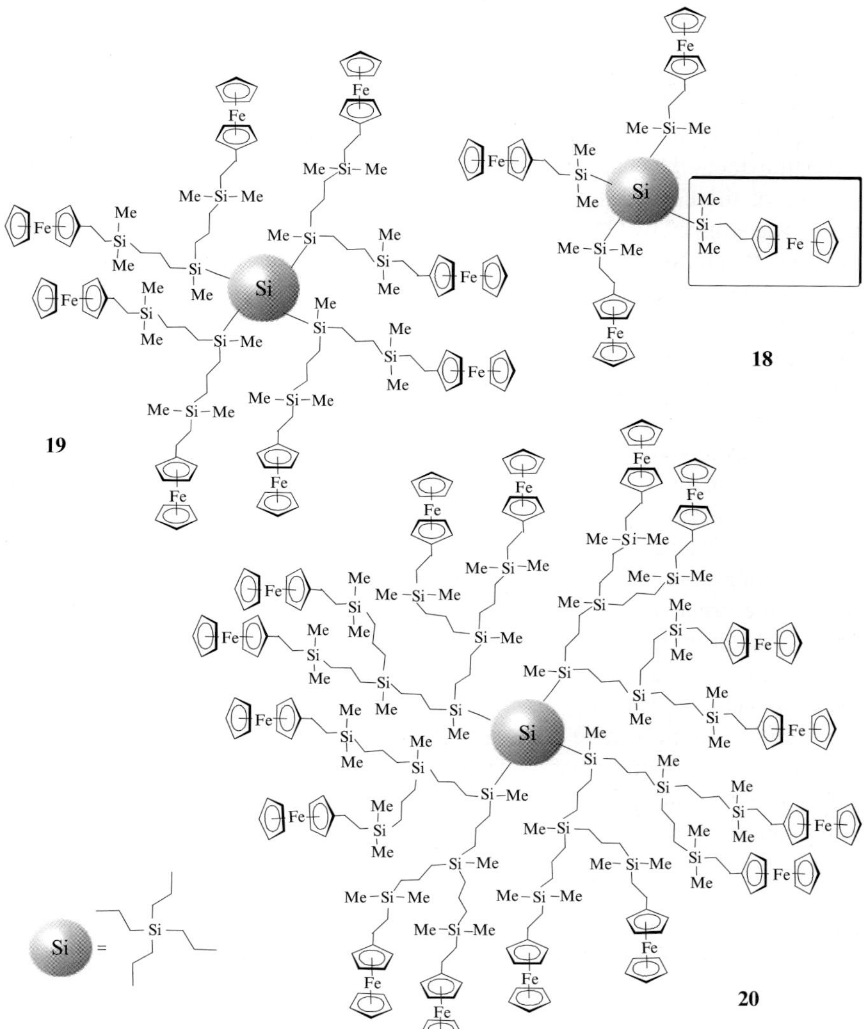

18

19

20

The three families of ferrocene-containing dendritic macromolecules described above, with ferrocenyl units attached to the end-groups of silicon-based dendrimers, were isolated as air-stable, red-orange, viscous materials. They were relatively easy to prepare and showed remarkably high solubilities in solvents such as dichloromethane, THF, and even non-polar hydrocarbons, such as hexane. Their structures were straightforwardly confirmed by a variety of spectroscopic and analytical techniques including NMR and IR spectroscopies, fast atom bombardment mass spectroscopy (FAB MS), matrix-assisted laser desorption/ionization time-of-flight mass spectroscopy (MALDI-TOF) MS, and elemental analysis. Interestingly, the high symmetry of these dendrimers made NMR spectroscopy a useful technique for their characterization as ^{1}H, ^{13}C, and ^{29}Si NMR spectra provided confirming evidence for the complete functionalization of their peripheral sites with ferrocenyl moieties.

We have also used computer-generated molecular models to gain further information about the shape and approximate dimensions of the three-dimensional structures of these dendritic materials. Figure 8.3 shows two energy-minimized structures determined by CAChe™ molecular mechanic calculations for the ferrocenyl dendrimers **9** and **20**, both containing 16 external ferrocenyl units but grown from different silicon-containing cores. We obtained approximate diameters of 2, 3 and 4 nm for the first, second, and third generation of these dendrimers, respectively. For comparison, the same figure also shows the molecular model of a nitrogen-based ferrocenyl dendrimer also prepared in our laboratory by the reaction of diaminobutane-based poly(propylene imine) (PPI) dendrimer with 16 amine groups with chlorocarbonyl ferrocene [46]. The three molecules shown in Fig. 8.3, all have sixteen peripheral ferrocenyl moieties which are linked to the silicon- or nitrogen-based dendritic scaffold through spacers of different chemical characteristics. It is evident from these structures that the shape of the ferrocenyl-containing dendrimers depends strongly on the chemical composition of the core, and on the nature and length of the organometallic end-branches. Thus, whereas the carbosilane-based ferrocenyl dendrimer **20** (in which the ferrocenyl groups are attached to the external silicon atoms through a two-methylene flexible spacer) adopts a nonextended, compact, globular structure (Fig. 8.3a), dendrimer **9** based on a tetramethylcyclosiloxane core is able to adopt a more extended, "butterfly-like" conformation (Fig. 8.3b), while the diaminobutane-based ferrocenyl dendrimer having 16 peripheral amido-ferrocenyl units shows a fairly open structure in which well-defined inner cavities can be observed (Fig. 8.3c).

8.2.1.2 Electrochemical Behavior of Silicon-Containing Dendrimers with Electronically Isolated Ferrocenyl Units

An important prerequisite for any kind of electrochemical application of the electroactive silicon-containing dendrimers is the chemical redox reversibility of organometallic moieties integrated into the dendritic structure. As noted earlier, the ferrocene moiety is undoubtedly the most attractive organometallic redox center to integrate into dendritic structures, not only because it is electrochemically well-behaved in most common solvents undergoing a reversible one-electron oxidation, but also because such electron removal usually does not involve fragmentation of the dendritic scaffold [33, 34, 62]. Likewise, the redox behavior of the ferrocene moiety can be tuned by substituents on the cyclopentadienyl rings when different

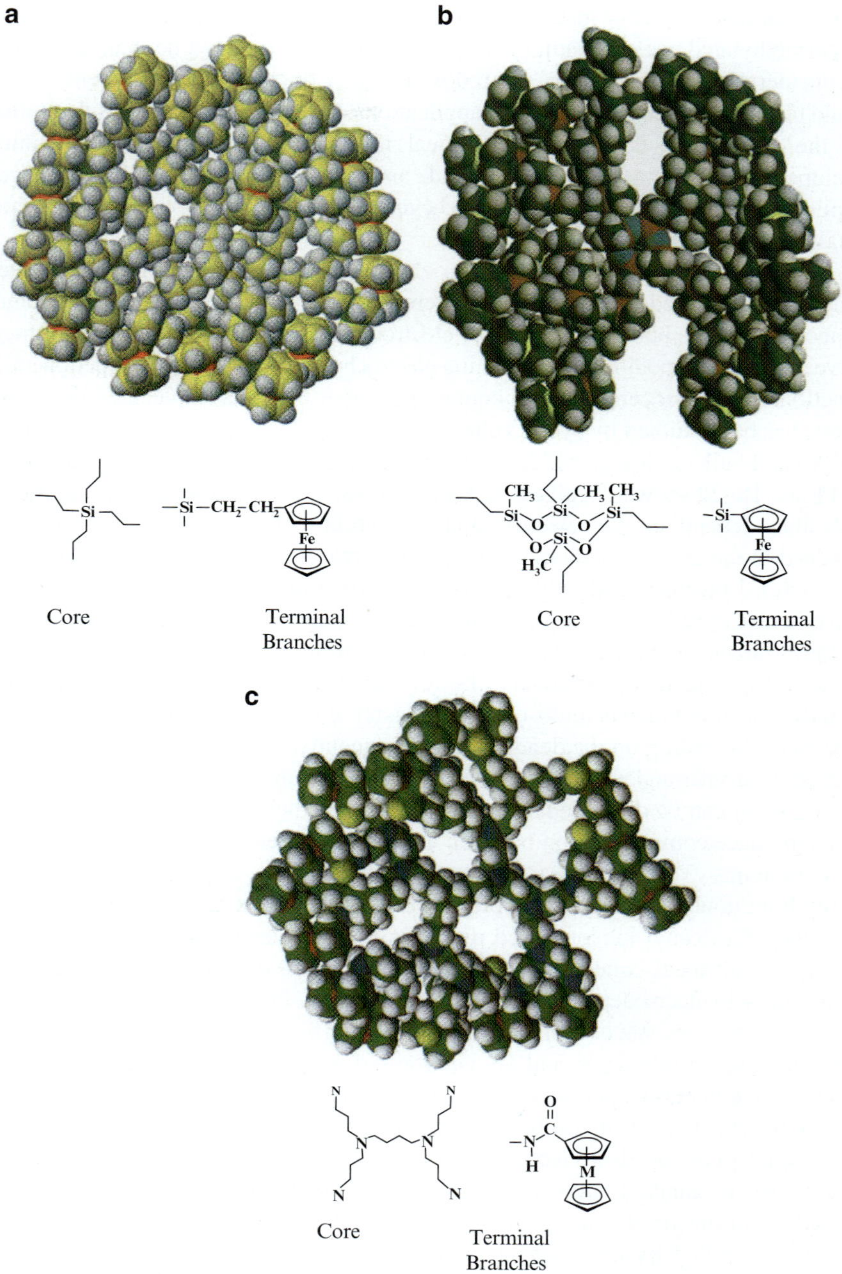

Fig. 8.3 Computer-generated structures of three third generation ferrocenyl dendrimers bearing 16 peripheral ferrocenyl units. (**a**) carbosilane-based dendrimer **9**; (**b**) cyclotetrasiloxane-based dendrimer **20**; (**c**) diaminobutane-based ferrocenyl-amido-terminated dendrimer DAB-dend-{NHCO(η^5-C_5H_4)Fe(η^5-C_5H_5)}$_{16}$ (DAB = DiAminoButane).

properties can result. For instance, as a result of enhanced electron-donating ability of permethylated cyclopentadienyl rings, polymethyl ferrocenyl derivatives exhibit a considerably lower value of the redox potential of the Fe^{II}/Fe^{III} system, which could be valuable in electrochemical applications. Indeed, effects of ring substitution on the structural, chemical and physical properties of compounds containing cyclopentadienyl rings can be profound, and they have dramatically influenced applications of metallocenes and related cyclopentadienyl compounds in synthetic, catalytic and materials chemistry [67].

A significant aspect of our work on electroactive metallodendrimers has been evaluation of the redox properties of different families prepared, not only in homogeneous solutions but also confined to electrode surfaces where metallodendrimers serve as electrode modifiers. The solution electrochemical behavior of the peripherally functionalized ferrocenyl silicon-containing dendritic macromolecules described above has been studied by cyclic voltammetry (CV), differential pulse voltammetry (DPV) and bulk coulometry [22, 23, 45, 50]. Cyclic voltammograms of dendrimers **4–11** and **18–20** showed in all cases, a single reversible oxidation process, at potentials that depend on the electron-donating ability of the different substituents attached to the cyclopentadienyl rings on the ferrocenyl moieties (see for example, Fig. 8.4a). Likewise, in the differential pulse voltammograms only one oxidation wave is observed (see Fig. 8.4b), while careful coulometry measurements resulted in the removal of the expected number of electrons for all peripheral ferrocenyl groups. These results showed that the reversible oxidation wave observed corresponds to a simultaneous multi-electron transfer at the same potential, of 4, 8 or 16 electrons, depending on the dendrimer. Consequently, multiple ferrocenyl moieties in each dendrimer undergo electron transfer independently, and no electronic coupling among them can be discerned. This electrochemical behavior clearly suggests that through-space communication between adjacent ferrocenes is not very effective in these systems.

Without doubt, the most noteworthy aspect of the redox behaviour of the above described ferrocenyl silicon-based dendritic macromolecules **4–6**, **10–11**, **18–20**, having a high local concentration of peripheral, non-interacting redox centres, is their ability to electrodeposit onto electrode surfaces. In this way, for the first time, electrode surfaces have been successfully modified with films of dendrimers containing reversible 4-, 8- and 16-electron redox systems, resulting in detectable electroactive materials persistently attached to the electrode surfaces [45].

The polyferrocenyl silicon-based dendrimers were immobilized in their oxidized forms onto platinum-disk electrodes by controlled potential electrolysis (at about +0.60 V versus standard calomel electrode, SCE) from degassed dendrimer-containing CH_2Cl_2 solutions. In the process, the amount of electrodeposited dendritic material can be controlled by the time interval during which the potential is held fixed. The electrodes thus coated were rinsed with CH_2Cl_2 to remove any adhering solution, dried in air and transferred to clean CH_2Cl_2 solution containing only the supporting electrolyte n-Bu_4NPF_6. The electrochemical response of an electrodeposited film of the octaferrocenyl carbosilane-based dendrimer **5** is shown in Fig. 8.4c as a

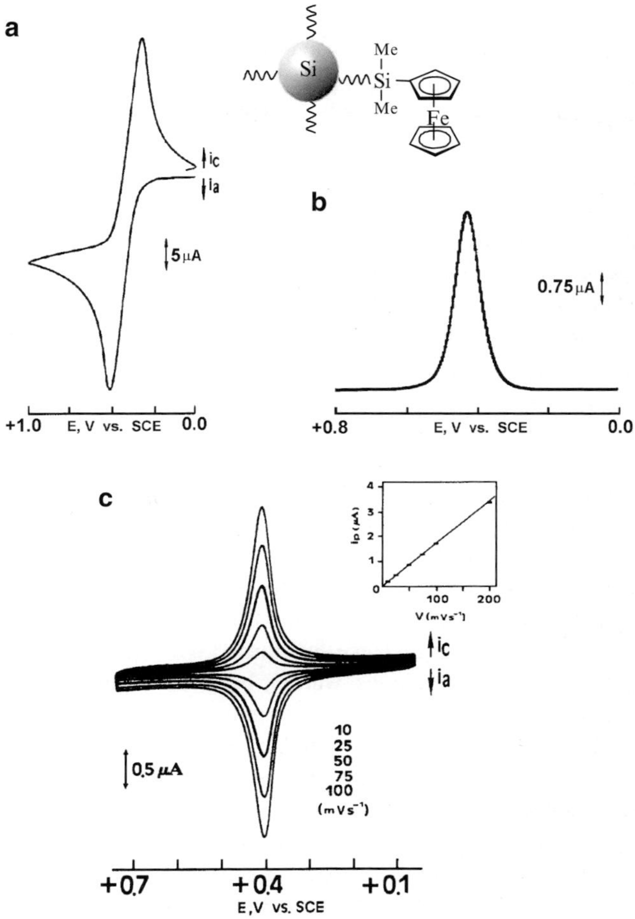

Fig. 8.4 (**a**) Cyclic voltammogram measured in CH_2Cl_2/0.1 M n-Bu_4NPF_6, at a Pt disk-electrode, of dendrimer **5**; (**b**) differential pulse voltammogram of **5** in CH_2Cl_2 solution; (**c**) cyclic voltammograms of a platinum disk-electrode modified with dendrimer **5** measured in 0.1 M n-Bu_4NPF_6/CH_2Cl_2. Inset: plot of peak current versus sweep rate (Adapted from [45]).

representative example. A well defined, symmetrical oxidation-reduction wave is observed, which is characteristic of surface-immobilized reversible redox couples, with the expected linear relationship of peak current with potential sweep rate v (see Fig. 8.4c inset) [68]. One remarkable feature of electrodes modified with films of these silicon-based ferrocenyl dendrimers is that they are very stable and reproducible. Indeed, cyclic voltammetric scans can be carried out in organic electrolyte solutions hundreds of times with no loss of electroactivity [45]. In fact, electroactivity of the modified electrodes is retained even after storage in air several weeks after preparation.

This is an important observation since many of anticipated electrochemical applications of modified electrodes will require extensive redox cycling.

Interestingly, the ability to prepare well-defined and stable ferrocenyl-containing dendrimer-modified electrode surfaces by electrodeposition, depends strongly on the chemical composition of the silicon-based skeleton. For instance, electrochemical studies indicated that, in contrast to the results discussed above, for the family of ferrocenyl dendrimers **7–9**, built from tetramethylcyclotetrasiloxane as a core, oxidation and reduction did not affect the solubility of the dendrimers, so that no oxidized dendritic material deposited on the electrode surface [50]. This feature can be attributed to the nature of the cyclosiloxane core which can modify the oxidized peripheral ferrocenes/solvent interaction in a manner that keeps the dendrimer in the solution.

8.2.1.3 Silicon-Based Dendrimers Functionalized with Carbonyl-Containing Organometallic Moieties

Stimulated by the results described above for the ferrocenyl silicon-based dendrimers, we attempted new synthetic strategies to build different families of organometallic dendritic macromolecules. For instance, we turned our attention to dendrimers in which tricarbonylchromium moieties were π-bonded to peripheral arene rings, directly attached to silicon atoms. Research into the synthesis and reactivity of $(\eta^6\text{-arene})Cr(CO)_3$ compounds is a significant and active area in organometallic chemistry because these compounds are key intermediates in a number of important stoichiometric and catalytic transformations.

The starting carbosilane dendrimers **21** and **22** functionalized at the periphery with aromatic rings shown in Reaction Scheme 8.3 were prepared using a divergent approach which starts from tetraallylsilane as a four-directional core [69]. Subsequent incorporation of $Cr(CO)_3$ moieties involves thermal replacement of CO from chromium hexacarbonyl by the peripheral dendrimer phenyl groups, in a donor solvent mixture of dibutylether/THF (9/1) at 140°C. This method was successful for the preparation of the first generation tetrametallic dendrimer **23** but failed for the functionalization of eight arms of the phenyl-terminated dendrimer **22** affording only the partially metalated organosilicon dendrimer **24** as the major reaction product. The dendrimer functionalized with eight $(\eta^6\text{-}C_6H_5)Cr(CO)_3$ peripheral units was not formed, and we believe that steric reasons were solely responsible for this. Cyclic voltammetric studies of the obtained $Cr(CO)_3$-containing metallodendrimers in a non-nucleophilic medium such as CH_2Cl_2/0.1 M n-Bu_4NPF_6, showed one single reversible oxidation wave at about $E_{1/2}$ = + 0.8 V versus SCE, which formally corresponds to oxidation of the zero valent chromium centers to the chromium(I) oxidation state. Thus, this indicates that the peripheral tricarbonyl chromium moieties are independent, non-interacting redox centers that are oxidized at the same potential.

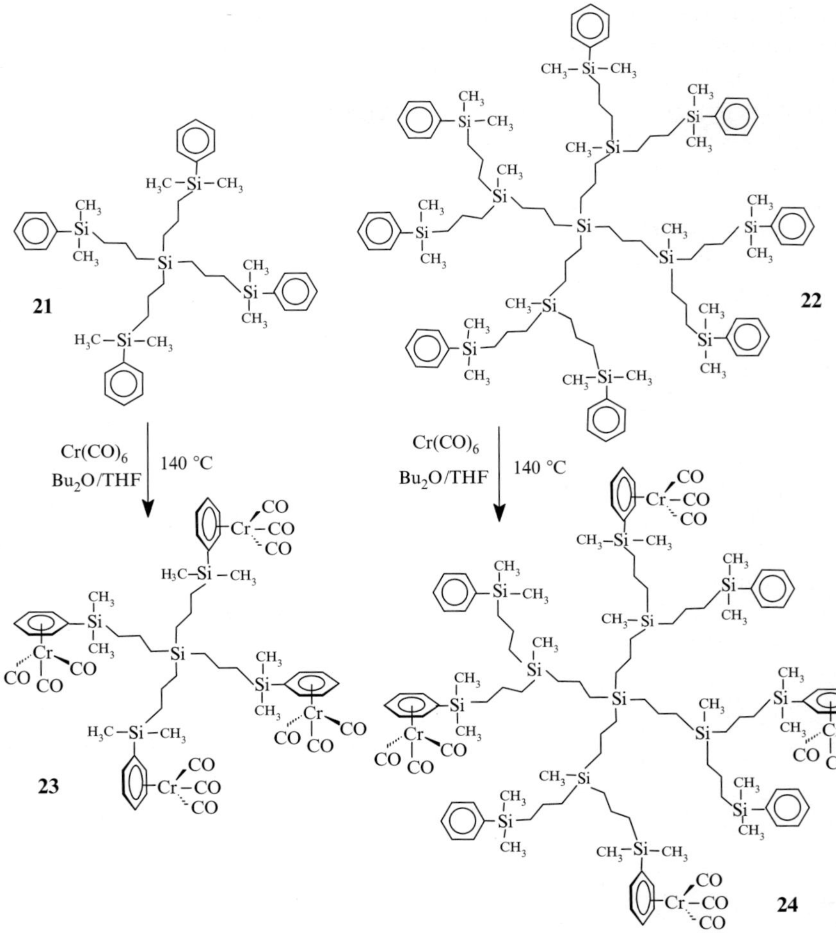

(8.3)

Cyclopentadiene is well known as one of the most important and widely used ligands in organometallic chemistry, and cyclopentadienyl derivatives, σ- or π-bonded to metal atoms, are known for all transition metals. Consequently, we attempted to incorporate this ligand into well defined carbosilane-based dendritic structures, because it offers enhanced variability in the designed generation of novel families of organometallic molecules. To pursue this goal, reaction of the sodium cyclopentadienide anion with a tetrafunctional Si–Cl dendrimer in THF was performed, affording the desired cyclopentadienyl-functionalized organosilicon dendrimer **25**, as shown in Reaction Scheme 8.4 [70]. The coordinating ability of these surface-located cyclopentadienyl ligands was assessed by the reaction with octacarbonyldicobalt where treatment of **25** with $Co_2(CO)_8$ in CH_2Cl_2 at reflux temperature, and in the presence of 1,3-cyclohexadiene, afforded the desired derivative **26**. In support of the η^5-coordination of cyclopentadienyl ligands to the $Co(CO)_2$ moieties, the characteristic transformations of the cyclopentadienyl ring resonances in both ^1H and ^{13}C NMR spectra, as well as the presence of two typical $\nu(C\equiv O)$ bands in the IR spectrum of **26** were observed, in accord with the proposed structure.

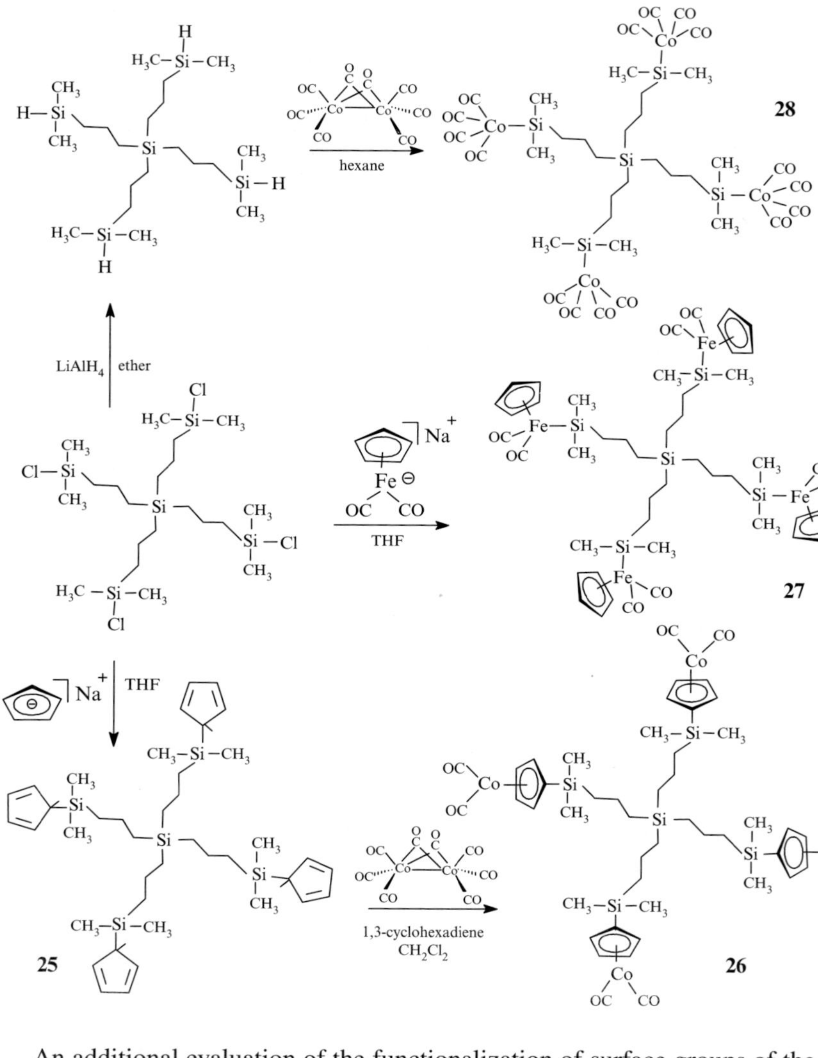

28

27

25

26

(8.4)

An additional evaluation of the functionalization of surface groups of the Si–Cl terminated carbosilane dendrimers was effected by the treatment with a THF solution of the carbonyl anion, Na[η^5-$C_5H_5Fe(CO)_2$] (prepared from [(η^5-C_5H_5) $Fe(CO)_2]_2$ and Hg/Na amalgam), to afford tetranuclear dendrimer **27**, which contained silicon-iron σ-bonds at the periphery (see Reaction Scheme 8.4). Similarly, the reactivity of octacarbonyldicobalt toward Si–H functionalized organosilicon derivatives afforded a novel tetrametallic dendrimer **28**, which bears organometallic moieties directly attached to the dendritic scaffold through silicon-metal σ-bonds (see also Reaction Scheme 8.4) [70].

Concerning the cobalt carbonyl-containing carbosilane dendrimers, note that Seyferth and his co-workers prepared early examples of this type of molecule [71]. The reaction of carbosilane dendrimers bearing peripheral ethynyl substituents with octacarbonyldicobalt, afforded the first and second generation dendrimers functionalized with acetylenedicobalt hexacarbonyl units, such as dendrimer **29**. The structure of the first generation cobalt-containing dendrimer $Si[CH_2CH_2SiMe_2C_2HCo_2(CO)_6]_4$ (**30**) was determined by X-ray diffraction studies. Later reports by Kim and Jung utilized closely related terminal phenylethynyl groups as the binding units for cobalt carbonyl-containing fragments, but, unlike Seyferth and co-workers, these authors reacted $Co_2(CO)_8$ with silicon-based dendrimers possessing two phenylethynyl functionalities on each silicon atom in the external dendritic surface [72]. Lang and co-workers were also successful in binding hexacarbonyldicobalt-containing moieties to ethynyl-terminated dendrimers having carbosiloxanes scaffolds [73].

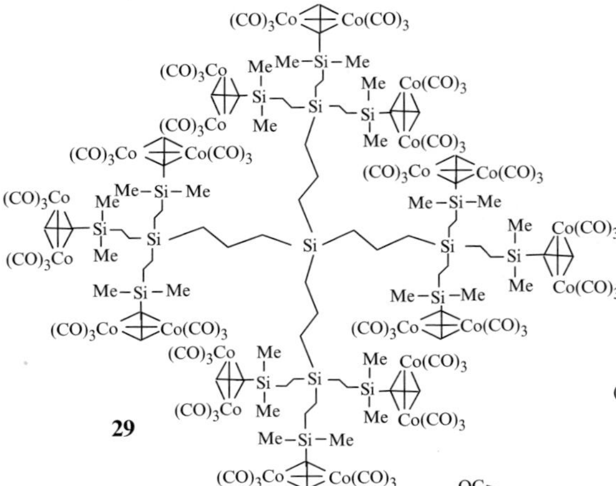

29

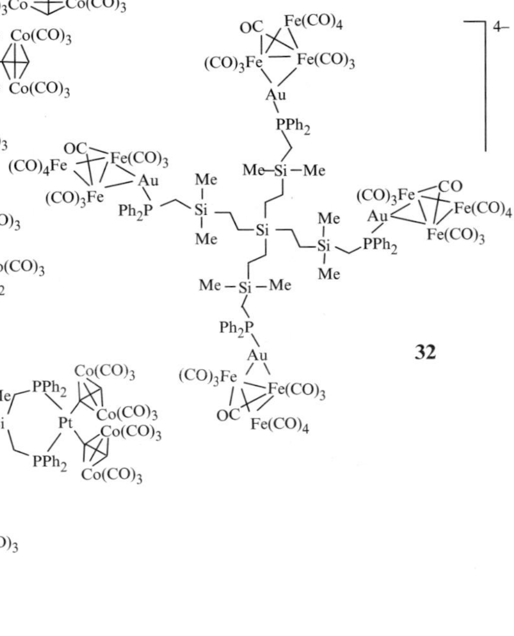

32

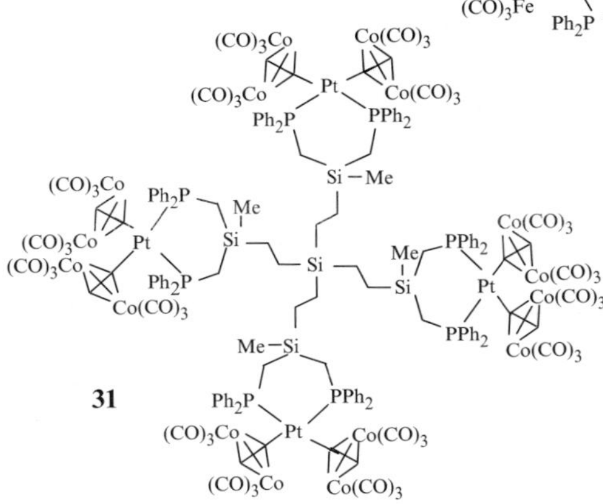

31

Interesting related examples of exterior functionalization of carbosilane dendrimers with $Co_2(CO)_6$ units have been reported by Rossell and coworkers [74]. In this case, carbosilane dendrimers containing peripheral platinum acetylide units were prepared first by the reaction of $[PtCl_2(C{\equiv}CPh)_2]$ with carbosilane dendrimers functionalized with peripheral PPh_2 ligands. The resulting dendrimers with $Pt(C{\equiv}CPh)_2$ end-groups are noteworthy because of their potential use as precursors for cluster-containing carbosilane dendrimers. For example, reaction of the platinum acetylide-functionalized dendrimers with $Co_2(CO)_8$ leads to heterometallic dendrimers with terminal $\{C_2Co_2(CO)_6\}$ cluster sites, such as the dendritic carbosilane **31**. The same authors have reported numerous interesting studies on carbosilane dendrimers functionalized with organometallic units [75]. For instance, remarkable heterometallic dendrimers terminated with iron-gold carbonyl-containing clusters, such as **32** were also prepared [75a]. Unfortunately, none of these systems has been characterized electrochemically.

Recently, we used hydrosilylation reactions between vinyl-functionalized carbonyl-containing organometallic fragments and Si–H functionalized scaffolds (including carbosilane dendrimers as well as cyclic and polymeric siloxanes) to prepare a series of novel multimetallic systems containing silicon-linked cyclopentadienyl dicarbonyl iron moieties [76]. For this purpose the silyliron organometallic fragment $(\eta^5\text{-}C_5H_5)Fe(CO)_2Si(CH_3)_2CH{=}CH_2$ (**33**), in which a silicon atom is directly bonded to a vinyl reactive group, was prepared by a salt elimination reaction between $Na[(\eta^5\text{-}C_5H_5)Fe(CO)_2]$ and dimethylvinylchlorosilane (see Reaction Scheme 8.5). Hydrosilylation of the monometallic **33** with appropriate Si–H functionalized cyclotetrasiloxane and dendritic carbosilane in the presence of Karstedt's catalyst afforded novel silyl carbonyl iron-functionalized tetrametallic molecules **34** and **35** in which organometallic units are attached to the silicon-containing scaffold through a two-methylene units flexible spacer.

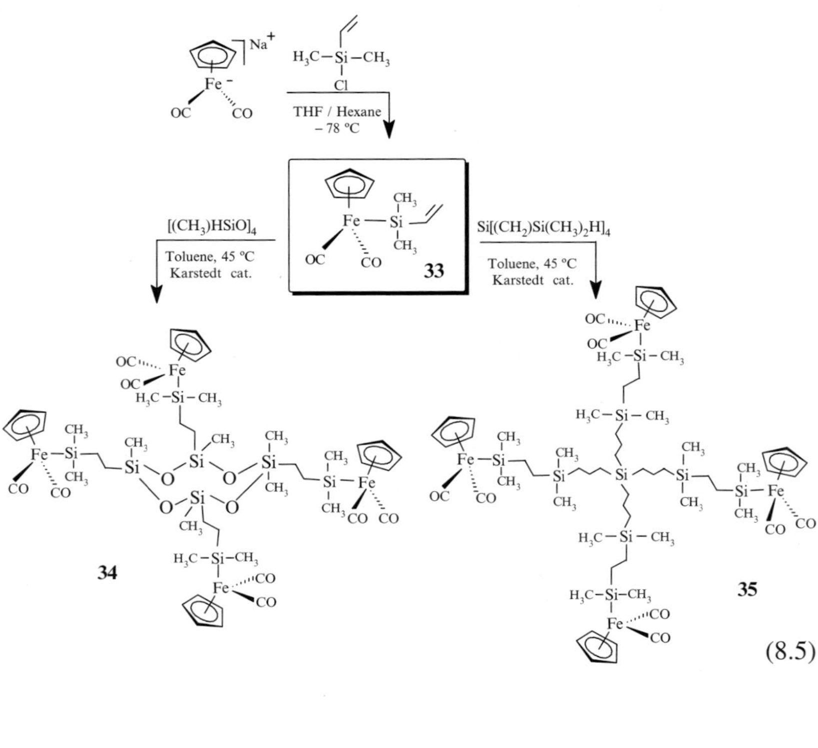

(8.5)

Electrochemical oxidation of compounds **33–35** was examined in dichloromethane, THF and acetonitrile solutions, containing n-Bu_4NPF_6 as supporting electrolyte. The cyclic voltammograms showed an anodic and cathodic peak, which were not associated with a reversible oxidation–reduction process. The separation of these peak potentials was very large (about 1.5 V) and it was found to depend on the scan rate. Likewise, as the sweep rate was increased, the reduction peak shifted in a negative direction, while the oxidation peaks shifted in a positive direction. These shifts in E_p and i_{pc}/i_{pa} values (much smaller than unity), clearly indicated that electrochemical oxidation of the iron carbonyl-containing compounds **33–35** was followed by a rapid chemical reaction (EC mechanism). Consequently, these compounds were oxidized to give the electron-deficient dicarbonyl-iron(III) species $[(\eta^5\text{-}C_5H_5)Fe(CO)_2Si(CH_3)_2CH{=}CH_2]^+$ ($[\mathbf{33}]^+$), $[\mathbf{34}]^{4+}$ and $[\mathbf{35}]^{4+}$. This oxidation process was followed by elimination of one CO ligand from organometallic moieties owing to a decreasing π-back bonding from electrogenerated Fe^{III} centers to the CO ligands. The elimination of CO was assisted by the nucleophilic attack of the solvent, yielding solvent adducts of the electron-deficient iron(III) species, such as $[(\eta^5\text{-}C_5H_5)Fe(CO)(\text{solvent})Si(CH_3)_2CH{=}CH_2]^+$ (for **33**), which were then reduced to yield neutral Fe^{II} monocarbonyl species, such as $[(\eta^5\text{-}C_5H_5)Fe(CO)(\text{solvent})Si(CH_3)_2CH{=}CH_2]$ (for **33**) (solvent = CH_2Cl_2, THF or CH_3CN) [76].

8.2.2 Silicon-Based Dendrimers from Organometallic Moieties

Traditionally, two complementary general synthetic growth strategies have been used for the construction of dendrimers: the divergent method, arising from the seminal work of Tomalia and Newkome [7, 8], and the convergent method, first reported by Hawker and Fréchet [77, 78] (see Chapter 1). Both strategies yield high generation dendrimers through repetitive activation and growth (i.e., protection/deprotection) steps.

In a similar way, organo-transition metal dendrimers can be constructed by preparing suitable organometallic functionalities, from which the build-up of the dendrimer structure is accomplished following either a divergent or convergent strategy. For example, a divergent approach to organometallic dendrimers involves growth from a polyfunctional organometallic core, where branching outward is accomplished via an increasing number of terminal branch transformations. Without doubt, the main disadvantage of such an approach is that a progressively larger number of organometallic monomers has to react successfully with the reactive functional groups of the precursor dendrimer surface (see also Chapter 1). As a consequence, and because of steric hindrance imposed by the sterically demanding organometallic units, it becomes extremely difficult at higher generation numbers to complete the reaction of all external functionalities of the reacting dendrimer. For this reason, a critical aspect in the divergent strategy is the choice of a suitable organometallic core, in which an appropriate number of reactive sites are conveniently placed, in order to assure that complete functionalization of the branches is accomplished easily, and in relatively high yields.

In the alternative convergent approach, the construction of organometallic dendrimers starts at what will ultimately become the outer surface of the dendrimer, and progresses inwardly. This approach first requires the synthesis of functionalized, progressively larger, organometallic dendritic wedges or dendrons which are then attached to a polyfunctional core to yield the multi-dendron organometallic dendrimer. A key requirement of the convergent approach is that the organometallic moiety be stable to the reaction conditions of the stepwise synthesis, since it is introduced at the beginning of the dendrimer preparation. The advantage of using the convergent approach, however, is that it allows precise control over the number and placement of functional organometallic groups (see also Chapter 1). The reactions are carried out in a stepwise fashion and the organometallics-containing dendrons can be isolated, purified and characterized by standard analytical and spectroscopic methods after each step. Since the products are generally highly soluble in spite of the large number of organometallic moieties present, the reactions can be monitored by multinuclear NMR spectroscopy, so that defects in branching can be detected prior to complexation with the molecule selected for the core. An additional attractive advantage of the convergent construction is that segment-block or layer-block organo-transition heterometallic dendritic macromolecules can be constructed. As is shown below, both divergent and convergent strategies have proved useful for preparing well-defined electroactive silicon-based organometallic dendrimers.

8.2.2.1 Dendrimers with Silicon-Linked Ferrocenyl Moieties Constructed via Convergent Approach

One of the richest areas of organometallic chemistry, which has grown enormously in the last few years, involves the synthesis and study of soluble molecular systems containing linked metallocenes [62, 63, 79, 80]. When two (or more) metallocene fragments are connected through a bridging system to obtain linked polymetallocenes, intermetallic electronic communication between the redox centers affords a wide range of potential new applications, and enables studies concerning intramolecular electron-exchange reactions. Interested in this subject, we focused on the construction of dendrimers possessing electroactive organometallic units linked together in close proximity to one another in order to enable electronic communication between the metal sites. This was a challenging synthetic goal which could, however, provide an access to new multimetallic dendrimers with appreciable electron mobility and consequently, with interesting electrical, redox and magnetic properties.

We utilized the convergent growth methodology to prepare dendrons and dendrimers based on silicon atoms as key components of the dendritic structures that possessed ferrocenyl units which could electronically communicate [47, 59]. Organometallic dendrons containing a carbon-carbon double bond at the focal point were constructed first, and then used in hydrosilylation chemistry to provide either higher-generation dendrons or final dendrimer molecules. The general procedure for this convergent synthesis, in which the ferrocenyl units are linked by a silicon bridge, is shown in Reaction Scheme 8.6. The key dendron was the silicon-bridged biferrocene **36**, which was prepared by the salt elimination reaction of ferrocenyllithium with vinylmethyldichlorosilane. A critical requirement for the first reaction step was to use extremely pure monofunctionalized lithioferrocene and to avoid the presence of any traces of dimetalated ferrocene species, $Fe(\eta^5-C_5H_4Li)_2$, which could produce undesirable and inseparable mixtures of dendritic products. For this purpose (tri-*n*-butylstannyl)ferrocene, $\{\eta^5-C_5H_4Sn(n-Bu_3)\}Fe(\eta^5-C_5H_5)$, was selected as the starting material since it had proved to be an excellent precursor for pure monolithioferrocene [81]. The transmetallation reaction of $\{\eta^5-C_5H_4Sn(n-Bu_3)\}Fe(\eta^5-C_5H_5)$ with *n*-buthyllithium was performed cleanly in THF at – 78°C and it gave pure monolithioferrocene, which was subsequently reacted with vinylmethyldichlorosilane to afford the targeted biferrocene **36**. The convergent growth of the organometallic dendron **36** was achieved by hydrosilylation reaction with phenylchlorosilane, in toluene and in the presence of catalytic amounts of Karstedt's catalyst, resulting in a new dendritic fragment **37** which contains a reactive chlorosilane functionality available for an ensuing alkenylation step. Treatment of its terminal Si–Cl moiety with allylmagnesium bromide in diethyl ether afforded the desired reactive dendron **38**, carrying four ferrocenyl units.

The availability of free olefinic substituents at the focal point of dendrons **36** and **38** facilitated incorporation of interacting organometallic redox centers in dendritic

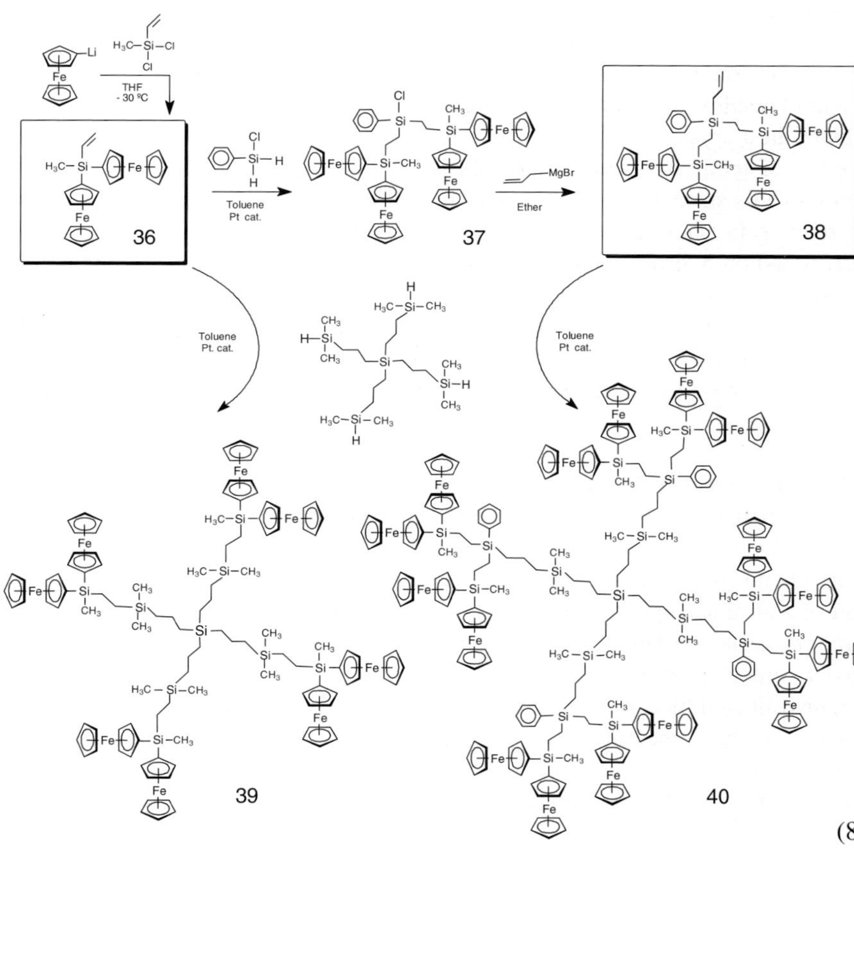

(8.6)

structures through hydrosilylation anchoring reactions with Si–H polyfunctional cores. The four-functional carbosilane, $Si[(CH_2)_3Si(Me)_2H]_4$, in which the reactive Si–H groups are located at the end of quite long silicon-containing chains, was selected as a core molecule, in order to prevent the possibility of steric congestion and provide the targeted dendrimers with a four-functional core (see Reaction Scheme 8.6). Completion of the anchoring reactions was easily monitored by ^1H NMR and IR spectroscopies, and it was established that full reaction of all four Si–H functionalities in the carbosilane core was easily achieved at room temperature, in toluene solution with Karstedt's catalyst. The ^1H NMR spectrum of the hydrosilylated dendrimer product proved that only the β-isomer was formed under these mild reaction conditions and that no Markovnikov addition (which would lead to the α-isomer) took place [82]. This assured a regular dendritic growth and generation of molecules of maximum symmetry, and in this way, we prepared dendrimers **39** and **40** (Reaction Scheme 8.6), containing on the external carbosilane surface 8 and 16 ferrocenyl moieties, respectively, linked together in pairs through a bridging silicon atom [47]. These were the first examples of organometallic dendritic molecules displaying significant metal-metal interactions in the dendritic structure.

All members of this family of organometallic dendrimers are stable to air and moisture and possess excellent solubilities in common organic solvents, an important characteristic which makes additional structural and electrochemical characterization quite straightforward. We were unable to obtain good single crystals of dendrimers **39** and **40**, although both dendrons **36** and **38** were successfully obtained in a crystalline form suitable for X-ray studies [47, 59]. The molecular structure of tetraferrocenyl dendron **38** is shown in Fig. 8.5 [83]. A remarkable feature of this structure is that silicon-linked ferrocenyl groups are oriented at ca. 90° relative to one another. The described convergent organometallic synthetic route, based on hydrosilylation and nucleophilic substitution steps, starting from a monofunctional vinyl organometallic moiety, represents a versatile method for accessing a potentially large number of multiredox dendritic macromolecules with organometallic moieties of different chemical nature, such as heterometallic dendrimers whose synthesis is described below.

8.2.2.2 Electrochemical Behavior of Silicon-Containing Dendrimers with Electronically Communicating Ferrocenyl Units

Electrochemical techniques have been among the most widely used tools to investigate metal-metal interactions in polymetallocene systems [79, 80]. Hence, the redox properties of dendritic molecules **36–40**, in which the peripheral ferrocenyl units are linked together in close proximity to one another by a silicon bridge, have been investigated by cyclic voltammetry and differential pulse voltammetry.

A key feature concerning the redox behaviour of dendritic molecules **36–40** is the presence of two successive, well separated, oxidation processes. Thus, for dendrons **36** and **38**, two reversible oxidation waves of equal intensity were observed in cyclic voltammograms in CH_2Cl_2 solution (see for example Fig. 8.6a), while differential pulse voltammetry (DPV) measurements showed two separated

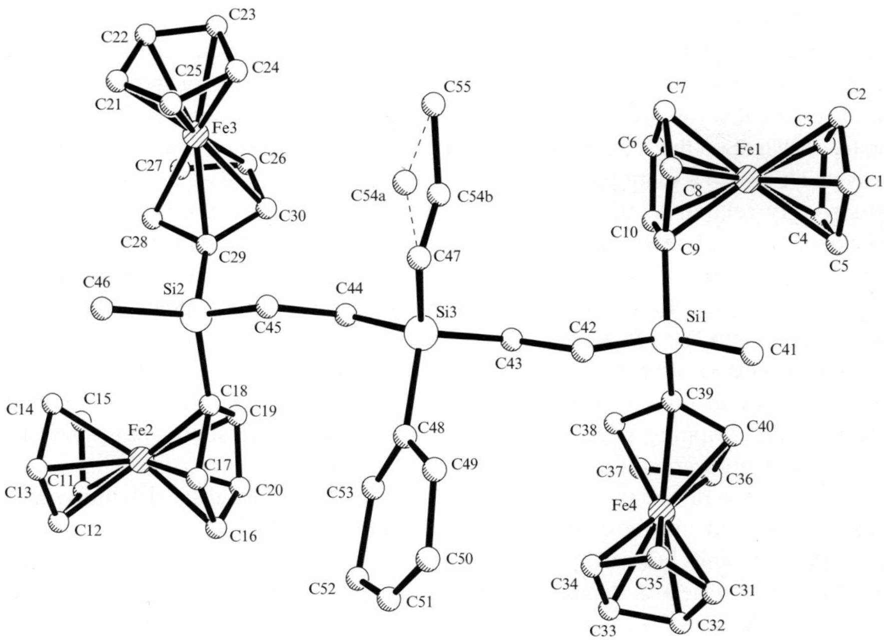

Fig. 8.5 Molecular structure of the tetrametallic dendritic fragment **38**. Hydrogen atoms have been omitted for clarity.

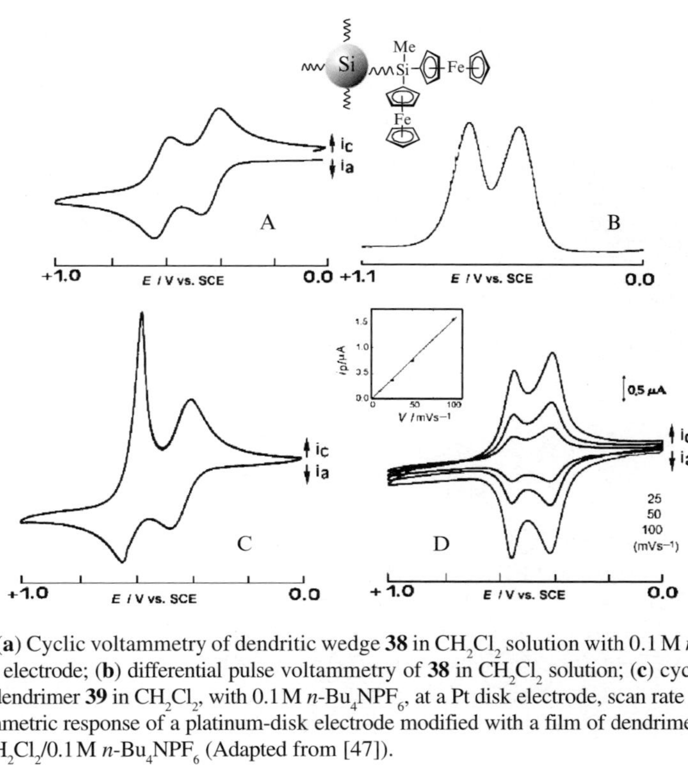

Fig. 8.6 (**a**) Cyclic voltammetry of dendritic wedge **38** in CH_2Cl_2 solution with 0.1 M *n*-Bu_4NPF_6, at Pt disk electrode; (**b**) differential pulse voltammetry of **38** in CH_2Cl_2 solution; (**c**) cyclic voltammetry of dendrimer **39** in CH_2Cl_2, with 0.1 M *n*-Bu_4NPF_6, at a Pt disk electrode, scan rate 100 mV s^{-1}; (**d**) voltammetric response of a platinum-disk electrode modified with a film of dendrimer **39**, measured in CH_2Cl_2/0.1 M *n*-Bu_4NPF_6 (Adapted from [47]).

oxidation waves of the same area (Fig. 8.6b). However, the voltammetric responses of polyferrocenyl dendrimers **39** and **40** in the same solvent were clearly different, as shown in Fig. 8.6c. For these dendrimers with a high loading of ferrocenyl units at the external surface, a change in solubility accompanied the change in oxidation state, so that upon the scan reversal after the second oxidation process, the reduction wave gave rise to a sharp peak for the reduction process. The sharpness of this peak is strongly related to the stripping nature of the electrochemical process, as dendritic material deposited onto the electrode surface is redissolved back into the solution as a result of the electron-transfer reaction. Therefore, the full oxidation of dendrimers **39** and **40** results in the precipitation of polycationic dendrimers $\mathbf{[39]}^{8+}$ and $\mathbf{[40]}^{16+}$ onto the electrode surface, and on the reverse scan, the dendrimers redissolve as they are reduced. However, if a small amount of CH_3CN is added to the electrolyte medium (for instance, in a system CH_2Cl_2/CH_3CN, 5:1 by volume), the cathodic stripping peak disappears, and the cyclic voltammograms become similar to that observed for dendritic fragments **36** and **38**.

The presence of two well separated oxidations processes for these dendritic molecules is consistent with the existence of significant metal-metal interactions between the two ferrocenyl units bridged by a silicon atom. The first oxidation of molecules **36** and **38–40** occurs in neutral dendrons at non-adjacent ferrocene sites (see for example, Reaction Scheme 8.7) which makes the subsequent removal of electrons from the remaining peripheral ferrocenyl centers, neighboring those already oxidized, more difficult. Consequently, the second oxidation process occurs at a higher potential.

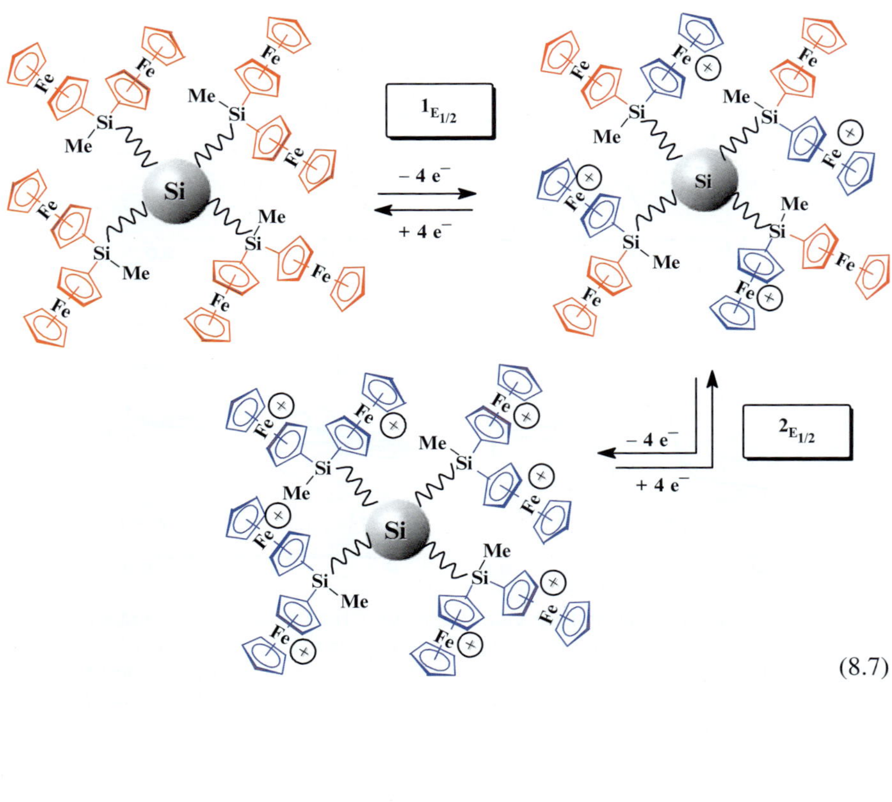

(8.7)

The observation of two oxidation processes in these ferrocenyl silicon-based dendritic molecules with linked ferrocenyl units, indicates a stabilization of the mixed valence species. Indeed, it is known that in linked metallocene systems the difference in the redox potentials observed for the two oxidation waves ($\Delta E = {}^2E_{1/2} - {}^1E_{1/2}$) provides information about the degree of electronic interaction between the metal centers [79, 80, 84], and this was used to calculate the comproportionation constant, $K_{com,}$ relative to the equilibrium shown in the following equations:

$$(\mathrm{Fe^{II}} - \mathrm{Fe^{II}}) + (\mathrm{Fe^{III}} - \mathrm{Fe^{III}}) \overset{K_{com}}{\rightleftarrows} 2(\mathrm{Fe^{II}} - \mathrm{Fe^{III}}) \qquad \text{(E.8.1)}$$

$$\Delta E = {}^2E_{1/2}(\mathrm{Fe^{III}} - \mathrm{Fe^{III}}) - {}^1E_{1/2}(\mathrm{Fe^{II}} - \mathrm{Fe^{III}}) \qquad \text{(E.8.2)}$$

$$K_{com} = \exp(F\Delta E / RT) = \exp(\Delta E/25.69) \text{ at } 25°\mathrm{C} \qquad \text{(E.8.3)}$$

From the wave splitting (ΔE) obtained for dendritic molecules **36** and **38–40**, which varies from 190 to 160 mV, the comproportionation constant K_{com} among three oxidation states of the two iron atoms in the dendrons was calculated according to Equation (E.8.3). Values of K_{com} = 1,630 to 507 were obtained, which indicates that the partially oxidized dendritic molecules **36**, **38**, **39** and **40** can be classified as the class II mixed-valence species, according to the Robin-Day classification, which is based on the degree of electronic delocalization in mixed-valence compounds [84]. These K_{com} values provide an approximate estimation of the electronic communication between the redox centers in dendrimers, and point out that, in the partially oxidized dendritic species, the charge is slightly delocalized between the two iron centers of the silicon-linked ferrocenyl units [85]. This was the first example of unambiguous electronic coupling of metal centers in a dendritic structure.

On the other hand, the tendency of dendrimers **39** and **40** to electrodeposit onto the electrode surfaces, observed in the above described voltammetric studies, allows the preparation of dendrimer-modified electrode surfaces. Consequently, dendrimers **39** and **40** were electrodeposited in their oxidized form onto platinum disk electrodes from degassed solutions in CH_2Cl_2 with 0.1 M n-Bu_4NPF_6, by maintaining the working electrode potential at about +0.70 V versus SCE. The voltammetric response of an electrodeposited film of octanuclear dendrimer **39**, studied in clean CH_2Cl_2 solution containing only supporting electrolyte, is shown in Fig. 8.6d. Two successive well-defined, reversible, oxidation-reduction waves were observed, with formal potential values of ${}^1E_{1/2} = 0.40$ and ${}^2E_{1/2} = 0.55$ V versus SCE, which are close to those observed for these dendrimers in solution. A linear relationship between peak current and potential sweep rate v was observed, and the potential difference between the cathodic and anodic peak was smaller than 10 mV at scan rates of 0.1 V s^{-1} or less. These voltammetric features unequivocally indicate the surface-confined nature of the electroactive dendrimer film [68]. As a consequence of strong interactions between the ferrocenyl units in these multiferrocenyl-functionalized dendrimer films, the first redox wave is broader than the second one. The surface coverage, Γ, of electroactive ferrocenyl moieties in dendrimer films on the electrode was determined by integration of the anodic wave in the cyclic voltammograms, and for the ferrocenyl dendritic film shown in Fig. 8.6d it was found to be 1.65×10^{-10} mol cm^{-2}. This was the first example of an electrode surface modified with redox-active dendrimer films possessing a controlled number of interacting metal centers.

8.2.2.3 Electroactive Heterometallic Silicon-Containing Dendrimers

As pointed out in Section 8.1, introduction of a controlled number of chemically different organometallic centers into predetermined sites of a dendritic structure is a challenging task. At the same time, the presence of two, or more, different metal centers within the same dendritic molecule can profoundly affect both physical properties and reactivity of the dendritic system. It is thus possible to construct heterometallic dendrimers capable of performing complex functions resulting from the integration of specific properties of their dissimilar constituent organometallic moieties. This may result in the development of a variety of novel and improved characteristics which can not be found in homopolymetallic dendritic molecules.

Of particular interest are metallodendrimers having two different redox-active organometallic units, since each of the chemically different organometallic moieties brings to the dendritic molecule its own ability to undergo redox processes at a certain potential. In addition, electroactive organometallic units of different types can be located in topologically equivalent or not equivalent sites of the dendritic structure to enable tailoring of redox patterns. Therefore, these heterometallodendrimers can be used to perform valuable functions such as reversible exchange (storage and release) of a large and predetermined number of electrons at different fixed potentials. Furthermore, such heterometallic dendrimers offer a unique opportunity to explore the dependence of electronic cooperativity on redox asymmetry and of chemical reactivity on the presence of a second, different redox center in the same metallodendritic molecule.

Motivated by these challenges, we targeted metallodendrimers containing electroactive organometallic fragments of different chemical nature. Our attention was first focused on the synthesis of heterogeneous metallodendrimers containing neutral ferrocene and cationic cobaltocenium units, the most stable organometallic redox systems, located simultaneously at the surface of a dendritic structure [51]. The preparation of such heterometallic dendrimers was accomplished by a combinatorial approach, in which the first four generations of commercially available amine-terminated diaminobutane-core poly(propylene imine) (PPI) dendrimers were reacted with an equimolar mixture of chlorocarbonyl ferrocene and the salt of chlorocarbonylcobaltocenium in CH_2Cl_2/CH_3CN (1:1) to yield a series of novel dendrimers terminated with 4, 8, 16 and up to 32 metallocene units. A representative member of this family of heterometallic dendrimers is molecule **41** shown in Fig. 8.7.

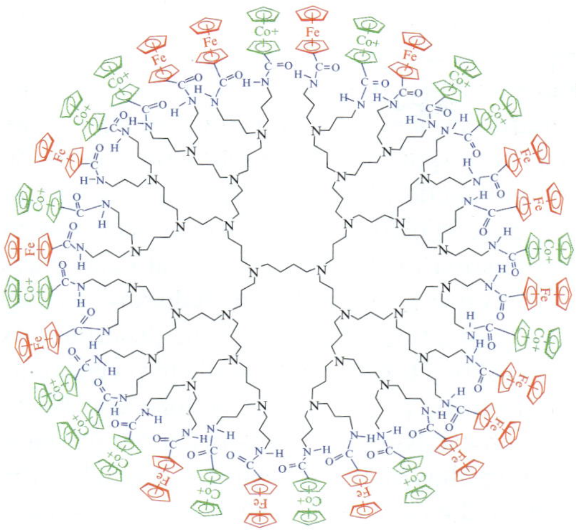

41

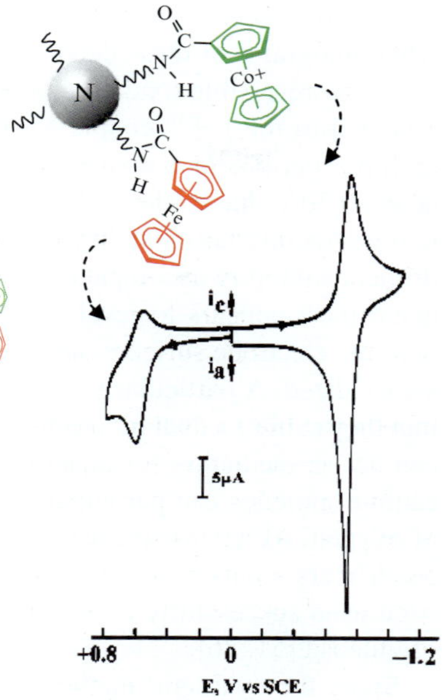

Fig. 8.7 Cyclic voltammogram recorded in acetonitrile solution at a glassy carbon electrode of the diaminobutane-based poly(propylene imine) heterometallic dendrimer **41** illustrated in the figure, containing both neutral ferrocene and cationic cobaltocenium units at the periphery (Adapted from [51]).

Ideally, every one of these metallodendritic molecules should have the same number of ferrocene and cobaltocenium units on the exterior surface. However, the experimental results showed that, as expected in accordance with the synthetic method we employed, several fractions with different ratios of the organometallic end-groups were isolated from a single reaction mixture. The mixed ferrocene-cobaltocenium dendrimers were characterized by NMR spectra, electrospray ionization mass spectrometry (ESI-MS) and total X-ray fluorescence (TXRF) analyses. The presence of neutral ferrocene and cationic cobaltocenium units resulted in an interesting solubility behaviour of these heteromultimetallic dendrimers, which could be modified by varying the ratio of the organometallic moieties on the periphery. For instance, dendrimers with a higher loading of cobaltocenium moieties were soluble in H_2O and insoluble in CH_2Cl_2, while for the structure loaded with neutral ferrocene units, solubility in water was decreased but solubility in CH_2Cl_2 was enhanced.

From the electrochemical point of view, these peripherally heterogeneous dendrimers are of special interest since they combine the reversible, one-electron electrochemical oxidation of ferrocene units to produce positively charged ferrocenium forms, with the fast, one-electron reversible reduction of cobaltocenium moieties to generate neutral cobaltocene. Indeed, as shown in Fig. 8.7, the cyclic voltammograms of these dendrimers exhibited two waves at very different potentials: one reversible oxidation at approximately +0.60 V (versus SCE) which was assigned to the $Fe^{II/III}$ couple for the simultaneous one-electron oxidation of the multiple peripheral noninteracting ferrocene and a single sharp reduction peak near −0.70 V due to the $Co^{III/II}$ couple corresponding to the redox process of the non-interacting terminal cobaltocenium moieties. Noteworthily, a change in dendrimers solubility accompanied the change in oxidation state, so that reduction of dendrimers appears to result in precipitation of the reduced neutral dendrimers onto the electrode surfaces, and on the reverse scan, dendrimers redissolve as they are oxidized. A particularly valuable feature of these heterometallic dendrimers is that they exhibit a dual bioelectrochemical function. Namely, while ferrocene units can act as mediators for anaerobic enzymatic processes, the peripheral cobaltocenium moieties can participate in electrocatalytic processess under the presence of oxygen. As a consequence, conducting films of electrodeposited heterometallic dendrimers, containing both ferrocene and cobaltocenium peripheral units, have been used successfully in a double way for aerobic and anaerobic determination of glucose [51, 86].

Since the synthetic methodology used for the preparation of heterometallic dendrimers with both ferrocene and cobaltocenium units does not allow precise control over the number of different metallocene units grafted to the dendritic surface, we attempted a different strategy to achieve this goal. In it, we combined on carbosilane dendritic scaffolds typical strong electron-donating fragments, such as ferrocene, with the electron-withdrawing η^6-aryltricarbonylchromium $(\eta^6\text{-}C_6H_5)Cr(CO)_3$ [55, 87]. The strategy was based on the convergent procedure described above for the homometallic dendritic structures possessing organometallic units that electronically communicated between themselves. It used organometallic dendrons containing a C = C group at the focal point that had been constructed first and then attached to a Si–H polyfunctional core via hydrosilylation chemistry in the final step of the dendrimer construction.

At this point, it should be remembered that dendrimer synthesis is inherently labour intensive process, involving lengthy and difficult synthetic steps and requiring relatively large quantities of readily accessible starting materials (see Chapter 1). Because of this, the redox-active organometallic compounds for the synthesis of heterometallic dendrimers need to: (a) be easy to prepare, (b) have a single reactive C=C group appropriate for the selected convergent growth scheme, (c) be sufficiently stable to tolerate the reaction conditions required for carbosilane chemistry (including hydrosilylation conditions and inertness toward the Grignard reagents), and (d) carry a potential ligand binding site for attachment of the second (different) organotransition metal fragment. In addition, it must also be possible to incorporate the second transition metal without decomplexing the already attached organotransition metal fragment.

All these prerequisites are fulfilled in the ferrocenylmethylphenylvinylsilane molecule **42** shown in Reaction Scheme 8.8, which was, therefore, selected as a new redox-active starting point for a convergent synthesis based on hydrosilylation reactions. Compound **42** was easily accessible by the salt elimination reaction of ferrocenyllithium with methylphenylvinylchlorosilane in THF at –30°C, and was isolated in high purity and reasonable high yield, as an air-stable orange crystalline solid.

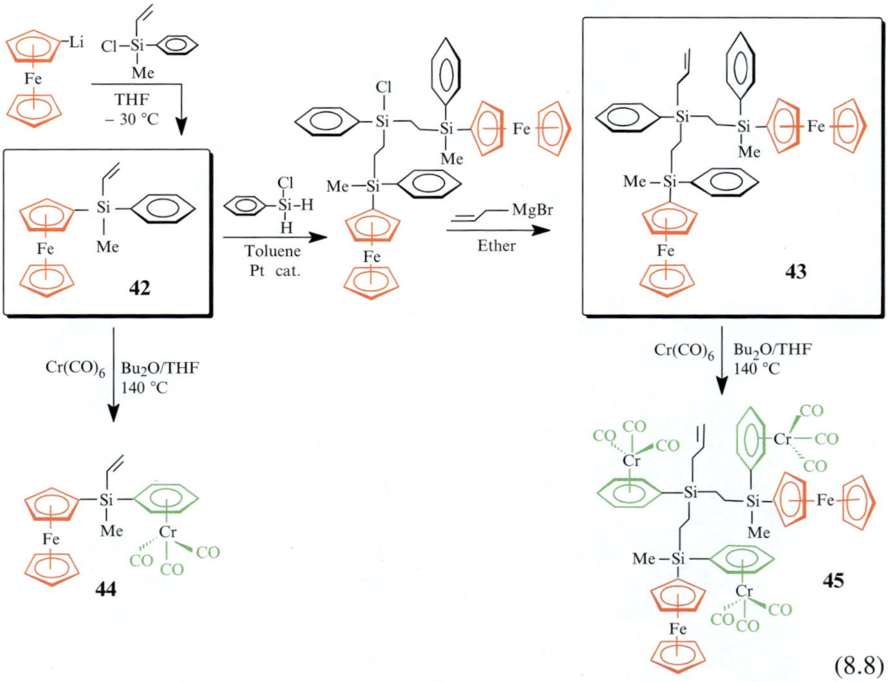

(8.8)

The convergent growth from **42** was achieved by hydrosilylation and alkenylation reaction steps, which led to the desired dendritic fragment **43**, carrying two ferrocenyl units linked to phenyl rings via a bridging silicon atom. The π-coordinating ability of the phenyl rings of **42** and **43** toward organometallic $Cr(CO)_3$ units allowed a simple synthetic access to the targeted heterobimetallic molecule **44** and heteropentametallic **45**, which were also isolated as crystalline solids. The molecular structures of vinyl-functionalized ferrocenylsilanes **42** and **44** were determined by single-crystal X-ray diffraction studies. For instance, the crystal structure of **44** showed that the $Cr(CO)_3$ group was disposed in a transoid configuration with respect to the ferrocenyl moiety (see Fig. 8.8 top), while the crystal structure of heterometallic **45** (see Fig. 8.8 bottom) showed association of similar organometallic moieties, a type of self-assembly of metals (Fe and Cr) which is apparently driving the packing into the crystal lattice. Thus, repeating chromium-rich and ferrocene-rich layers, can be clearly distinguished.

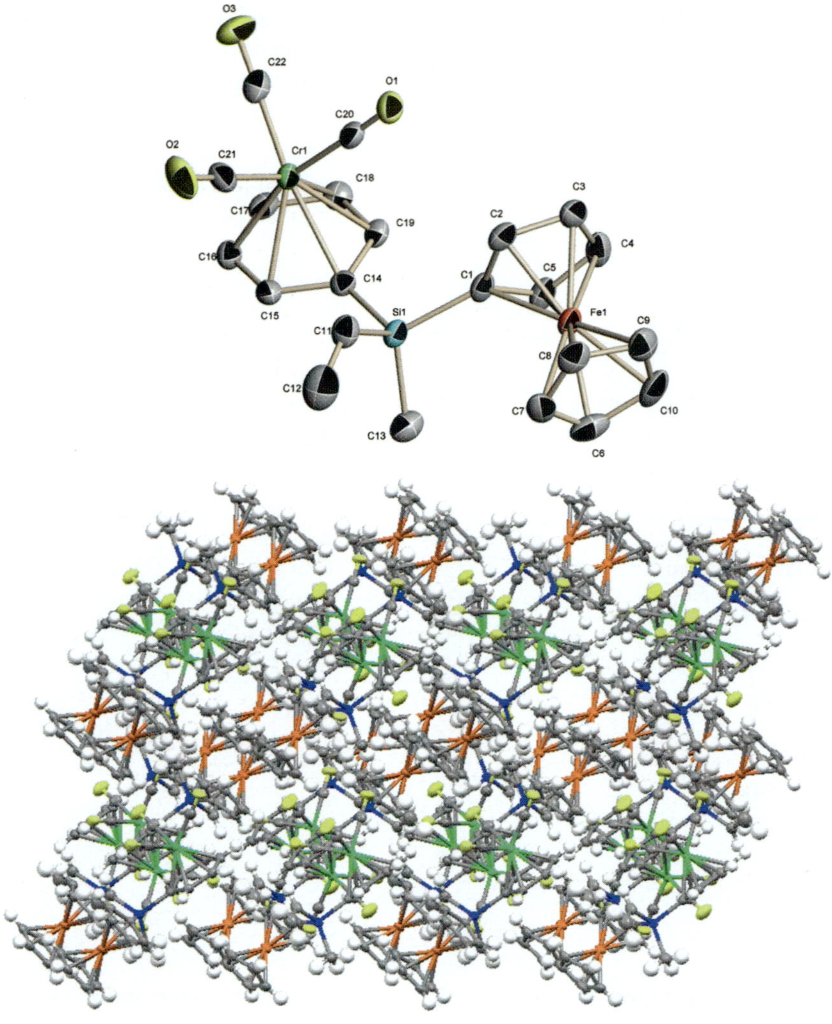

Fig. 8.8 (**a**) Molecular structure of heterometallic **44**; (**b**) crystal-packing diagram of **44** showing the iron-rich and chromium-rich pseudolayers (Adapted from [55]).

The accessibility of vinyl or allyl functionalities at the focal points of dendrons **42–45** enabled their attachment to Si–H polyfunctional cores via hydrosilylation chemistry, in the final step of the dendrimer construction. From two different vinyl-functionalized molecules **42** and **44**, we investigated two synthetic pathways to the targeted first generation heterooctametallic dendrimer

46, shown in Reaction Scheme 8.9. The advantage of route A (which involved coupling of the preformed heterometallic fragment **44** to the four-directional carbosilane core $Si[(CH_2)_3Si(Me)_2H]_4$), should have been that the heterometallic dendrimer **46** was produced in a single hydrosilylation reaction step after the complexation of the phenyl rings with $Cr(CO)_3$ units in **44**. However, all attempts to graft the $Cr(CO)_3$ complexed **44** to the carbosilane core by hydrosilylation failed to afford the expected heterooctametallic dendrimer **46**. This was most likely due to the presence of $Cr(CO)_3$ units in **44**, which induced electron-deficiency in the phenyl rings bound to the Si–CH=CH_2 group, which, in turn, clearly decreased the effectiveness of hydrosilylation with Si–H functionalized molecules [88, 89]. Consequently, we developed another synthetic route, B (see Reaction Scheme 8.9), in which hydrosilylation of Si–CH=CH_2 focal groups with tetrafunctional core was performed before the coordination of phenyl ligands to the tricarbonylchromium fragments. Although this route involves two reaction steps, it allowed successful synthesis of both the first-generation tetraferrocenyl dendrimer **47** and its targeted heterooctametallic counterpart **46**. Similarly, the second generation octaferrocenyl dendrimer $Si\{(CH_2)_3Si(Me)_2(CH_2)_3PhSi[(CH_2)_2MePhSi(\eta^5\text{-}C_5H_4)Fe(\eta^5\text{-}C_5H_5)]_2\}_4$ (**48**) was synthesized by hydrosilylation of dendrons **43** and the same carbosilane core $Si[(CH_2)_3Si(Me)_2H]_4$. Subsequent thermal treatment of dendrimers **47** and **48** with $Cr(CO)_6$ led to the desired first- and second-generation heterometallic dendrimers **46** and **49**, bearing 8 and 20 organometallic moieties on their periphery, respectively (see Reaction Scheme 8.9).

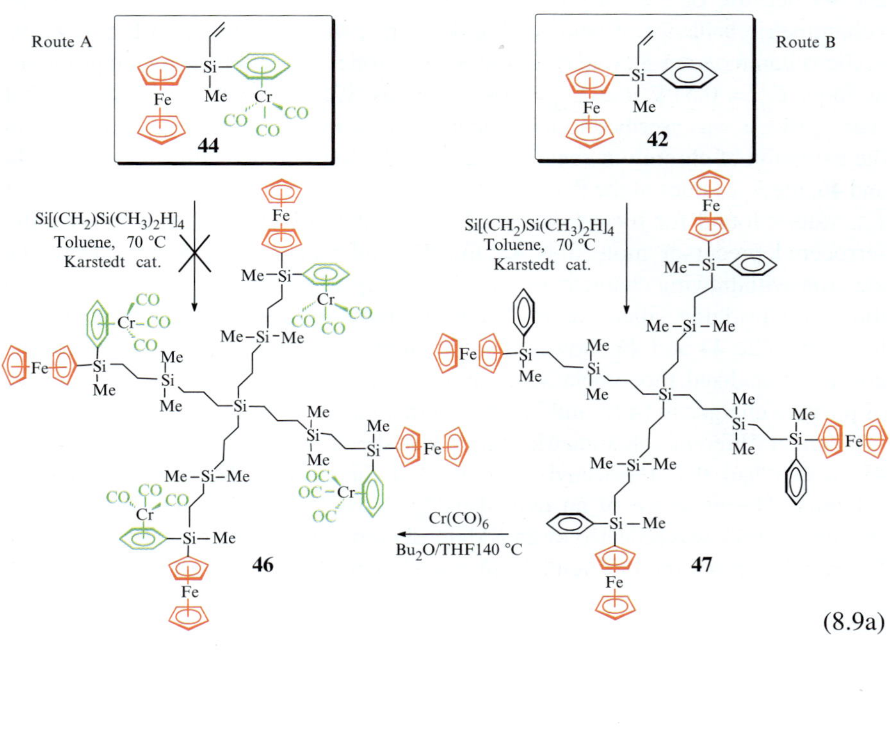

(8.9a)

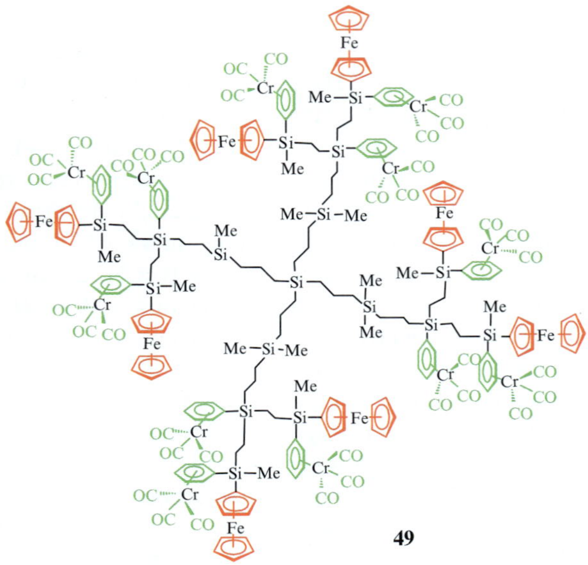

49

(8.9b)

8.2.2.4 Electrochemical Behavior of Dendrimers with Peripheral Silicon-Bridged Ferrocenyl and Aryltricarbonyl-Chromium Moieties

Of special interest is the electrochemical behavior of dendritic molecules **44**, **45**, **46** and **49** because of their heterometallic nature. As a representative example, the voltammetric behavior of octametallic dendrimer **46** is illustrated in Fig. 8.9a. Its cyclic voltammogram showed two diffusion controlled, reversible oxidation processes at about $E_{1/2}$ = 0.51 V and $E_{1/2}$ = 0.93 V versus SCE, respectively [55]. The first redox process was ascribed to the oxidation of iron centers, and the second one to the oxidation of chromium centers. Interestingly, for heterometallic compounds **44** and **46**, the $E_{1/2}$ values of the first oxidation process are higher than the corresponding $E_{1/2}$ values found for the oxidation of iron centers of the related homometallic ferrocenyl precursor molecules **42** and **47**, and this is most likely due to the electron-withdrawing character of the adjacent (η^6-C_6H_5)Cr(CO)$_3$ moieties, bonded through a bridging silicon atom. Similarly, the chromium centred oxidation of heterometallic **44** and **46** showed a slight anodic shift with respect to $E_{1/2}$ for the non-functionalized monometallic compound (η^6-C_6H_6)Cr(CO)$_3$ as a consequence of positive charges in **[44]**$^+$ and **[46]**$^{4+}$ generated after the first oxidation process.

A rather different voltammetric feature was shown by the pentametallic molecule **45**, which had two ferrocenyl moieties and three (η^6-aryl)chromiumtricabonyl subunits. The presence of an extra (η^6-C_6H_5)Cr(CO)$_3$ moiety, non bonded to a ferrocenyl unit, was reflected by an additional chromium-centered oxidation process. Figure 8.9b shows the differential pulse voltammogram of **45** in CH_2Cl_2 solution.

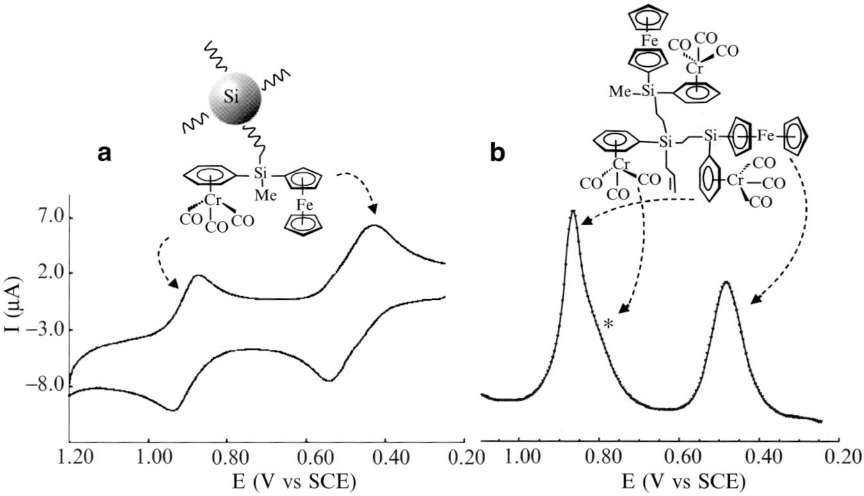

Fig. 8.9 (**a**) Cyclic voltammogram of the heterometallic dendrimer **46** in CH_2Cl_2, with 0.1 M n-Bu_4NPF_6; (**b**) differential pulse voltammogram of the heterometallic dendron **43** (recorded at a glassy carbon electrode in CH_2Cl_2/n-Bu_4NPF_6).

Two separated peaks of different heights can be seen; the second peak being broader than the first one and displaying a shoulder (marked with an asterisk) suggesting two overlapped redox processes. Integration of the peak areas gives a first:second waves ratio of 2:3. The first oxidation peak (at $E_{1/2}$ = 0.51 V) was attributed to the simultaneous two-electron transfer of two electrons removed from the two ferrocenyl subunits in the molecule resulting in the formation of dicationic species $[\mathbf{45}]^{2+}$. The central, almost unresolvable, oxidation peak (at about $E_{1/2} \approx$ 0.80 V) is presumably centered in the chromium center of the isolated (η^6-C_6H_5) $Cr(CO)_3$ unit linked to the Si–allyl group. Finally, two electrons are removed (at $E_{1/2}$ = 0.91 V) from the remaining chromium centers of (η^6-C_6H_5)$Cr(CO)_3$ moieties neighboring on the ferrocenyl moieties already oxidized. For the second generation heterometallic dendrimer **49** the electrochemical behavior was rather complicated since the second oxidation wave was broadened and the redox process displayed electrochemical irreversibility.

8.2.2.5 Poly(methylsiloxane) Backbones with Pendant Organometallic Dendrons: Silicon-Based Ferrocenyl Dendronized Polymers

Dendronized polymers constitute a novel class of macromolecules which merges concepts of dendrimers and linear polymers. They are composed of a linear polymer backbone to which dendrons of increasing size (i.e., generations) are appended. At high degrees of dendronization, dendronized polymers gain stiffness and adopt a cylindrical shape [90–93]. This exclusive architecture makes dendronized polymers unique macromolecules and interesting candidates for a variety of nanoscale applications, such as molecular wires, self-adaptive materials and functional scaffolds for catalysis. In spite of the fact that a variety of dendronized polymers has been synthesized and numerous metallodendrimers have been reported, only very few studies have addressed the synthesis of dendronized polymers cont-aining metallic moieties [94].

Our long standing interest in siloxane-based organometallic polymers and in silicon-based organometallic dendrimers, naturally led to the investigation of novel organometallic macromolecules in which dendritic building-blocks (dendrons) containing electroactive organometallic units are grafted onto multifunctional, flexible poly(methylsiloxane) backbones. The combination of unique architectural features of dendritic molecules with electroactivity of organometallic functions and the well-known, remarkable properties of poly(siloxanes) (such as chemical stability, high permeability to gases and low toxicity) [95] is very attractive for physical and redox properties of the resulting hybrid macromolecules. Hence, we prepared siloxane-based dendronised molecules functionalized with either electronically communicating ferrocenyl moieties (**50** and **51**) [53], or electronically isolated ferrocenyl units (**52** and **53**) [96] (see Reaction Scheme 8.10) which represent the first members of a new family of silicon-based organometallic dendronized polymers.

Once again, the chemistry of silicon opened new perspectives in this field. Thus, we employed hydrosilylation chemistry for the construction of these novel dendronized poly(siloxanes) by using dendrons such as **36** and **43** bearing a single reactive C=C functionality in their focal groups. With Karstedt's catalyst, these wedges were appended to a poly(methylhydrosiloxane-co-dimethylsiloxane) copolymer $(Me_3SiO)(Me_2SiO)_m(MeSiHO)_n(SiMe_3)$ (m = 70–75%, n = 25–30%; $M_w \approx 1{,}900–2{,}000$), and to a poly(methylhydrosiloxane) $(Me_3SiO)(MeSiHO)_n(SiMe_3)$ ($n \approx 35$) in toluene solutions at temperatures between 40°C and 60°C. In general, the reactions were run overnight, which in all cases resulted in complete disappearance of the ν(Si–H) IR band (near 2,155 cm^{-1}) of the starting linear siloxane chains. The resulting hydrosilylated products **50–53** were purified by repeated dissolution in dichloromethane and precipitation into ethanol or hexane. The obtained compounds were isolated as air-stable, orange-brown, tacky, oily materials which hardened on standing and were soluble in solvents such as dichloromethane and THF. All dendronized polymers formed free-standing amber films when cast from dichloromethane solutions. Structural characterization of poly(ferrocenyl) polymers **50–53** was achieved by FTIR, ^{1}H, ^{13}C and ^{29}Si NMR spectroscopies, and MALDI-TOF-MS.

In dichloromethane solution ferrocenyl dendronized polysiloxanes **50** and **51** exhibited well-defined, two-step, reversible oxidations similar to that encountered in their dendritic precursor **36**. This electrochemical pattern was consistent with the existence of appreciable interactions between the two iron centers linked together by a silicon bridge in the pendant dendrons. Cyclic voltammograms

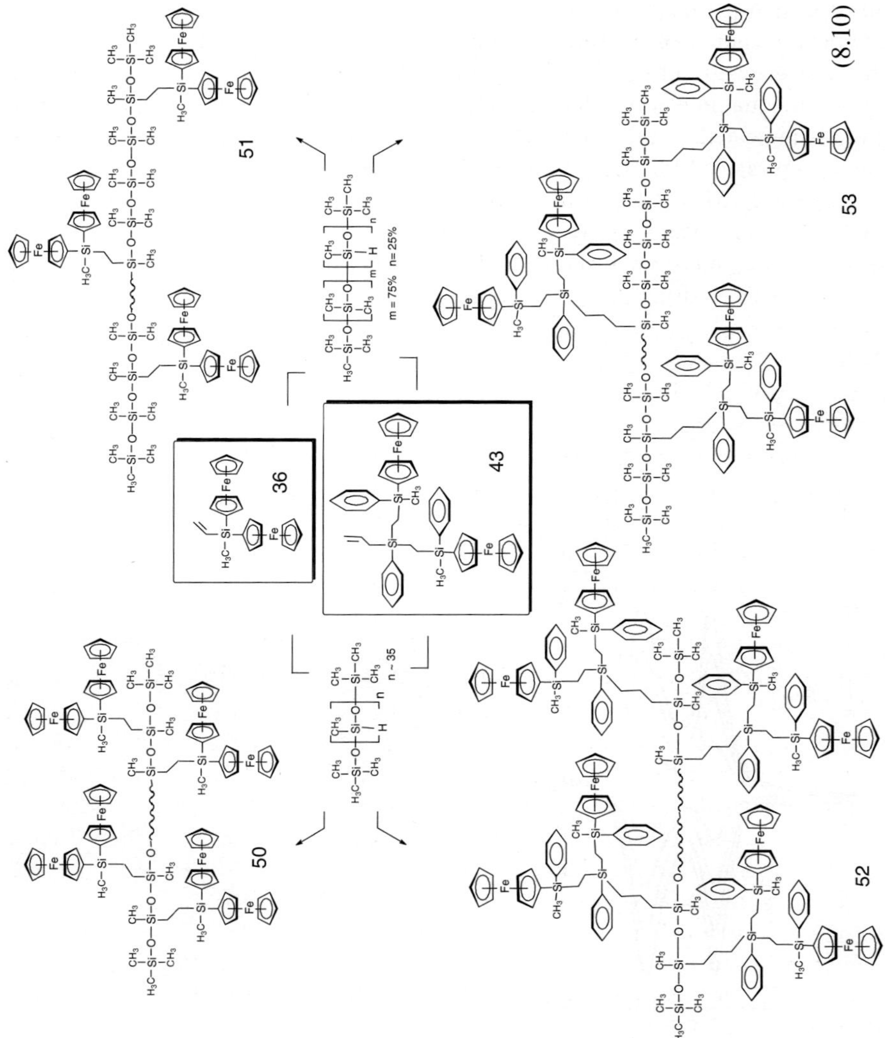

(8.10)

showed that poly(ferrocenylsiloxanes) **50** and **51** exhibited a clear tendency to deposit onto the electrode surfaces, as shown by the narrow shape of the waves [53]. This electrodeposition was performed either by controlled potential electrolysis at +0.9 V or alternatively, by repetitive cycling between 0 and +1.0 V potential limits. Noticeably, in the latter case, a regular increase in both current peaks upon subsequently repeated scans was observed, clearly indicating that an electroactive polymer film was growing on the electrode surface. The electrochemical response of the film of copolymer **51** shown in Fig. 8.10a exhibits two successive, well-defined, highly symmetrical, reversible oxidation-reduction waves of equal intensity, which are characteristic of surface-confined redox couples, as indicated by the linear relationship of both peak currents with the potential sweep rate *v* [68]. The microstructure of films from ferrocenyl-dendronised polysiloxanes **51** and **53** electrochemically deposited on platinum wire working electrodes was examined by scanning electron microscopy (SEM). As can be seen from the micrograph shown in Fig. 8.10b, a film of polymer **51** exhibited a "coral-reef-like" structure.

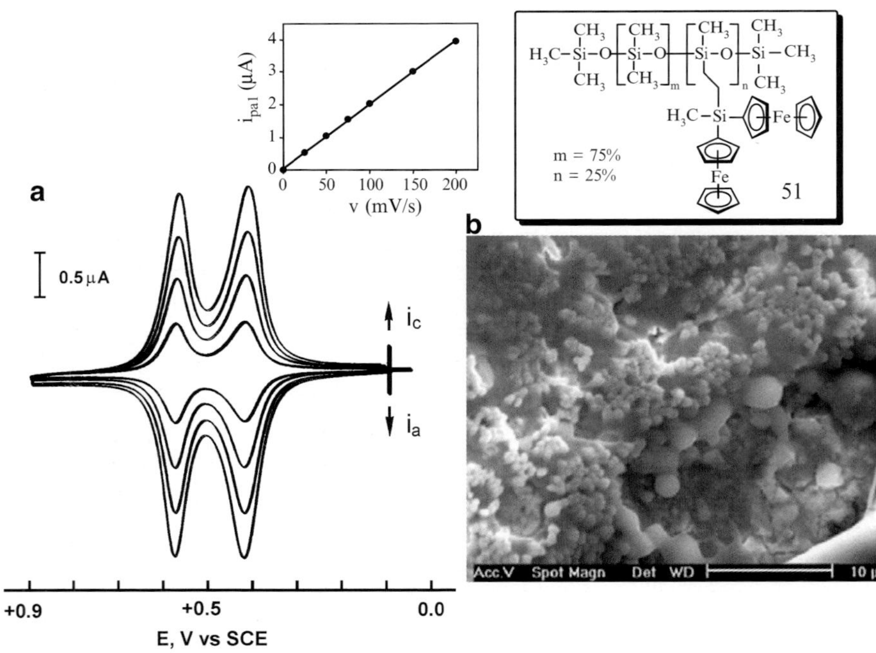

Fig. 8.10 (**a**) Voltammetric response of a platinum-disk electrode modified with a film of poly(ferrocenylsiloxane) copolymer **51**, measured in CH_2Cl_2/0.1 M *n*-Bu_4NPF_6. Scan rates: 25, 50, 75 and 100 mV s^{-1}. Inset: Scan rate dependence of the first anodic peak current; (**b**) scanning electron micrograph of a film of polymer **50**, electrochemically deposited on a platinum wire electrode (0.5 mm of diameter) (Adapted from [53]).

8.2.2.6 Ferrocene as a Core for Divergent Construction of Organometallic Silicon-Containing Dendritic Molecules

In our quest for new dendritic molecules possessing ferrocenyl moieties at predetermined sites in their organosilicon scaffolds, we reasoned that metallocene derivatives polyfunctionalized with olefinic substituents might be a good starting point for divergent syntheses, since their reactive C=C functionalities should allow attachment of many different reactive organometallic moieties (e.g., through hydrosilylation chemistry) to afford the desired multimetallic dendritic systems. Our first attempts to grow such dendritic molecules from a vinyl-functionalized ferrocenyl-containing core were made by using the commercially available 1,1′-divinylferrocene, $\{\eta^5\text{-}C_5H_4(CH=CH_2)\}_2Fe$, in a reaction with methyldichlorosilane, Cl_2MeSiH. Unfortunately, this hydrosilylation failed, either completely or to a large extent, and mainly afforded oxidation products of the starting difunctional ferrocene derivative, and other presumably polymeric materials, which were not characterized. Believing that steric reasons were important for the failure of this divergent synthesis, we decided to try a different divinyl-ferrocene, and for this purpose prepared 1,1′-bis(dimethylvinylsilyl)ferrocene (**54**) (see Reaction Scheme 8.11). In this compound, the steric strain imposed by the sterically-demanding metallocene units is released since the reactive vinyl sites are separated from the cyclopentadienyl rings by the $SiMe_2$ groups. The key organometallic molecule **54** was prepared by the low temperature reaction of 1,1′-dilithioferrocene´TMEDA (TMEDA = N,N′-tetramethyl-ethylenediamine) with dimethylvinylchlorosilane. After appropriate purification by column chromatography, compound **54** was isolated in a 65% overall yield as a reddish-orange liquid.

Indeed, the divinyl derivative **54** was found to be an effective, two-directional, readily functionalizable, core for the divergent construction of homo and heterometallic dendritic systems. For example, successful hydrosilylation of **54** with methyldichlorosilane, was evidenced by multinuclear NMR spectroscopy and afforded the versatile ferrocene-based molecule **55** possessing four Si–Cl reactive end-groups. The high reactivity of **55** was exploited by its subsequent conversion to pentametallic **56,** by a salt-elimination reaction with monoferrocenyllithium in THF at −30°C. After purification by column chromatography, the desired compound **56**, possessing a central ferrocene and four outer ferrocenyl units linked together in pairs through a bridging silicon atom, was isolated as an air-stable orange crystalline solid and was fully characterized by NMR (^{1}H, ^{13}C and ^{29}Si) and IR spectroscopies, mass spectrometry and electrochemical techniques.

The very reactive tetrafunctional **55** is also an excellent starting point for further syntheses, particularly for alkenylation and hydrosilylation reactions as growth stages. For instance, reaction of the terminal Si–Cl functionalities of **55** with an excess of allylmagnesium bromide, in diethyl ether at reflux, afforded the desired dendritic molecule **57** carrying two allyl functions linked by a bridging silicon atom. Treatment of **57** with methyldichlorosilane, afforded a novel ferrocene-based derivative containing eight reactive Si–Cl bonds which, after subsequent treatment with 1-lithioferrocene, led to the second generation dendrimer **58** having an internal

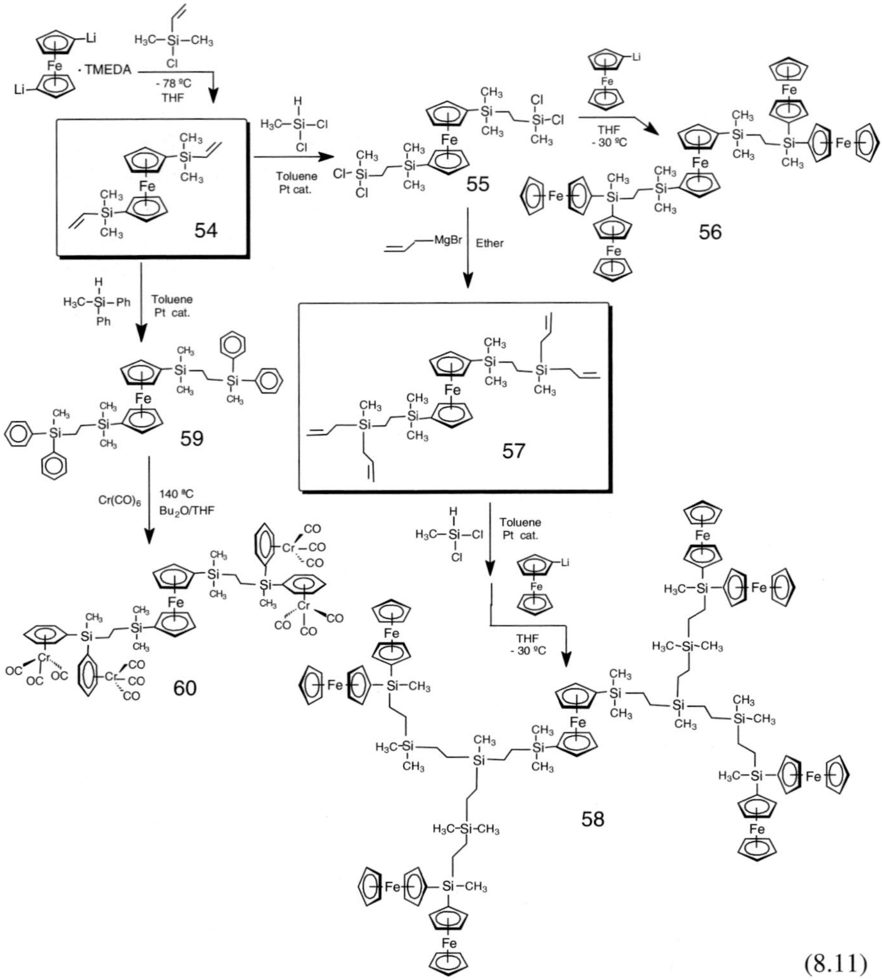

(8.11)

ferrocenyl moiety and eight peripheral units. On the other hand, the hydrosilylation reaction of **54** with methyldiphenylsilane yielded a tetraphenyl-functionalized molecule **59** which, after subsequent thermal treatment with an excess of $Cr(CO)_6$, yielded a first generation heterometallic pentanuclear dendritic molecule possessing a central ferrocene and four outer electroactive chromium-containing centres. As is clearly shown in Reaction Scheme 8.11, the high reactivity and versatility of both lithiated ferrocenyl intermediates $(\eta^5\text{-}C_5H_4Li)_2Fe$ and $(\eta^5\text{-}C_5H_4Li)Fe(\eta^5\text{-}C_5H_5)$ proved to be synthetically very useful for divergent construction of silicon-based electroactive dendritic molecules.

Regarding the electrochemical properties of these molecules, note that the degree of interaction between the electroactive metal-based moieties linked by a bridging silicon atom, depends on their chemical nature. Thus, the cyclic voltammogram of hetero-pentametallic **60** showed a reversible oxidation wave at $E_{1/2}$ = 1.05 V, corresponding to a simultaneous multielectron transfer of four electrons removed from the four terminal chromium centers. The fact that pentanuclear **60** exhibits a single-step oxidation process at this potential value is of relevance since it implies that peripheral chromium tricarbonyl moieties, linked together by a silicon bridge, do not essentially communicate electronically with each other. The lack of interaction between the $(\eta^6\text{-}C_6H_5)Cr(CO)_3$ units linked by a bridging silicon atom contrasts interestingly with the case of the corresponding pentaferrocenyl molecule **56** for which significant electronic interactions between the terminal silicon-bridged ferrocenyl moieties were observed. This indicates that the nature of arene rings imposes a significant influence on the degree of interaction between metallic centers.

8.3 Electrochemical Applications of Ferrocenyl Silicon-Containing Dendritic Molecules

8.3.1 Ferrocenyl Dendrimers with Si–NH Linkages as Redox Sensors for Recognition of Inorganic Anions

Dendritic molecules offer attractive possibilities in molecular recognition processes because they may engage in multivalent host-guest interactions, acting as either host (receptors) or guest (substrate) systems [33, 99, 100]. Molecular recognition of non-electroactive anionic guest species by redox-active receptor molecules is an area of intense current research activity [101, 102]. This is due to the fundamental roles that anions play in biological and chemical processes, as well as to the importance of developing novel sensor devices for environmentally relevant anions, of which phosphate is one of the most significant examples. Redox-responsive receptor molecules are able to selectively bind and recognize guest species through the perturbation of their redox systems provoked by the host-guest interactions. The corresponding sensor systems aim to convert such interactions at the molecular level into measurable electrochemical signals.

The ferrocene system has been extensively used in recent years as an electrochemical sensor since it exhibits reversible electron transfer kinetics [33, 34, 101]. The oxidized and reduced states of ferrocene-based receptors, display different degrees of affinity to anionic guest species, and the oxidation state determines the thermodynamic stability of the complex formed between the receptor and the guest [33]. The basis of this differential affinity is only electrostatic and perturbation of the charge in electroactive ferrocene host can result in increased anion binding affinity. When the binding interaction is strong, the electrochemistry may clearly reflect two different redox states, since the complexed and the free forms typically exhibit separate redox waves. Therefore, the half-wave potential ($E_{1/2}$) of the ferrocene/ferrocenium (Fe^{II}/Fe^{III}) system is different in the presence and in the absence of a substrate whose recognition is being sought. In addition, the binding constant of the guest and the host containing the ferrocene unit is not the same in the neutral Fe^{II} redox form of ferrocene and in its Fe^{III} oxidized form.

Dendrimers containing multiple peripheral ferrocene units are of interest as supramolecular redox-active receptors for molecular recognition processes because they may engage in multiple interactions with suitable anionic guest species. The ferrocenyl dendrimers **10** and **11**, that contain multiple peripheral Si–NH groups as well as inner cavities which can be occupied by small molecules, can potentially bind and sense anionic guests through cooperative effects involving NH–anion hydrogen bonding interactions in the neutral state, and electrostatic attractions with the positively charged ferrocenium moieties, after electrochemical oxidation of terminal ferrocenyl units in the neutral dendritic receptors. As an example, a schematic representation of the possible binding interactions for dihydrogenphosphate anion in shown in Fig. 8.11a. Electrochemical anion-sensing properties of these silicon-based ferrocenyl dendrimers were investigated by cyclic voltammetry, by titration of the ferrocenyl dendrimers with $[n\text{-}Bu_4N]^+$ salts of the anion guests $H_2PO_4^-$, HSO_4^-, Cl^- and Br^- [22, 23, 103]. Here, we describe the results obtained with dendrimer **11**, which contains eight ferrocenyl peripheral units linked to the dendritic framework through Si–NH groups.

Before the beginning of titration, the cyclic voltammogram of the octaferrocenyl dendrimer **11** in CH_2Cl_2/0.1 M $n\text{-}Bu_4NPF_6$ solution showed a unique reversible oxidation wave at 0.41 V versus SCE (see Fig. 8.11b solid line), corresponding to a simultaneous multielectron transfer of eight-electrons (as expected for

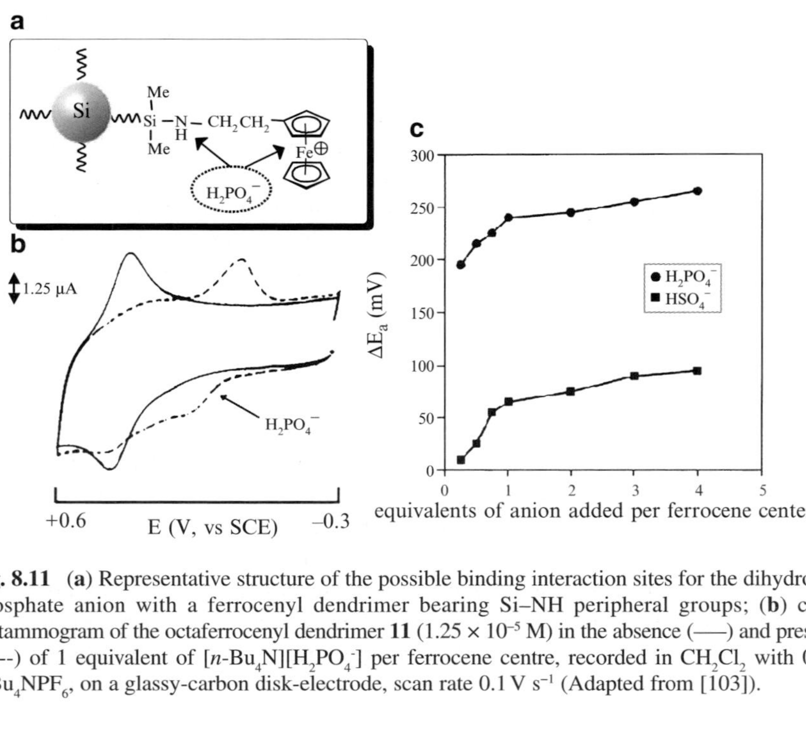

Fig. 8.11 (**a**) Representative structure of the possible binding interaction sites for the dihydrogen-phosphate anion with a ferrocenyl dendrimer bearing Si–NH peripheral groups; (**b**) cyclic voltammogram of the octaferrocenyl dendrimer **11** (1.25×10^{-5} M) in the absence (——) and presence (-----) of 1 equivalent of $[n\text{-}Bu_4N][H_2PO_4^-]$ per ferrocene centre, recorded in CH_2Cl_2 with 0.1 M $n\text{-}Bu_4NPF_6$, on a glassy-carbon disk-electrode, scan rate 0.1 V s^{-1} (Adapted from [103]).

independent, reversible, one-electron processes) at the same potential as the eight peripheral ferrocenyl moieties. However, once the titration began, dendrimer cyclic voltammograms showed significant anion induced cathodic perturbations, indicating that the addition of $H_2PO_4^-$ or HSO_4^- caused a decrease in the intensity of the redox wave, along with the progressive appearance of a new redox wave, of increasing intensity, at a less positive potential (see Fig. 8.11b dashed line). The current associated with the new redox couple increased linearly with the anion concentration until the original wave disappeared completely and the new redox couple reached full development. Interestingly, the largest magnitude of cathodic shift was observed with the $H_2PO_4^-$ anion (see Fig. 8.11c). However, the addition of Cl^- and Br^- anions to the ferrocenyl dendrimer-containing solution did not give rise to a new wave but, instead, only produced a progressive cathodic shift of the initial wave until one equivalent of the anion was added per dendrimer branch. The $\Delta E_{1/2}$ values reflect the difference in the substrate binding strength between the two redox states of the ferrocenyl-containing dendrimer. The binding of the anion effectively stabilized the positive charge of the oxidized ferrocenium units causing the redox couple to shift to more negative potentials. These electrochemical results clearly indicate that the shape of the oxidation wave changes from a reversible redox process to an EC mechanism (electron transfer followed by a chemical reaction) especially in the presence of the anionic guest $H_2PO_4^-$ and HSO_4^-. The fact that the most remarkable changes of voltammograms were obtained for large and tetrahedral anions such as $H_2PO_4^-$ and HSO_4^- could be related to the presence of cavities in the receptor with dimensions complementary to these anions.

Of significant interest to the development of sensor technology are the results of electrochemical competition experiments. Hence, in order to establish the selectivity of the dendritic receptors in electrochemical recognition, such experiments were carried out. It was found that the cyclic voltammogram recorded when one equivalent of $H_2PO_4^-$ per ferrocene unit was added to a CH_2Cl_2/electrolyte solution of dendrimer **11** in the presence of tenfold excess of HSO_4^-, Cl^- and Br^- was similar to that recorded for the dendrimer **11** in the presence of $H_2PO_4^-$ alone. The selectivity of the ferrocenyl dendrimer functionalized with peripheral Si–NH linkages to dihydrogenphosphate was good as judged by the potential shifts observed in the presence of an anion mixture. In a similar manner, HSO_4^- could be detected unambiguously, in the presence of Cl^- and Br^-. The results obtained showed that this dendrimer displayed the following selectivity trend: $H_2PO_4^- > HSO_4^- > Cl^- > Br^-$.

A promising new technique for electrochemical anion sensing devices relies on immobilization and/or preorganization of redox active receptors on the electrode surface. Concentration of active recognition sites on solid surfaces is known to greatly enhance the sensing and binding properties of the resulting material compared to those of the same individual molecules in solution. In this context, it is important to highlight that electrode surfaces can be readily modified with thin films of the dendrimer receptor **11** by electrooxidation on a glassy-carbon or a platinum electrode [103]. The voltammetric response of electrodes modified with electroactive films of octametallic dendrimer **11** was found to be sensitive to the presence and concentration of the $H_2PO_4^-$ guest anion. This result, which is an example of a successful transfer of electrochemical sensing properties of a redox probe from the homogeneous phase to the electrode surface, presents encouraging prospects for the development of novel types of molecular sensory devices.

Concerning the applications of organometallic dendrimers in anion recognition, it should be further noted that in collaboration with the group of Professor Angel E. Kaifer, at the University of Miami, we also investigated the anion binding interactions of a different family of dendrimers, this one based on a PPI scaffold terminated with 4, 8, 16 and up to 32 ferrocenyl-urea moieties, respectively [52]. These dendrimers proved to be very sensitive to the presence of dihydrogenphosphate anion, even at submillimolar concentration levels, in DMSO solution. Electrochemical anion sensing experiments were carried out using square wave voltammetry (SWV) because this technique has a lower detection limit than CV. A large shift in the half-wave oxidation potential of 112 mV to less positive values was found for $H_2PO_4^-$, whereas HSO_4^- and Cl^- caused much smaller potential shifts (40 and 16 mV, respectively). The electrochemical data obtained also suggest that, in these dendrimers, two ferrocene–urea arms bind a single dihydrogenphosphate anion. Evidence of a "dendritic effect" was observed in the redox response in the change from the first (four ferrocene units) to the third (16 ferrocene units) generation in the presence of $H_2PO_4^-$. As the number of ferrocenyl-urea end-units increased, so did the magnitude of the cathodic shift in the ferrocene/ferrocenium couple. In similar work, Astruc and coworkers made use of dendrimers containing amido-metallocene units as exoreceptors for the electrochemical recognition of anions [25, 104].

8.3.2 Ferrocenyl Silicon-Containing Dendrimers as Electron-Transfer Mediators in Amperometric Biosensors

One of the most remarkable applications of ferrocene derivatives is their use as redox mediators in the so-called enzymatic electrodes or biosensors [105, 106]. Biosensors are unique analytical tools, in terms of specificity and sensibility, which are able to determine, in a simple and rapid way, the concentration of substances of clinical and physiological interest. The methodology is based on the fact that, in the presence of enzyme-catalysed reactions, the electrode currents are considerably amplified. One of the most significant applications of this technique is determination of the small concentrations of glucose in the blood, particularly to help type II diabetes mellitus patients to monitor their daily sugar levels.

Among many protocols available for measuring glucose, the most commonly used method is based on the oxidation reaction catalyzed by the enzyme glucose oxidase (GOx) shown in Reaction Scheme 8.12. In this system, the common electron-acceptor oxygen generates the product hydrogen peroxide, which is measured by electrochemical oxidation at a platinum electrode. Historically, the first determination of glucose in the blood was based on this reaction, by titration of the generated H_2O_2.

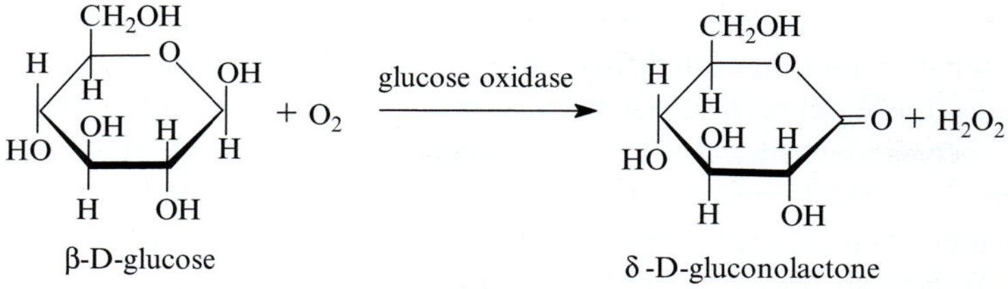

(8.12)

In the initial step of the biocatalytic reaction, the flavin adenine dinucleotide (FAD) group of the oxidised enzyme $GOx_{(Ox)}$ oxidizes β-D-glucose and converts it into δ-D-gluconolactone forming the $FADH_2$-containing reduced enzyme $GOx_{(Red)}$ (Reaction Scheme 8.13). Because these redox centres are essentially electrically insulated within the enzyme molecule, direct electron transfer to the surface of a conventional electrode does not occur to any measurable degree. In nature, the abstraction of the two electrons from the $GOx_{(Red)}$ is usually performed by dioxygen, which in turn is reduced by $FADH_2$ into hydrogen peroxide (Reaction Scheme 8.14). H_2O_2 may then diffuse out of the enzyme and can be detected electrochemically. However, a major problem with this approach lies in the sensitivity to many common, interfering, easily-oxidizable species present in the biological fluids. In order to overcome this, many artificial redox mediators have been investigated as electron acceptors in the development of amperometric glucose sensors.

$$\beta\text{-D-glucose} + GOx_{(Ox)} \longrightarrow \delta\text{-D-gluconolactone} + GOx_{(Red)} \quad (8.13)$$

$$GOx_{(Red)} + O_2 \longrightarrow GOx_{(Ox)}) + H_2O_2 \quad (8.14)$$

Amperometric enzyme electrodes with electroactive ferrocene-containing species acting as mediators, replacing the natural electron-acceptor, dissolved oxygen, are of special interest. When the molecular oxygen is substituted for an electrogenerated ferrocenium Fe^{III} cation, the catalytic reaction produces the corresponding neutral ferrocene Fe^{II} species (rather than H_2O_2), which in turn is re-oxidized at the electrode surface, thus triggering the glucose → gluconolactone conversion following the catalytic mechanism illustrated in Fig. 8.12.

Monomeric ferrocene mediators were initially used as electron-shuttling redox couples. Nevertheless, in these ferrocene-based mediators the solubility of the

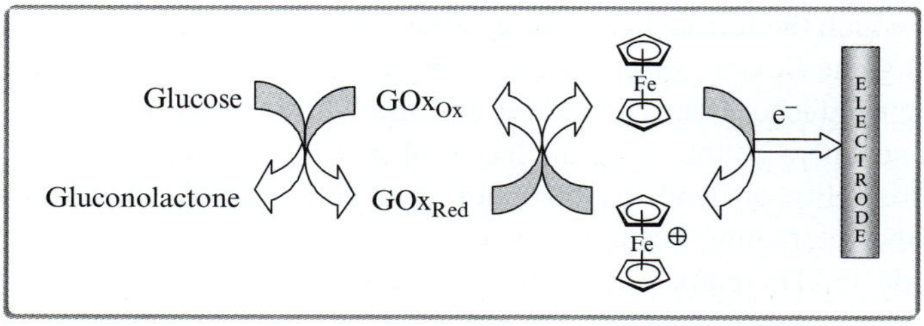

Fig. 8.12 Schematic illustration of the oxidation reaction of glucose catalysed by the enzyme glucose oxidase (GOx) in the presence of an electroactive ferrocene-containing species acting as mediator.

oxidized form would inevitably be a source of long-term instability, due to the loss of ferrocenium ions that can diffuse away from the electrode surface. For this reason, over the last years there has been growing interest in systems where the mediating species is chemically bound in a manner that allows close contact between the FAD/$FADH_2$ redox centers of the enzyme and the mediator yet prevents the latter from diffusing away from the electrode surface. Increasing the molecular weight of the ferrocene-containing mediator should significantly decrease this possible loss.

Potentially, our polyferrocenyl-containing dendritic macromolecules described in Section 8.2, with relatively high molecular weights and in which the ferrocenyl units are at the end of long flexible silicon-containing branches, can serve to electrically connect the enzyme, facilitating a flow of electrons from the enzyme to the electrode. For this reason, in order to test the ability of ferrocenyl dendrimers to act as electron mediating species, a study of the efficiency of dendrimer/glucose-oxidase/carbon paste electrodes was undertaken [23, 107] and some of the most significant results of this study for dendrimers **4**, **5**, **18**, and **19** are illustrated in Fig. 8.13. The dendrimer/glucose oxidase/carbon paste electrodes were constructed by mixing the ferrocene-containing dendrimers (previously dissolved in CH_2Cl_2) and graphite powder. After evaporation of the solvent, the appropriate amounts of glucose oxidase and paraffin oil were added, and the resulting mixture was blended into a paste and then placed in a hole at the end of a carbon paste electrode.

Figure 8.13a shows as a representative example a cyclic voltammogram of a carbon paste electrode containing glucose oxidase and silicon-containing ferrocenyl dendrimer **5** as mediator, before and after the addition of glucose. Clearly, the addition of glucose led to the enhancement of the oxidation current, while cathodic current was not observed. This fact is indicative of enzyme-dependent catalytic reduction of the ferrocenium cations. In addition, the electrodes are clearly sensitive to small changes in glucose concentration, and display a good response over long periods of time.

It is also worth mentioning that the glucose sensor responses were found to be dependent on the number of peripheral ferrocenyl units and on the nature of the silicon-based dendritic framework. For instance, for equimolar amounts of ferrocene moieties, the octaferrocenyl dendrimers **5** and **19**, possessing longer organosilicon branches (see Figs. 8.4 and 8.7), proved to mediate electron transfer more efficiently than the relay systems based on tetrametallic dendrimers **4** and **18** (see Figs. 8.19b and 8.19c). On the other hand, it is clear that dendrimers **18** and **19**, in which the ferrocenyl units are attached to the dendritic scaffold through a two methylene flexible spacer, were more effective at mediating electron transfer between reduced glucose oxidase and the carbon paste electrode, than the related dendrimers of the same nuclearity **4** and **5**. This result clearly suggests that the flexibility of dendritic mediator is an important factor in its ability to facilitate the interaction between the mediating species and the flavin adenine dinucleotide (FAD) redox centres of glucose-oxidase. Studies in which monomeric and polymeric ferrocene-containing mediators were compared, showed that sensors based on octanuclear ferrocenyl dendrimers displayed a similar response to glucose as sensors based on monomeric ferrocene mediators but showed better operational stabilities because their oxidized forms were less soluble

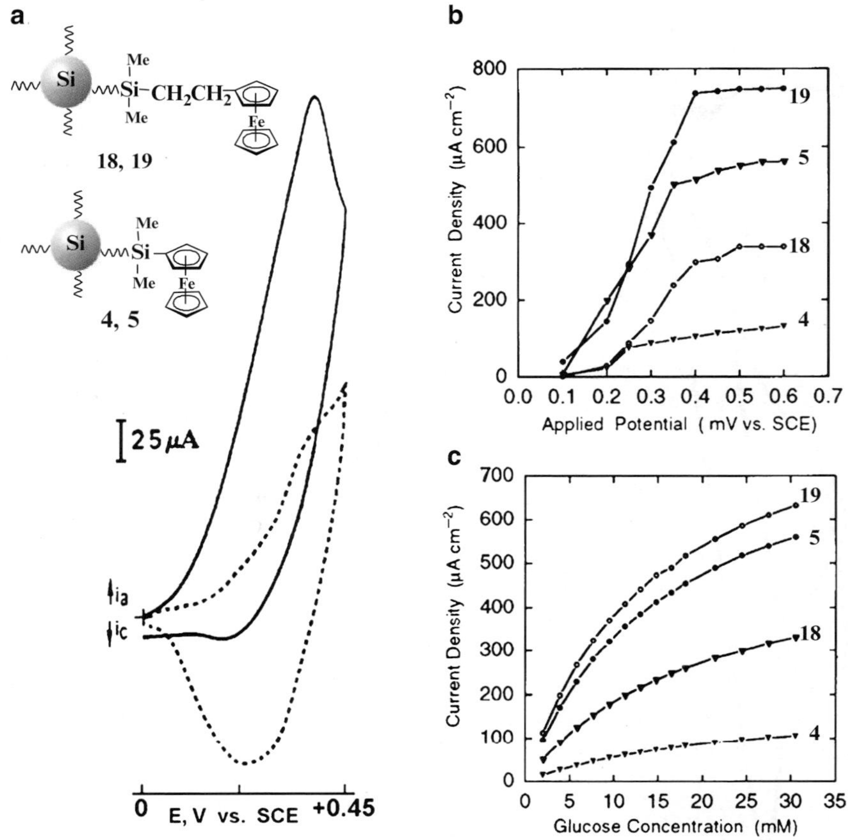

Fig. 8.13 (**a**) Cyclic voltammograms for the dendrimer **5**/glucose oxidase/carbon paste electrode, recorded at 5 mV/s in pH 7.0 sodium phosphate buffer (with 0.1 M KCl) solution, with no glucose present (- - - -), and in the presence of 0.1 M glucose (——); (**b**) steady-state polarization curves of the dendrimer/glucose oxidase/carbon paste electrodes (for dendrimers **4**, **5**, **18** and **19**) in the presence of 25 μM glucose; (**c**) variation of the steady-state current of the dendrimer/glucose oxidase/carbon paste electrodes with glucose concentration (at +350 mV versus SCE) (Adapted from [107]).

than those of the freely diffusing mediators. Furthermore, ferrocenyl dendrimer-based sensors exhibited a higher sensitivity than ferrocene-modified polymer mediated electrodes.

8.3.3 *Electrocatalytic Oxidation of Ascorbic Acid Mediated by a Ferrocenyl Siloxane-Based Network Polymer*

It is well-known that ascorbic acid exists extensively in fruits and plays an important role as an antioxidant agent in many biological reactions. Therefore, electrochemical methods have been employed for determination of ascorbic acid in biological systems

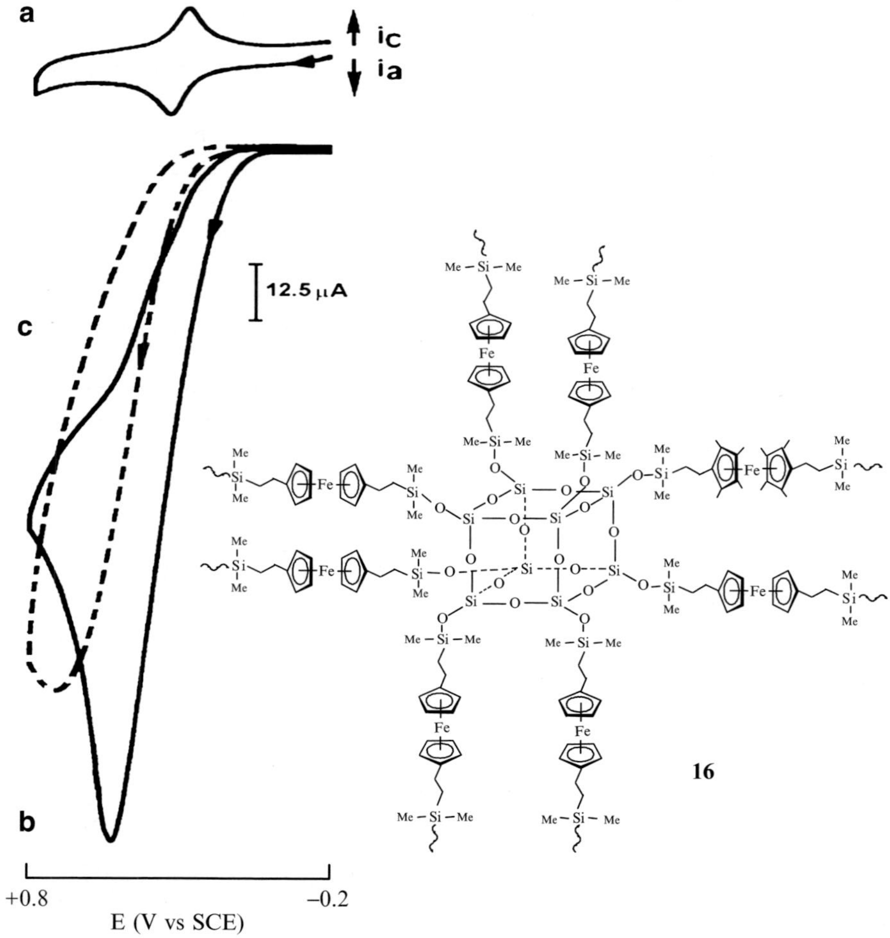

Fig. 8.14 Cyclic voltammograms in 0.1 M acetate buffer with 0.1 M $LiClO_4$ of: (**a**) a glassy-carbon electrode modified with a film of the octasilsesquioxane-based ferrocenyl polymer **16**; (**b**) same electrode as in **a** recorded with 10^{-3} M ascorbic acid; (**c**) a bare (unmodified) glassy-carbon electrode measured in 10^{-3} M ascorbic acid (Adapted from [50]).

and in foods. However, electrochemical oxidation of ascorbic acid at metallic or carbon electrodes proceeds at high overpotential and consequently, suffers from interference from other oxidizable compounds that exist in biological solutions. A number of electrocatalytic oxidations of ascorbic acid systems with chemically modified electrodes have been proposed, which, for instance, comprise a polyvinylferrocene film covalently bound to a glassy carbon [108]. Therefore, we decided to examine the ability of electrodes modified with ferrocenyl octasilsesquioxane-based polymer **16** films to catalyze ascorbic acid oxidation [50].

The coating of platinum, glassy-carbon and gold electrode surfaces with silsesquioxane-based ferrocenyl films of polymer **16** was readily accomplished by repeated cycling (in CH_2Cl_2 with Bu_4NClO_4) over the 0 to +1 V versus SCE potential range, so that the amount of electroactive polymeric material electrodeposited was controlled by the number of scans performed. As shown in Fig. 8.14a, the resulting modified electrodes transferred from clean aqueous electrolyte solution (0.5 M acetic acid, 0.1 M $LiClO_4$) displayed a well defined, stable electrochemical response for the ferrocenyl units at $E_{1/2}$ = +0.32 V versus SCE (ΔE_p = 30 mV at 10 mV s^{-1}), characteristic of surface-immobilized reversible redox couples, with the expected linear relationship between the peak current and potential sweep rate v [68].

To test electrocatalytic activity of the octasilsesquioxane-based ferrocenyl polymer film toward ascorbic acid oxidation, we studied the cyclic voltammetric responses in acetic buffer solution, in the absence and in the presence of ascorbic acid. The results shown in Fig. 8.14b and c indicate that electrodes modified with films of polymer **16** were effective. Namely, curve B corresponds to the voltammetric response to the ascorbic acid (10^{-3} M) in 0.1 M acetic buffer of a glassy-carbon electrode modified with a film of polymer **16**. It can be seen that the anodic peak current of the ferrocenyl polymer film is greatly enhanced while the corresponding cathodic wave disappears, which is consistent with a catalytic process. Catalytic activity of the ferrocenyl polymer modified electrode is further confirmed by the decrease in the oxidation potential (225 mV) and by the increase of the anodic peak current relative to the oxidation of the ascorbic acid at a bare electrode in the same medium (Fig. 8.14c).

8.4 Concluding Remarks

This chapter highlights the rich structural diversity of dendritic molecules with silicon atoms in the branch junctures and electroactive organometallic entities in their structures. It is shown that organometallic silicon-containing dendritic molecules combine the unique features of dendrimer architecture with the extraordinarily versatile chemical behaviour of silicon and the rich chemistry of organotransition metal compounds. Beside the synthetic challenges and pleasant aesthetics of these molecules, organometallic dendrimers based on silicon atoms have already been shown to be useful for a range of interesting applications such as dendritic catalysts (see also Chapter 9), electron-transfer mediators in redox catalysis and also as redox-active receptors for molecular recognition processes. Likewise, silicon-based dendrimers having a high local concentration of peripheral redox centres, regardless of whether they are electronically isolated or communicating, have proved to be highly valuable materials for developing chemically modified electrodes with films of organometallic dendritic molecules.

Moreover, the opportunities for future progress in the design of novel organometallic dendrimers based on silicon atoms are almost unlimited considering the great diversity of synthetic organometallic routes that are as yet unexplored, the numerous variations in possible topologies, the remarkable versatility of organosilicon chemistry and the large number of organo-transition metal moieties that can be incorporated into dendritic structures. Most of the synthetic work described here was based on the functionalization of silicon dendritic skeletons with the stable, chemically accessible ferrocenyl moiety. The integration of redox-active metallocene units other than ferrocene such as cobaltocenium, nickelocene, vanadocene or chromocene, as well as other related sandwich and half-sandwich compounds is a synthetic challenge which will enable construction of novel organometallic silicon-containing dendritic molecules with remarkable electrochemical properties, distinctly different from all those prepared so far. Attachment of such new types of redox-active silicon-containing dendrimers to electrode surfaces will undoubtedly result in interesting novel metallocene-based electron relay systems. Likewise, more elaborate synthetic routes, such as those leading to heterometallic dendrimers **46–49**, and to ferrocenyl-based dendronized poly(siloxanes) **50–53** may yield electroactive macromolecules with more complex structures containing different organometallic moieties and with yet unexplored and, hopefully, interesting electrochemical properties. Exploration of these novel structures and their electrochemical properties is one of the major goals of future research in this field.

Acknowledgements This work would not have been possible without the effort and dedication of all researchers working in the group over the past years, whose names appear in the reference list. A special mention should go to Dr. Beatriz Alonso and Dr. Magdalena Zamora, who have worked in this area with great enthusiasm and have synthesized most of the silicon-containing organometallic dendritic molecules reported in this chapter. I would also like to acknowledge the collaboration of Dr. Carmen Mª Casado who prepared the majority of the series of ferrocenyl polymers and has contributed many of the results summarized in this review. I am equally, deeply indebted to Prof. José Losada (E.T.S.I.I. Universidad Politécnica de Madrid) who helped me take my first steps in the application of electrochemistry to the study or organometallic compounds and has contributed for many years his dedication and invaluable scientific insight shown in his discussions of the development of electrochemical applications of electroactive organometallic molecules prepared in our group. I wish to warmly express my gratitude to Prof. Angel E. Kaifer, at the University of Miami, who was so kind as to revise the manuscript and contribute valuable comments to its contents. Finally, I gratefully acknowledge the Spanish Ministerio de Educación y Ciencia's support to this research (Projects CTQ2005-02282/BQU, BQU-2001-0210 and PB97-0001).

References

1. Newkome GR, Moorefield CN, Vögtle F (2001) Dendrimers and dendrons–concepts, syntheses, applications. Wiley-VCH, Weinheim
2. Fréchet JMJ, Tomalia DA (eds) (2001) Dendrimers and other dendritic polymers. Wiley, Chichester
3. Newkome G (ed) Advances in dendritic macromolecules. JAI Press, Greenwich, CT (1994) vol 1 (1995) vol 2 (1996) vol 3 (1999) vol 4 (2002) vol 5
4. (a) Tomalia DA, Fréchet JMJ (2002) J Polym Sci Part A: Polym Chem 40:2719; (b) Vögtle F, Gesternmann S, Hesse R, Scwierz H, Windisch B (2000) Prog Polym Sci 25:987; (c) Tully DC, Fréchet JMJ (2001) Chem Commun 1229; (d) Fréchet JMJ (2003) J Polym Sci Part A: Polym Chem 41:3713
5. Some recent reviews on dendrimers: (a) Tomalia D A (2005) Prog Polym Sci 30:294; (b) Liang C, Fréchet JMJ (2005) Prog Polym Sci 30:385; (c) Boas U, Christensen JB, Heegaard PMH

(2006) J Mater Chem 16:3785 (d) Majoral JP (2007) N J Chem 31:1039; (e) Smith DK (2006) Chem Commun 34; (f) Dvornic PR (2006) J Polym Sci Part A: Polym Chem 44:2755
6. Buhleier E, Welmer W, Vögtle F (1978) Synthesis 78:155
7. Newkome GR, Yao ZQ, Baker GR, Gupta K (1985) J Org Chem 50:2003
8. Tomalia DA, Baker H, Dewald JR, Hall M, Kallos G, Martin S, Roeck J, Ryder J, Smith P (1985) Polym J 17:117
9. For some recent reviews on biomedical applications of dendrimers, see (a) Lee CC, MacKay JA, Fréchet, JMJ, Szoka FC (2005) Nat Biotechnol 23:1517; (b) Svenson S, Tomalia DA (2005) Adv Drug Deliv Rev 57:2106; (c) Boas U, Heegaard PMH (2004) Chem Soc Rev 33:43; (d) Duncan R, Izzo L (2005) Adv Drug Deliv Rev 57:2215
10. Boas U, Christensen JB, Heegaard PMH (2006) Dendrimers in medicine and biotechnology–new molecular tools. RSC, Cambridge
11. (a) Denti G, Campagna S, Serroni S, Ciano N, Balzani V (1992) J Am Chem Soc 114:2944; (b) Newkome GR, Cardullo F, Constable EC, Moorefield CN, Cargill Thompsom AMW (1993) Journal of the Chemical Society-Chemical Communications 925–927
12. (a) Liao Y-L, Moss JRJ (1993) Journal of the Chemical Society-Chemical Communications 1774–177; (b) Knapen JWJ, van der Made AV, Wilde de JC, van Leeuwen PWNM, Wijkens P, Grove DM, van Koten G (1994) Nature 372:659 (c) Moulines F, Djakovitch L, Boese R, Gloaguen B, Thiel W, Fillaut J-L, Delville M-H, Astruc D (1993) Angew Chem Int Ed Engl 32:1075; (c) Achar S, Puddephatt RJ (1994) Angew Chem Int Ed Engl 33:847
13. For general reviews on metallodendrimers see references 13–21: Hwang S-H, Schreiner CD, Moorefield C, Newkome GR (2007) New J Chem 31:1192
14. Newkome GR, He E, Moorefield CN (1999) Chem Rev 99:1689
15. Majoral JP, Caminade AM (1999) Chem Rev 99:845
16. Berger A, Gebbink RJM, van Koten G (2006) Top Organomet Chem 20:1
17. Mery D, Astruc D (2006) Coordin Chem Rev 250:1965
18. van Manen H-J, van Veggel FCJM, Reinhoudt DN (2001) Top Curr Chem 217:121
19. Serroni S, Campagna S, Puntoriero F, Pietro CD, McClenaghan ND, Loiseau F (2001) Chem Soc Rev 30:367
20. Gorman C (1998) Adv Mater 10:295
21. (a) Manners I (2004) Metallodendrimers. In: Synthetic metal-containing polymers. Wiley-VCH, Weinheim, p 237; (b) Hang S-H, Newkome GR (2007) Metallodendrimers and their potential utilitarian applications. In: Abd-El-Aziz ASA, Manners I (eds) Frontiers in transition metal-containing polymers, Chapter 10. Wiley-Interscience, Hoboken, NJ
22. For reviews on organometallic dendrimers see references 22–30: Cuadrado I, Morán M, Losada J, Casado CM, Pascual C, Alonso B, Lobete F (1996) In: Newkome G (ed) Advances in dendritic macromolecules, vol 3. JAI Press, Greenwich, CT, pp 151–195
23. (a) Cuadrado I, Moran M, Casado CM, Alonso B, Losada J (1999) Coord Chem Rev 193:395; (b) Casado CM, Cuadrado I, Morán M, Alonso B, García B, Gonzalez B, Losada J (1999) Coord Chem Rev 185:53
24. (a) Hearshaw MA, Moss JR (1999) Chem Commun 1; (b) Hearshaw MA, Moss JR (1999) In: Newkome G (ed) Advanced in dendritic macromolecules, vol 4. JAI Press, Stamford, CT, pp 1–60
25. (a) Astruc D, Blais J-C, Cloutet E, Djakovitch L, Rigaut S, Ruiz J, Sartor V, Valerio C (2000) Top Curr Chem 210:230
26. Chase P A, van Koten G (2004) J Organomet Chem 689:4016
27. Rossell O, Seco M, Angurell I (2003) C R Chimie 6:805
28. Kreiter R, Kleij AW, Gebbink RJMK, van Koten G (2001) Top Curr Chem 217:163
29. Alonso B, Alonso E, Astruc D, Blais J-C, Djakovitch L, Fillaut JL, Nlate N, Moulines F, Rigaut S, Ruiz J, Valerio C (2002) In: Newkome G (ed) Advances in dendritic macromolecules, vol 5. JAI Press, Greenwich, CT, pp 89–127
30. Kiyotaka O, Shigetoshi T (2003) Top Curr Chem 228:39
31. For reviews on bioorganometallic chemistry see, for example: (a) Fish RH, Jaouen G (2003) Organometallics 22:2166; (b) Fouda MFR, Abd-Elhazer MM, Abdelsamaia RA, Labib AA (2007) Appl Organometal Chem 21:613

32. Balzani V, Campagna S, Denti G, Juris A, Serroni S, Venturi M (1998) Accounts Chem Res 31:26
33. Kaifer AE, Gómez-Kaifer M (1999) Supramolecular electrochemistry. Wiley-VCH, Weinheim/New York
34. Kaifer AE (2007) Eur J Inorg Chem 5015
35. Venturi M, Ceroni P (2003) C R Chimie 6:935
36. Hecht S, Fréchet JMJ (2001) Angew Chem Int Ed 40:74
37. Nierengarten J-F (2003) C R Chimie 6:725
38. (a) Cameron CS, Gorman CB (2002) Adv Funct Mater 12:17; (b) Gorman CB, Smith JC (2001) Accounts Chem Res 34:60; (c) Gorman CB (2003) C R Chimie 6:911
39. Flanagan JB, Margel S, Bard AJ, Anson FC (1978) J Am Chem Soc 100:4248
40. (a) Morán M, Cuadrado I, Losada J (1987) Organometallics 6:2341; (b) Morán M, Cuadrado I, Losada J (1988) J Chem Soc Dalton Trans 833; (c) Morán M, Cuadrado I, Pascual C, Casado CM, Losada J (1992) Organometallics 11:1210; (d) Morán M, Pascual C, Cuadrado I, Losada J (1993) Organometallics 12:811
41. Casado CM, Morán M, Losada J, Cuadrado I (1995) Inorg Chem 34:1668
42. (a) Morán M, Casado CM, Cuadrado I, Losada J (1993) Organometallics 12:4327; (b) Casado CM, Cuadrado I, Morán M, Alonso B, Lobete F, Losada J (1995) Organometallics 14:2618
43. Cuadrado I, Casado CM, Lobete F, Alonso B, González B, Losada J (1999) Organometallics 18:4960
44. Alonso B, Cuadrado I, Morán M, Losada J (1994) J Chem Soc Chem Commun 2575
45. Alonso B, Morán M, Casado CM, Lobete F, Losada J, Cuadrado I (1995) Chem Mater 7:1440–1442
46. Cuadrado I, Morán M, Casado CM, Alonso B, Lobete F, Garcia R, Ibisate M, Losada J (1996) Organometallics 15:5278
47. Cuadrado I, Casado CM, Alonso B, Morán M, Losada J, Belsky V (1997) J Am Chem Soc 119:7613
48 (a). Castro R, Cuadrado I, Alonso B, Casado C, Morán M, Kaifer AE (1997) J Am Chem Soc 119:5760; (b) González B, Casado CM, Alonso B, Cuadrado I, Morán M, Wang Y, Kaifer AE (1998) Chemical Communications 2569–2570
49 (a). Takada T, Diaz DJ, Abruña HD, Cuadrado I, Casado C, Alonso B, Morán M, Losada J (1997) J Am Chem Soc 119:10763
50. Casado CM, Cuadrado I, Morán M, Alonso B, Barranco M, Losada J (1999) Appl Organomet Chem 13:2345
51. Casado CM, González B, Cuadrado I, Alonso B, Morán M, Losada J (2000) Angew Chem Int Ed 39:2135
52. Alonso B, Casado CM, Cuadrado I, Morán M, Kaifer AE (2002) Chemical Communications 1778–1779
53. Alonso B, González B, Ramírez E, Zamora M, Casado CM, Cuadrado I (2001) J Organomet Chem 637–639:642
54. Zamora M, Herrero S, Losada J, Cuadrado I, Casado CM, Alonso B (2007) Organometallics 26:2688
55. Zamora M, Alonso B, Pastor C, Cuadrado I (2007) Organometallics 26:5153
56. (a) Frey H, Schlenk C (2000) Top Curr Chem 210:70; (b) Frey H, Lach C, Lorenz K (1998) Adv Mater 10:279
57. Lang H, Lühmann B A (2001) Adv Mater 20:1523
58. Son DY (2001) In: Rappoport Z, Apeloig Y (eds) The chemistry of organic silicon compounds, vol 3. Wiley, New York, pp 745–803
59. Alonso B (1997) Ph.D. thesis, Universidad Autónoma de Madrid
60. Togni A, Hayashi T (eds) (1995) Ferrocenes. VCH, Weinheim
61. Long NJ (1998) Metallocenes: an introduction to sandwich complexes. Blackwell Science, London
62. Zanello P (2003) Inorganic electrochemistry. Theory, practice and application. Royal Society of Chemistry: Cambridge

63. For reviews on metallocene-containing polymers see, for example: (a) Peckham TJ, Gómez-Elipe P, Manners I (1998) Metallocene-based polymers. In: Togni A, Halterman R (eds) Metallocenes. Wiley-VCH: Weinheim, pp 723–773; (b) Nguyen P, Gómez-Elipe P, Manners I (1999) Chem Rev 99:1515; (c) Abd-El-Aziz A, Manners I (2005) J Inorg Organomet Polym Mater 15:157; (d) Hudson RDA (2001) J Organomet Chem 637–639:47
64. van der Made AW, Van Leeuwen PWNM (1992) Journal of the Chemical Society-Chemical Communications 1400–1401
65 (a). Roovers J,Toporowski PM, Zhou LL (1992) Polym Prepr Am Chem Soc Div Polym Chem 33:182; (b) Zhou LL, Roovers J (1993) Macromolecules 26:963
66. Seyferth D, Son DY, Rheingold AL, Ostrander RL (1994) Organometallics 13:2682
67. For interesting reviews on the substituent effects of the cyclopentadienyl ring in metallocene and related cyclopentadienyl compounds see: (a) Hays ML, Hanusa TP (1996) Adv Organomet Chem 40:117; (b) Okuda J (1992) Top Curr Chem 160:97
68. (a) Murray RW (ed) (1992) Molecular design of electrode surfaces. Wiley-Interscience: New York; (b) Abruña HD (1988) Electrode modification with polymeric reagents. In: Skotheim TA (ed) Electroresponsive molecular and polymeric systems, vol 1, Chapter 3. Marcel Dekker, New York
69. Lobete F, Cuadrado I, Casado CM, Alonso B, Morán M, Losada J (1996) J Organomet Chem 509:109
70. Cuadrado I, Morán M, Moya A, Casado CM, Barranco M, Alonso B (1996) Inorg Chim Acta 251:5
71. Seyferth D, Kugita T, Rheingold AL, Yap GPA (1995) Organometallics 14:5362
72. (a). Kim C, Jung I (1998) Inorg Chem Commun 1:427; (b) Kim C, Jung I (1999) J Organomet Chem 588:9
73. Brüning K, Lang H (1999) J Organomet Chem 592:147
74. Benito M, Rossell O, Seco M, Muller G, Ordinas JI, Font-Bardiá M, Xolans X (2002) European Journal of Inorganic Chemistry Issue: 9:2477–2487
75. (a) Benito M, Rossell O, Seco M, Segalés G (1999) Inorg Chim Acta 291:247; (b) Benito M, Rossell O, Seco M, Segalés G (1999) Organometallics 16:5191; (c) Benito M, Rossell O, Seco M, Segalés G (2001) J Organomet Chem 619:245;(d) Angurell I, Muller G, Rocamora M (2003) Dalton Transactions (2003) 6:1194–1200 (e) Angurell I, Lima JC, Rodríguez L-I, Rodríguez L, Rossell O, Seco M, New J Chem (2006) 30:1004; (f) L-I, Rodríguez L, Rossell O, Seco M, Muller (2007) J Organomet Chem 692:851
76. Rámirez-Oliva E, Cuadrado I, Casado CM, Losada J, Alonso B (2006) 691:1131
77 (a). Hawker CJ, Fréchet JMJ (1990) J Am Chem Soc 112:7638; (b) Hawker CJ, Fréchet JMJ (1990) J Chem Soc Chem Commun 15:1010–1013
78. For an excellent recent review on the convergent construction of dendrimers see: Grayson SK, Fréchet JMJ (2001) Chem Rev 101:3819
79. Barlow S, O'Hare D (1997) Chem Rev 97:637
80. Ceccon A, Santi S, Orian L, Bisello A (2004) Coord Chem Rev 683:724
81. Guillaneux D, Kagan HB (1995) J Org Chem 60:2502
82. (a) Marciniec B (ed) (1992) Comprehensive handbook on hydrosilylation. Pergamon, Oxford; (b) Ojima I (1989) The hydrosilylation reaction. In: Patai S, Rappoport Z (eds) The chemistry of organic silicon compounds, part 2. Wiley, New York, pp 1479–1526
83. Alonso B, Cuadrado I, unpublished results
84. (a) Robin MB, Day P (1967) Adv Inorg Chem Radiochem 10:247; (b) Creutz C (1983) Prog Inorg Chem 30:1
85. It has been recently shown that the combination of CH_2Cl_2 and $[n\text{-}Bu_4N][B(C_6F_5)_4]$ as solvent and supporting electrolyte, respectively provides close-to optimal conditions for electrochemical oxidation studies of multiferrocenyl compounds by minimizing nucleophilic attack by the electrolyte anion and improving product solubilities. See for example: (a) Camire N, Mueller-Westerhoff UT, Geiger WE (2001) J Organomet Chem 637–639:823; (b) Barrière F, Kirss RU, Geiger WE (2005) Organometallics 24:48
86. Alonso B, García P, Losada J, Cuadrado I, Casado CM (2004) Biosens Bioelectron 19:1617
87. Zamora M (2006) Ph.D. thesis, Universidad Autónoma de Madrid

88. It is well-known that electron-withdrawing substituents on the vinyl group decrease the rate of hydrosilylation processes compared to the more electron-donating groups. On the other hand, steric factors due to substituents in both silane or vinyl groups, may also affect the rate and completeness of hydrosilylation reactions. See, for example: (a) Stein J, Lewis LN, Smith KA, Lettko KX (1991) J Inorg Organomet Polym 1:325; (b) Liu HQ, Harrod JF (1990) Can J Chem 68:1100
89. For a recent work illustrating the effect of the substituents of the Si–H and C = C groups on hydrosilylation reactions see: Hilf S, Cyr PW, Rieder DA, Manners I, Ishida T, Chujo Y (2005) Macromol Rapid Commun 26:950
90. (a). Schlüter AD (1998) Top Curr Chem 197:165 139; (b) Schlüter AD, Rabe JP (2000) Angew Chem Int Ed 39:864
91. Schlüter AD (2005) Top Curr Chem 245:151
92. Frauenrath H (2005) Prog Polym Sci 30:325
93. Hawker CJ, Wooley KL (2005) Science 309:1200
94 (a). Suijkerbuijk BMJM, Shu L, Klein Gebbink RJM, Schlüter AD, van Koten G (2003) Organometallics 22:4175; (b) Kim Y, Mayer MF, Zimmermman SC (2003) Angew Chem Int Ed 42:1121; (c) Chow H-F, Leung C-F, Li WL, Wong K-W, Xi L (2003) Angew Chem Int Ed 42:4919; (d) Zhang Y, Xu Z, Li X, Chen Y (2005) J Polym Sci Part A Polym Chem 45:3303
95. See for example: (a) Clarson SJ, Semlyen JA (eds) (1993) Siloxane polymers. Prentice Hall, Englewood Cliffs, NJ; (b) Noll W (1968) Chemistry and technology of silicones. Academic, New York; (c) Mark JE, Allcock HR, West (eds) (1992) Inorganic polymers. Prentice-Hall, Englewood Cliffs, NJ, p 141.
96. Zamora M, Alonso B, Cuadrado I, Manuscript in preparation
97. García B, Casado CM, Cuadrado I, Alonso B, Morán M, Losada J (1999) 18:2349
98. García B, Casado CM, Alonso B, Cuadrado I, unpublished results
99. (a) Zimmerman SC, Lawless LJ (2001) Top Curr Chem 217:96; (b) Zeng F, Zimmerman SC (1997) Chem Rev 97:1681
100. Ong W, Gómez-Kaifer M, Kaifer AE (2004) Chemical Communications 15:1677–1683
101. See for example: (a) Beer PD, Bayly SR (2005) Top Curr Chem 255:125 and references therein; (b) Beer PD, Hayes EJ (2003) Coord Chem Rev 240:167; (c) Beer PD, Gale PA (2001) Angew Chem Int Ed Engl 40:486; (d) Beer PD (1998) Accounts Chem Res 31:71
102. Mc Quade DT, Pullen AE, Swager TM (2000) Chem Rev 100:2537
103. Casado CM, Cuadrado I, Alonso B, Morán M, Losada J (1999) J Electroanal Chem 463:87
104. Valerio C, Fillaut J-L, Ruiz J, Guittard J, Blais J-C, Astruc D (1997) J Am Chem Soc 119:2588
105. (a) Wang J (2008) Chem Rev 108:814; (b) Wang J (2001) Electroanalysis 13:983
106. Ryabov AD (2004) Adv Inorg Chem (2004) 55:201–269
107. Losada J, Cuadrado I, Morán M, Casado CM, Alonso B, Barranco M (1997) Anal Chim Acta 338:191
108. For some examples of electrocatalytic oxidation of ascorbic acid by ferrocene-based modified electrodes see, for example: (a) Raoof J-B, Ojani R, Kiani A (2001) J Electroanal Chem 515:45; (b) Wang Wu, Z Tang J, Teng R, Wang E (2001) Electroanalysis 13:1093; (c) Kazakeviien B, Valincius G, Niaura G, Talaikyt Z, Kaemkait M, Razumas V, Plauinaitis D, Teierskien A, Lisauskas V (2007) Langmuir 23:4965

Chapter 9
Carbosilane Dendrimers: Molecular Supports and Containers for Homogeneous Catalysis and Organic Synthesis

Maaike Wander, Robertus J.M. Klein Gebbink, and Gerard van Koten

9.1 Introduction

The attachment of catalytic species to support materials is a widely applied method to combine the advantages of homogeneous and heterogeneous (supported) catalysis. The commonly used organic supports are insoluble polymeric materials, which have been developed with great success for solid phase organic synthesis and have a long history and importance. Obvious difficulties with these materials are their restricted loading capacity, the wettability issues, the often restricted accessibility of active (supported) sites, their reactivity or incompatibility towards reactive reagents, such as organometallics, and last but not least their high polydispersity. The use of soluble support materials can solve some of these problems, and for this reason soluble dendrimers have been explored as supports for homogeneous catalysts. Some of the advantages of dendrimers over many other types of macromolecules are their well defined structures and low polydispersity, good solubility in common organic solvents, and the presence of well-defined end-groups for the anchoring of catalytic species (see Chapter 1), all of which facilitate analysis of the (loaded) dendrimers often with atomic precision.

During the last decade, several reviews appeared describing the use of dendrimers as soluble supports for catalysts [1–11]. Among these, the silicon-based carbosilane dendrimers (see Chapter 3) assume a special position because of their structural robustness and stability towards highly reactive reagents. These are important prerequisites for any derivatization of the dendritic structure as well as for the introduction of catalytic metal sites, *vide infra*. Carbosilane dendrimers derive their kinetic and thermodynamic stability from the relatively high dissociation energy (306 kJ/mol) and low polarity of the Si–C bond [1]. The first demonstration of the potential of these unique properties was the successful synthesis of a carbosilane dendrimer **1** functionalized at its periphery with catalytically active NCN-pincer

M. Wander, R.J.M. Klein Gebbink, and G. van Koten
Chemical Biology & Organic Chemistry, Debye Institute for Nanomaterials Science, Faculty of Science, Utrecht University, Padualaan 8, 3584 CH Utrecht, The Netherlands
E-mails: m.wander@uu.nl; r.j.m.kleingebbink@uu.nl; g.vankoten@uu.nl

P.R. Dvornic and M.J. Owen (eds.), *Silicon-Containing Dendritic Polymers*, Advances in Silicon Science 2, nickel catalysts (NCN = $[C_6H_3(CH_2NMe_2)_2\text{–}2,6]^-$) [12]. Due to their molecular size of about 2 nm, such catalytic species can be separated from the reaction solutions by nanofiltration, which in principle opens the way for recycling of the catalyst as well as for continuous use of such catalysts in membrane reactors.

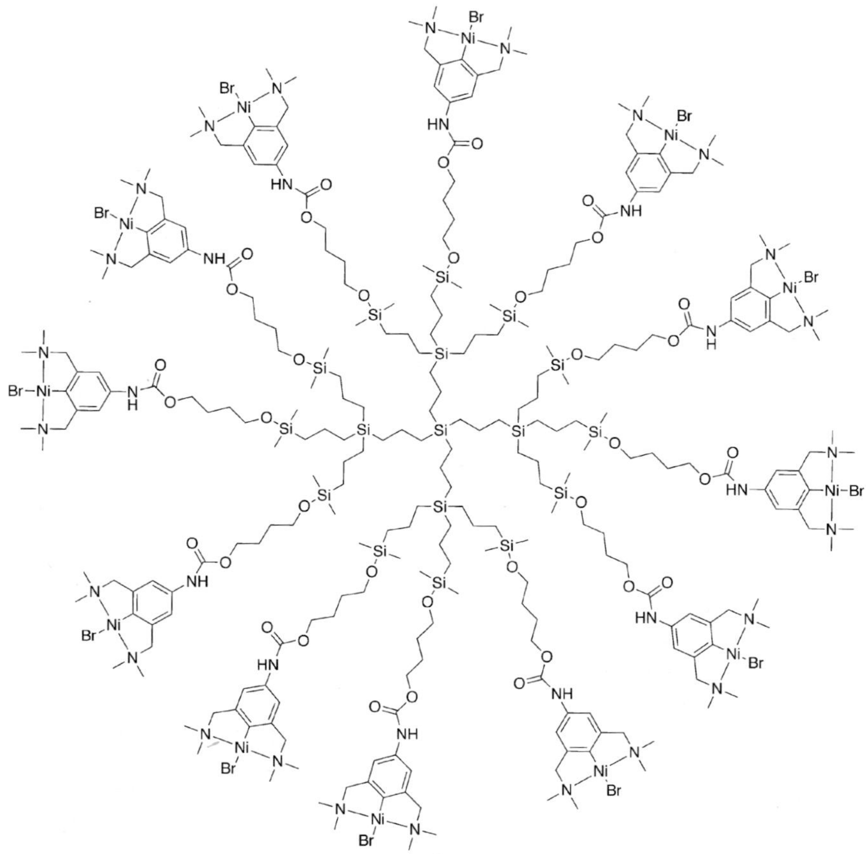

1

Carbosilanes are particularly attractive dendrimers because of the degree of control one has over the sites where the catalytic species can be attached to the dendrimer scaffold. Figure 9.1 shows various possibilities for this attachment: (A) on the periphery of a dendrimer or dendron; (B) at the core or at the focal point; (C) at the branching points, or (D) as dendrimer-encapsulated nanoparticles [3]. However, the last possibility is not applicable to carbosilane dendrimers, since interactions between the catalyst and the scaffold are unlikely because of the absence of internal metal-coordinating moieties (compare with Section 11.5).

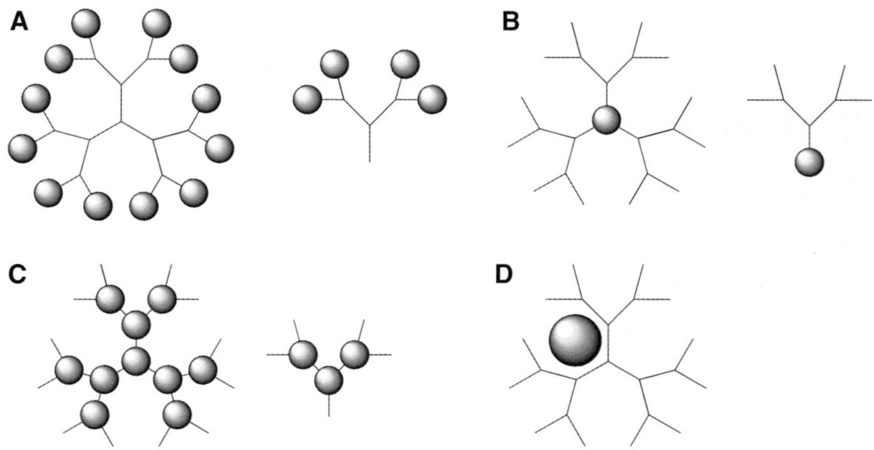

Fig. 9.1 Schematic representation of different binding modes of catalytic species to a (carbosilane) dendritic (i.e., dendrimers or dendrons) architecture. (**a**) At the periphery; (**b**) at the core; (**c**) at the branching points; (**d**) as encapsulated nanoparticles [3].

However, the high degree of branching which originates from the use of silicon as the core atom and branching points, makes these carbosilane dendrimers very suitable for use as carriers for catalytic species.

Since dendritic architecture can be designed with a high degree of control (see Chapter 1), features such as mutual proximity of catalytic sites (e.g., mutual deactivation or cooperativity effects) and secondary coordination (sphere effects) by the catalyst-surrounding dendritic organic scaffold can be implemented in a controlled fashion [13–15]. The sizes of the carbosilane dendrimers (larger than 2 nm) and their low polydispersities enable application of (nano)filtration methods with commercially available membranes. In fact, dendrimers with sizes in the range of 2.5–3.0 nm are already suitable for nanofiltration [15]. Other types of catalytic dendrimers that have been used in nanofiltration set-ups, include, for example, porphyrin-functionalized pyrimidine dendrimers, which were used as recyclable photosensitizers for the oxidation of various olefinic compounds [16]. For applications in continuous catalytic processes retentions of at least 99.99% are required in order to obtain a catalyst system that remains in the reactor for a prolonged period of time [17].

Several reactor designs and membrane types have been developed for the separation of soluble, catalytic, carbosilane dendrimers from the product stream [3, 18–26]. Membrane technology can be performed either batch-wise or in continuous-flow membrane reactors, using micro-, ultra-, nano-, diafiltration or reverse osmosis. Passive dialysis is used batch-wise, since it utilizes the difference in concentrations of the reactants and products on either side of the membrane (see Fig. 9.2). In continuous-flow membrane reactors, reactants and product, but not the

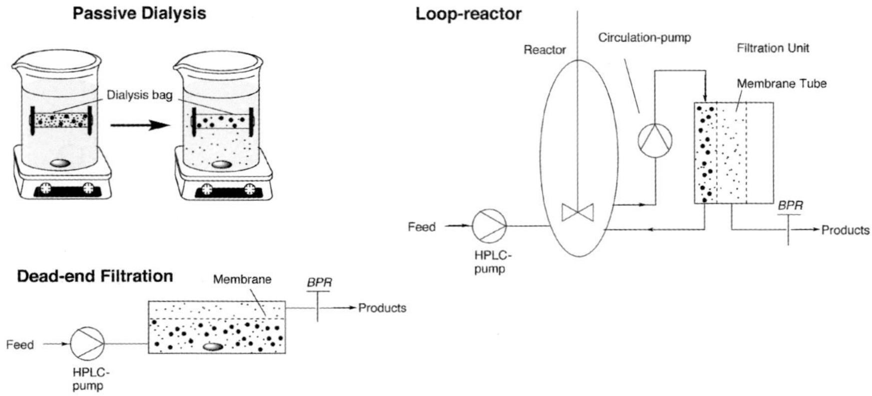

Fig. 9.2 Schematic representation of different combinations of reactors and (nano)filtration methods (*BPR* = Back-pressure regulator) [19].

dendrimer catalyst, are transported through the membrane by applying an external force like air pressure. Dijkstra et al. reviewed the different types of continuous-flow membrane reactors used with dendrimer catalysts [19].

Different kinds of membranes, prepared from polymeric as well as ceramic materials, have been developed. They need to be solvent resistant, and preferably able to withstand industrially important solvents such as tetrahydrofuran (THF), dimethylformamide (DMF), dimethylsulfoxide (DMSO), *N*-methylpyrrolidone (NMP) and dichloromethane (DCM). Most currently available polymeric membranes, however, still show some incompatibility problems with these conditions. Ceramic membranes, on the other hand, overcome some of these problems, but suffer from brittleness and are still more expensive. Vankelecom et al. nicely reviewed the field of solvent resistant nanofiltration techniques [26].

In this chapter, we review the use of carbosilane dendrimers in catalysis research, including different derivatives with covalently bonded metal catalysts and different types of catalysis in which they have been used. Multiple uses of such relatively expensive carrier molecules, whose synthesis is labor intensive, will be of importance when it comes to industrial applications. Most of the catalytic species that have been covalently attached to dendrimers are simple mimics of the catalysts used in the corresponding homogeneous catalytic processes and are based on either phosphine or other ligand donor systems. In each section of this chapter, a brief overview of the synthesis and structural aspects of the respective dendrimer catalysts is presented first, followed by a discussion of the catalytic processes in which they have been applied. Special attention is given to filterability, cooperativity effects, solvent compatibility, recyclability, secondary coordination sphere effects, etc. The chapter closes with a concise overview of the use of newly developed, soluble carbosilane dendrimers as carrier molecules for multistep supported organic synthesis. It is particularly noted that many requirements that are so important for the successful use of soluble dendrimer catalysts in homogeneous catalysis are encountered in this application again. These include the number and proximity of active sites, solvent and reagent compatibility, as well as inertness and filterability of the dendrimer supports.

9.2 Carbosilane Dendrimers with Covalently Bound Catalysts

The common method for the introduction of catalytic species onto a dendrimer scaffold is attachment of the catalyst directly to the periphery of the preformed dendrimer. In this way, well-defined multi-catalyst structures can be obtained, in which the activity of each site can be easily compared to that of the corresponding monomeric analogue. The effects of the presence of dendritic mass around each active metal center, as well as proximity effects of neighboring metal centers, are often reflected in the catalytic reactivity that is observed. This reactivity (selectivity) can be either unchanged, increased or lowered with respect to its monomeric analogue, i.e., either a positive or negative "*dendritic effect*" is observed. In addition to the peripherally-functionalized dendrimers, dendrimers that are core-functionalized with a catalyst have also been reported. The large dendritic mass surrounding the single catalytic site in such dendrimers can cause a site-isolation effect, leading to the formation of a microenvironment around the metal center that influences its catalytic properties. Two main groups of catalysts covalently bonded to dendrimer scaffolds include those with phosphine-based dendrimer ligands and those with "other" (non-phosphine-based) dendrimer ligands.

9.2.1 Synthesis and Structural Aspects of Phosphine-Based Dendrimer Catalysts

Phosphine ligands are widely used in homogeneous catalysis to stabilize complexes and to direct activity and selectivity. Many mono- and polydentate phosphine ligands have been synthesized and used in, for example, palladium and platinum complexes, yielding homogeneous catalysts for a plethora of reactions. One way to make these catalysts recyclable and applicable in continuous catalytic processes is to connect them to a soluble support, i.e. to create dendrimers with sizes in the range of 2–3 nm which allows their separation from the reaction solutions by nano- or diafiltration techniques.

For the synthesis of phosphine-based dendrimer catalysts, mainly carbosilane dendrimers comprising silicon atoms at the core and at the branching points with ethane-diyl- or propane-1,3-diyl-spacers have been used, although other types of connectivities between the silicon centers are also known. The synthesis of alkyl-based carbosilane dendrimers of low polydispersities has been extensively studied [27] and these are now available for many applications. As these dendrimers have relatively well-defined structures, spectroscopic techniques such as NMR are applicable for the study of their structural features in solution. In addition to these, other dendritic species have also been developed in which a carbosilane dendron is connected to a non-carbosilane core. Examples are the silsesquioxane cores with carbosilane dendrons studied by Cole-Hamilton et al. and Morris et al. (see also Chapter 7) [28–33].

In most cases, the phosphine ligands, most frequently diphenylphosphino groups, are first anchored covalently to the periphery of the dendrimer scaffold and then converted into the corresponding phosphine-metal complexes. The phosphine groups can be connected to the carbosilane dendrimer either directly, or via a linking moiety, a tether, such as a benzylic group or aliphatic chain [34–36]. The generally applied method for direct attachment involves the use of a lithiated phosphine, which is quenched with the chlorosilane dendrimer end-groups (see Reaction Scheme 9.1) [35]. In some cases (not shown), a hydrosilylation reaction was used to connect the phosphine ligands to the allylic or vinylic end-groups of the dendrimer scaffold [29].

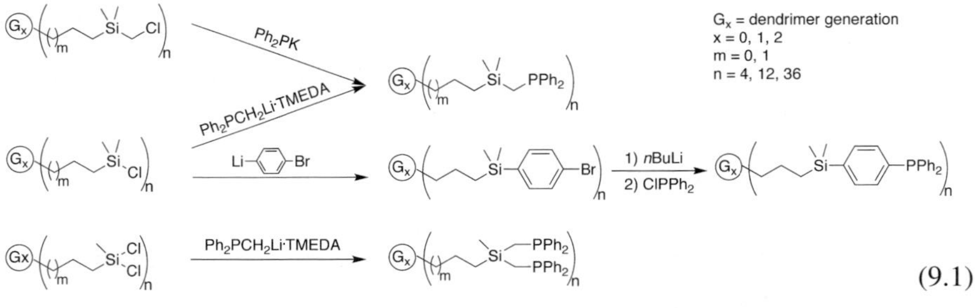

(9.1)

In general, introduction of phosphine ligands in the last step of the synthetic sequence reduces the chance of losing or damaging the sensitive and expensive phosphanyl functions during the synthesis of the dendrimer scaffold. The attachment of a phosphine group via a tether to a dendrimer periphery can be performed in two ways. Either the tether is first connected to the dendrimer and then loaded with the phosphine groups, or it is first attached to the phosphine, followed by connecting the tether to the dendrimer surface. In the latter case the tether-phosphine unit can be perfectly tuned for the formation of both the ligand and the phosphine-metal complex site, since dendrimer effects are less likely to play a role.

9.2.2 *Catalytic Reactivity of Phosphine-Based Dendrimer Catalysts*

Phosphine-based dendrimer catalysts have been tested in a wide variety of catalytic reactions. This section deals with a comparison between their reactivity and selectivity with those of their monomeric analogues.

9.2.2.1 Allylic Substitution

Palladium phosphine complexes are excellent catalysts of allylic substitution or alkylation reactions. Van Leeuwen et al. reported on the synthesis and catalytic activity of several carbosilane dendrimers functionalized with phosphine-palladium units [37–39]. Peripheral phosphine-functionalized carbosilane dendrimers up to the second generation were synthesized by the reaction of peripheral chlorosilyl-functionalized dendrimers with [(diphenylphosphino)methyl]lithium·TMEDA [37, 38], yielding dendrimers containing one or two diphenylphosphanyl groups per end-group (see Reaction Scheme 9.1). The palladium complexes were formed by subsequent reaction of the phosphine functionalized dendrimers with either [PdClMe(COD)] or $[PdCl[(\eta^3\text{-}C_3H_7)]_2$. Dendrimers with one PPh_2 end-group per Si-branching point formed trans-complexes when reacted with [PdClMe(COD)], whereas dendrimers with two PPh_2 end-groups per Si-branch point formed cis-complexes, i.e. the latter end-groups were acting as bidentate ligands. The resulting metallodendrimers were tested as catalysts in allylic alkylation and amination reactions (see Reaction Schemes 9.2A and B, respectively), both in batch and in continuous processes [37]. It was found that the number of phosphine units per dendrimer end-group had some influence on the activity and selectivity. The dendrimer catalyst containing bidentate ligands appeared to be more active in both the allylic amination reaction between crotyl acetate and piperidine and the allylic alkylation reaction between crotyl acetate and sodium 2-methylmalonate. The selectivity in the allylic alkylation reactions was hardly influenced by the dendrimer size and number of phosphine end-groups, whereas in the allylic amination reaction the selectivity induced by bidentate ligands was slightly different from that of the monodentate ligands.

The size of the dendrimer scaffold hardly affected the catalytic activity, which makes these dendrimers suitable supports. It also enables their application in continuous-flow membrane reactors, as indicated by the retention of up to 99.7% obtained for the second generation dendrimer catalyst in dichloromethane during the allylic alkylation reaction. The preservation of activity with increasing dendrimer generation also indicates that all catalytic sites act as independent catalysts even when steric crowding at the periphery increases, as is the case for higher generation dendrimers. However, the test of dendrimer catalysts in continuous processes, resulted in a rapid drop of the yield of allylic substitution reactions, rather than in leaching of the dendrimer catalyst (as demonstrated by previous retention studies [38]). Since this drop in yield was ascribed to catalyst deactivation, dendrimers with an ethanediyl tether between the terminal silicon and the phosphorus atom, i.e. with $SiCH_2CH_2PPh_2$ end-groups, were prepared and tested. This seemingly subtle change in the tether length positively influenced the stability of the catalytic systems and although the yield of the reaction products still decreased to some extent, decomposition of the catalysts was not observed in continuous allylic amination reactions. Combining these results with those obtained from previous retention studies, the retention of the dendrimer catalyst with ethanediyl tether was estimated to be 98.5–99%.

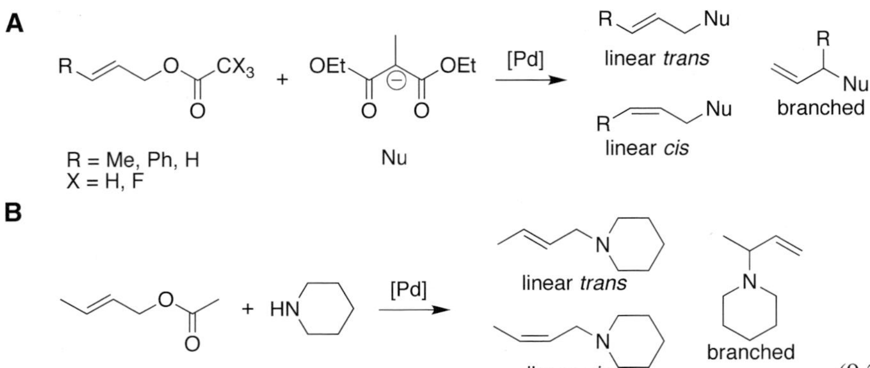

(9.2)

The synthesis of 1,1′-bis(diphenylphosphino)ferrocene (dppf) ligands having dendrons connected to the para-positions of the phenyl groups yielded carbosilane dendrimers with a bidentate phosphorus ligand at the core **2** [39]. The dendrons of up to the third generation, were converted to the corresponding bisphosphine palladium complexes by reaction with $[PdCl_2(MeCN)_2]$. The complexation of the ligand in a bidentate cis fashion, forming similar palladium complexes as dppf allowed comparison of the catalytic activity between dendritic and palladium-dppf complexes. For catalytic experiments, the dendritic ligands were first reacted with crotylpalladium chloride dimer to form dppf-like palladium complexes, which were then used in the palladium-catalyzed allylic alkylation of 3-phenyl-1-allyl acetate with diethyl sodium-2-methylmalonate (Reaction Scheme 9.2A; R = Ph; X = H) and were all found to be active and producing mainly the linear *trans*-product. A decrease in reaction rate was observed, when using the higher generation dendrons, especially on going from the second to the third generation. This was ascribed to a decreased mass transport of the reagents through the dendrimer shell surrounding the catalytic site. However, the selectivity for the branched product increased with increasing generation, probably because the increased steric bulk of the dendrimer shell caused hindering of the nucleophile attack on the palladium-allyl. As another reason for these different selectivities a gradual change in the apolar microenvironment created by the carbosilane dendrons was also suggested.

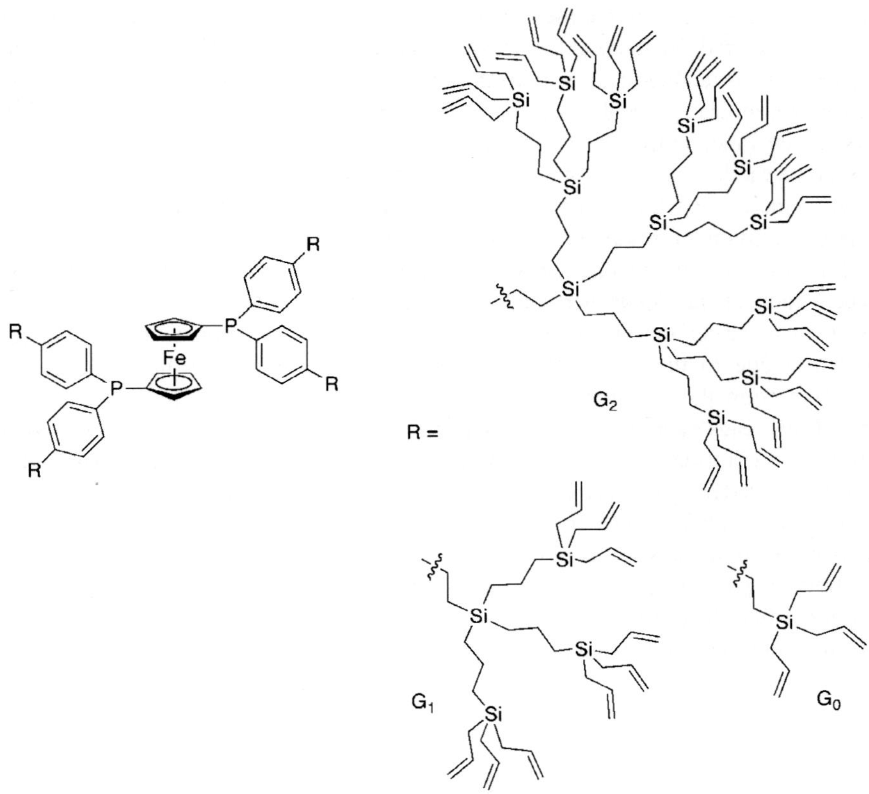

2

9.2.2.2 Hydrovinylation

Carbosilane dendrimers with phosphine palladium complexes at the periphery were also tested in hydrovinylation processes. Van Koten's and Vogt's groups described the synthesis of monomeric and dendrimeric carbosilanes functionalized with various ω-(diphenylphosphino) carboxylic acid ester end-groups (see Reaction Scheme 9.3) [40]. The palladium complexes of these ligands were successfully applied in the palladium-catalyzed hydrovinylation of styrene. The monomeric model compounds showed higher activity than the corresponding 0G dendrimer catalysts and became more active with increasing Pd-P,O ring size. However, while almost complete isomerization occurred in the batch-wise processes, when hydrovinylation reactions with dendrimer complexes were carried out in a continuously operated nanofiltration membrane reactor with 0G-Pd_4 catalyst, hardly any isomerization or formation of other side products were observed. This can be ascribed to the shorter contact times between the catalyst, reagents and products in the continuous set up.

In addition to the easy separation when using dendrimer catalysts, this decreased amount of isomerized product in the product stream is an important advantage over the non-supported or batch-wise reactions. However, as the 0G-Pd_4 catalyst showed only modest retention in the membrane reactor, a loss of catalytic species during the reaction occurred, resulting in lower catalysts concentrations and corresponding lower conversions. In addition, formation of palladium black, which was observed on the membrane surface, probably accelerated the decrease in catalyst activity. Consequently, higher generations of these dendrimer catalysts were prepared and tested in order to overcome the low retention of 0G-Pd_4 [41]. As expected, the retention of dendrimer catalyst increased when going from the zeroth to the first generation, but the time dependent product formation resembled the one found for the 0G catalyst.

Furthermore, although the selectivity of the hydrovinylation reaction also increased, deactivation of the dendrimer catalysts was again observed. This deactivation was ascribed to easier double or multiple phosphine complexation to the palladium, probably due to the increased proximity of neighboring phosphino end-groups resulting from the flexibility of dendrimer arms. Formation of these diphosphine complexes leads to a lack of free (available) ligand in the catalytic solution, which results in the formation of palladium black and thereby catalyst deactivation. NMR studies suggested that deactivation of the dendrimer palladium catalysts took place during the catalysis, since freshly prepared solutions contained only pure monophosphine complexes. Since the deactivation process is similar in both batch and continuous processes, but different total turnover numbers are reached, it can be excluded that deactivation is correlated to the amount of converted styrene or to the amount of passed solvent [41].

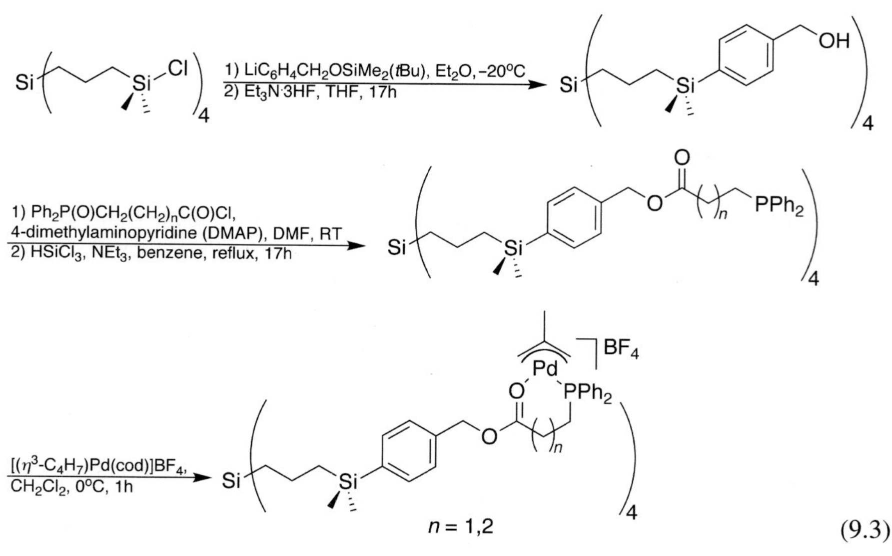

(9.3)

Rossell et al. synthesized palladium-functionalized carbosilane dendrimers of up to the third generation with peripheral P-stereogenic monophosphines **3** [42]. These dendrimer catalysts were formed by a reaction of dinuclear $[Pd(\mu\text{-}Cl)(\eta^3\text{-}2\text{-}MeC_3H_4)]_2$ with carbosilane dendrimers having chiral monophosphines (with either 2-biphenylyl or 9-phenanthryl substituents) at their periphery, thereby grafting the dendrimer surface with [PdCl(η^3-2-MeC$_3$H$_4$)] units. The activity of these catalysts was tested in the asymmetric hydrovinylation of styrene to give 3-aryl-1-butenes and related derivatives, and compared to those of two chiral, monomeric model compounds. It was found that activity, product selectivity, and enantiomeric excesses (ee) depended strongly on both the nature of the phosphine and the halide abstractor used. For the first generation dendrimer carrying 2-biphenylylphosphine ligands, the best results were obtained by using Na[BArF] (BArF = $\{B[3,5\text{-}(CF_3)_2C_6H_3]_4\}^-$) as activator instead of $AgBF_4$. The best results in terms of ee (79% towards the S-isomer) were obtained with the third generation dendrimer-palladium catalyst. Furthermore, all generations of dendrimer catalysts with 9-phenanthryl substituents (activated with $AgBF_4$) were extremely active but did not induce any enantioselectivity. Surprisingly, the second generation of this dendrimer catalyst, when activated with Na[BArF], yielded mainly the (R)-3-phenyl-1-butene isomer, in contrast to the first generation which mainly produced the (S)-3-phenyl-1-butene, similar to all other dendrimer systems studied. When, in the case of dendrimer catalysts with 2-biphenylylphosphine ligands, the solvent was changed from CH_2Cl_2 to supercritical CO_2, selectivities and ee values comparable to those obtained in the previous studies were observed, although the activities were somewhat lower [43]. In general, no clear dendrimer effect was observed for these systems. The hydrovinylation reactions were performed in autoclaves, not membrane reactors, so that no conclusions about the retention of dendrimer catalysts could be provided.

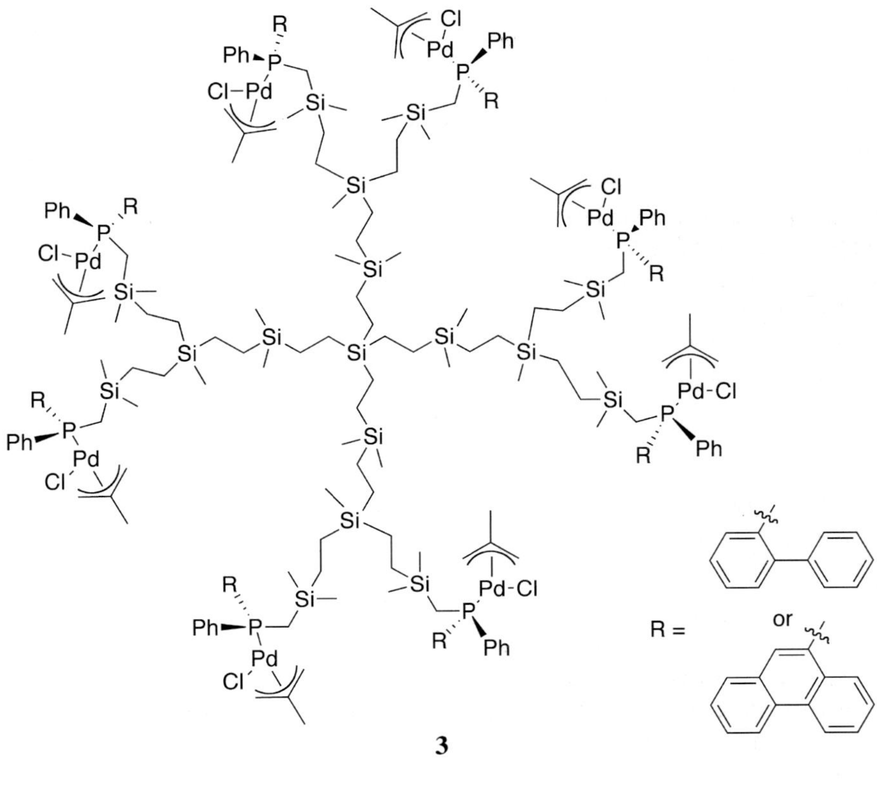

3

In 2002, the group of Benito and Rossell used the same ligand system as Van Leeuwen and his coworkers, with diphenylphosphino-terminated carbosilane dendrimers containing one or two PPh_2 groups per dendrimeric arm (see Reaction Scheme 9.1). These ligands were palladated and platinated and the palladium complexes were tested in the batch-wise hydrovinylation of styrene (see Reaction Scheme 9.4) [44]. Comparable monomeric systems were also synthesized, in which one or two phosphine groups were present per end-group. The activity of the dendrimer catalysts appeared to be lower than that of the monomeric analogues and the polynuclear complexes containing four or eight catalytic end-groups were more active than the dendrimer catalysts of Van Koten/Vogt containing 8 or 12 terminal bidentate P,O-coordinating atoms at the periphery [41]. Moreover, a good selectivity towards 3-phenyl-1-butene was observed.

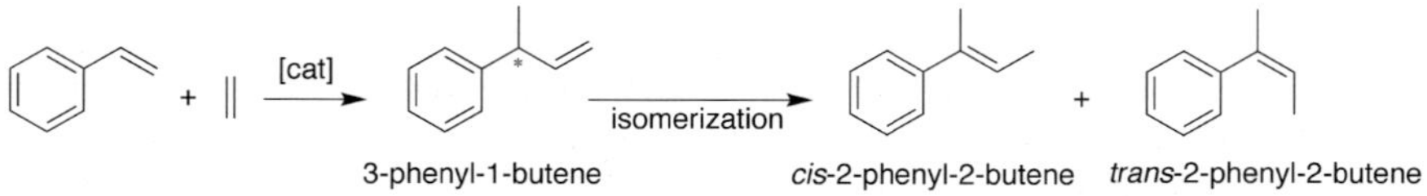

(9.4)

9.2.2.3 Hydrogenation and Transfer Hydrogenation

Recently, rhodium and ruthenium catalysts with carbosilane dendrimers containing P-stereogenic monophosphine ligands prepared by Rossell et al., were used in hydrogenation reactions [45]. The same ligand system was also used in the hydrovinylation reaction, but for this purpose it was metallated with rhodium or ruthenium using either [RhCl(cod)] (cod = cyclooctadiene) or [$RuCl_2$(p-cymene)], respectively (see Reaction Scheme 9.5 where P^* = carbosilane dendrimer, as depicted in Reaction Scheme 9.3, functionalized with (S)-CH_2PPh(2-biphenylyl). The rhodium complexes were tested in the hydrogenation of dimethyl itaconate and the relationship between the size/generation and its catalytic properties was investigated. Going from the model compound to the first generation dendrimer the activity decreased, probably due to the decreased accessibility of the metal centers in the dendrimer species. The ee value was zero in all cases.

The ruthenium complexes were tested in the transfer hydrogenation of acetophenone. A positive dendrimer effect was observed, since the zeroth generation dendrimers were slightly more active than the model compounds, although the enantiomeric excesses were low and appeared to be hard to reproduce.

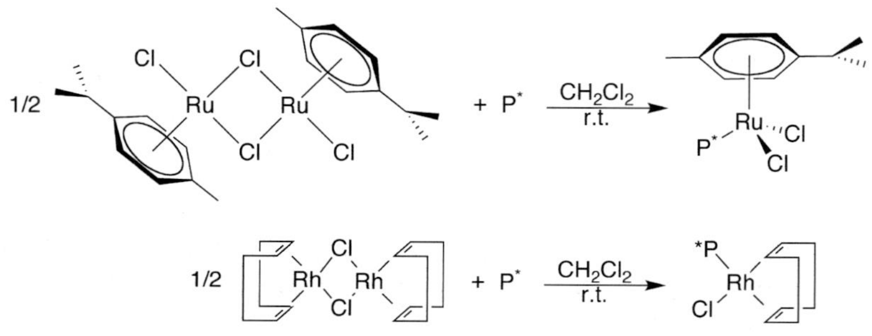

(9.5)

In 2003, the same group reported on the synthesis of neutral and cationic diphenylphosphino rhodium- and iridium-containing dendrimers up to the second generation [46]. The dendrimer ligands contained one or two diphenylphosphino groups per dendrimer arm and were reacted with $[MCl(cod)]_2$ (M = Rh, Ir) to form neutral or cationic catalytic species. The products were tested as catalysts in the hydrogenation of 1-hexene and the results obtained with the neutral dendrimer rhodium(I) were compared to those obtained with mononuclear and dinuclear rhodium complexes, showing a higher activity for the dendrimer catalysts. However, there was also a slight decrease in turnover frequency (TOF) for the higher generations of dendrimer species. The cationic rhodium(I) metallodendrimer was less active than the cationic monomeric analogue, but it showed higher activity than the neutral species of the same generation. The first generation iridium(I) metallodendrimer showed comparable results to those of the rhodium one. However, these results were irreproducible, which was attributed to the low solubility of iridium(I) dendrimers under the reaction conditions applied.

Dendrimer catalysts bearing phosphine ligands on the periphery, synthesized by the groups of Van Leeuwen and Rossell (see Reaction Scheme 9.1), were also tested in the transfer hydrogenation reactions, e.g. reduction of cyclohexanone to cyclohexanol [47]. "Single" and "double" metallic-layered dendrimers containing ruthenium complexes were prepared from diphenylphosphino ligands (see structures **4** and **5**) where in the double layered dendrimers one metal site functioned as a branching point. In preliminary catalytic studies with mononuclear ruthenium complexes an increase in activity was observed when the reaction was performed in refluxing propan-2-ol compared to the reaction at room temperature. It also appeared that neutral complexes were more active than the corresponding cationic ones. Under the same conditions, the single-layered dendrimer complexes showed lower activities than the mononuclear analogues, pointing to a negative dendrimer effect. Furthermore, among the first generation compounds, the neutral compounds were more active than the corresponding cationic analogues, following the same trend that was found for the mononuclear complexes. The double metallic-layered dendrimers were unstable and could not be tested.

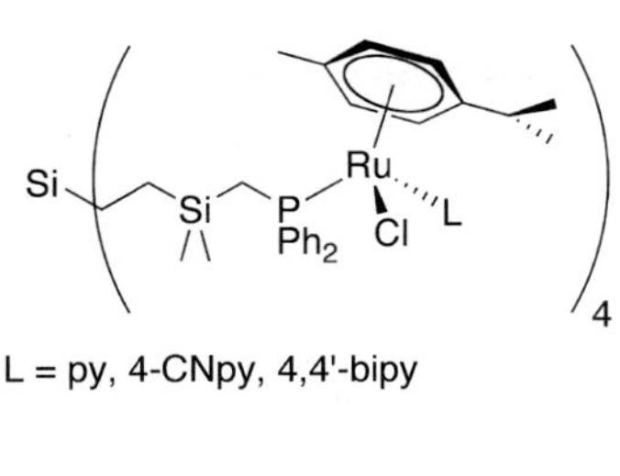

4

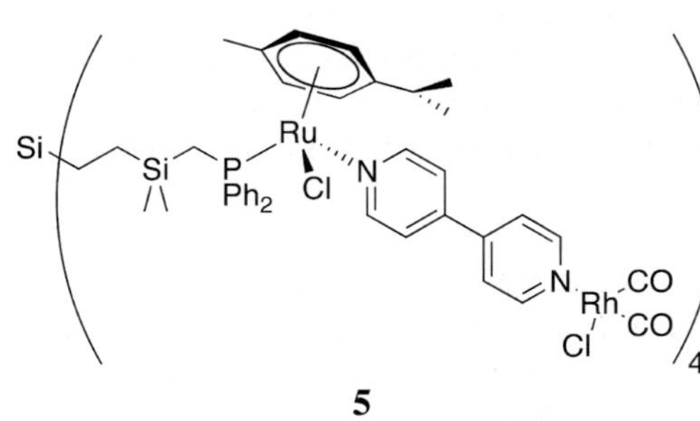

5

A different system was developed by Kakkar and co-workers, who synthesized a dendrimer with a dimethylsilyl-linked 3,5-dihydroxybenzyl alcohol scaffold, containing diphenylphosphino end-groups [48]. By reaction with [RhCl(cod)], catalytic species **6** was obtained that appeared to be active in the hydrogenation

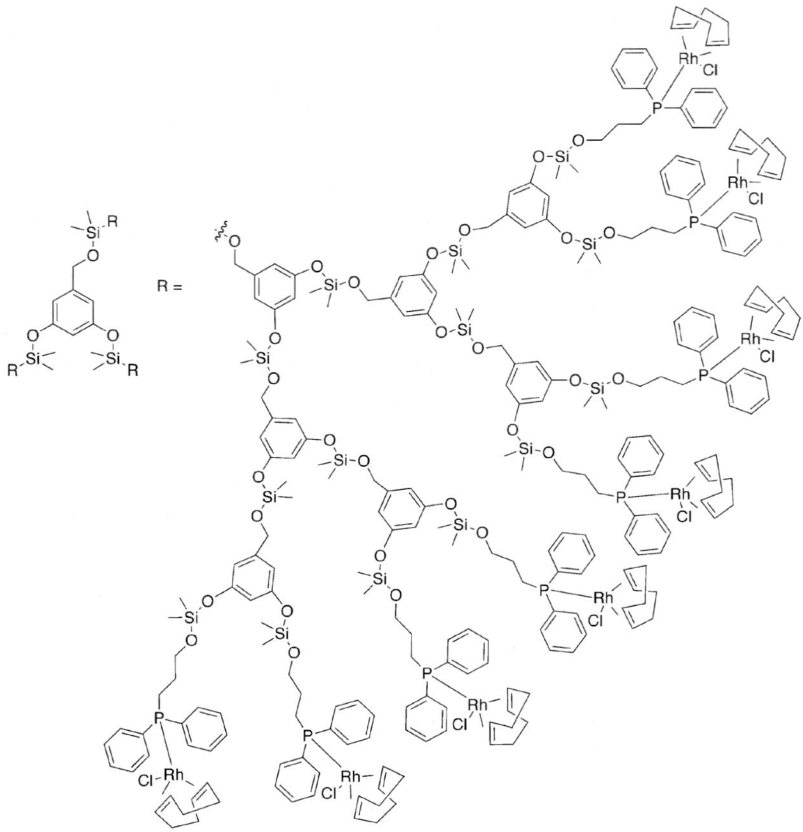

6

of 1-decene. The catalytic activity was found to be slightly dependent on the dendrimer generation and reaction time, giving higher activity for higher generation, i.e. showing a positive dendrimer effect [48, 49] All dendrimer catalysts showed an increasing activity with time. In addition, while with the first generation dendrimer nearly full conversion was reached after 5 h, the same was achieved with the third generation already after 2 h. However, it should be noted that the number of Rh(I) centers at the periphery of metallodendrimers increases with increasing generation, and therefore causes the higher TOFs per dendrimer catalyst, not per single Rh-center. Furthermore, when these results were compared with those obtained for Rh(I)-supported tri(alkyl)phosphine dendrimers which were prepared earlier [50] and in which the catalytic sites were distributed throughout the dendrimer scaffold, higher TOFs were observed for metallodendrimers with peripherally-supported catalytic sites. These differences can be explained by both steric factors (diphenylphosphine ligands are shielding the Rh-center more than trialkylphosphine ligands) and differences in electronic density at the metal centers.

Recovery of metallodendrimer catalysts and their suitability for recycling were also investigated. It was found that recycled catalysts retained their efficiency and, surprisingly, an increase in activity was observed upon recycling [49]. In addition to this, it was also found that during the dendrimer synthesis, the use of uncontrolled reaction conditions (i.e. a multi-step [two-, four- or six-step procedures], reaction without dropwise addition or control of temperature) led to the formation of hyperbranched carbosilane polymers (see Chapters 12 and 13) [48]. These polymers were also tested in hydrogenation of 1-decene, giving good conversion rates, with reaction times as low as 0.5 h for the two- and four-step hyperbranched polymers [49]. In the case of the six-step hyperbranched polymer, the catalytic activity increased gradually during the catalysis, as in the case of the metallodendrimers. This difference in behavior was explained by suggesting that the large six-step hyperbranched polymer was almost a single species organometallic dendrimer, since its MALDI-TOF spectrum showed only one extremely dominant peak (in contrast to the lower generation hyperbranched polymers that consisted of mixtures of organometallic macromolecules present in equal amounts). As a consequence, it behaved more like a typical metallodendrimer catalyst [49].

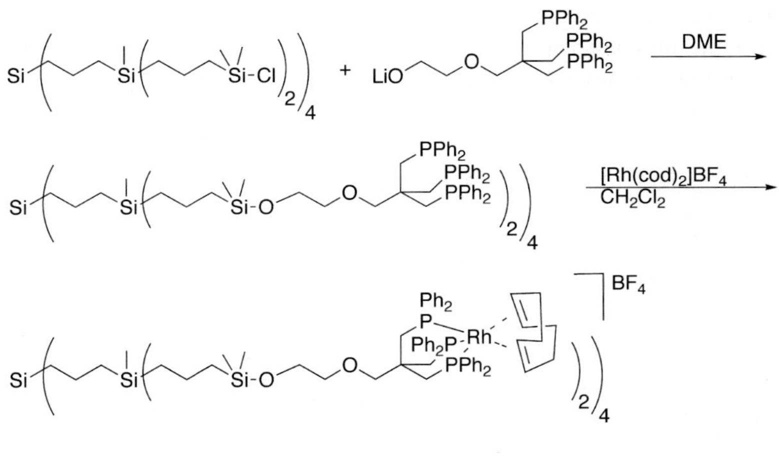

(9.6)

Findeis and Gade synthesized mononuclear model compounds and the corresponding dendrimers with tripodal trisphosphine ligands at their peripheries, which were then transformed into the corresponding rhodium complexes (see Reaction Scheme 9.6) [51]. Mononuclear molybdenum complexes were also prepared, but these were not tested as catalysts. In contrast to the dendrimers described earlier, these tripodal trisphosphine ligands were first connected to a tether moiety, forming a ligand-tether combination, which was then connected to the periphery of the dendrimer scaffold and subsequently metallated using $[Rh(cod)_2]BF_4$. The catalytic properties of these metallodendrimers in hydrogenation of styrene and 1-hexene were the same as those of the monomeric catalysts. However, since the catalysts were connected to the dendrimer scaffold, they were robust enough to withstand several sequential recycling steps. This robustness also allowed a complete analysis of these dendrimer catalysts, which confirmed their uniformity (i.e., low polydispersity).

9.2.2.4 Hydroformylation and Hydrocarbonylation

The diphenylphosphino-functionalized carbosilane dendrimers of Reaction Scheme 9.1, were also applied as catalysts in hydroformylation reactions by both Van Leeuwen and Cole-Hamilton et al. [28–31, 35]. Van Leeuwen investigated dendrimers comprising silicon atoms at the core and branching points, and ethanediyl- or propane-1,3-diyl- spacers, which were functionalized with diphenylphosphino end-groups at the periphery. These ligands were metallated with [Rh(acac)$(CO)_2$] and the resulting dendrimer-phosphino rhodium complexes were tested in the hydroformylation of 1-octene (A) and hydrocarbonylation of 1-alkenes (B) in Reaction Scheme 9.7 [35, 52]. These dendrimer catalysts showed the same selectivity as the mononuclear analogues, however, their activity depended on their size and flexibility. Those with more flexible C_3-spaces in the backbone generally gave higher conversions than the more compact ones containing the C_2-spacers. Furthermore, the C_3-dendrimer monophosphine rhodium catalysts showed a decrease in activity with increasing dendrimer generation. The difference between the catalysts derived from the dendrimers with monodentate and bidentate end-groups was apparent, since the latter gave slower reactions. A similar difference was also observed for the mononuclear rhodium parent compounds. In all cases similar proportions of linear-to-branched products (7:3) were obtained.

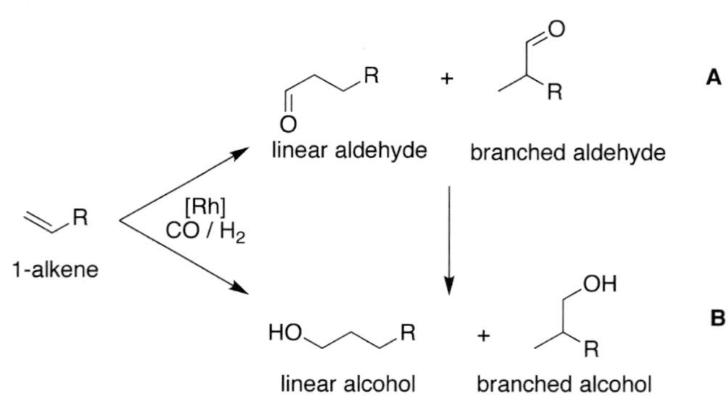

(9.7)

Cole-Hamilton et al. described the synthesis of diphenyl- and dialkyl-phosphino-containing carbosilane dendrimers based on a polyhedral silsesquioxane core, with up to 48 phosphino end-groups (**7**) [28–32]. The first and the second generation diphenyl- and diethyl-phosphino-containing dendrimers were successfully applied as ligands for the synthesis of metallodendrimer rhodium catalysts which were then tested in hydroformylation and hydrocarbonylation reactions of 1-octene. In the hydroformylation reaction, the use of the first and the second generation diphenylphosphino dendrimer ligands resulted in high linear-to-branched ratios (up to 14:1). However, on changing from the first to the second generation ligand a drop in reaction rate by a factor of 2 was noted, probably due to increased steric hindrance at the periphery of the more bulky second generation ligand.

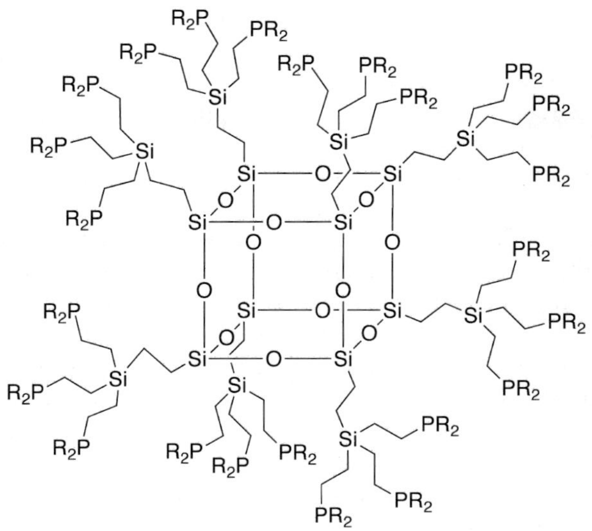

R = alkyl, phenyl

7

In the hydrocarbonylation reactions of alkenes (hex-1-ene, oct-1-ene, non-1-ene, prop-1en-2-ol (see Reaction Scheme 9.7, *vide supra*), the linear-to-branched ratios of the resulting products were slightly higher (3.1:1) with the dendrimer ligands than with free triethylphosphine (2.4:1) [29, 32]. Depending on the nature of the phosphine end-groups and the complexity of the dendrimer scaffold, differently functionalized dendrimers showed different properties during the catalytic experiments. For example, hexyl and ethyl functionalized phosphines (compared to methyl phosphines) and longer alkanediyl spacers between the branching points in the dendrimer scaffold gave higher solubilities and therefore more homogeneous systems. The generation of the dendrimers did not seem to affect the selectivity of the hydroformylation reaction. However, the branching pattern did show a large influence on the reaction rate, in a way that the amount of peripheral functional groups and the length of the bridges between the phosphines in the dendrimers appeared to be the determining factors for the reactivity of the corresponding rhodium catalysts.

9.2.3 Synthesis and Structural Aspects of Non-phosphine-Based Dendrimer Catalysts

In addition to carbosilane dendrimers with phosphine ligands, the synthesis and reactivity of many other dendrimer systems functionalized with ligands with different donor atoms, e.g. nitrogen containing ligands, have been investigated. The first prepared and tested metallodendrimer catalyst was reported by Van Koten et al. in 1994 [12]. This system contained the so-called NCN-pincer metal catalysts connected to the periphery of a carbosilane dendrimer via spacer moieties. The ATRA (atom-transfer radical addition) catalysis, such as mononuclear [NiX(NCN)] for addition of CCl_4 to activated alkenes had been developed earlier by the Van Koten group [12, 53, 54] and was found to function equally well on the dendrimer scaffold. Following this report, many other metallodendrimer catalysts containing multiple catalytic sites at the periphery of carbosilane as well as non-carbosilane dendrimers have been described [1–11].

In general, two approaches to the introduction of catalytic metal-ligands are known. They follow strategies similar to those described earlier for the introduction of phosphine-based ligands to a carbosilane dendrimer scaffold. According to one approach, the metal can be introduced at the end of the synthesis of the carbosilane poly-ligand scaffold, which requires novel ways for polyfunctionalization of the carbosilane dendrimer periphery since many of these donor atoms have weaker coordination power than the phosphorus-donor atom in a triorganophosphine. In the other approach, the metal-ligand unit can be synthesized before connecting the complex to the periphery of the dendrimer, which has the advantage that the most sensitive step in the synthesis can be carried out separately from the synthesis of the dendrimer support which allows working with pure metal-ligand units.

As with the chemistry of phosphine-based ligands, metal-ligand units have been either put onto the periphery of dendrimers or modified by attachment of one or various dendrons to the ligand(s) of the metal-ligand complex. For example, cyclopentadienyl ring(s) of metallocenes or half-sandwich complexes have been functionalized with dendrons [55], resulting in the formation of dendritic catalysts that are functionalized either at the dendrimer core or at the focal point of the dendrons (see for example structures **10A** and **B**). The post-synthesis modification of the focal point of a dendron allows for incorporation of more reactive cores, which can be useful for the chemistry of both dendrimers and hyperbranched polymers [55]. Furthermore, when a dendron is attached to a well-known ligand system, the stability of the catalyst can be increased, making the catalyst more robust towards water and air.

In most cases, dendrimers are used as supports for catalytic species themselves. However, examples are also known in which the periphery of a dendrimer is functionalized with anionic cocatalytic species, such as $–[B(C_6F_5)_3]^-$ units, which during catalysis are applied as counteranions of actual cationic catalytic species (see Structure **9**) [56]. Another possibility is attachment of species that function as initiators, e.g. in radical or atom transfer radical (ATR) polymerization reactions, which in certain cases can lead to the formation of multi-arm star polymers [57–61] (see also Chapters 3, 10 and 11).

9.2.4 *Catalytic Reactivity of Non-phosphine-Based Dendrimer Catalysts*

Various non-phosphine-based metallodendrimer systems described above have been applied as catalysts in a variety of different reactions. In this section, the activity of these systems is described and compared to their mononuclear analogues.

9.2.4.1 Aldol Condensation

The catalytic activity of metal complexes of the so-called ECE-pincer ligands (ECE = $[C_6H_3(CH_2E)_2\text{-}2,6]^-$ where E = NR_2, PR_2, AsR_2, OR, SR) has been explored intensively (see, for example, [20, 62–65]). The first example in which 12 NCN-pincer Ni halide moieties were connected via linkers to a carbosilane dendrimer scaffold [12] and tested as a catalyst in a Kharasch addition reaction involving atom-transfer radical addition (ATRA) was described in 1994. Moreover, it was also shown that the resulting metallodendrimer had a suitable size for use in a membrane reactor. Later, NCN-pincer palladium halide units were connected directly, without a linker, to hyperbranched-polycarbosilane supports (see Fig. 9.3) and tested in the aldol condensation of benzaldehyde and methyl isocyanoacetate as a model reaction [66]. The activities observed were comparable to those of the analogous single-site palladium catalyst, while the size of the structures allowed their purification by dialysis to obtain a metallodendritic polymer with a low enough polydispersity to allow its application in a continuous membrane reactor.

In 2001, the same group reported on the synthesis of macrocyclic structures in which a dendrimer scaffold was functionalized with monoanionic C,N-chelating

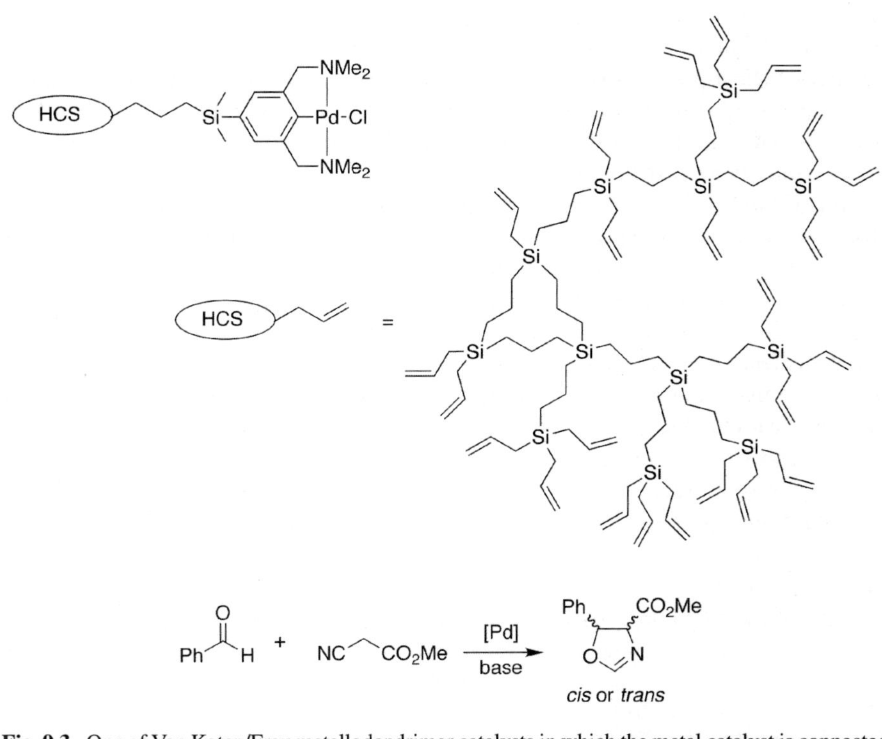

Fig. 9.3 One of Van Koten/Frey metallodendrimer catalysts in which the metal catalyst is connected to a hyperbranched carbosilane dendritic scaffold and used for aldol condensation of benzaldehyde and methyl isocyanoacetate [66].

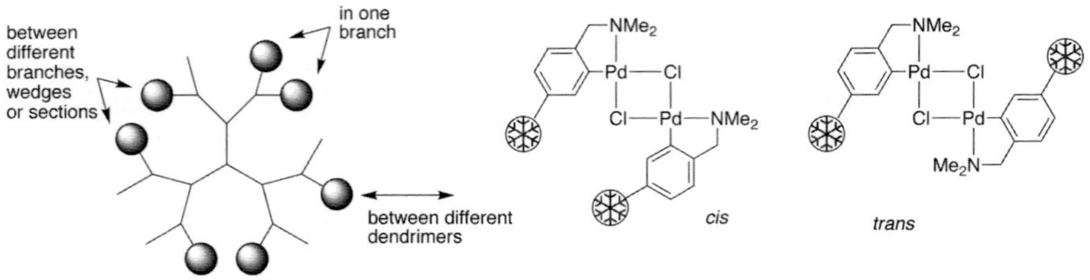

Fig. 9.4 Schematic representation of the formation of stereoisomeric dendrimer dimers in cyclopalladated CN-derivatized carbosilane dendrimers [67].

ligands [67]. On palladation of these ligands the formation of dimeric structures occurred which in the case of the first and second generation dendrimers led to the formation of several stereoisomers. These could originate from dimer formation between [CN-PdCl] complexes of different branches, sections or dendrons (intramolecular), or even different dendrimers (intermolecular) (see Fig. 9.4).

Upon addition of pyridine, the single-site, mononuclear pyridine adducts of these macromolecular constructs were formed and used as catalysts for the aldol condensation of benzaldehyde and methyl isocyanoacetate. Abstraction of the halide anion yielded the polycationic analogues, which were used in subsequent catalytic experiments without palladium black formation. This indicated an increased stability of the polycationic palladium-dendrimer species as compared to the mononuclear palladium complexes. It was also observed that a metallodendrimer system which had a more rigid skeleton or a higher degree of peripheral crowding was less active in the aldol condensation reaction, also causing a slight change in the *cis/trans* ratio of the oxazoline product.

In addition to these, other mononuclear and dendrimer systems with cage-like structures were also prepared [68]. In these structures, the catalytic NCN-pincer palladium complexes were encapsulated within carbodiazasilane cages, forming mononuclear macrocycles. These cages could be connected to each other via a central core molecule, yielding a multicage dendrimer structure containing three macrocycles. Interestingly, in the same catalytic aldol condensation reaction, the dendrimer (multicage) cationic derivative appeared to be more active than the mononuclear analogues.

Pyrenoxy-based NCN-pincer palladium molecular tweezers were attached to carbosilane dendrimers of up to the second generation (see Structure **8**) [69]. Aldol condensation reactions between methylisocyanoacetate and aromatic aldehydes, catalyzed by the monomeric pincer palladium species with and without pyrenoxy ligands and the metallodendrimers were compared. For the monomeric series it was found that the presence of pyrenoxy groups in the catalyst increased initial reaction rates by a factor of 2. One of the reasons for this could be that pyrenoxy groups stabilize one or more of the transition states or intermediates in the catalytic cycle. On the other hand, dendrimer catalysts caused a dramatic fourfold drop in the reaction rate compared to the monomeric analogue, probably due to steric crowding, which makes the catalytic sites less accessible. However, the *cis/trans* ratio of the resulting products was not dependent on the nature of the catalyst used. Although these metallodendrimers are large enough to be used in the continuous processes, no details on this aspect have been reported.

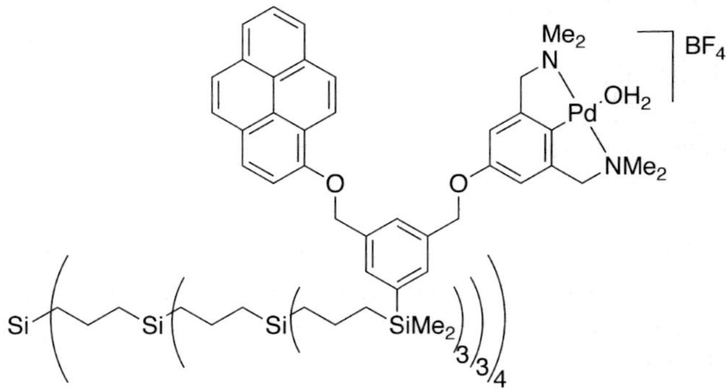

8

9.2.4.2 Polymerization

The metallocene-catalyzed polymerization of α-olefins allows production of polymers with new architectures and properties. The active species during this catalysis is a cationic metallocene stabilized by a non-coordinating anion. This contact ion pair is formed by treating the neutral metallocene with an activating cocatalyst, such as methyl aluminoxane (MAO) or perfluorophenylborane, $B(C_6F_5)_3$. In most cases, the metallocene (in the catalytic site) is connected to the supporting dendrimer. However, Mager et al. described a study of the polymerization properties of metallocene-cation-anion pairs in which the anions were embedded in an extremely sterically crowded environment by using a carbosilane dendrimer's periphery [56]. The resulting steric hindrance led to weaker coordination of the anions that were less nucleophilic thus leading to their increased stability. This made the system suitable for use even in cases of very sterically demanding ion pairs.

Polyanionic carbosilane dendrimers functionalized with $-[B(C_6F_5)_3]^-$ groups on their periphery up to the second generation were also synthesized (see Structure **9**). The stable ammonium salts of these dendrimers were tested in olefin polymerization with various zirconocene catalysts and they all showed high activities. Other advantages of these systems are their high stability during polymerization (no decrease in activity was observed) and the possibility to use aliphatic solvents such as hexane.

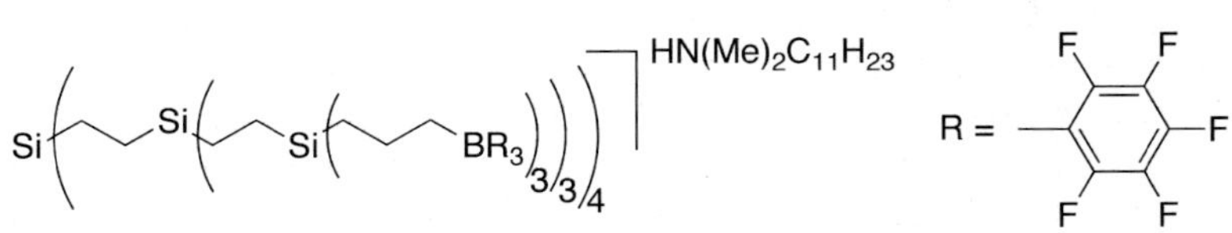

9

The De Jésus group described several dendrimer-catalyst systems that were active in the polymerization of α-olefins. The difference between these systems and those described above was in the way the dendrimer scaffold was connected to the catalytic center, which either involved the focal point of a dendron, or the periphery of the dendrimer. The system described by Andrés et al., contained organometallic complexes at the focal point of a dendron [55, 70, 71], for example, the first generation carbosilane dendrons (with C_3-linkers) with cyclopentadiene groups at focal points. Subsequent reaction of potassium hydride with these Cp groups yielded the cyclopentadienyl salt that upon reaction with $TiCl_4$ or $ZrCl_4$·2THF in toluene afforded the corresponding metallocene dendrimer **10A** [55]. These metallocene dendrimers were activated with MAO and tested in the polymerization of both ethylene and propylene. Compared to the simple metallocene catalyst [MCp_2], the activity of the titanium Cp-dendron catalyst was lower by one order in the polymerization of ethylene, whereas the corresponding zirconium catalyst showed no decrease in activity. Comparable results were obtained earlier for carbosilanes with C_2 linkers, although for these systems a decrease in activity of about 30% was observed for the zirconocene compound [71].

The dendrimer metallocene zirconium catalysts were also used for propylene polymerization where their activity was found to be comparable to that of the monomeric analogue. The second and the third generation catalysts could not be synthesized, probably because of steric issues or because appropriate reaction conditions were not found [55]. In another study, a β-diketiminato ligand was chosen as the focal point of dendrons [70] and the reaction of dendritic (and non-dendritic) β-diketimines with half-sandwiched titanium and zirconium complexes (η^5-$CpMCl_3$) in the presence of triethylamine as Lewis acid yielded mixed cyclopentadienyl(β-diketiminato)metal complexes **10B**. Both non-dendritic and dendritic complexes were activated with MAO and applied in ethylene polymerization experiments where the metallodendrimer catalysts showed a slightly increased activity compared to their non-dendritic analogues. Nevertheless, their performance was still far from that of the well-established metallocene complexes (η^5-Cp_2MCl_2; M = Ti, Zr) [55, 71].

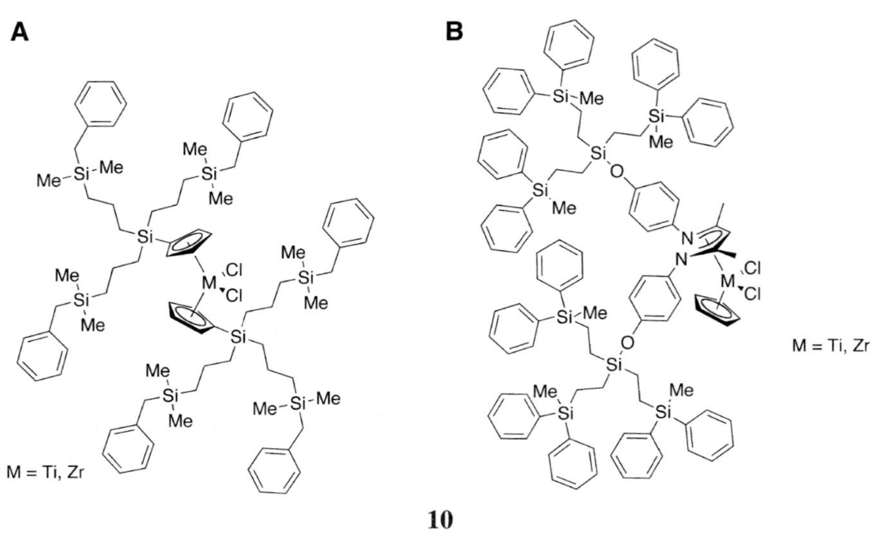

10

Arévalo et al. reported on carbosilane dendrimers with monometallic cyclopentadienyl-aryloxy metal complexes at their periphery [72, 73]. Various titanium and zirconium derivatives of 4-allyl-2-methoxyphenol or 4-allyl-2,6-dimethoxyphenol precursors **11**, were synthesized and tested in the polymerization of ethylene. In addition to dendrimers in which [–$OTiCl_2Cp$] groups were placed para to the aliphatic (dendrimer) substituent, systems with the same groups positioned ortho to the aliphatic chain were also prepared. However, the catalytic activity of these complexes was found to be very low or negligible after activation with MAO. In general, systems having OMe group(s) ortho to the [–$OTiCl_2Cp$] group yielded high molecular weight polyethylene with low polydispersity. Probably the proximity of ortho-OMe groups allows for additional electronic stabilization of active species, resulting in a higher propagation rate. The use of either monometallic or metallodendrimer catalysts did not affect the polymerization activity, but metallodendrimer catalysts yielded polymers with higher crystallinity.

Furthermore, a difference in activity was observed between fresh and aged toluene solutions of the dendrimer titanium complexes. When activated with MAO, the fresh solutions behaved as moderately active systems, while the aged ones were very active. It was suggested that this could be ascribed to the occurrence of dendrimer aggregation processes. The aggregates showed high hydrodynamic volumes and low polydispersities, which was confirmed by the relaxation times distribution that showed only one dynamic process with a narrow distribution. When peripheral [$TiCl_2Cp$] units were replaced with [$TiCl_2Cp^*$] or [$MClCp_2$] (M = Ti, Zr) groups or when 2-allyl-6-methylphenol was incorporated as a ligand (yielding complexes with peripheral units ortho to the aliphatic chain) no aggregation of the resulting metallodendrimer was observed on solution aging. These results show that steric and/or electronic effects induced by different ligands or organometallic units can have a large effect on the synthesis and catalytic activity of the resulting metallodendrimeric catalysts.

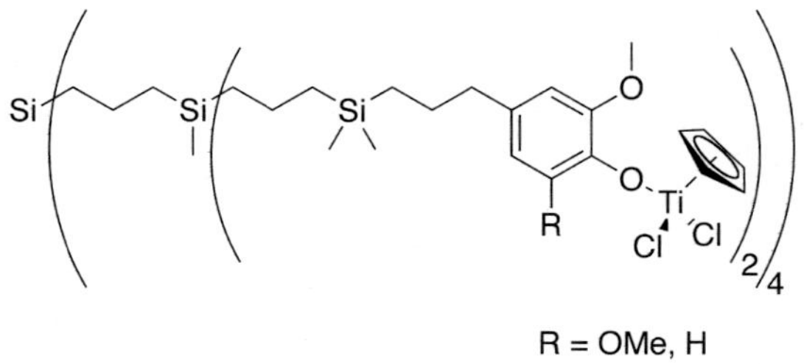

11

Benito et al. synthesized a series of monomeric complexes and dendrimers with peripheral N,N′-iminopyridine chelating ligands, **12** [74–76]. Neutral and cationic palladium derivatives of these dendrimer ligands were prepared and the cationic complexes were tested in copolymerization reactions [74]. The cationic palladium compounds were found to be active catalysts for the alternating syndiospecific copolymerization of CO and 4-tert-butylstyrene, producing mainly syndiotactic polyketones due to a chain-end-controlled mechanism. Changing the ligand by adding methyl substituents to the aryl ring increased activity of the palladium catalysts, whereas addition of a methyl group ortho to the coordinating N-atom of the pyridine ring deactivated the catalyst completely by steric hindrance. In contrast to other dendrimer catalysts, the presence of a dendrimer support appeared hardly relevant in terms of the stability of catalytic species. The catalytic performance was, however, found to be dependent on the dendrimer generation, which influenced the microstructure of the copolymerization products. Higher generations showed superior activities, i.e. a positive dendrimer effect, but at the same time produced shorter and less stereoregular copolymer chains.

Nickel derivatives of the same N,N′-iminopyridine chelating dendrimer-ligand system **12**, were studied as catalysts for the polymerization of ethylene [75, 76]. When activated with MAO under mild reaction conditions, these catalytic systems were active for both oligomerization and polymerization of this monomer. Their catalytic activity and selectivity depended on the ligand structure and dendrimer generation similar to the previously described palladium catalysts. Steric protection of the axial coordination sites by addition of methyl substituents to the aryl ring increased the polymerization and oligomerization activities, while a methyl group in the six-position of the pyridine ring hindered the equatorial coordination sites and thereby reduced activities. It turned out that production of ethylene insertion products (oligomer for high generation vs. polymer for low generation), as well as the oligomer chain-length distribution, the branching density, molecular weight and polydispersity of the polymers could be regulated by selecting the dendrimer generation. Catalysts derived from higher generation dendrimers produced polyethylenes with lower degrees of branching but with higher molecular weights in cases where polymers were formed. In addition, they showed higher oligomerization activities. A combination of steric pressure on the growing chains and microenvironmental protection of the polymerization catalytic species by higher generation dendrimers, may explain these results [76].

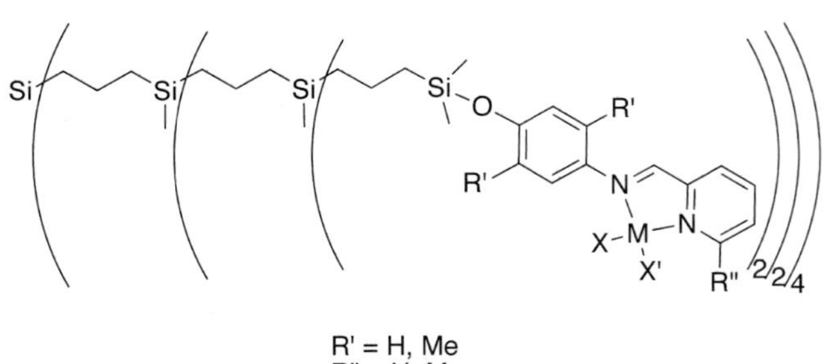

R' = H, Me
R'' = H, Me
M = Pd; X = Cl, Me; X' = Cl, NCMe
M = Ni; X = Br; X' = Br

12

Iron containing metallodendrimer catalysts for polymerization of ethylene were synthesized and tested by Li and coworkers [77]. Metallodendrimers, containing four or eight (2,6-bis(imino)pyridyl)iron(II) dichloride end-groups were prepared starting from a Si–H terminated carbosilane dendrimer. The ligands were connected to the dendrimer periphery by hydrosilylation, followed by complexation with $FeCl_2 \cdot 4H_2O$ (**13**). After activation with modified MAO, catalysts with a low Al/Fe molar ratio (e.g. Al/Fe = 500) showed much higher activity in ethylene polymerization and produced higher molecular weight polymers than the mononuclear analogues. However, activities of mononuclear and dendrimer catalysts were comparable for the system with a higher Al/Fe ratio (e.g. Al/Fe = 1,500), although dendrimer catalysts still produced higher molecular weight polyethylenes with higher melting temperatures. Thus, it seems that steric crowding in dendrimer iron catalysts controls the chain transfer mechanism during ethylene polymerization, indicating a positive dendrimer effect.

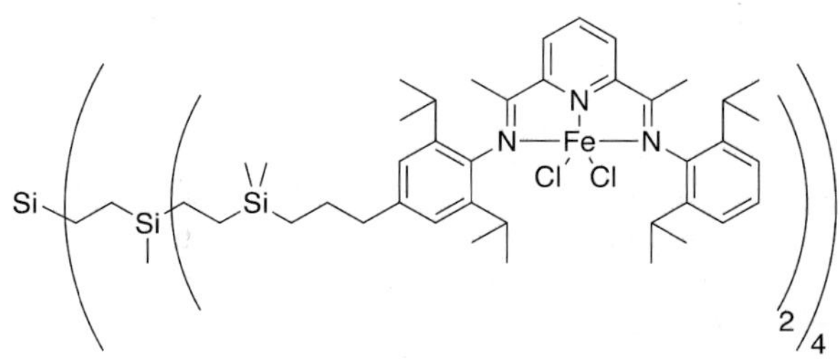

13

A similar (2,6-bis(imino)pyridyl)iron(II) dichloride complex containing different dendritic wedges in the *para*-positions of aryl rings has been developed by Moss et al. [78, 79]. Both carbosilane and poly(benzylphenylether) wedges were attached via ether linking groups to the ligand system, followed by complexation with $FeCl_2 \cdot 4H_2O$ (**14**). The resulting complexes were activated with MAO (Al/Fe = 400) and tested in the catalytic oligomerization of ethylene to higher 1-alkenes. The activity of the catalysts appeared to be unrelated to the type of dendritic wedges attached, however, the size of the wedges slightly influenced the catalyst activity in a positive way.

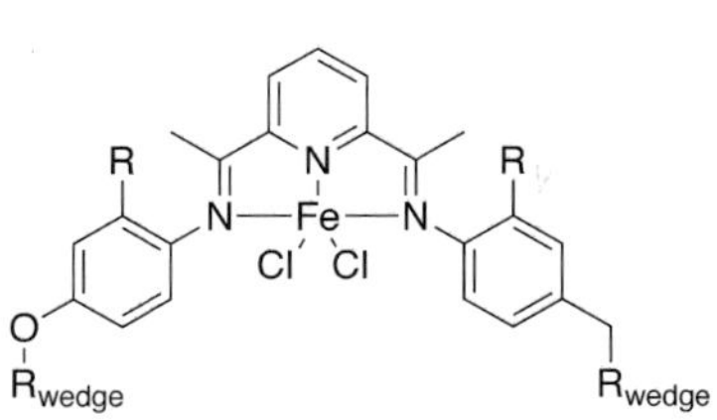

R = H, Me
R_{wedge} = $Br(CH_2)_4SiMe_2Ph$, $Br(CH_2)_4Si((CH_2)_3SiMe_2Ph)$

14

The use of the zeroth and the first generation carbosilane dendrimers functionalized with 2-bromoisobutyryl end-groups as initiators in the copper(I) bromide/N-(*n*-octyl)-2-pyridylmethanimine-mediated living-radical polymerization of methyl methacrylate (MMA) was studied by Van Koten et al. (see Reaction Scheme 9.8) [60, 61]. The star polymers formed during this polymerization had narrow molecular weight distributions (PDI < 1.3) and M_n close to theoretical values predicted by the amount of monomer consumed. The polymerization rates were lower than those produced by the monomeric analogue tested, probably because of initial intramolecular termination reactions (star-star couplings).

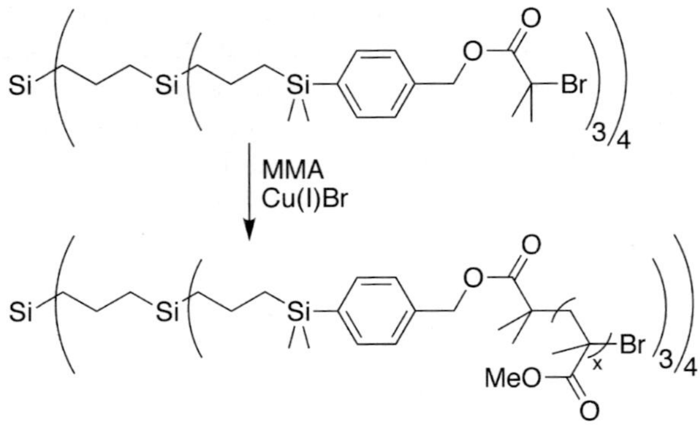

(9.8)

9.2.4.3 Michael Addition

Soluble copper(I) catalysts were developed for the 1,4-addition of diethyl zinc to 2-cyclohexenone by Van Koten's group [80]. ortho-Aminoarenethiolatocopper(I) catalyst was prepared and attached to a zeroth generation carbosilane dendrimer **15**, obtained by a convergent synthetic method. The resulting catalyst was tested in both polar and apolar solvents and showed excellent activity. Its solubility in apolar solvents opens possibilities for a wider variety of substrates that are not soluble in conventional (polar) solvents to be used in this type of reaction. Compared to the unsupported analogue, a clear positive dendrimer effect was observed, since the supported catalyst was more robust towards water and air, had increased solubility in most common organic solvents, and comparable (or even higher) catalytic activity. Furthermore, the increased stability and size of the dendrimer catalyst will allow its separation by nanofiltration, for which higher generations homologues will have to be synthesized.

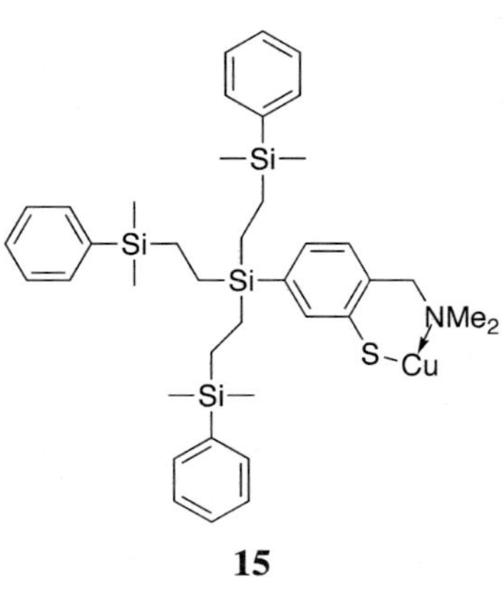

15

9.2.4.4 Ring-Opening Metathesis Polymerization (ROMP)

Beerens et al. synthesized metallodendrimers for catalytic ring-opening metathesis polymerization (ROMP) of norbornene by coupling of ruthenium complexes to low generation carbosilane dendrimers via an olefin metathesis reaction [57–59]. Phosphine ligands were used for stabilization of these complexes, but since these ligands were not a part of the dendrimer support, these systems are considered as non-phosphine-based dendrimer ligands. The obtained initiators showed very high activities for the ROMP of norbornene (see Reaction Scheme 9.9) yielding multi-arm star polymers in a controlled manner. The activity and selectivity of the catalysts were comparable to those of their mononuclear analogues [57].

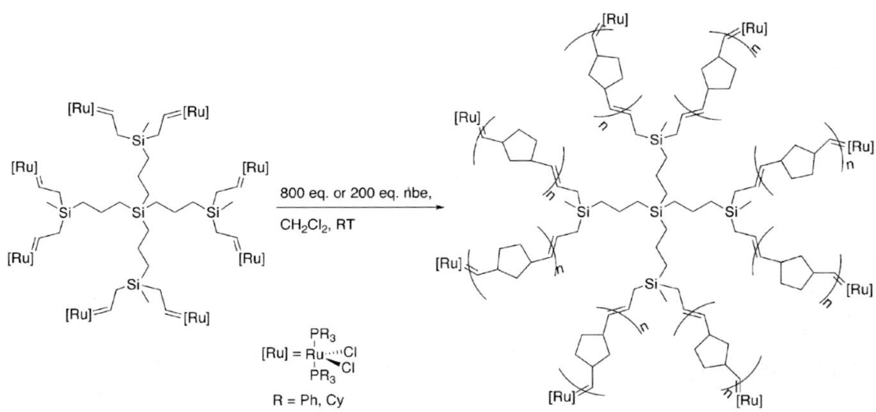

(9.9)

In addition to several different ruthenium complexes, aryloxy-tungsten complexes were also anchored to small carbosilane dendrimers [59]. These complexes also showed high activities, yielding, after complete conversion of monomer, high molecular weight branched star polymers by a further dismutation reaction. This reaction, which caused the coupling of two dendrimeric units, was followed by elimination of two metal centers from the dendrimer surface and formation of unidentified multi-tungsten species. A probable reason for this phenomenon may be the high activity of tungsten-alkylidene dendrimer relative to that of the corresponding ruthenium systems.

9.2.4.5 Ring-Closure Metathesis

In addition to the ring opening metathesis polymerization, ring-closure metathesis can also be catalyzed by metallodendrimer catalysts. In the Van Koten group, Si–Cl terminated carbosilane dendrimers were functionalized by using organo-lithium or organomagnesium reagents [34]. Furthermore, polylithiation of the 4-bromophenyl-functionalized dendrimers yielded valuable starting materials for further functionalization, e.g. to pyridyl alcohols, which could be used as ligands for the formation of ruthenium complexes **16**. Dendrimer ligands were prepared using both the zeroth and the first generation carbosilane dendrimers, of which only the smallest ones were metallated and tested in the ring-closure metathesis of diethyl diallylmalonate. The activity of the dendrimer catalyst was found comparable to that of the unimolecular catalyst system, so that after 30 min 100% conversion was achieved. In another experiment, the catalyst solution was separated from the reaction mixture by nanofiltration through a membrane (molecular weight cut-off 400), the reaction was stopped after reaching a 20% conversion and catalyst decomposition was observed, which was ascribed to deactivation by the membrane surface. This clearly shows that the development of membranes that are resistant to organic solvents and reagents is still an interesting and very important issue.

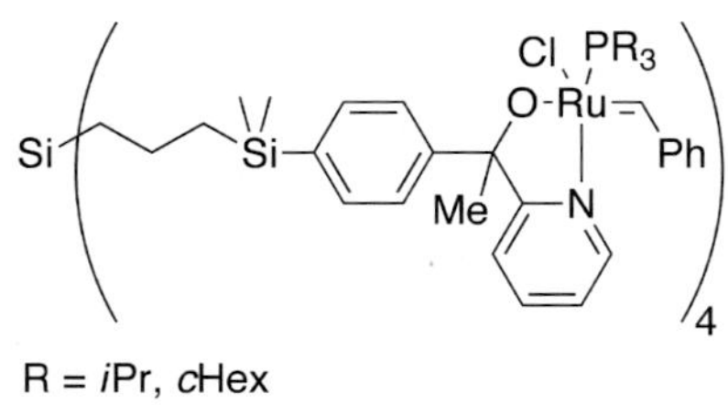

16

9.2.4.6 Atom-Transfer Radical Addition (Kharasch Addition)

The first attempt to close the gap between homogeneous and heterogeneous catalysis by the use of dendritic supporting scaffolds involved the synthesis of a nanosized dendrimer-supported homogeneous catalyst that would allow performance of a reaction under homogeneous conditions and separation of the catalyst by nanofiltration from the resulting product. Toward this end, various NCN-pincer nickel(II) halide catalysts [14, 15, 81] were grafted onto carbosilane dendrimer scaffolds (see for example catalyst **1**) and successfully applied in the Kharasch addition reaction. Initially the NCN-pincer ligands were connected to the dendrimer periphery via a relatively long linker (**A**), but later the grafting was directly onto the Si-centers (**B**) (see Structures **17**). An obvious difference between these two types of catalysts is in the accessibility of the nickel centers.

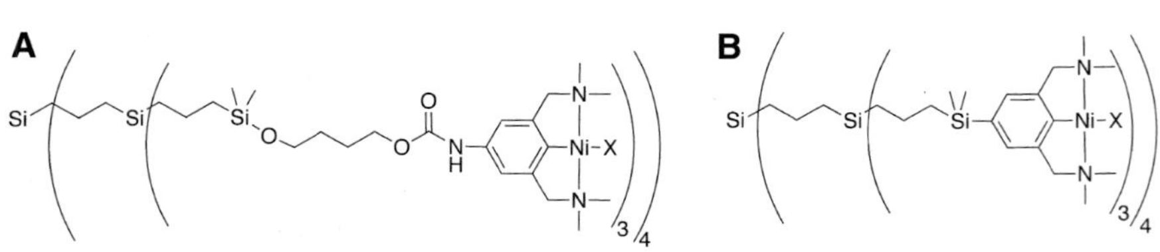

17

Both types of catalysts were tested in the Kharasch addition of CCl_4 to methyl methacrylate and it was observed that catalytic activity per nickel site of higher generation dendrimers **17B** (up to 36 NCN-pincer nickel halide units) decreased dramatically compared to the same generation of dendrimer **17A** and the mononuclear analogues. This was explained by suggesting that in the more congested metallodendrimer **17B**, more densely packed nickel centers enabled an intramolecular redox deactivation reaction shown in Reaction Scheme 9.10. This "proximity effect" is more pronounced in the dendrimer catalysts of the type **17B**, since in this design, in the absence of extended linker, the nickel centers are in closer proximity to each other.

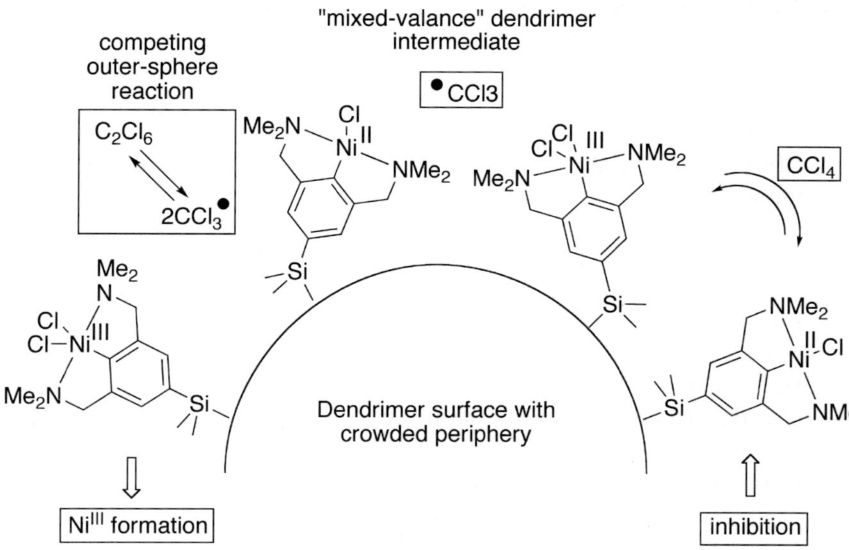

(9.10)

To overcome this deficiency, dendrimers with more elongated arms have been synthesized by changing the degree of branching within the dendrimer scaffold. In these species (see Structure **18** [15]), the distance between the nickel centers was considerably increased, which resulted in improved catalytic efficiencies (less intramolecular deactivation) that ultimately approximated the activity of the mononuclear analogue. Consequently, in this study a case of negative cooperation of the catalytic sites in a metallodendrimer could be studied in great detail.

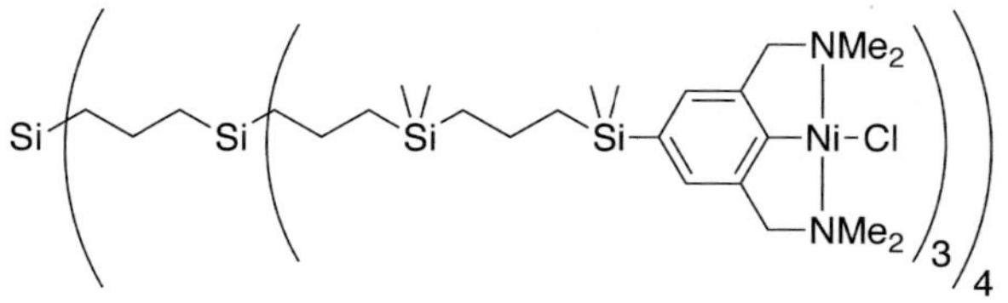

18

The metallodendrimer catalysts for the Kharasch addition were also applied in membrane reactors. Already the first generation catalysts (without linker; diameter ~2.5–3 nm) showed high retentions in ultrafiltration membrane reactors and could be applied in continuous-flow reactors with no significant leaching of catalyst through the membrane. However, the formation of purple nickel(III)-containing precipitates revealed the proximity effect as well as effects arising from the membrane surface. This leads to lower catalytic efficiencies and irreversible formation of inactive, very stable NCN-pincer Ni(III) sites.

9.2.4.7 Hydrosilylation

A series of dendrimers capped with $–C_6F_4B(C_6F_5)_2$ end-groups was synthesized by Piers et al. [82]. In contrast to the borate dendrimers discussed above, these borane dendrimers have not been fully investigated and applied yet. They were prepared via a self-catalyzed silylation reaction of the hydrosilane dendrimer end-groups with appropriate aryl ether containing the $–C_6F_4B(C_6F_5)_2$ group (see Reaction Scheme 9.11), in which the borane functions as its own catalyst. The dendrimers differed in the number of end-groups at the periphery from 4 for the zeroth generation to 8 or 12 for the second generation dendrimers. They were tested as catalysts for the hydrosilylation reaction of acetophenone using triethylsilane and were found to be only slightly less active than $B(C_6F_5)_3$ itself. The more crowded second generation dendrimer containing 12 end-groups showed a somewhat lower reaction rate compared to the other two dendrimer systems, but it was still an effective catalyst under the conditions applied. These results suggest that all boron centers of these dendrimers acted independently.

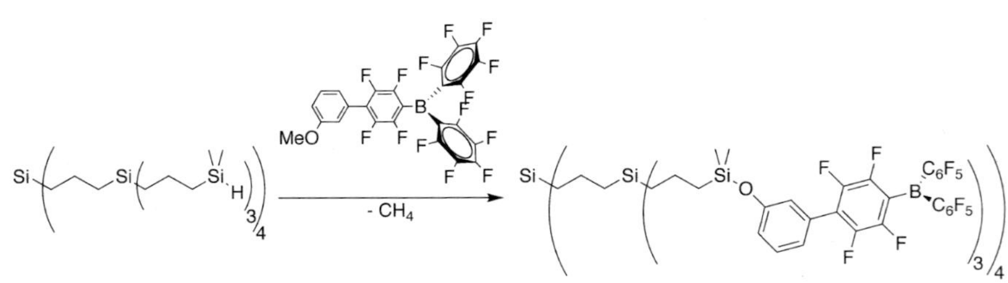

(9.11)

9.2.4.8 Enantioselective Addition of Dialkylzincs to Aldehydes

Sato et al. synthesized carbosilane dendrimers loaded with up to 12 units of chiral β-amino alcohols, yielding a chiral, dendrimer catalyst **19** [83]. These catalysts were tested in the enantioselective addition of various dialkylzinc compounds to aldehydes, and were found to efficiently catalyze the formation of enantiomerically enriched *sec*-alcohols with ee values of up to 93%. In earlier work, analogues, more rigid dendrimeric chiral ligands were synthesized and tested [84, 85]. The ee values obtained in these studies were somewhat lower or comparable to the ee values obtained with the carbosilane dendritic catalysts, but in the more rigid dendrimer catalysts the chiral sites were more isolated and working independently. The authors proposed that flexibility of the carbosilane dendrimer catalysts enables interaction between the chiral sites and thereby the formation of products with high enantioselectivities [83]. The activities of both the zeroth and the first generation dendrimer catalysts and of a dimeric model compound (with a –Si(Me)$_2$(CH$_2$CH$_2$)(Me)$_2$Si– spacer between the two chiral β-amino alcohol units) were reported. However, no comparison or benchmarking with reactions of the chiral β-amino alcohol itself has been made.

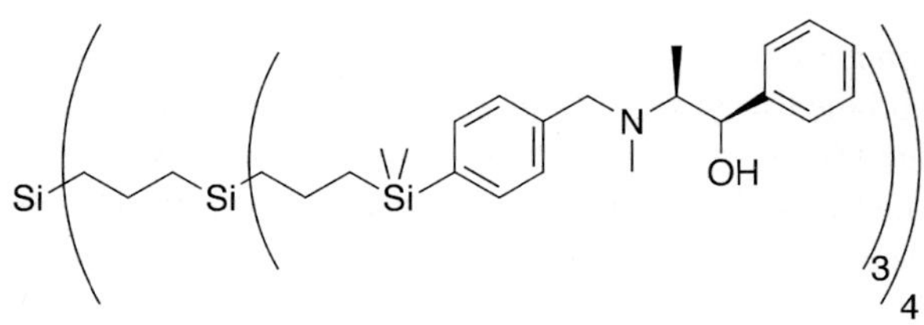

19

9.2.4.9 Hydrogenation

Metal-phosphine complexes are commonly used as catalysts for hydrogenation of alkenes. Recently, Gade et al. reported the synthesis of a new, non-phosphine based class of molecular hydrogenation catalysts, which were also connected to different types of dendron and dendrimer scaffolds (**20**) [86, 87]. The monomeric palladium complexes of these so-called BPI-ligands (1,3-bis(2-pyridylimino)isoindolate) were applied to the hydrogenation of olefins. Unfortunately, catalytic studies with dendritic catalysts have not been reported yet.

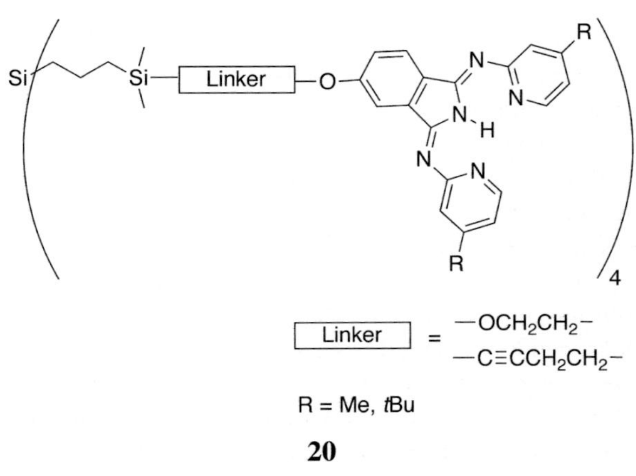

20

One of the problems encountered with phosphorus-ligand-based metal complexes is their sensitivity towards oxidation when exposed to air, which results in limited recycling possibilities. However, the dendrimer versions of the BPI-palladium complexes appeared to be thermally and kinetically more stable than the well-established phosphine containing catalysts. This increased stability of dendrimer BPI-palladium complexes opens the possibility for application of these catalytic species in, for example, continuous hydrogenation processes.

Feng et al. described the synthesis of a carbosilane dendrimer with peripheral aminopropyl groups (**21**) and the corresponding platinum and palladium complexes (prepared by reaction with $H_2PtCl_6{\cdot}6H_2O$ or $PdCl_2{\cdot}2H_2O$) [88, 89]. The palladium complex (no information on the Pd to N molar ratio was provided) was tested in hydrogenation reactions of several organic compounds (e.g. styrene, allyl alcohol and acetophenones) and appeared to be an effective catalyst for the reduction of both the C=C and the C=O bonds of the substrates tested. The palladium dendrimer complex appeared to be far more active than monomeric $PdCl_2{\cdot}2H_2O$ at the same palladium concentration. In addition to this, the monomeric catalyst caused palladium plating during the reaction. An explanation for this increase in activity of the dendrimer catalyst may be the possible formation of coordinatively unsaturated palladium sites during the catalysis, which does not occur with the monomeric catalyst. It was found that dendrimer catalysts could be reused without any loss in activity. However, no details were given on their recyclability.

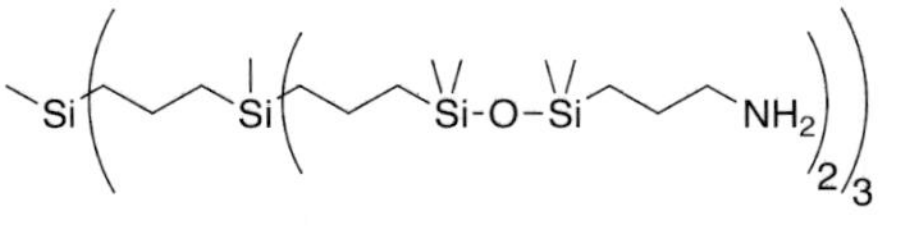

21

9.3 Supported Organic Synthesis on Soluble Carbosilanes

In addition to the use of carbosilane dendrimers as soluble supports for homogeneous catalysts, it is also possible to apply them as supports for supported organic synthesis (SOS). Since the introduction of the well-known Merrifield resins in the 1960s [90], insoluble solid supports or soluble polymeric supports have been widely used in Solid Phase Organic Synthesis (SPOS). For this, precisely defined soluble supports will have several advantages, such as more homogeneous reaction mixtures leading to more linear reaction kinetics and higher reaction rates. Furthermore, standard spectroscopic methods (e.g., NMR-spectroscopy) can be applied to make monitoring of single reaction steps easier. Carbosilane dendrimers are perfect candidates for this purpose, since they are chemically inert, good anchors for several reactive molecules, can be separated from small molecules by simple filtration, i.e. reused after isolation by various kinds of separation techniques, and can withstand harsh reaction conditions [91]. As a consequence, the development of peripherally-functionalized carbosilane dendrimers opened various possibilities for the application of dendrimer ligands as soluble supports for SOS (see for example, [34, 92]).

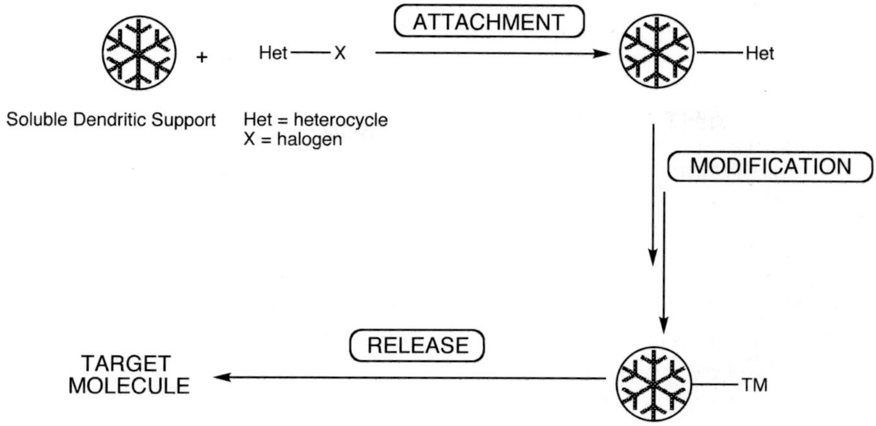

Fig. 9.5 General scheme of the process for use of dendrimers as soluble supports in supported organic synthesis [93].

A general approach to SOS consists of three subsequent steps: attachment of the substrate to the dendrimer support, modification of the substrate into the target molecule, and release of the product from the support (see Fig. 9.5). In addition to this, subsequent recycling of the soluble dendrimer support is possible after its separation from the product solution (see Fig. 9.2).

Recently, in the group of Klein Gebbink, the use of carbosilane dendrimers as soluble supports for organic synthesis has been further advanced [93]. Already second generation dendrimers were shown to be large enough to separate from reaction mixtures by dialysis. This enabled purification after each step of the reaction sequence and thus provided cleaner products. Different types of reactions have been performed, including lithiations and palladium catalyzed C–C coupling reactions. With the loading of dendrimers with pyridine moieties, stepwise modification of the periphery, and release of the products from it, a proof of principle of this technique has been provided (see Reaction Scheme 9.12).

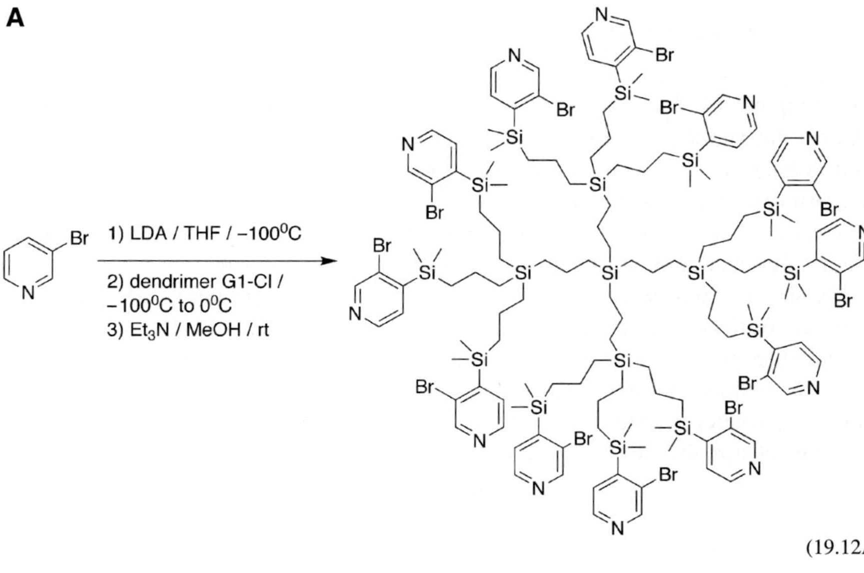

(19.12A)

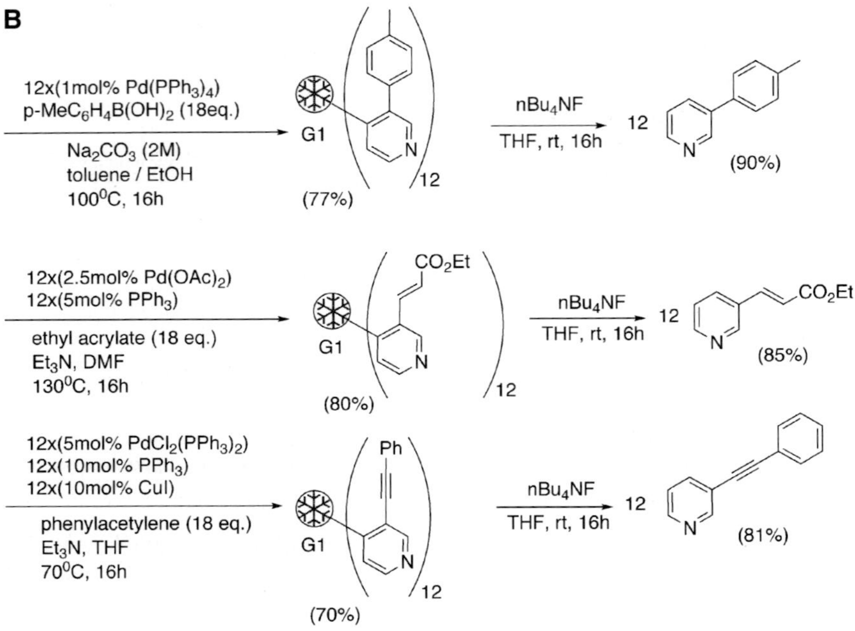

(9.12B)

Furthermore, in order to obtain an even better diafiltration performance, carbosilane dendrimers with more rigid or shape-persistent cores have been developed as soluble supports. With increased rigidity of the core, membranes with larger pores sizes can be used, which allows higher filtration rates, throughput of products and reactants without loss of dendrimeric support [19].

9.4 Conclusions and Future Outlook

This chapter describes the current state of the use of carbosilane dendrimers and dendrons as supports for homogeneous catalysts, with active sites located either in their cores or focal points, or at the periphery in the end-groups. The inertness of the carbosilane dendritic scaffold to most of the reagents that are required for the synthesis of metal-ligand complexes as well as its solubility makes carbosilane dendrimers excellent carrier molecules for homogeneous catalytic systems. Via a proper choice of alkane-diyl chains connecting the Si-branching points, the rigidity and solubility of these scaffolds can be tailored, i.e. C_2-linkers lead to a more rigid, often less soluble (more crystalline) carbosilane backbone than C_3-linkers. Furthermore, lower generation dendrimers already have sizes that make them suitable for separation from the product solutions by nanofiltration techniques and for use in continuous reaction set-ups. Due to this property, the use of dendrimer catalysts in automated processes is a research area that is explored by many groups (for reviews on this topic, see also [4, 94]). In addition, the development of membranes that are resistant to organic solvents and reagents is still an interesting and very important field.

Fundamental studies have revealed that with a proper not too dense coverage of the dendrimer periphery, each catalytic site can function as if it were an independent catalyst. It was also found that a too high catalyst site density in various types of reactions can lead to severe lowering of the catalytic performance of the metallodendrimer catalysts (negative cooperative effect), although the opposite (positive cooperative effects) has been observed as well. Approaches aimed at reducing this site density by decreasing the degree of branching of the dendrimer scaffold near the periphery to regain the activity of the single catalyst site are also described. It should be noted that for the future development of this field it will be necessary that the homogeneous catalyst on the dendrimer scaffold has considerable robustness to make multiple use possible. It goes without saying that recycling is only economical for processes that require higher concentrations of homogeneous catalysts, which is often the case in the synthesis of fine chemicals and special products. Many of the catalytic species described here make use of phosphine ligands, which are connected directly or via linker moieties to the dendrimer scaffold. The performance of these dendrimer catalysts is often comparable to that of the mononuclear analogues.

With these developments, systems are now at hand that combine favorable properties of homogeneous catalysts, such as activity, selectivity and well defined structural features, with some of those properties that are specific for heterogeneous catalysts, such as recyclability, easy engineering of complex systems, high total turn over numbers and last but not least an easy, eventually continuous, separation of catalyst from the product solution. Accurate and easy characterization combined with easy purification and catalyst recycling are advantages that were not achieved with other catalytic systems yet. A possible application of these properties might be in membrane bags to compartmentalize the dendrimer catalyst in the reaction mixture. Through permeation (driven by reverse osmosis) the catalyst and reactants can interact with each other but the catalyst can still be removed easily from the reaction mixture. This approach, which can be extended to applications in, for example, mini reactors, is currently being investigated by the research group of Klein Gebbink and others. This method opens a route to the so-called tandem- or cascade-catalysis with one, two or a number of nano-sized, compartmentalized dendrimer catalysts present in a single reaction mixture.

The developments described for the application of dendrimer supports in catalysis can also be applied to supported organic syntheses. The requirements needed for catalyst supports, e.g. recyclability, nanofiltration, inertness, robustness, are also required for supported organic synthesis with soluble supports. This new field of application of carbosilane dendrimers has been opened only recently but it already shows promise for the near future.

Finally, it should be also noted that almost all described dendrimer catalysts consist of catalytic species covalently bonded to the carbosilane scaffolds. However, recent developments have provided systems, in which the catalyst is non-covalently bonded to the core of a core-shell dendrimer species [9, 95–102]. The systems developed by Van Koten et al. comprise a non-covalently bonded metal catalyst which is charged with an anionic tether and a core-shell dendrimer carrier/container that has a polycationic core. The dendrimer container-metal catalyst assembly forms itself by ion exchange and is primarily held together by Coulombic forces. The dendrimers used for this purpose are based on carbosilane core molecules, functionalized with polybenzyl aryl ether dendrons. One of the advantages of this system lies in the fact that catalytic species are bonded strongly enough to prevent leaching, but at the same time can easily be removed from the dendrimer container. These recent developments open new ways for application of carbosilane dendrimers as recyclable supports for various catalysts for a wide variety of reactions [95–101].

References

1. Schlenk C, Frey H (1999) Monatsh Chem 130:3
2. Andrés R, de Jesús E, Flores JC (2007) New J Chem 31:1161
3. Berger A, Klein Gebbink RJM, van Koten G (2006) Top Organomet Chem 20:1
4. Méry D, Astruc D (2006) Coord Chem Rev 250:1965
5. Helms B, Fréchet JMJ (2006) Adv Synth Catal 348:1125
6. Astruc D, Chardac F (2001) Chem Rev 101:2991
7. Kreiter R, Kleij AW, Klein Gebbink RJM, van Koten G (2001) Top Curr Chem 217:163
8. Reek JNH, Arévalo S, van Heerbeek R, Kamer PCJ, van Leeuwen PWNM (2006) Adv Catal 49:71
9. Reek JNH, de Groot D, Oosterom GE, Kamer PCJ, van Leeuwen PWNM (2002) Rev Mol Biotech 90:159
10. Twyman LJ, King ASH, Martin IK (2002) Chem Soc Rev 31:69
11. Oosterom GE, Reek JNH, Kamer PCJ, van Leeuwen PWNM (2001) Angew Chem Int Ed 40:1828
12. Knapen JWJ, van der Made AW, de Wilde JC, van Leeuwen PWNM, Wijkens P, Grove DM, van Koten G (1994) Nature 372:659
13. Liang C, Fréchet JMJ (2005) Prog Polym Sci 30:385
14. Kleij AW, Gossage RA, Jastrzebski JTBH, Boersma J, van Koten G (2000) Angew Chem Int Ed 39:176
15. Kleij AW, Gossage RA, Klein Gebbink RJM, Brinkmann N, Reijerse EJ, Kragl U, Lutz M, Spek AL, van Koten G (2000) J Am Chem Soc 122:12112
16. Chavan SA, Maes W, Gevers LEM, Wahlen J, Vankelecom IFJ, Jacobs PA, Dehaen W, de Vos DE (2005) Chem Eur J 11:6754
17. Müller C, Nijkamp MG, Vogt D (2005) Eur J Inorg Chem 4011
18. Vankelecom IFJ (2002) Chem Rev 102:3779
19. Dijkstra HP, van Klink GPM, van Koten G (2002) Acc Chem Res 35:798
20. Albrecht M, van Koten G (2001) Angew Chem Int Ed 40:3750
21. Albrecht M, Schlupp M, Bargon J, van Koten G (2001) Chem Commun 1874
22. Albrecht M, Lutz M, Spek AL, van Koten G (2000) Nature 406:970
23. Albrecht M, Gossage RA, Lutz M, Spek AL, van Koten G (2000) Chem Eur J 6:1431
24. Albrecht M, van Koten G (1999) Adv Mater 11:171
25. van Heerbeek R, Kamer PCJ, van Leeuwen PWNM, Reek JNH (2002) Chem Rev 102:3717
26. Vandezande P, Gevers LEM, Vankelecom IFJ (2008) Chem Soc Rev 37:365

27. van der Made AW, van Leeuwen PWNM (1992) J Chem Soc Chem Commun 1400
28. Ropartz L, Morris RE, Schwarz GP, Foster DF, Cole-Hamilton DJ (2000) Inorg Chem Commun 3:714
29. Ropartz L, Morris RE, Foster DF, Cole-Hamilton DJ (2002) J Mol Catal A Chem 182–183:99
30. Ropartz L, Morris RE, Foster DF, Cole-Hamilton DJ (2001) Chem Commun 361
31. Ropartz L, Haxton KJ, Foster DF, Morris RE, Slawin AMZ, Cole-Hamilton DJ (2002) J Chem Soc Dalton Trans 4323
32. Ropartz L, Foster DF, Morris RE, Slawin AMZ, Cole-Hamilton DJ (2002) J Chem Soc Dalton Trans 1997
33. Jaffrès PA, Morris RE (1998) J Chem Soc Dalton Trans 2767
34. Wijkens P, Jastrzebski JTBH, van der Schaaf PA, Kolly R, Hafner A, van Koten G (2000) Org Lett 2:1621
35. de Groot D, Emmerink PG, Coucke C, Reek JNH, Kamer PCJ, van Leeuwen PWNM (2000) Inorg Chem Commun 3:711
36. Reek JNH, de Groot D, Oosterom GE, Kamer PCJ, van Leeuwen PWNM (2003) CR Chim 6:1061
37. de Groot D, Reek JNH, Kamer PCJ, van Leeuwen PWNM (2002) Eur J Org Chem 1085
38. de Groot D, Eggeling EB, de Wilde JC, Kooijman H, van Haaren RJ, van der Made AW, Spek AL, Vogt D, Reek JNH, Kamer PCJ, van Leeuwen PWNM (1999) Chem Commun 1623
39. Oosterom GE, van Haaren RJ, Reek JNH, Kamer PCJ, van Leeuwen PWNM (1999) Chem Commun 1119
40. Hovestad NJ, Eggeling EB, Heidbüchel HJ, Jastrzebski JTBH, Kragl U, Keim W, Vogt D, van Koten G (1999) Angew Chem Int Ed 38:1655
41. Eggeling EB, Hovestad NJ, Jastrzebski JTBH, Vogt D, van Koten G (2000) J Org Chem 65:8857
42. Rodríguez LI, Rossell O, Seco M, Grabulosa A, Muller G, Rocamora M (2006) Organometallics 25:1368
43. Rodríguez LI, Rossell O, Seco M, Orejón A, Masdeu-Bultó AM (2008) J Organomet Chem 693:1857
44. Benito M, Rossell O, Seco M, Muller G, Ordinas JI, Font-Bardia M, Solans X (2002) Eur J Inorg Chem 2477
45. Rodríguez LI, Rossell O, Seco M, Muller G (2007) J Organomet Chem 692:851
46. Angurell I, Muller G, Rocamora M, Rossell O, Seco M (2003) Dalton Trans 1194
47. Angurell I, Muller G, Rocamora M, Rossell O, Seco M (2004) Dalton Trans 2450
48. Bourrier O, Kakkar AK (2003) J Mater Chem 13:1306
49. Bourrier O, Kakkar AK (2004) Macromol Symp 209:97
50. Petrucci-Samija M, Guillemette V, Dasgupta M, Kakkar AK (1999) J Am Chem Soc 121:1968
51. Findeis RA, Gade LH (2003) Eur J Inorg Chem 99
52. Cheliatsidou P, White DFS, Cole-Hamilton DJ (2004) Dalton Trans 3425
53. Grove DM, Van Koten G, Mul P, Zoet R, Van Der Linden JGM, Legters J, Schmitz JEJ, Murrall NW, Welch AJ (1988) Inorg Chem 27:2466
54. Gossage RA, van de Kuil LA, van Koten G (1998) Acc Chem Res 31:423
55. Andrés R, de Jesús E, de la Mata FJ, Flores JC, Gómez R (2005) Eur J Inorg Chem 3742
56. Mager M, Becke S, Windisch H, Denninger U (2001) Angew Chem Int Ed 40:1898
57. Beerens H, Wang W, Verdonck L, Verpoort F (2002) J Mol Catal A Chem 190:1
58. Beerens H, Verpoort F, Verdonck L (2000) J Mol Catal A Chem 151:279
59. Beerens H, Verpoort F, Verdonck L (2000) J Mol Catal A Chem 159:197
60. Hovestad NJ, Jastrzebski JTBH, van Koten G, Bon SAF, Waterson C, Haddleton DM (1999) Polym Preprints (Am Chem Soc Div Polym Chem) 40:393
61. Hovestad NJ, van Koten G, Bon SAF, Haddleton DM (2000) Macromolecules 33:4048
62. van der Boom ME, Milstein D (2003) Chem Rev 103:1759

63. Dupont J, Consorti CS, Spencer J (2005) Chem Rev 105:2527
64. Solin N, Kjellgren J, Szabo KJ (2003) Angew Chem Int Ed 42:3656
65. Singleton JT (2003) Tetrahedron 59:1837
66. Schlenk C, Kleij AW, Frey H, van Koten G (2000) Angew Chem Int Ed 39:3445
67. Kleij AW, Klein Gebbink RJM, van den Nieuwenhuijzen PAJ, Kooijman H, Lutz M, Spek AL, van Koten G (2001) Organometallics 20:634
68. Rodríguez G, Lutz M, Spek AL, van Koten G (2002) Chem Eur J 8:45
69. Slagt MQ, Jastrzebski JTBH, Klein Gebbink RJM, van Ramesdonk HJ, Verhoeven JW, Ellis DD, Spek AL, van Koten G (2003) Eur J Org Chem 1692
70. Andrés R, de Jesús E, de La Mata FJ, Flores JC, Gómez R (2005) J Organomet Chem 690:939
71. Andrés R, de Jesús E, de la Mata FJ, Flores JC, Gómez R (2002) Eur J Inorg Chem 2281
72. Arévalo S, de Jesús E, de La Mata FJ, Flores JC, Gómez R, Rodrigo MM, Vigo S (2005) J rganomet Chem 690:4620
73. Arévalo S, de Jesús E, de la Mata FJ, Flores JC, Gómez R, Gómez-Sal MP, Ortega P, Vigo S (2003) Organometallics 22:5109
74. Benito JM, de Jesús E, de la Mata FJ, Flores JC, Gómez R (2006) Organometallics 25:3045
75. Benito JM, de Jesús E, de la Mata FJ, Flores JC, Gómez R (2005) Chem Commun 5217
76. Benito JM, de Jesús E, de la Javier Mata F, Flores JC, Gómez R, Gómez-Sal P (2006) Organometallics 25:3876
77. Zheng ZJ, Chen J, Li YS (2004) J Organomet Chem 689:3040
78. Overett MJ, Meijboom R, Moss JR (2005) Dalton Trans 551
79. Blom B, Overett MJ, Meijboom R, Moss JR (2005) Inorg Chim Acta 358:3491
80. Arink AM, van de Coevering R, Wieczorek B, Firet J, Jastrzebski JTBH, Klein Gebbink RJM, van Koten G (2004) J Organomet Chem 689:3813
81. van Koten G, Jastrzebski JTBH (1999) J Mol Catal A Chem 146:317
82. Roesler R, Har BJN, Piers WE (2002) Organometallics 21:4300
83. Sato I, Kodaka R, Hosoi K, Soai K (2002) Tetrahedron Asymmet 13:805
84. Sato I, Shibata T, Ohtake K, Kodaka R, Hirokawa Y, Shirai N, Soai K (2000) Tetrahedron Lett 41:3123
85. Sato I, Kodaka R, Shibata T, Hirokawa Y, Shirai N, Ohtake K, Soai K (2000) Tetrahedron Asymmet 11:2271
86. Meder MB, Haller I, Gade LH (2005) Dalton Trans 1403
87. Meder M, Galka CH, Gade LH (2005) Monatsh Chem 136:1693
88. Li CF, Li DX, Zhang ZJ, Feng SY (2005) Chin Chem Lett 16:1389
89. Li CF, Li DX, Feng SY (2005) Polym Int 54:1041
90. Merrifield RB (1963) J Am Chem Soc 85:2149
91. Klein Gebbink RJM, Kruithof CA, van Klink GPM, van Koten G (2002) Rev Mol Biotechnol 90:183
92. Hovestad NJ, Ford A, Jastrzebski JTBH, van Koten G (2000) J Org Chem 65:6338
93. Le Nôtre J, Firet JJ, Sliedregt LAJM, van Steen BJ, van Koten G, Klein Gebbink RJM (2005) Org Lett 7:363
94. Fan QH, Li YM, Chan ASC (2002) Chem Rev 102:3385
95. van de Coevering R, Klein Gebbink RJM, van Koten G (2005) Prog Polym Sci 30:474
96. van de Coevering R, Bruijnincx PCA, Lutz M, Spek AL, van Koten G, Klein Gebbink RJM (2007) New J Chem 31:1337
97. van de Coevering R, Bruijnincx PCA, van Walree CA, Klein Gebbink RJM, van Koten G (2007) Eur J Org Chem 2931
98. van de Coevering R, Alfers AP, Meeldijk JD, Martínez-Viviente E, Pregosin PS, Klein Gebbink RJM, van Koten G (2006) J Am Chem Soc 128:12700
99. van de Coevering R, Kreiter R, Cardinali F, van Koten G, Nierengarten JF, Klein Gebbink RJM (2005) Tetrahedron Lett 46:3353

100. van de Coevering R, Kuil M, Klein Gebbink RJM, van Koten G (2002) Chem Commun 1636
101. Kleij AW, van de Coevering R, Klein Gebbink RJM, Noordman AM, Spek AL, van Koten G (2001) Chem Eur J 7:181
102. Ribaudo F, van Leeuwen PWNM, Reek JNH (2006) Top Organomet Chem 20:39

Chapter 10
Liquid Crystalline Silicon-Containing Dendrimers with Terminal Mesogenic Groups

Valery Shibaev and Natalia Boiko

10.1 Introduction

It is well-known that one of the main features of low-molar-mass liquid crystals and liquid-crystalline (LC) polymers is the presence of anisometric molecular fragments (*mesogenic groups*) responsible for LC phase (*mesophase*) formation. The majority of mesogenic groups consist of rigid rod-, board- (or lath-) and disk-shaped molecular moieties, which play a role of specific "building blocks", a spontaneous ordering of which leads to the formation of different LC phases. A compound that under suitable conditions of temperature, pressure, and concentration can exist as a mesophase is usually called a mesogen or mesogenic compound.

Figure 10.1 shows various types of the best known and wide-spread mesophases. Depending on the orientational and positional organization of molecules, these mesophases may be roughly divided into *nematic*, *smectic*, and *columnar* LC phases. All these types of mesophases are usually formed by melting of crystalline organic solids (or cooling of an isotropic melt) and are, therefore, called *thermotropic liquid crystals*. The temperature at which the transition between the mesophase and the isotropic phase occurs is called the *clearing* (T_{cl}) or *isotropization temperature* (T_{iso}). Detailed information relating to low-molar-mass and polymer liquid crystals and their nomenclature may be found in a comprehensive three-volume handbook [1] and in the IUPAC Recommendations of basic terms associated with liquid crystals [2].

Although the low-molar-mass liquid crystals have been discovered about 120 years ago [3], for a relatively long time afterward the studies of liquid crystals were confined to purely academic research. However, in the mid 1960s, the accumulated knowledge in the LC field led to practical application of liquid crystals

V. Shibaev and N. Boiko
Chemistry Department, Moscow State University, Moscow 119991, Russia
E-mail: lcp@genebee.msu.su

P.R. Dvornic and M.J. Owen (eds.), *Silicon-Containing Dendritic Polymers*, Advances in Silicon Science 2, © Springer Science+Business Media B.V. 2009

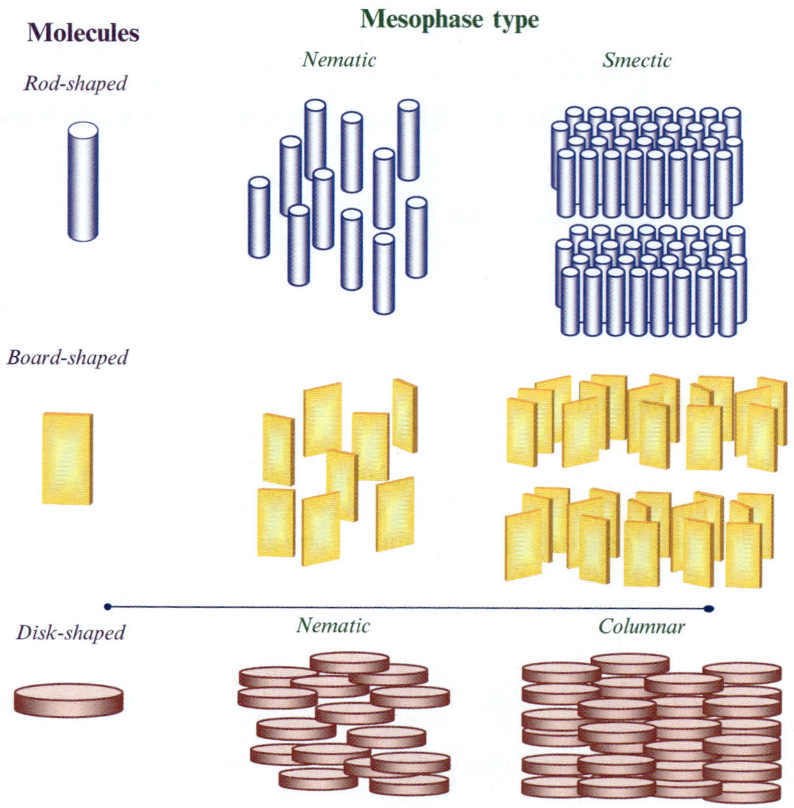

Fig. 10.1 Various types of mesophases formed by rod-, board-, and disk-shaped molecules of low-molar-mass liquid crystals.

in electronic and optoelectronic devices, telecommunication systems, mobile cell-phones, and display technologies [1, 4]. The unique combination of ordered arrangement of molecules in different LC phases, high molecular mobility, and fast response of liquid crystals to changes in external fields (electric, magnetic and mechanical) pushed liquid crystals into the foreground as a novel class of practically-important materials capable of providing rapid and reliable response to the control signals transmitted by electromagnetic fields. The liquid crystal indicators provided a basis for the modern display technology as well as for mobile communication devices, optical discs, smart optical cards, and other optically active materials.

The peak of the practical achievements and successes of LC research came with the creation of LC polymers in mid 1970s. These studies resulted in the synthesis of the so-called mesogenic LC polymers incorporating mesogenic groups (simulating the constitution of low-molar-mass liquid crystals) (see

Fig. 10.1) either in the main chains of their macromolecules (*main chain LC polymers*) or as the pendant side branches (*side-chain or comb-shaped*) *LC polymers* (see Fig. 10.2) [5, 6]. The latter closely resemble the constitution of dendrimers with terminal mesogenic groups. The comb-shaped polymers offered a convenient matrix for the creation of the first thermotropic LC polymers synthesized by the Russian [7, 8] and German groups [9] independently of each other. In such systems, the side mesogenic groups were chemically linked to the macromolecular backbone by flexible (e.g. aliphatic) *spacers* (see Fig. 10.2). This concept of spacers, originally introduced in the above-mentioned works, inspired the development of a new direction of research aimed at creation of thermotropic comb-shaped LC polymers. The flexible aliphatic (or any other) spacers separating mesogenic groups enable a high degree of independence for the anisometric mesogenic fragments with regard to the backbone which makes possible their cooperative interaction that results in the formation of a mesophase. Using the spacer concept, several tens of thousands of LC comb-shaped polymers were synthesized and the number of them continues to increase [10–16].

In the 1990s, extensive investigations of LC polymers expanded into the field of branched and dendritic compounds, and we implemented the same "spacer concept" to the synthesis of dendrimers with mesogenic groups [17–21]. The general approach to the synthesis of LC dendrimers was based either on the incorporation of mesogenic groups into the whole volume of their interiors or on the introduction of mesogenic fragments as terminal groups into the periphery (see Fig. 10.3). Thus, the 1990s can be considered as the period of the "liquid-crystalline dendrimer boom". Many new LC dendrimers containing various mesogenic groups and different dendritic interiors such as polyesters, poly(propylene imine), polycarbosilane, poly(amidoamine) and polysiloxane were synthesized in various laboratories in many countries.

The first major, comprehensive review of liquid-crystalline dendrimers was published by us in *Polymer Science* in 2001 [22]. It summarized both the data

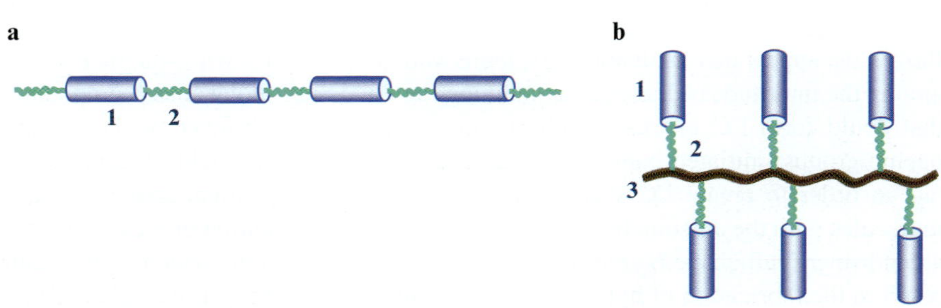

Fig. 10.2 Schematic representation of macromolecules of (**a**) main chain and (**b**) side chain (comb-shaped) LC polymers. (1) mesogenic group; (2) spacer; (3) backbone.

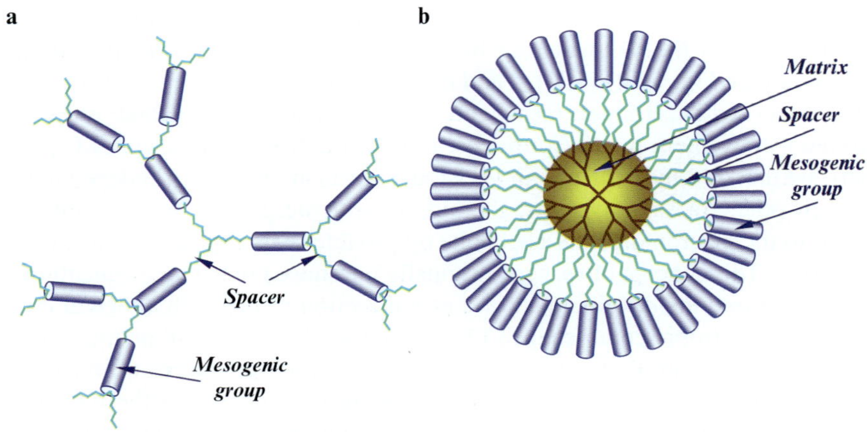

Fig. 10.3 Schematic representation of two types of LC dendrimers with different arrangement of mesogenic groups. (**a**) Mesogenic groups in the whole volume of a molecule; (**b**) molecule of LC dendrimer with terminal mesogenic groups.

available in the literature and the results of original investigations performed by our group at the Chemistry Department of Moscow State University. Ever since, LC dendrimers have attracted a growing scientific interest from researchers working in such diversified fields as the chemistry and physics of liquid crystals, physical chemistry of macromolecular compounds, and supramolecular chemistry. At present, compounds with dendritic architecture, including LC dendrimers, are being incorporated as principal building blocks into nanomaterials and nanotechnology.

10.2 Peculiarities of the Molecular Structure of LC Silicon-Containing Dendrimers

Before the appearance of dendrimers, there were no molecules with spherical shape among the multitude of mesogen-containing low- and high-molar-mass compounds that could form LC phases. Synthesis and studies of dendrimers (without mesogenic groups) initiated interest of researchers working in the field of liquid crystals in order to create LC systems which combine the spherical shape of these molecules with the anisometry of rigid rods. In fact, incorporation of mesogens into a dendrimer (ball-shaped) architecture is an intringuing design concept, which can result in the fabrication of hybrid structures combining LC properties and dendritic functionalities and shapes. The vast majority of experimental data regarding silicon-containing LC dendrimers is from the study of LC dendrimers containing terminal mesogenic groups considered in this chapter.

The interest in LC dendrimers is dictated by their unusual, exotic structure (see Fig. 10.3b). Each such dendrimer can be thought of as a sphere, the internal part of which is a highly branched dendritic interior, while the external part or periphery consists of rigid mesogenic fragments chemically linked to the interior by flexible spacers of different lengths. The main features of LC dendrimers that distinguish them from linear polymers are their regular highly branched topology, low polydispersity, absence of entanglements, and large number of terminal mesogenic groups. Similar to the two-faced Janus, the LC dendrimers are clearly characterized by the *dual nature* of their physico-chemical properties. On the one side, their dendritic molecules have spherical symmetry and tendency to acquire isotropic spatial distribution in space, which is a typical *entropic factor* inherent to macromolecules, while on the other, their rigid rod-like mesogens tend to form anisotropic phases due to a large gain in enthalpy, which is a typical *energetic factor* inherent to liquid crystals. As a result of these contradictory tendencies, a phase-separated and self-assembled structure is formed, a type of structure that results from *microsegregation* followed by *self-assembly* of the mesogenic fragments. In fact, LC dendrimers are original hybrids combining macromolecular properties of dendritic interiors with those of low-molar-mass liquid crystals.

As mentioned above, all silicon-containing LC dendrimers with terminal mesogens are comprised of a dendritic matrix (or interior), aliphatic spacers, and mesogenic groups (see Fig. 10.3b). Therefore, the problem of establishing the main regularities and features of LC mesophase formation in such systems is directed towards determining the role of the dendrimer generation number, the spacer length, and the chemical nature of mesogens on their structure and phase behavior. To date, many silicon-containing LC dendrimers based on siloxane, carbosilane and some other cyclic silicon-containing oligomers have been synthesized and studied. In this chapter, we first consider general approaches to their synthesis and then discuss the main relationships between their molecular constitution, structural peculiarities, and physico-chemical behavior.

10.3 Polyorganosiloxane Dendrimers with Terminal Mesogenic Groups

The general principle for the synthesis of LC dendrimers with terminal mesogenic groups was originally developed by the Moscow State University research team headed by one of us (V.S.), in close collaboration with the research group of the Institute of Synthetic Polymeric Materials of the Russian Academy of Sciences headed by A. Muzafarov (see Chapter 2) [17–21]. According to this approach, mesogens are attached via a spacer to a dendrimer obtained by a divergent synthesis and containing reactive Si–Cl end-groups. Reaction Scheme 10.1 shows the synthesis of polysiloxane LC dendrimers containing terminal cholesterol groups:

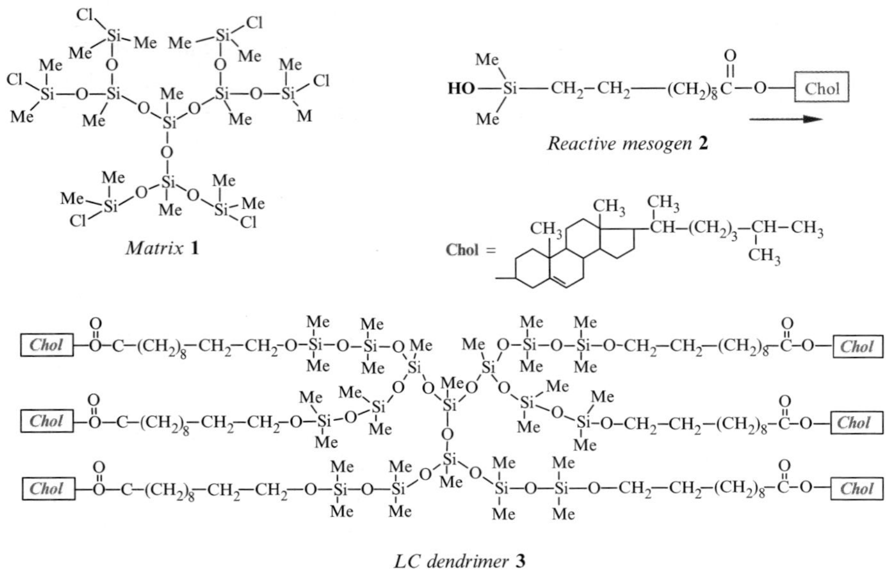

(10.1)

The first stage of this three-step synthesis involved the synthesis of a polychloromethylsilsesquioxane dendrimer precursor, **1**, containing six terminal Si–Cl groups. The second stage, consisting of several reactions, included preparation of the mesogenic compound **2**, containing cholesterol mesogenic groups and dimethylhydroxysilyl reactive centers separated by long undecanoate fragments. It was suggested that the structure of compound **2** with its long spacer would significantly decrease the possible steric hindrance during the grafting (attachment) of this mesogen to the dendritic siloxane compound **1** and would also promote the formation of a LC phase in the desired product **3**, a siloxane-based LC dendrimer of the first generation.

This cholesterol-containing dendrimer **3**, bearing six terminal mesogenic (cholesterol) groups, formed a smectic A (SmA) type mesophase in a broad temperature range from the glass transition temperature (T_g = −1.5°C) to the isotropization temperature (T_{iso} = 120°C). Figure 10.4 shows a scheme for the packing of these dendrimers into a SmA mesophase, based on the results of X-ray diffraction measurements and molecular modeling [17–21]. It can be seen that this model represents a layer smectic packing of the cholesterol mesogens. The spherical form of the methylsesquioxane dendrimer **1** is disrupted due to the ordering of the cholesterol mesogenic groups from different dendrimer molecules forming the layered packing. To achieve this, the cholesterol mesogens with aliphatic undecylene spacers (L = 25 Å) must adopt a "stretched" conformation in the smectic layer (d = 39 Å) and the flexible dendritic interior **1** must be somewhat "flattened" in the direction perpendicular to the layer plane. At the same time, the LC dendrimer molecule as a

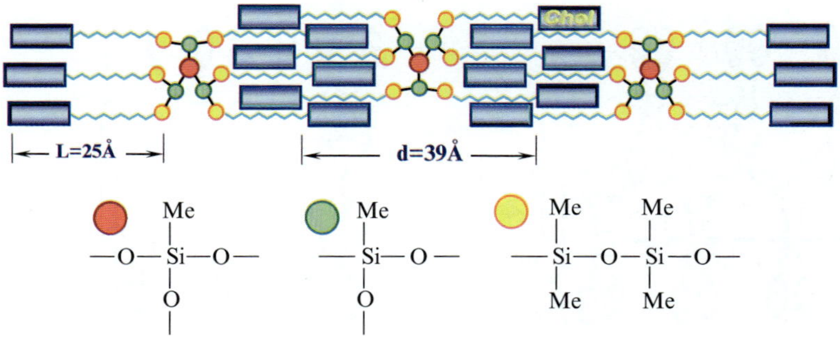

Fig. 10.4 Schematic representation of molecular packing of LC dendrimer **3** in S_A mesophase.

whole must become "prolate" rather than spherical which distinguishes this type of dendrimer from those without mesogenic groups.

Following this, Mehl and Goodby [23] described LC siloxanes **4–6** containing four cyanobiphenyl mesogens, which can be considered as siloxane dendrimers of the zeroth generation:

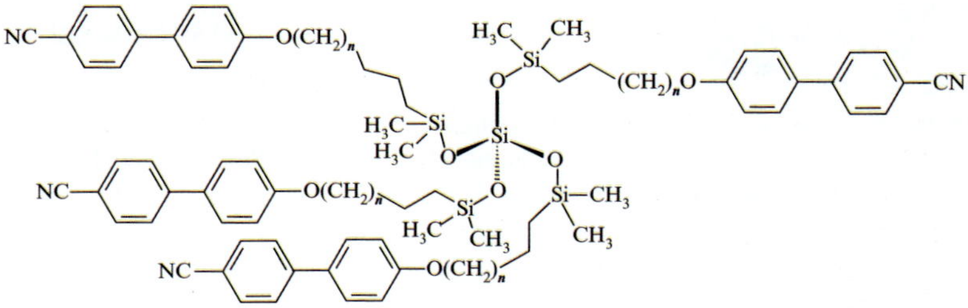

n= 2 **4**

n= 4 **5**

n= 9 **6**

They also used the spacer concept to chemically link the mesogenic groups to the silicon-containing matrix. As the aliphatic spacer length increased from 4 to 11 methylene units, these compounds displayed LC phases of SmA and SmC types and their isotropization temperatures increased from 89°C to 130°C. All these siloxane dendrimers exhibited glass transitions in the temperature range from −5°C to 15°C and, based on small-angle X-ray data, several possible variants for the packing of these molecules in SmA mesophase were suggested (see Fig. 10.5) Subsequently, a series of siloxanes with different matrix structures (linear, cyclic and dendritic) and various numbers of identical terminal cyanobiphenyl mesogens (from 1 to 8) were described [24]. It was found that the structure of the central siloxane matrix in these LC dendrimers did not affect their properties.

More complicated dendritic structures comprising an octasilsesquioxane core **7** and different mesogens attached end-on (dendrimer **8**) and laterally (dendrimer **9**) were described by Goodby et al. [25, 26].

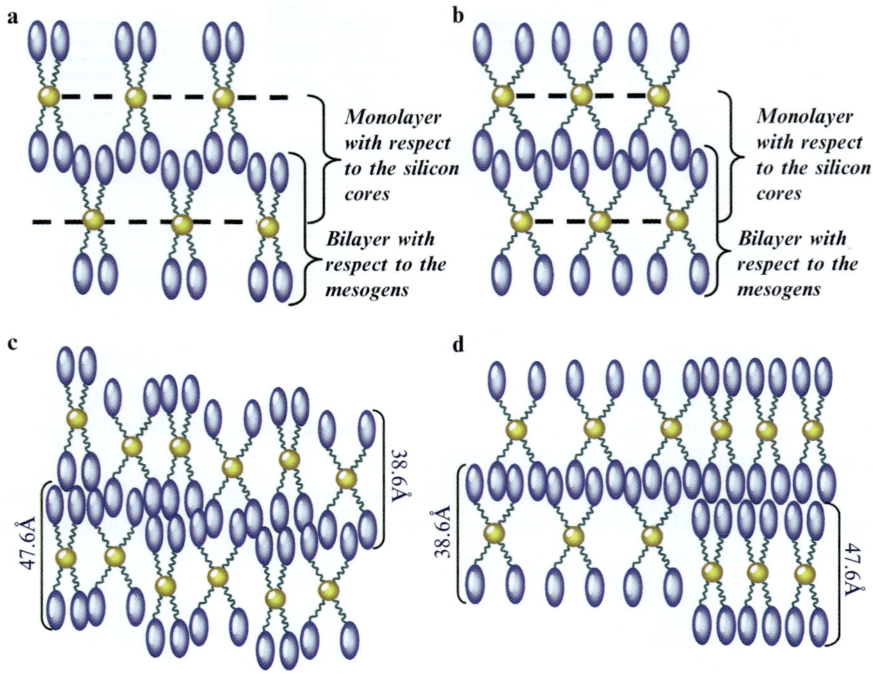

Fig. 10.5 Molecular packing in the smectic A mesophases of compounds (**a**) **4**, (**b**) **5**, and (**c, d**) **6** (two-phase and incommensurate, respectively) (Reprinted from [23]. ©1996 Wiley-VCH Verlag GmbH & Co KG. With permission).

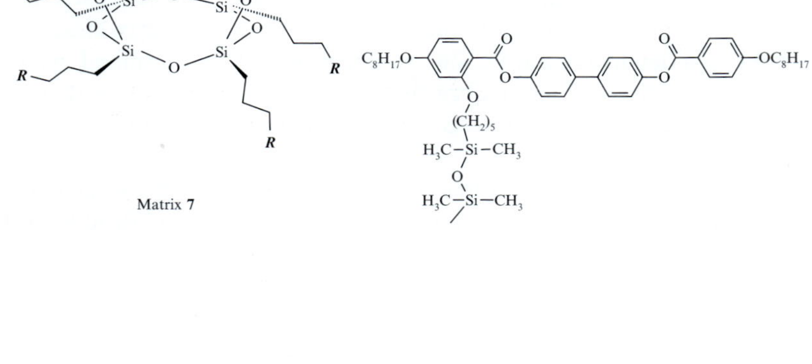

Matrix **7**

8

9

These dendrimers could be classified as carbosilane-siloxane compounds because their structures contain both carbosilane and siloxane units, although the latter dominate. The LC dendrimer **8** was synthesized by a hydrosilylation reaction where the cyanobiphenyl mesogens containing terminal Si–H groups were attached to an octasilsesquioxane core containing terminal allyl groups. This process is similar to the scheme used to obtain carbosilane LC dendrimers considered in the next section.

Dendrimer **8** contains 16 cyanobiphenyl mesogens attached end-on to a dendritic interior via undecylene aliphatic spacers, and forms the tilted SmC phase (SmC 63,1 SmA) and orthogonal SmA phase (SmA 91,7 I) having low glass transition temperature ($T_g = -17.5°C$). In the authors' opinion [25], the molecules of dendrimer **8** must have a rod-like shape in order to pack in layers and, moreover, the molecules within the layers should be disordered. Computer simulation showed that all of the mesogens can not lie on one side of the octasilsesquioxane core; therefore, the mesogenic fragments must be split into two groups with the core unit sandwiched in between as shown in Fig. 10.6. If mesogenic groups are attached laterally to the silsesquioxane core **7** (e.g. LC dendrimer **9**), then nematic phases dominate and their stability is only influenced by the number of mesogenic units.

Moreover, linking 16 laterally-attached chiral mesogenic groups **10** to an octasilsesquioxane core **7** leads to formation of *chiral enantiotropic mesophases*, such as the rectangular columnar (Col^*_{rec}), disordered hexagonal columnar (Col^*_{hd}), and chiral nematic (N^*) phases: Col^*_{rec} 86 Col^*_{hd} 105 N^* 110 I.

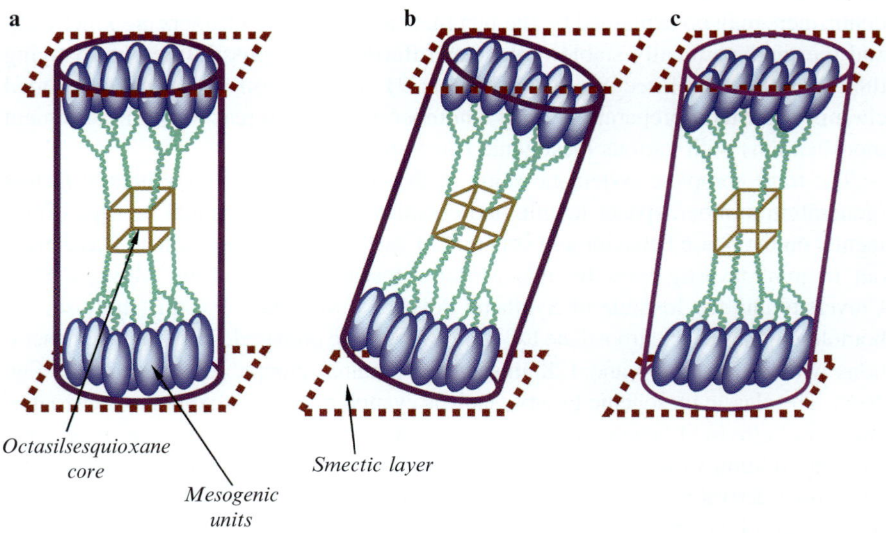

Fig. 10.6 Possible molecular topologies enforced by mesogenic fragments of dendrimer **8** in (**a**) SmA and (**b, c**) SmC mesophases (Reprinted from [25]. ©1996 Taylor & Francis. With permission).

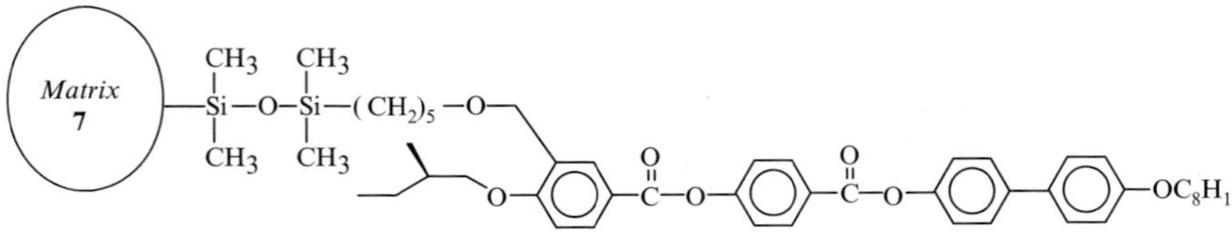

10

Possible packing of dendritic molecules with laterally-attached mesogens **10** in mesophases is shown in Fig. 10.7a and b. It was suggested, that the octasilsesquioxane cube lies approximately at the center, floating in "a sea of flexible siloxane and aliphatic chains", with mesogenic units forming a rod-like "cloud" around them (Fig. 10.7c). Thus, the formation of columnar phases for dendrimers with a silsesquioxane core connected with mesogenic fragments indicates that the rigid core exerts primary control over the occurrence of such disordered structures. It follows that the lateral or side-on attachment of the mesogens to the interior is a key design feature in this particular mesophase type formation.

10.4 Carbosilane LC Dendrimers

Among LC dendrimers of different chemical structures a considerable amount of works has been devoted to carbosilanes. These dendrimers have been extensively studied by Russian [17–22, 28, 29], German [36–41], and Japanese groups [42, 43] from the beginning of 1990s. The wide use of these cores for the synthesis of LC dendrimers may be explained by the fact that carbosilane dendrimers are kinetically and thermodynamically stable molecules offering broad possibilities of changing their dendritic architecture (see Chapter 3). In addition, the well-developed chemistry for their preparation offers a number of chemical reactions for convenient modifications with various mesogenic fragments.

The most complete systematic study of the effects of the main structural factors (generation number, spacer length, and chemical nature of terminal mesogen fragments) on the phase behavior and structure of carbosilane LC dendrimers was carried out in joint investigations by two Moscow research groups from Moscow State University and the Institute of Synthetic Polymeric Materials. For this study several homologous series of carbosilane LC dendrimers were prepared, spanning five generations with 8, 16, 32, 64 and 128 anisometric terminal groups, respectively (see Fig. 10.8). The classic mesogenic groups, such as cyanobiphenyl, alkoxyphenylbenzoates, and chiral ethyl-(*S*)-lactate derivatives containing two or three benzene rings, were used in addition to photochromic mesogenic fragments based on azobenzene- and cinnamoyl-derivatives, as well as hydrophobic aliphatic linear chains and phenol-based hydrophilic fragments. In all cases, the mesogens were attached to dendrimers via long undecylene (Und) or decylene aliphatic spacers. The influence of the aliphatic spacer length, containing three, four, and five CH_2 groups, was studied in a series of LC dendrimers of the fourth generation and with cyanobiphenyl mesogens [22].

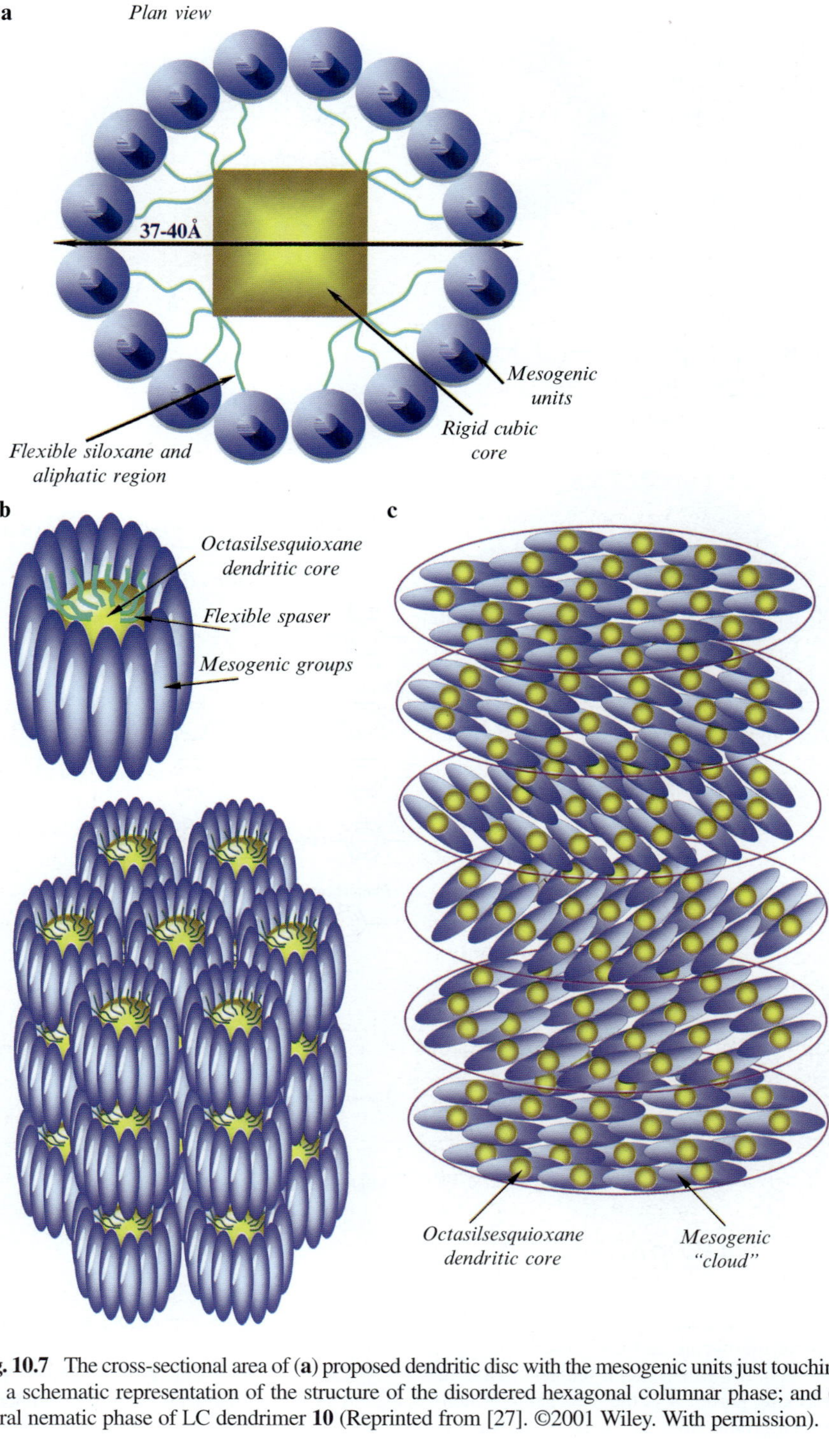

Fig. 10.7 The cross-sectional area of (**a**) proposed dendritic disc with the mesogenic units just touching; (**b**) a schematic representation of the structure of the disordered hexagonal columnar phase; and (**c**) chiral nematic phase of LC dendrimer **10** (Reprinted from [27]. ©2001 Wiley. With permission).

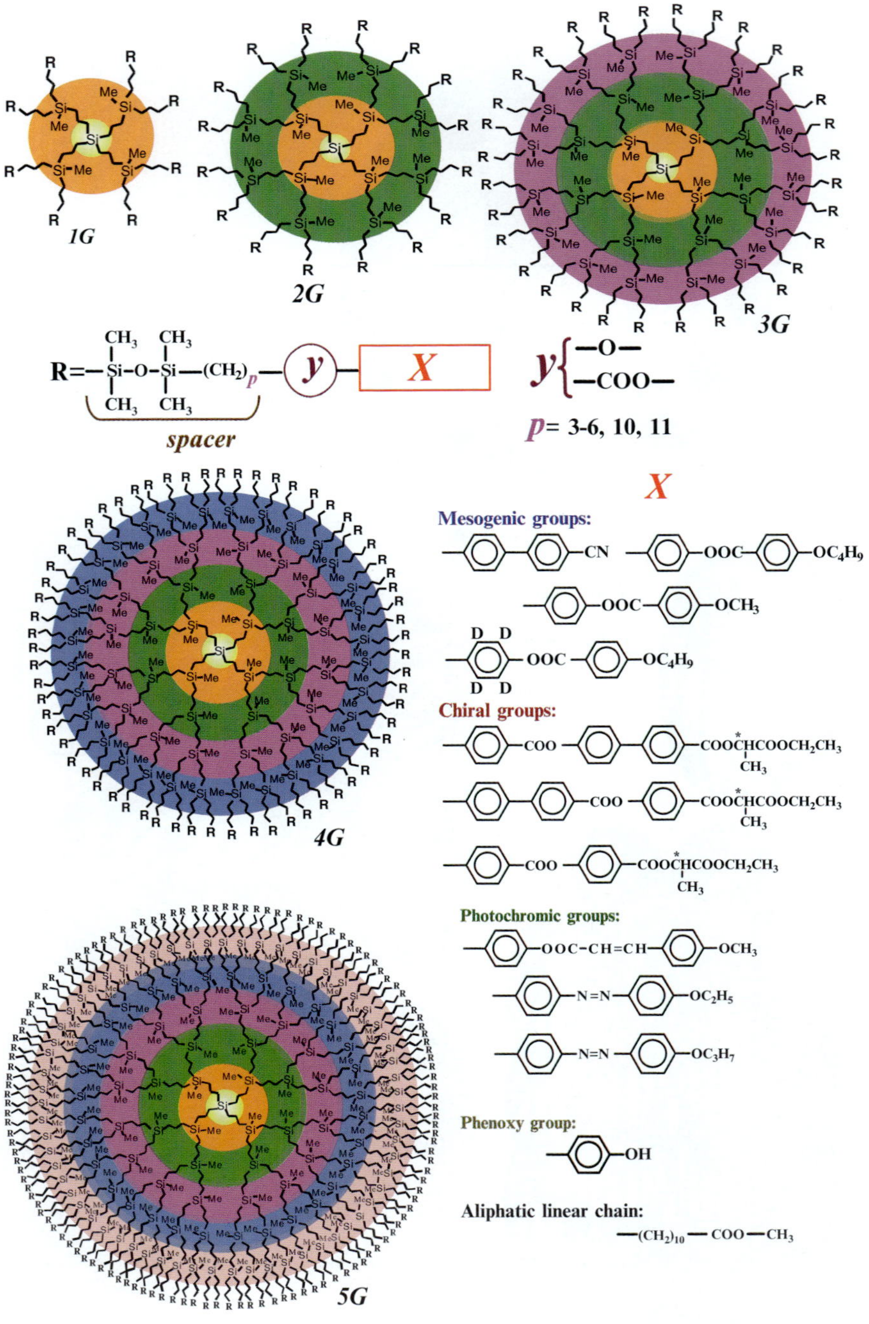

Fig. 10.8 The general structure of LC carbosilane dendrimers of all five generations with different terminal mesogenic and non-mesogenic groups. nG-(spacer-y-X)m corresponds to the structural formula of dendrimer; nG is the generation number; spacer is the methylenic groups number; **y**- is the linking group; X is the mesogenic or non-mesogenic terminal fragments; m is the number of terminal groups (m = 8, 16, 32, 64, 128).

Apart from *homodendrimers*, the carbosilane *codendrimers* and *block-codendrimers* containing spacers of different lengths and terminal mesogenic groups have also been recently synthesized and studied (see below).

10.4.1 Synthesis of Carbosilane LC Dendrimers with Different Molecular Architectures

The synthesis of carbosilane LC dendrimers with terminal mesogenic groups is not a trivial task because the requirements for monodispersity are very stringent. The general strategy consists of three main steps [28, 29].

The first step is the controlled synthesis of a carbosilane dendrimer precursor with terminal allyl groups. Starting with tetraallylsilane and using methyldichlorosilane as a branching agent, five generations of these dendrimers were synthesized according to Reaction Scheme 10.2 (see also Chapter 3).

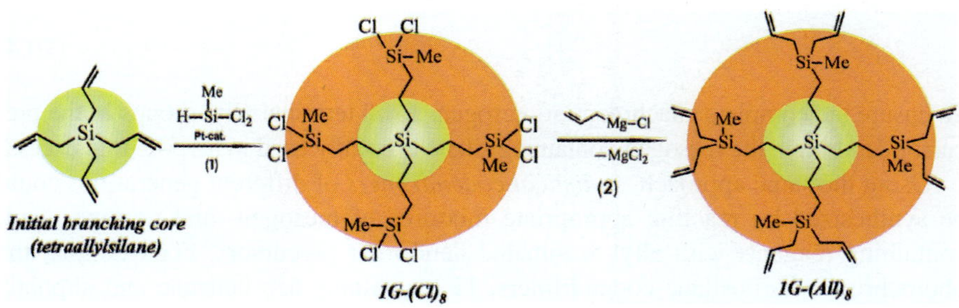

(10.2)

The second step includes the synthesis of functionalized mesogenic groups with a spacer. Reaction Scheme 10.3 shows the synthesis of such a mesogenic group with an undecylenic spacer containing a terminal Si–H group capable of reacting with the carbosilane dendrimer having allyl end-groups. The final third step is the hydrosilylation attachment of these two reagents, as illustrated by an example of a LC dendrimer of the third generation containing 16 mesogenic groups in Reaction Scheme 10.4.

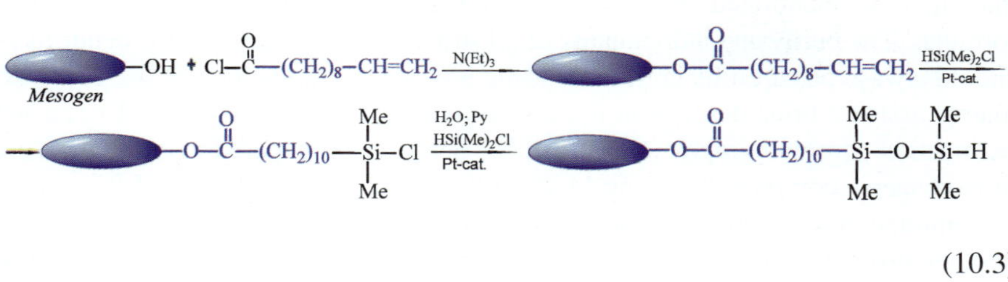

(10.3)

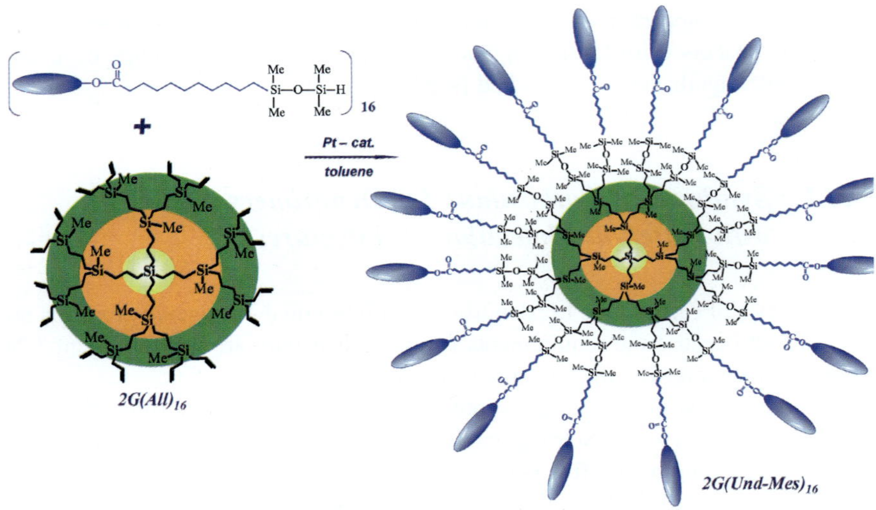

(10.4)

To ensure the complete attachment of mesogens to all terminal allyl groups of the precursor dendrimer, the mesogen-containing silane is usually used in a 1.5–2-fold excess.

Using the same approach, *statistical codendrimers* of different generations could be synthesized by reacting appropriate mixtures of mesogen- and non-mesogen-containing reagents with allyl-terminated dendrimer precursors. For example, the photochromic carbosilane codendrimers, **11**, containing azo-benzene and aliphatic fragments of different compositions (see Fig. 10.9a) have been synthesized and studied [44–46]. Another way of obtaining statistical codendrimers is based on partial substitution of the terminal functional groups of non-liquid crystalline homodendrimers by mesogenic fragments. For example, Fig. 10.9b shows a recently reported amphiphilic LC codendrimer of the third generation, **12**, containing butoxyphenylbenzoate and phenolic terminal groups [47, 48]. In order to obtain LC *block-codendrimers* a complex multistep synthetic method was developed by our colleagues at the Institute of Synthetic Polymer Materials in Moscow [49] and used by us to obtain the block-codendrimers, **13**, containing the same photochromic azo-benzene and aliphatic decylenic linear fragments (see Fig. 10.9c) [46].

In conclusion of this "synthetic part" of this chapter it is very important to stress that all above-mentioned reactions perform well and yield products with presented formulas. The purity and individuality of all structures of synthesized LC dendrimers should always be carefully confirmed using physical methods such as IR, ^{1}H NMR, matrix assisted laser desorption ionization-time of flight (MALDI-TOF) and gel permeation chromatography (GPC). All LC dendrimers prepared in our work were monodisperse compounds having M_w/M_n ratios of no more than 1.01–1.03.

Simultaneously with our early publications [18, 19, 28, 29], the results of another investigation of carbosilane dendrimers with terminal mesogenic groups

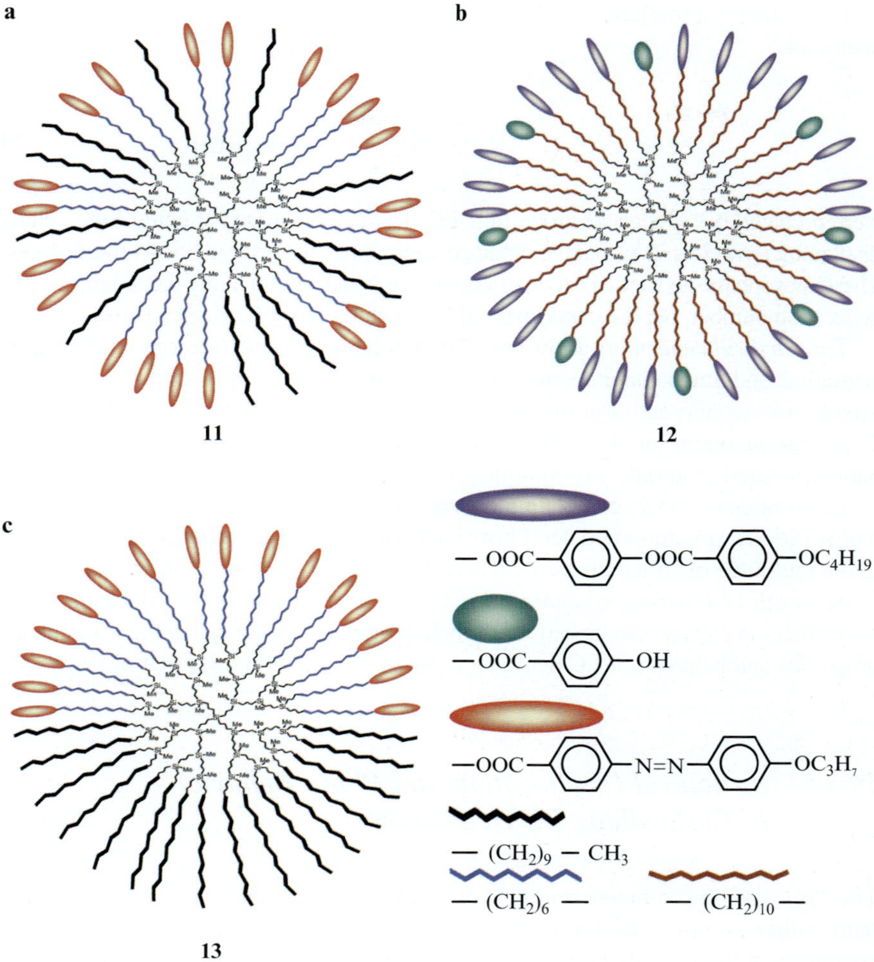

Fig. 10.9 Statistical (**a**) photochromic, (**b**) amphiphilic LC carbosilane co-dendrimers of the second generation, and (**c**) LC block-codendrimer containing photochromic and aliphatic terminal groups (second generation).

were reported by the German research group headed by Frey [36–41]. These authors also synthesized three generations of dendrimers with cholesterol and cyanobiphenyl groups; however, in contrast to our structures, their terminal mesogenic groups were attached to the dendrimers via shorter aliphatic spacer of only five CH_2 units. In addition, their LC dendrimers were synthesized via reaction of dendrimer polyols with cholesteryl-chloroformate or the reactive derivative of cyanobiphenyl, and they observed the formation of a SmA structure. The isotropization temperatures were independent of the spacer length, but increased from the first to the second dendrimer generation.

Interesting carbosilane LC dendrimers with 3, 9 and 27 terminal *chiral fragments*, **14**,

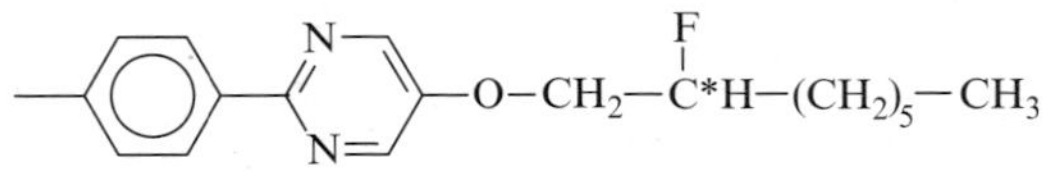

14

were prepared by a Japanese group [42, 43]. The authors aimed at chiral SmC* phase displaying *ferroelectricity*, but the obtained compounds formed only SmA mesophases. These mesophases existed in a temperature range which significantly increased with the generation number, but the dendrimers did not exhibit ferroelectric properties.

Thus, *hybrid dendrmers* allow one to combine in the same macromolecule liquid crystalline and functional properties of their constituents: hydrophilic and hydrophobic groups, mesogenic and non-mesogenic fragments, photochromic and chiral units. This feature opens up interesting new opportunities for the creation of novel nanostructured materials with unique properties.

As mentioned above, competing tendencies of mesogenic fragments to arrange in highly ordered structures and dendrimer interiors to form symmetrical shapes can fuse into a multitude of structural forms. Therefore, *structure formation in LC dendrimers is the result of microsegregation and organization processes.* Such dualism of LC dendrimers is clearly manifested in all their properties. In the following sections the properties and behavior of LC dendrimers are discussed from this point of view.

10.4.2 Structural Organization and Phase Behavior of Carbosilane LC Dendrimers

The "internal micro-heterogeneity" of individual LC dendrimers predetermines their rather complicated hierarchy of structural ordering. Despite the spherical symmetry of their highly branched interiors having a tendency to an isotropic space distribution, the interactions of their mesogenic rigid rod fragments break the "ball symmetry" and lead to anisotropic arrangement of the terminal mesogens. Such a situation is the main characteristic of low generations (1G-4G) of LC dendrimers which are considered in the next section.

10.4.2.1 Low Generations

Figure 10.10 shows a molecular model of a LC dendrimer of the first generation containing eight terminal cyanobiphenyl groups with an undecenyl spacer 1G-(Und-CB)$_8$. It can be clearly seen from this figure, that the mesogenic groups of the same molecules can be packed more or less densely forming a typical layered structure. The neighboring dendrimers can be superimposed on one another to create the smectic structure.

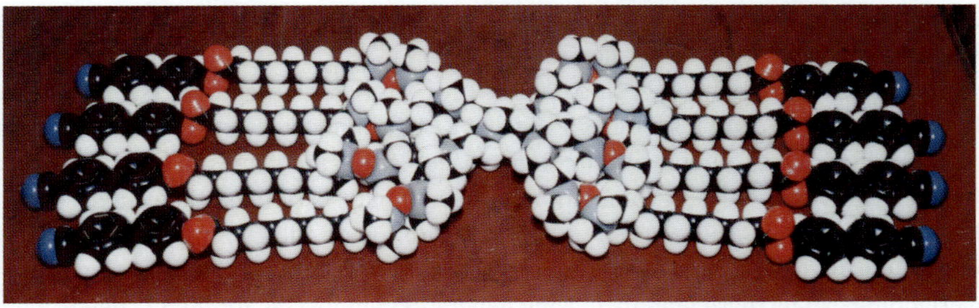

Fig. 10.10 Molecular model of LC dendrimer containing eight cyanobiphenyl mesogenic groups with undecylenic spacer 2G-(Und-CN)$_8$.

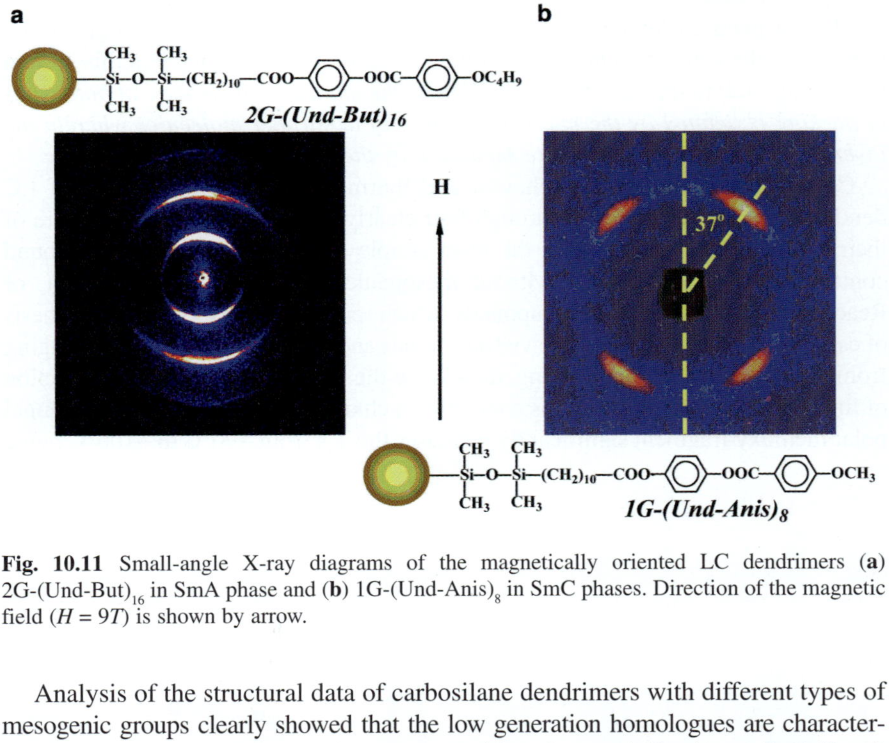

Fig. 10.11 Small-angle X-ray diagrams of the magnetically oriented LC dendrimers (**a**) 2G-(Und-But)$_{16}$ in SmA phase and (**b**) 1G-(Und-Anis)$_8$ in SmC phases. Direction of the magnetic field (*H* = 9*T*) is shown by arrow.

Analysis of the structural data of carbosilane dendrimers with different types of mesogenic groups clearly showed that the low generation homologues are characterized by the formation of smectic structure with orthogonal (SmA) or tilted (SmC) disposition of mesogens with respect to the planes in which the flexible interior is located. The detailed X-ray analysis of a number of LC carbosilane dendrimers at small and wide angles allowed calculation of their layer spacings and suggested several variants of mesogens packing [21, 28–30, 33–35, 46, 47, 50–52]. Figure 10.11 shows typical small angle X-ray diffraction patterns for two samples of LC dendrimers oriented in the strong magnetic field (*9T*). Both X-ray diffraction patterns are characterized by a diffuse halo at wide angles close to 4.8–5.0 Å, demonstrating the liquid-like order in the direction normal to the mesogenic long axis. The small angle reflexes are split into two arcs for the orthogonal SmA mesophase and into four arcs for the tilted SmC structure.

The layer spacings have been calculated for various LC dendrimers from X-ray diffraction data. It can be seen from Fig. 10.12 that in most cases for three series of LC carbosilane dendrimers with the same spacer length the values of layer spacings, *d*, increased with increasing generation number. Analysis of these data suggests that dendrimer molecules pack in the orthogonal (SmA) and tilted (SmC) modifications with full or partial overlap of mesogenic groups. Therewith, the degree of mesogen overlapping can increase with increasing generation number, as for the LC dendrimers with cyanobiphenyl mesogenic groups (Fig. 10.12 curve 1); the *d*-values for the first three generations of these series of dendrimers are in close agreement.

General schemes for the molecular packing of LC dendrimers in SmA and SmC mesophases with full and partial mesogen overlap are shown in Fig. 10.13. The *d*-values depend on the chemical structure of mesogenic groups, the length of the spacer, and the generation number. For the LC dendrimers considered in this section they usually fall in the 35–55 Å range. *Hence, the main specific feature of both types of packing is defined by the microsegregation of dendrimer molecules wherein the layers of carbosilane interiors are separated by the layers of mesogenic groups.*

Comparison of the phase behavior and thermal properties of a series of LC dendrimers of generations one through four clearly demonstrates the dual nature of their behavior. Let us start with the most simple carbosilane dendritic compound containing only *allyl groups* without mesogenic units, such as 1G-$(Allyl)_8$ of Reaction Scheme 10.2. These compounds, which serve as precursors for the synthesis of carbosilane LC dendrimers, are viscous liquids and have very low T_g values ranging from −90°C to −100°C, depending slightly on the generation number. Introduction of linear aliphatic terminal groups containing a chain of $–CH_2–$ units with a terminal polar methoxy fragment significantly increases the T_gs from −90°C to −100°C range

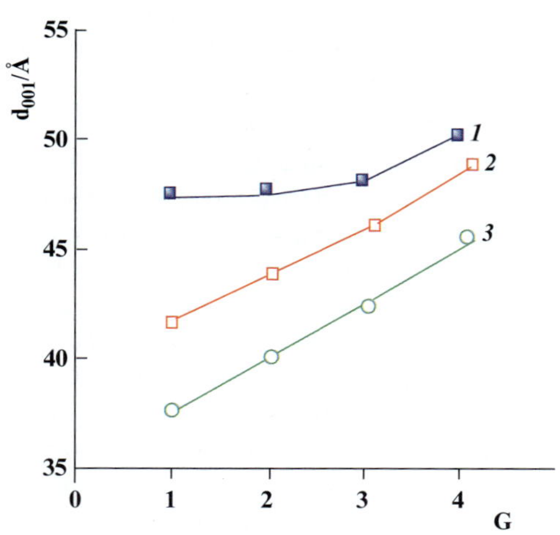

Fig. 10.12 Influence of the generation number on the long spacings d_{001} for three series of LC dendrimers with the same spacer length at 50°C: (**1**) nG-(Und-CN); (**2**) nG-(Und-Anis); (**3**) nG-(Und-But).

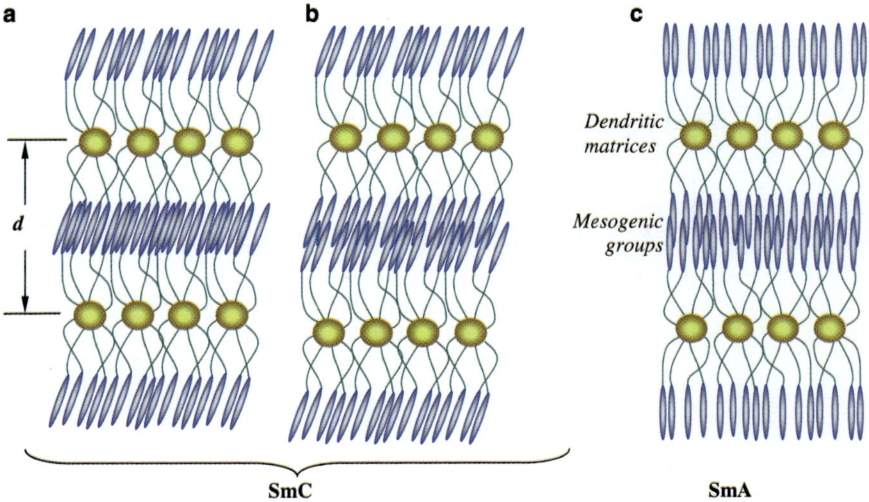

Fig. 10.13 Schemes of molecular packing of LC dendrimers in (**a, b**) tilted smectic C and (c) orthogonal smectic A phases with the full (**a**) and partial (**b, c**) mesogenic groups overlapping (Reprinted from [52]. ©2006 Ivanovo State University. With permission).

to −70°C (see Fig. 10.14). However, no LC phases are formed, but dendrimers have a tendency to crystallize forming two crystalline modifications K1 and K2 with low melting points that depend only slightly on the generation number.

Analysis of enthalpies of melting showed that their values decreased as a function of generation number. Increase in generation number led to a reduction of ordering and crystallization of terminal aliphatic flexible groups. In the next section we shall see that introduction of rigid-rod mesogenic groups into high generation dendrimers promotes formation of "disc" or "torus"(doughnut)-shaped dendritic molecules which become "building blocks" for the formation of columnar structures.

The above-described thermal behavior shows very clearly the *dual nature of dendrimers* even without mesogenic groups. While flexible dendrimer molecules do not crystallize and are not responsible for low T_gs, aliphatic terminal groups do crystallize and are responsible for melting points, but their interactions also slightly increase the T_gs. When terminal groups are sufficiently independent in their behavior to demonstrate their autonomous character; phase separation takes place and the long aliphatic chains crystallize (for other types of dendrimer-based multiarm star polymers see also Chapter 3, Section on Star Polymers with Carbosilane Dendrimer Core and Section 11.3). The energetic interactions of terminal groups overcome the entropic forces associated with the existence of flexible amorphous dendrimer interiors.

Figure 10.15 shows the influence of the structure of mesogenic groups on T_gs of carbosilane LC dendrimers. It can be seen that all dendrimers have similar T_gs lying within a low temperature range from −20°C to −5°C, and indicating that T_g values do not depend on the generation number within each series of these LC dendrimers. Similarly, T_g values change only slightly from one type of mesogenic group to another, demonstrating that chemical structure of mesogens also has only a slight effect on this

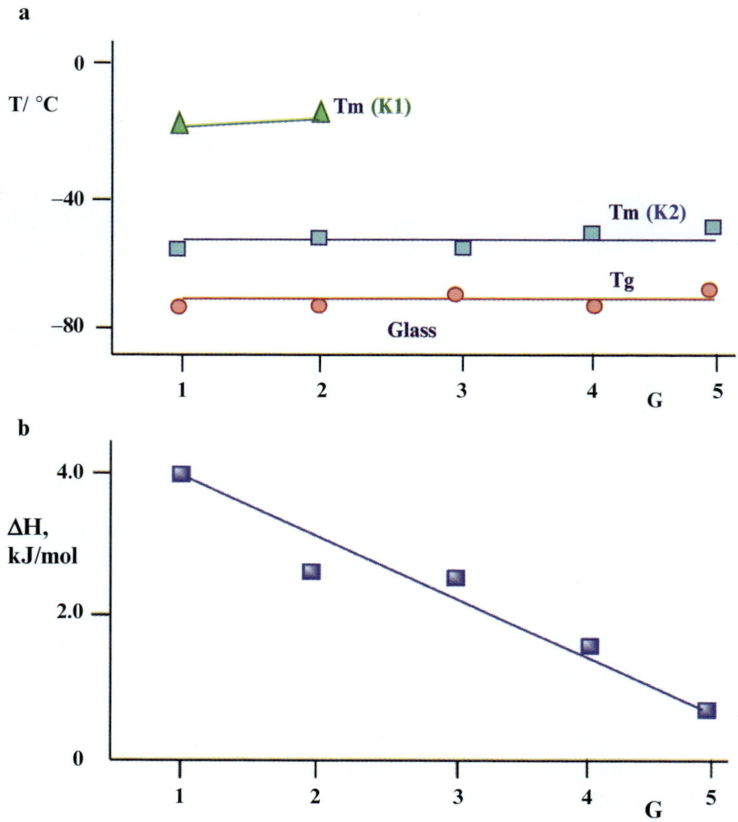

Fig. 10.14 Dependences of (**a**) a melting point (T_m) and glass transition temperatures (T_g) as well as (**b**) enthalpy of crystalline modification K1 and K2 melting as a function of generation number for carbosilane dendrimers containing —$(CH_2)_{10}$–COO–CH_3 terminal group (Reprinted [52]. ©2006 Ivanovo State University. With permission).

property. Thus, it follows that glass transition temperatures of these LC dendrimers are mainly determined by the chemical structure of their carbosilane interiors.

Thermal properties and phase behavior of these three series of LC dendrimers of five generations are summarized in Fig. 10.16. In all cases, the glass transition temperatures as well as melting points of crystalline modifications of LC dendrimers lie mainly below ambient temperature (see Fig. 10.16c). At the same time, all carbosilane dendrimers from generation one to four show layered smectic SmC and/or SmA structures with a rather high isotropization temperatures (T_{iso}, from 90°C to 120°C). These temperatures are mainly determined by the generation number and the spacer length. The higher the generation number and the longer the spacer length, the higher is the isotropization temperature.

The influence of spacer length on the formation of LC phase, namely, the glass transition, the clearing point, and the enthalpy of transition is shown in Fig. 10.17

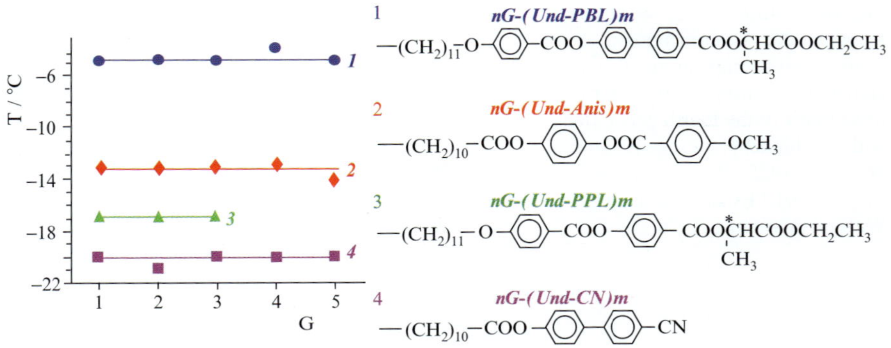

Fig. 10.15 Influence of generation number on glass transition temperature for carbosilane LC dendrimers with the same dendritic core but with different mesogenic groups.

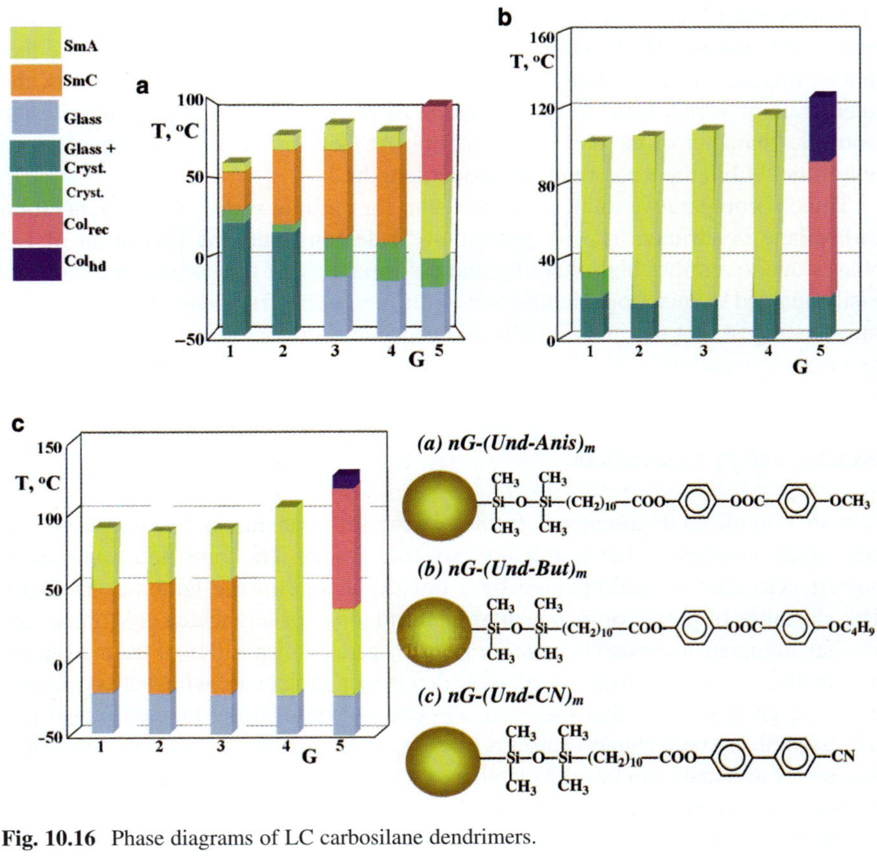

Fig. 10.16 Phase diagrams of LC carbosilane dendrimers.

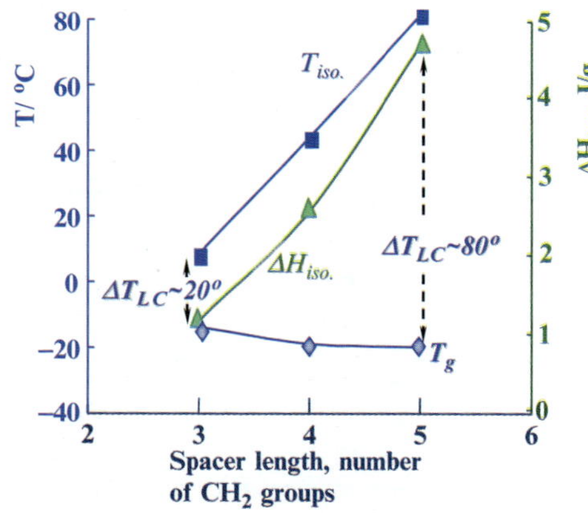

Fig. 10.17 Influence of the spacer length on the glass transition (T_g), isotropization temperature (T_{iso}) and enthalpy of transition (ΔH_{iso}) for LC dendrimers of the fourth generation with CN-biphenyl mesogenic groups 4G-(Spacer-CB) (Reprinted from [22]. ©2001 Pleiades Publishing. With permission).

for an example of the fourth generation carbosilane LC dendrimer with cyanobiphenyl mesogenic groups. It can be seen that the increase in spacer length from three to five $-CH_2-$ groups led to a virtually linear increase in both temperature and enthalpy of isotropization. As a result, the temperature interval of the LC mesophase existence extended five times (from 18°C to 98°C). This phenomenon is related to the fact that an increase in spacer length favors greater mobility of terminal mesogens, the interaction of which leads to the formation of anisotropic mesophases despite the isotropic character of the dendrimer interior. At the same time, T_g values changed only little within the same range of spacer lengths.

Thus, a comparative analysis of the data for various series of low generation carbosilane dendrimers (up to generation 5) demonstrates the formation of LC phases due to a rather high mobility and deformability of the carbosilane interiors easily adapted to anisotropic interaction of the mesogenic fragments. It is reasonably safe to suggest that the individuality of dendrimer interiors is strongly suppressed by microsegregation in which the dominant role belongs to mesogens.

10.4.2.2 High Generations

A more complicated structure is formed in the fifth generation of carbosilane LC dendrimers containing 128 terminal mesogenic groups. The transition from low to high generations is accompanied by a drastic change in the lamellar structure. The existence of plenty of mesogen groups and long spacers which are forced into a radial arrangement around the interior prevents parallel disposition of the mesogenic fragments. A molecular model of such dendrimer molecules with butoxyphenylbenzoate groups shows that they can adopt the shape of a flat tapered object and self-assemble into supramolecular disks (see Fig. 10.18). In other words, each individual dendrimer molecule can be considered as a laterally compressed nanoparticle. This leads to self-assembly into supramolecular columns, which can further self-organize into columnar mesophases.

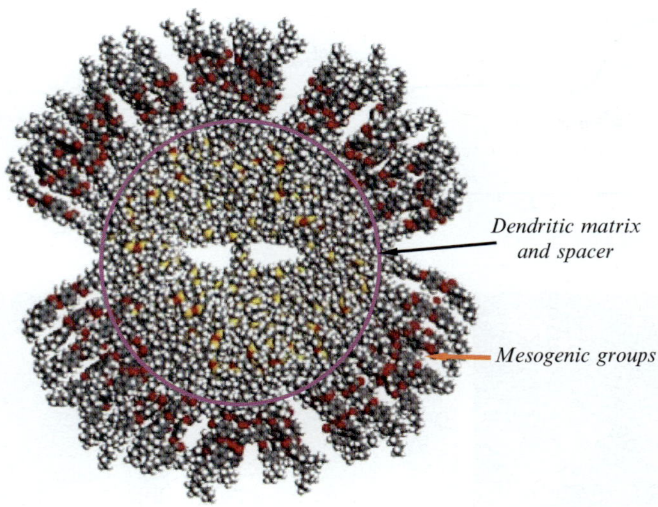

Fig. 10.18 Molecular model of the fifth generation dendrimer 5G-(Und-But)$_{128}$ in the form of a flattened nanoparticle (Reprinted from [48]. ©2005 American Chemical Society. With permission).

Phase diagrams (see Fig. 10.16) illustrate the existence of columnar phase for all LC dendrimers of the fifth generation at elevated temperatures. Note that this type of mesophase can be characterized by two different columnar modifications with rectangular (Col_{rec}) and hexagonal disordered (Col_{hd}) column arrangements. Based on small- and wide angle X-ray and neutron scattering data, obtained for oriented (in magnetic field) samples at various temperatures [22, 30, 33–35, 50–52], generalized structural models of mesophase modifications formed by carbosilane LC dendrimers are presented below.

The most extensively studied carbosilane LC dendrimer of the fifth generation [30, 33] containing cyanobiphenyl mesogenic groups demonstrates the most prominent example of structural polymorphism. The differential scanning calorimetry (DSC) trace of this dendrimer clearly shows the existence of three transition points: T_g and two first order transitions with different heats of transitions (see Fig. 10.19). At high temperatures, just below the clearing point, a grey (non coloured) mosaic texture is observed. At slightly lower temperatures, just after the second transition, a very bright coloured mosaic texture is seen. After cooling of the sample below 50°C, the mosaic texture is gradually destroyed and a sort of lamellar texture is formed. This structure corresponds to a SmA phase. Two types of mosaic texture are assigned to the different types of columnar mesophase.

The results of detailed X-ray measurements of LC dendrimer of the fifth generation at different temperatures allowed calculation of the parameters *a* and *c* of a two-dimensional rectangular unit cell and suggested structural models of dendrimer molecules shown in Fig. 10.20. At still lower temperatures the SmA lamellar structure is formed. Heating of the sample leads to gradual change of the lattice parameters:

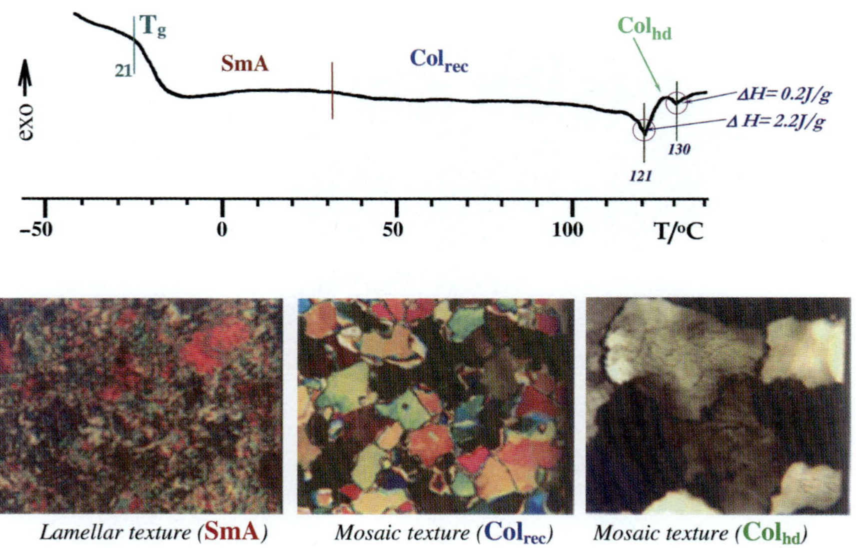

Fig. 10.19 DSC curve and optical textures formed by carbosilane dendrimer of the fifth generation containing cyanobiphenyl mesogenic groups 5G-(Und-CB)$_{128}$ at various temperatures (Reprinted from [33]. ©2000 American Chemical Society. With permission).

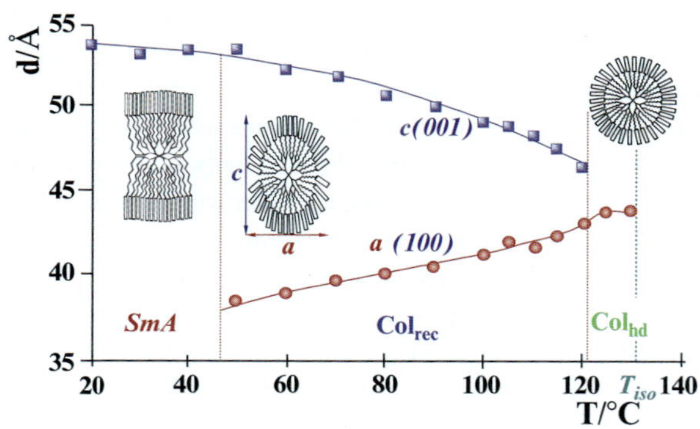

Fig. 10.20 Temperature dependence of the parameters *a* and *c* of unit cells in the different mesophases of LC dendrimer 5G-(Und-CB)$_{128}$ and schemes of molecular models of LC dendrimer at different temperatures.

a values increase, and *c* values decrease. In this temperature interval, the rectangular columnar phase Col$_{rec}$ appears, while at higher temperatures, only one slightly broad peak in the X-ray difractogram remains, which means that dendrimer molecules become more spherical with increasing temperature.

Generalized diagrams of polymorphic modifications of the same LC dendrimer of the fifth generation are shown in Fig. 10.21. The SmA structure formed at low temperatures assumes the alternations of layers consisting of mesogenic groups with layers containing carbosilane dendritic interiors, and probably also parts of the spacers. The dendrimer molecules are significantly elongated to enhance the interaction between mesogenic groups. They are also flattened and arranged in layers (see Fig. 10.21a).

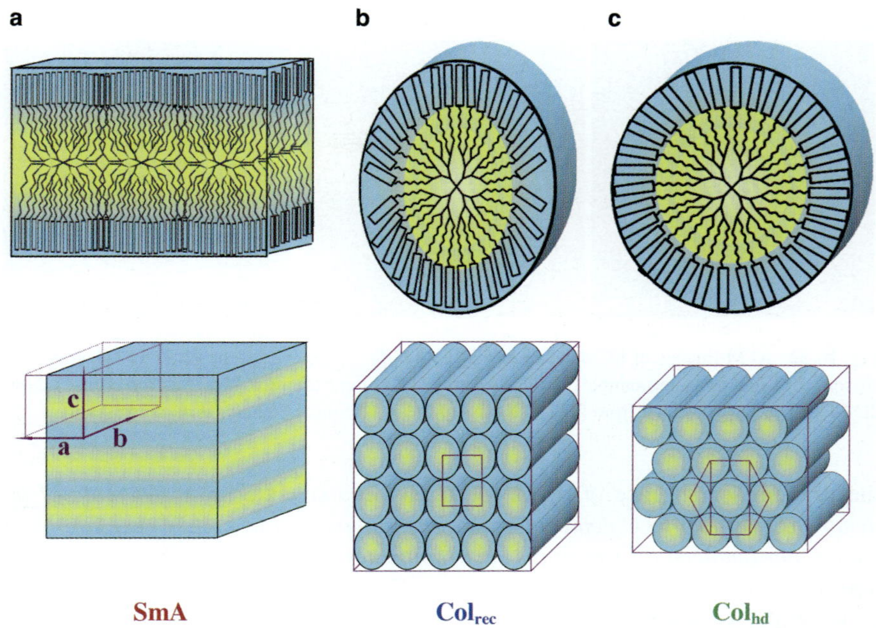

Fig. 10.21 Schematic diagrams of a possible structures and molecular packings formed by carbosilane LC dendrimer 5G-(Und-CB)$_{128}$ at various temperatures. (**a**) Lamellar SmA mesophase, 40°C; (**b**) columnar structure with rectangular ordering of ellipsoidal columns, 70°C (Col$_{rec}$); (**c**) hexagonal disordered columnar phase (Col$_{hd}$) with rounded-shaped dendrimer molecules, 130°C (Reprinted from [52]. ©2006 Ivanovo State University. With permission).

Heating reduces strong interactions between mesogens and they tend to orient in all directions. Each molecule becomes less elongated and more symmetrical, but anisotropy is preserved as molecules adopt ellipsoidal shapes. As a result, the lamellar structure is transformed into the columnar phase with rectangular lattice Col$_{rec}$ (see Fig. 10.21b). At still higher temperatures, the interactions between mesogenic groups decrease and the hexagonal-disordered columnar structure is formed (see Fig. 10.21c). No strong interactions between mesogens take place in this modification, which is composed entirely of dendrimer interiors due to microphase separation between the aliphatic dendritic units and mesogenic fragments. This is confirmed by a rather large enthalpy of the Col$_{rec}$→Col$_{hd}$ phase transition (2.1 J/g) in comparison to a tenfold smaller value for the Col$_{hd}$→I transition (0.2 J/g). Each column is composed of globular dendrimer molecules, with mesogens residing exclusively on the external surface and the inner part comprising amorphous dendrimer interiors (see Fig. 10.21b).

Thus, we may conclude that the highly branched topology of LC dendrimers induces microphase separation that leads to the formation of lamellar or columnar supramolecular structures depending on the generation number. The role of the chemical nature of terminal mesogens is simply to determine particular characteristics of a given mesophase, such as the temperatures and enthalpies of transitions, and

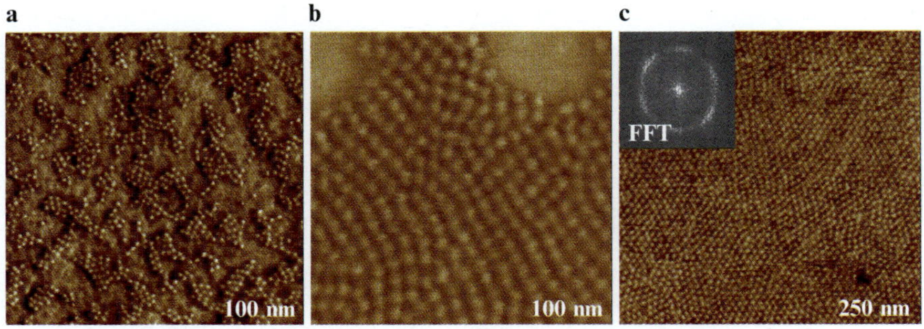

Fig. 10.22 AFM images of LC dendrimer 5G-(Und-CB)$_{128}$ prepared on the different substrates: (**a**) silicon; (**b**) a non-annealed sample on mica; (**c**) annealed sample on mica. Inset (*left top corner*) shows 2D FET spectra (Reprinted from [31]. ©2000 American Chemical Society. With permission).

the presence or absence of tilt of mesogens in a smectic layer. *The role of the dendritic interior of LC dendrimers is most pronounced in higher generations, in which supramolecular structures of a columnar type begin to form within the amorphous dendritic layer.*

Recently, we observed the process of supramolecular ordering in carbosilane LC dendrimers using atomic force microscopy (see Fig. 10.22) [31]. The samples of the fifth generation LC dendrimers with cyanobiphenyl groups were prepared by spin-coating of dilute chlorophorm solution (10^{-4} wt %) on silicon wafers and mica plates. Figure 10.22a clearly shows individual dendrimer molecules as monodisperse spherical nanoparticles. The dimensions of these particles are about 5 nm, in good agreement with X-ray diffraction data. Figure 10.22b shows aggregates of dendrimer particles with rectangular packing. Thermal annealing of dendrimer domains on mica led to a transition from rectangular to hexagonal packing as is clearly shown in Fig. 10.22c. Thus, the AFM data confirmed the structural models obtained from the small-angle X-ray scattering data described above.

10.5 Chiral Carbosilane LC Dendrimers with Ferroelectric Properties

As shown above, most if not all, carbosilane LC dendrimers of low generations (1G–4G) form the simplest smectic phases: SmA and SmC. However, it should be noted, that among the various known smectic phases, the chiral smectic C phase (SmC*) is the most interesting, because it can show ferroelectric properties and very fast switching between two stable states of the mesogenic units.

It is well-known that low-molar-mass ferroelectric liquid crystals show great potential for electro-optical applications [1, 4]. The synthesis of ferroelectric LC side chain polymers was first achieved by us, using comb-shaped polymers in 1984 [53, 54]. Considering significant scientific and technological interests in the SmC* phase, synthetic methods have been explored [55–58] to obtain chiral LC dendrimers capable of forming the SmC* phase in order to create ferroelectric liquid crystals with unique dendritic structure. The terminal mesogenic fragments, ethyl-(*S*)-lactate derivatives, containing three phenyl-biphenyl-lactate (PBL) and biphenyl-phenyl-lactate (BPL) as well as two benzene rings phenyl-phenyl-lactate (PPL) with chiral fragments, were coupled to the carbosilane dendrimers via an aliphatic spacer consisting of 11 methylene groups (Und = undecylenic group):

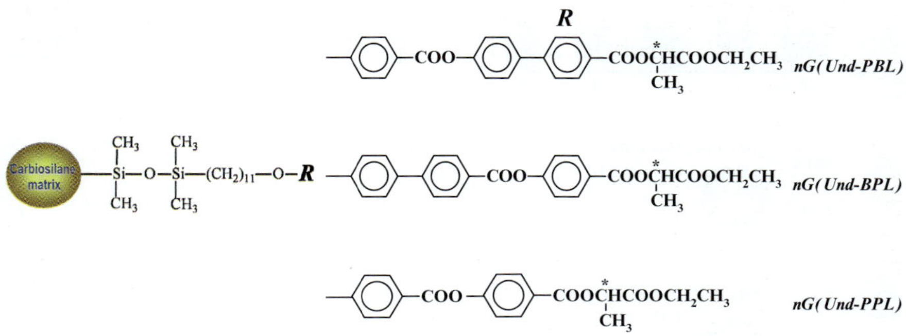

It was shown [53, 54, 59], that combination of four factors: (1) a long spacer, (2) the extended rigid mesogenic fragments, (3) a transverse dipole moment, and (4) the chiral groups, leads to the formation of the SmC* phase possessing ferroelectric properties. Three generations of these chiral LC dendrimers displayed chiral SmC* phases over a very broad temperature range (see Fig. 10.23). Their T_g values were in the low temperature range (from about −5°C) and the LC phases were stable up to 170–180°C. The SmC* structure was preserved in the glassy state and the chiral dendrimers of the fourth and fifth generations displayed only the rectangular columnar phase. This was probably caused by steric hindrance in the arrangement of rigid rod mesogens and layered packing of these first three generations being replaced by a columnar structure (compare with Section 10.4.1.1).

As mentioned above, the main feature of the smectic packing of LC dendrimers is microsegregation of dendrimer molecules, which in turn is caused by the alternation of carbosilane dendrimer interiors and mesogenic layers (see Fig. 10.24). According to this scheme, the chiral SmC* phase forms a *helicoidal structure* in which mesogens are placed in layers (Fig. 10.24a), and tilted at an angle Θ from the normal. The change in tilt angle from layer to layer gradually describes a helix characterized by a pitch depending on the chemical structure of the compound. The existence of a dipole moment and chiral center in each mesogen generates in each layer a polarization *p* (shown by arrows in Fig. 10.24a). However, since the directions of dipole moments change from one layer to another, the spontaneous polarization P_S is reduced to zero throughout the bulk of the sample. In order to obtain a true ferroelectric phase, the chiral smectic C* phase should be untwisted and a macroscopic nonvanishing polarization can be generated (see Fig. 10.24b).

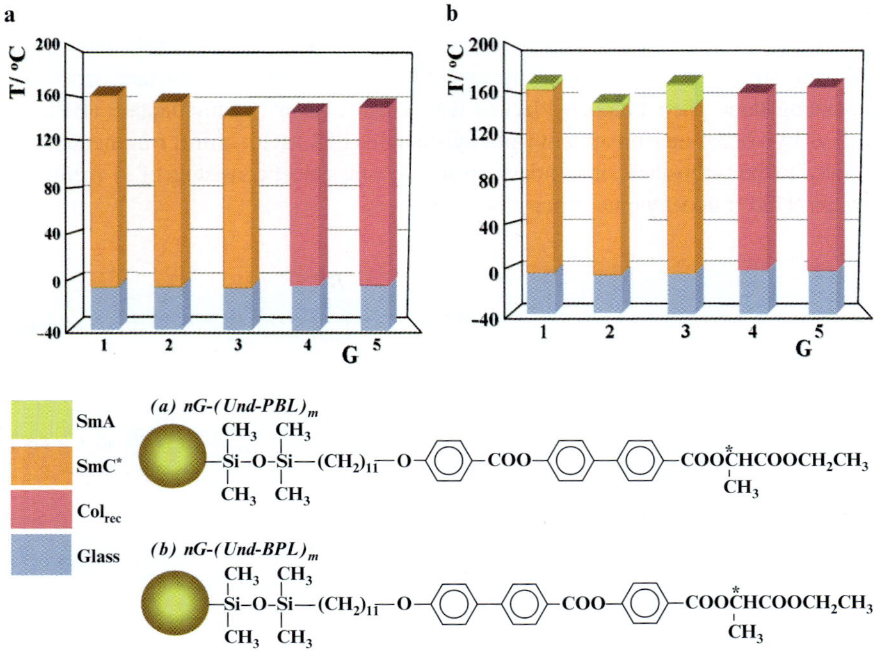

Fig. 10.23 Phase diagrams of two series of LC carbosilane dendrimers with chiral mesogenic groups.

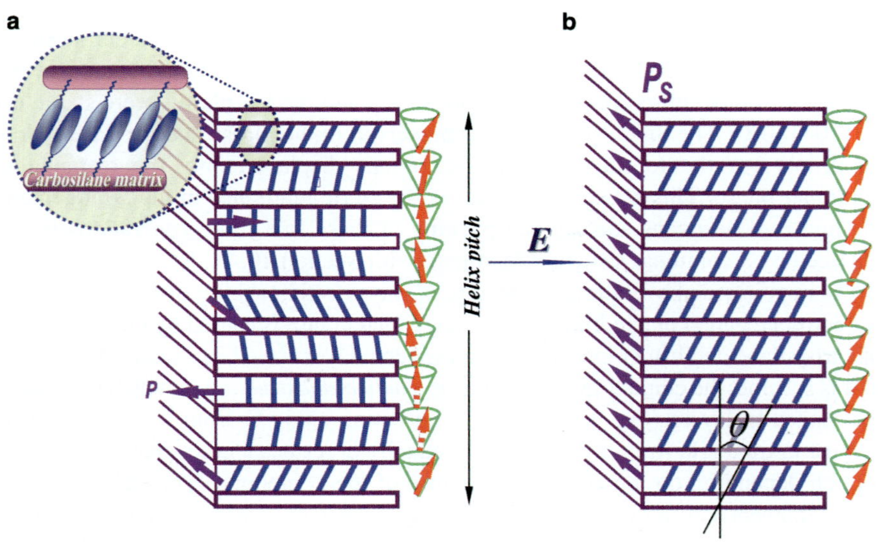

Fig. 10.24 Schematic illustration of the arrangement of mesogens in a chiral SmC* mesophase of LC dendrimers (**a**) before and (**b**) after helix untwisting under an electric field. The upper inset at the left shows the mesogenic group packing in the tilted smectic layer. P_S – spontaneous polarization. Red arrows show mesogens orientation in each layer.

Electrooptical studies of chiral LC dendrimers of the first three generations of Fig. 10.23 revealed effects of untwisting the helixes by the action of electric field. Ferroelectric properties were investigated and values of spontaneous polarization were calculated for several series of these chiral carbosilane dendrimers [55, 56]. In order to make a comparison of the effect of the generation number on ferroelectric properties, the measurements were performed for different dendrimers at the same reduced temperature, T_{red}, given by the ratio

$$T_{red} = T - T_{tr} \qquad \text{(E. 10.1)}$$

where: T_{tr} corresponds to the Isotropic→SmC* transition temperature for the carbosilane dendrimers of the series nG-(Und-PBL), and to the SmA–SmC* transition temperature for the dendrimers of the series nG-(Und-BPL).

As shown in Fig. 10.25a, the maximum value of P_S was observed for the first generation dendrimer ($P_S \sim 140\,\mathrm{nC/cm^2}$). The higher the generation number and the higher the temperature, the more significant was the disordering influence of the dendrimer interior. Thus, the increase in generation number increased disorder in the packing and led to a decrease in the average tilt angle in the chiral SmC* phase. This might be one of the reasons for the decrease in P_S values with increasing generation number.

Of highest importance for technological applications is the switching time, τ, of the electrooptical effect. As shown in Fig. 10.25b, the switching times for the LC dendrimers were in the range 1–10 ms, similar to that for side chain ferroelectric LC polysiloxanes. Since the switching time is related to rotational viscosity, η, applied electric field, E, and, spontaneous polarization by the simple relation

$$\tau = \eta / E \cdot P_S \qquad \text{(E. 10.2)}$$

one can assume that the rise of τ with increasing generation number is associated with the decrease in P_S and increase in viscosity (see Fig. 10.25b).

According to Equation (E.10.2) fast switching times are characteristic of LC compounds with high P_S values. Therefore, an active search is on for novel chemical compounds possessing the highest P_S values. Our above mentioned publications [55–58] provide a good basis for the development of novel types of ferroelectric LC carbosilane compounds with dendritic architecture.

Similarly, carbosilane LC dendrimers with *bent-core mesogenic* units at the periphery are of significant current interest in LC research. It should be pointed out that molecules with a bent rigid core have also been included into the mesogenic fragments and named *banana-shaped mesogens* [60–62]. Some time ago it was predicted that chirality of liquid crystals should not be necessary for ferroelectricity [62]. The bent-core molecules provide the possibility of polar order in smectic layers leading to soft matter with ferroelectric or antiferroelectric properties. One example of a banana-shaped molecule is shown in structure **15**.

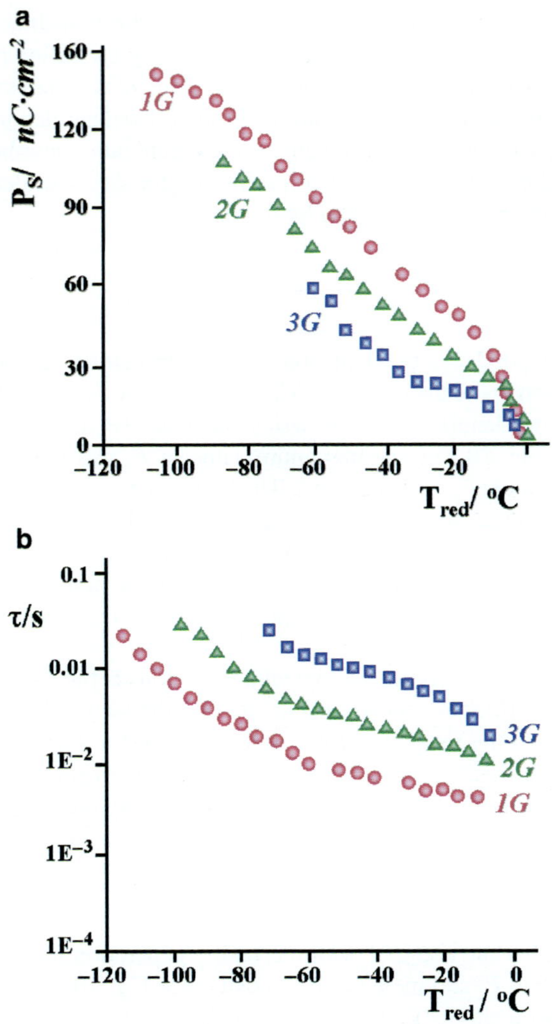

Fig. 10.25 Temperature dependences of (**a**) the spontaneous polarization, P_S, and (**b**) switching time, τ, for the ferroelectric LC dendrimers nG-(Und-PBL) of the first three generations.

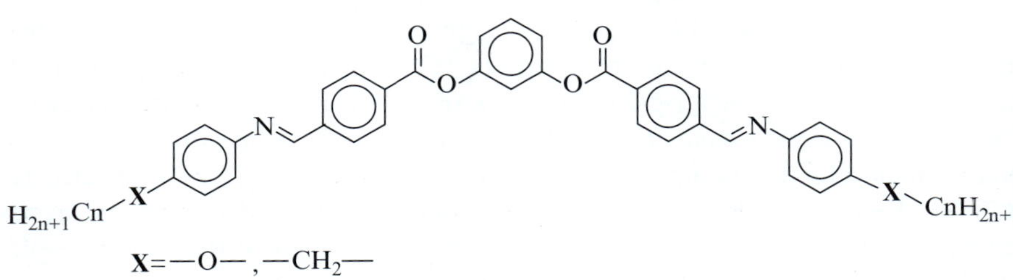

15

Such molecules are non-chiral, but their polar phase is the result of anticlinic ordering of two SmC layers in a coupled bilayer structure, as shown in Fig. 10.26a. The understanding of the relationships between the chemical structure and properties of banana-shaped compounds is still at a very early stage. Nevertheless, it was found that materials with banana-shaped molecules can display ferroelectric and antiferroelectric properties with very high values of spontaneous polarization (P_S = 200–700 nC/cm^2) [61, 62].

The first carbosilane LC dendrimers with banana mesogens, **16**, were described by Tschierske et al. [63]:

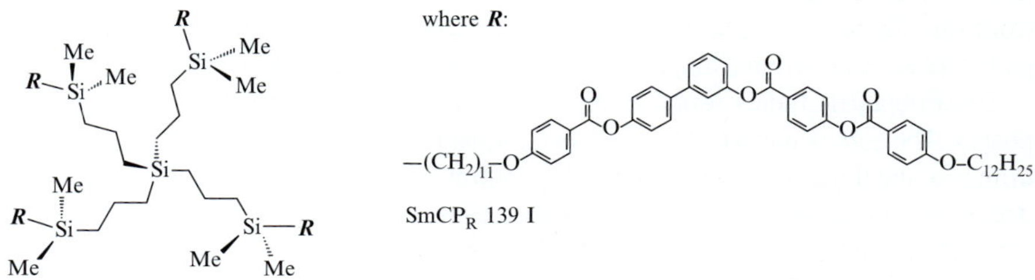

16

The dendrimer of the first generation formed an interesting, new, randomly tilted polar smectic C phase (SmCP$_R$) (R-random) which was stable up to 139°C and transformed into a glassy LC state on cooling (see Fig. 10.26b). On applying an electric field, the polar directors of the layers aligned parallel to the field, as shown in Fig. 10.26c. As a result, "a ferroelectric state" (SmCP$_F$) appeared and a ferroelectric switching process with a very high value of spontaneous polarization (P$_S$ = 1,400 nC/cm^2) was observed.

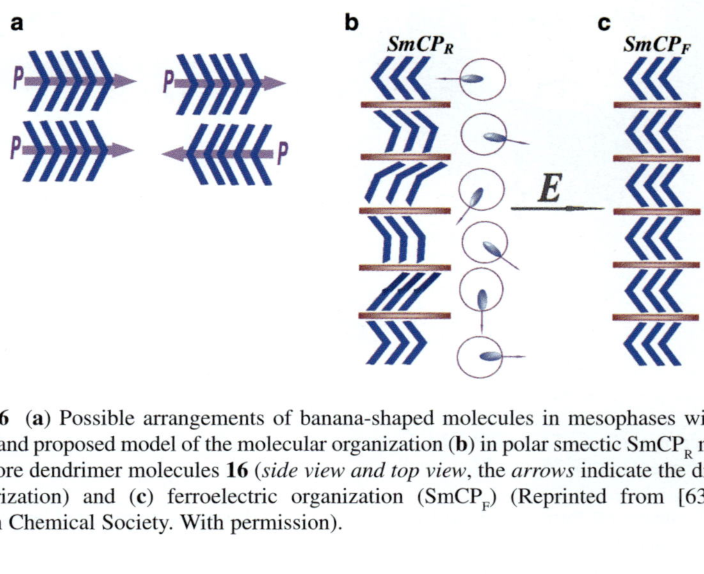

Fig. 10.26 (**a**) Possible arrangements of banana-shaped molecules in mesophases with a polar order (***p***) and proposed model of the molecular organization (**b**) in polar smectic SmCP$_R$ mesophase of bent-core dendrimer molecules **16** (*side view and top view*, the *arrows* indicate the direction of the polarization) and (**c**) ferroelectric organization (SmCP$_F$) (Reprinted from [63]. ©2002 American Chemical Society. With permission).

More recently the Halle research group headed by Tschierske [64] synthesized the next three generations of LC dendrimers with the same banana mesogen (as in **16**) containing one oxygen atom in the aliphatic spacer (see Fig. 10.27a). Detailed X-ray analysis and optical polarization measurements showed formation of crystalline and fluid smectic phases without an in-plane order. The layer distance was only slightly dependent on the generation number and much smaller than the size of the dendrimers, suggesting that the layer is formed by the organization of individual bent (banana) aromatic cores. These layers are separated by disordered layers of alkyl chains and dendrimer interiors. Organization of dendrimer molecules in antipolar smectic phase in the form of triply segregated layers is schematically presented in Fig. 10.27b. It can be seen from this figure, that dendrimer interiors, bent-core (banana) units, and the aliphatic parts (spacer and terminal alkyl chains) are organized in separate sub-layers.

Electrooptical studies revealed very high *Ps* values corresponding to the $SmCP_A$ phases. The bent dendrimer 1G displays P_S equal to 700 nC/cm^2, whereas the bent dendrimer of the third generation 3G had $P_S = 600$ nC/cm^2. In other words, the value of P_S decreases with increasing generation number, as was observed for the above-mentioned carbosilane dendrimers of the nG-(Und-PBL) and nG-(Und-BPL) series (see above).

Appearance of spontaneous polarization in the chiral and non-chiral bent-shaped carbosilane dendrimers provides another support for the dual nature of LC dendrimer

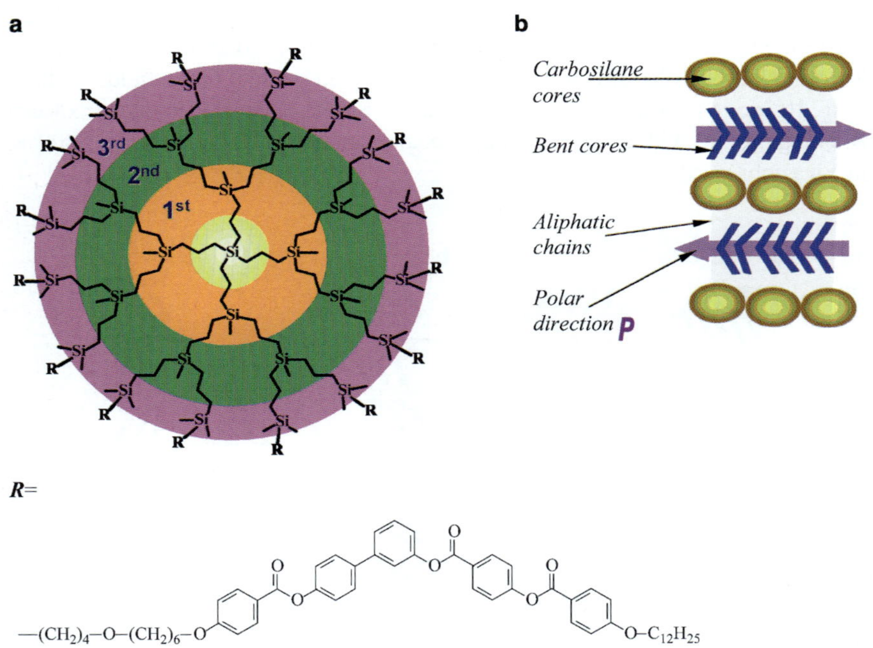

Bent-dendrimer (*1G*) *Cr 73 SmCP$_A$ 111 I*
Bent-dendrimer (*3G*) *Cr 79 SmCP$_A$ 113 I*

Fig. 10.27 Structural formulas of (**a**) LC carbosilane dendrimers with terminal bent-core mesogenic groups and (**b**) arrangement of these dendrimers in the antipolar smectic phase $SmCP_A$ (polar directions are indicated by *arrows*) (Reprinted from [64]. ©2006 Wiley. With permission).

molecules. The flexible carbosilane interior does not prevent formation of chiral helical smectic phases for low dendrimer generations. At the same time, dendritic structure clearly manifests itself in higher generations by the disappearance of smectic structure and formation of the supramolecular columnar phase. In some cases, the banana-shaped terminal groups of carbosilane LC dendrimers induce formation of crystalline and LC phases without any polar ordering. Such an example was recently described [65] for the first generation of a carbosilane dendrimer containing bent-shaped mesogenic groups **17**.

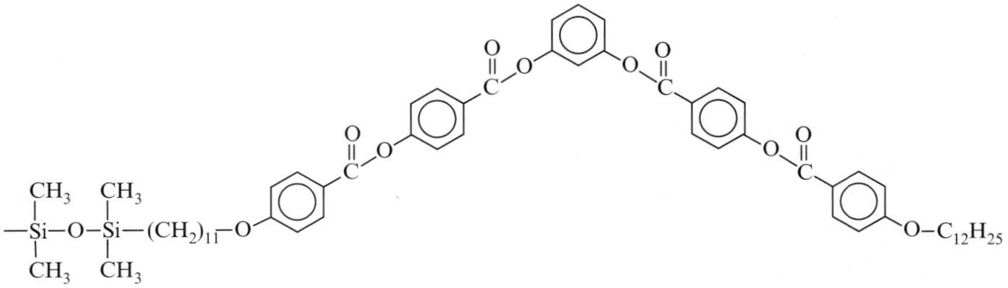

Cr 75 SmA 143

All these peculiarities of ferroelectric LC dendrimers stimulate intensive research of the silicon-containing dendrimers with mesogenic fragments. The banana-shaped dendrimers attract a great interest because such molecules exhibit a variety of novel mesophase types, unknown in conventional mesogenic compounds.

10.6 Photochromic LC Carbosilane Dendrimers

Recent years have brought an ever-growing interest in new photosensitive materials which can change their optical properties under light irradiation. This interest has been generated by new challenging applications of such materials in the design of various devices for optical recording and data storage. In 2003 we published a review of photoactive LC polymer systems including photosensitive dendrimers with light controllable structure and optical properties [66]. In the literature, there is a number of publications on dendrimers with photoactive groups (usually azobenzene fragments) localized either within the dendrimer interior or grafted onto their surface, as shown in Fig. 10.3. However, there is almost no information on photochromic dendrimers forming LC phases. In our opinion, the introduction of mesogenic photoactive groups into dendrimer molecules capable, for example, of isomerization, is of great interest from different points of view.

First, while the geometrically progressive increase in photocromic mesogen molecular content drastically changes many physico-chemical properties, such as free volume, density and viscosity, the effect of the generation number of LC dendrimers on their photochemical and photooptical properties is still not clear. Second, in so far as the LC dendrimers are intermediate in their physical properties between macromolecular compounds and low-molar-mass liquid crystals, the photochromic LC dendrimers are of most interest as new potential photooptical media for optical application and photonics. Indeed, the low viscosity inherent to all dendrimers, low glass transition temperatures typical for carbosilane dendrimers, and capability for responding to action of external fields, can be combined in one and the same material. Combination of these factors is very important because such compounds could also perform specific functions such as in photoinduced processes (photoisomerization, photocyclization, photoorientation) or antenna effects, leading to promising new LC photochromic materials with dendritic topology.

The first photochromic LC carbosilane dendrimer with terminal cynnamoyl photosensitive mesogenic groups that undergoes configurational changes in response to light irradiation was described by us in 2001 [67]. This photochromic LC dendrimer **18** was obtained by the reaction of a 1G dendrimer containing 8 terminal hydroxyl groups with 4-methoxycinnamoyl chloride, and it formed a SmA phase with low T_g of −28°C and isotropization temperature at 58°C.

Under UV-irradiation this dendrimer **18** undergoes *E–Z* isomerization of the cynnamoyl groups (see Reaction Scheme 10.5) and a {2 + 2} photocycloaddition reaction leading to the formation of a three-dimensional network, insoluble in organic solvents. Because of this, in our studies of photoresponsive LC dendrimers, we focused on a series of photochromic LC azobenzene-containing dendrimers which remain soluble after the light irradiation and can undergo reversible isomerization processes:

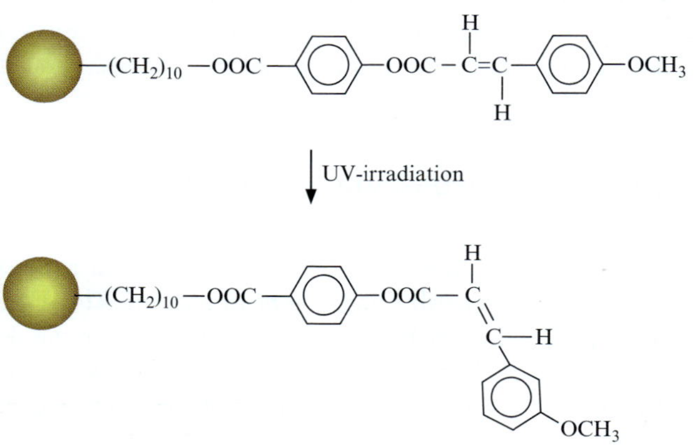

18 18

18a 18a

(10.5)

The first investigation of these azo-containing LC carbosilane homodendrimers was performed with the first, third and fifth generation containing ethoxyazobenzene terminal fragments (see, for example, Structure **19**) linked to the carbosilane interior with a flexible spacer (see also Fig. 10.28) [68]. These dendrimers form only crystalline phases which melt into isotropic liquids within the 81–91°C range, depending on the generation number. Because of this, their photochemical and photooptical properties have been studied in solution and in amorphous transparent films prepared by spin-coating in order to prevent crystallization. The photochemical results convincingly revealed that dendritic interior does not affect the *trans-cis* (*E–Z*)-isomerization of azobenzene groups, as shown in Fig. 10.29. During UV-irradiation, *E–Z*-isomerization takes place and the absorbance peak corresponding to the π–π^* electron transition (at 360 nm) of azobenzene groups (*E*-isomer) decreases (Fig. 10.29a) in a process which is thermally

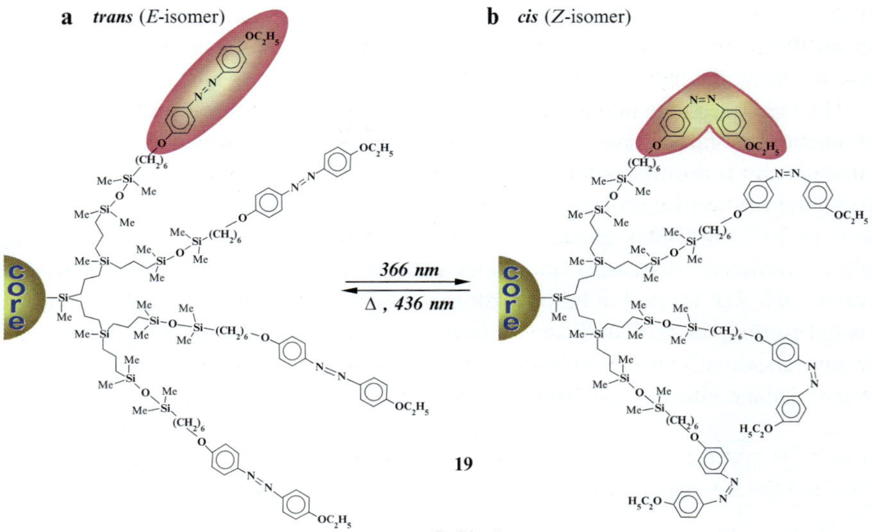

19

Cr 91 I

Fig. 10.28 LC photochromic homodendrimer **19** of the first generation with terminal ethoxy-azobenzene fragments linking to carbosilane matrix by six CH_2 groups (only one half of dendrimer core is shown): (**a**) before and (**b**) after UV-irradiation.

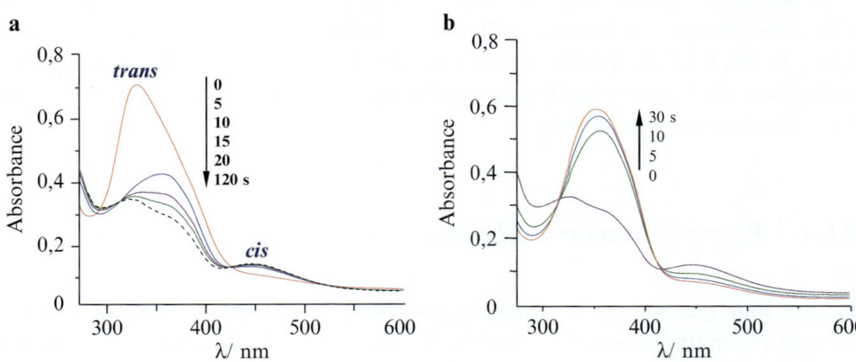

Fig. 10.29 Changes of absorbance spectra during (**a**) UV irradiation (λ = 360 nm) and (**b**) visible light irradiation of spin-coated amorphous dendrimer film **19** (see Fig. 10.28). Dashed line corresponds to the photostationary state (Reprinted from [69]. ©2002 American Chemical Society. With permission).

and photochemically reversible. Irradiation of the same sample with visible light leads to an increase of the absorption in the spectral region of the π–π^* electron transition over several seconds (Fig. 10.29b). (The small shift of the maximum in Fig. 10.29a to shorter wavelengths during UV-irradiation was explained by the breakdown of aggregates of azobenzene groups in the initial amorphous dendrimer film [69]). Consequently, UV- and visible light irradiation enable induction of a reversible photochemical isomerization of photochromes despite the presence of the highly branched carbosilane dendrimer interior. Thus, the dendrimer interior does not influence photochemical reactions of azobenzene end-groups, which again demonstrates the dual nature of the photochromic LC dendrimers.

The next questions in our continuing research on the synthesis and properties of photochromic carbosilane LC dendrimers concerned the influence of dendrimer topology and generation number on their phase behavior and photooptical properties. To begin to answer some of these questions, a series of novel photosensitive carbosilane dendrimers of different generations and various molecular architectures including homodendrimers, statistical and block-codendrimers was prepared and studied [44–46, 69–71]. Instead of ethoxyazobenzene groups (see Fig. 10.28) which led to crystallization, in this study we selected the photochromic propoxyazobenzene groups, chemically linked by hexamethylene spacers (see below) to five generations of carbosilane interiors nG-(Hex-Azo)homo:

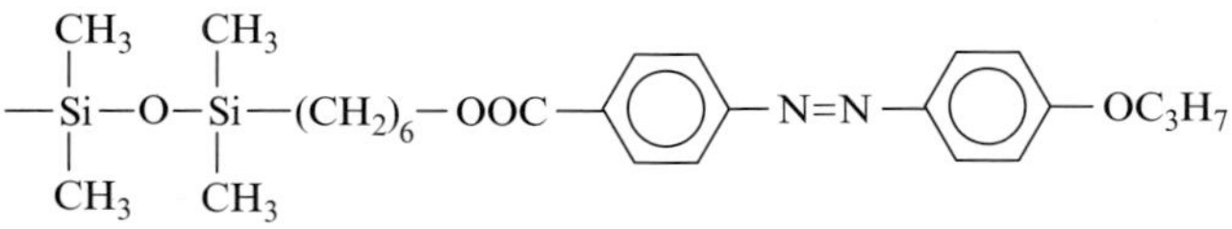

(Hex-Azo)homo

These homodendrimers, together with a series of statistical nG-(Hex-Azo)stat (see LC codendrimers **11** of Fig. 10.9) and block codendrimers nG-(Hex-Azo)block (see LC block-codendrimer **13** of Fig. 10.9), made up a specific series of photochromic LC carbosilane dendrimers of different molecular topology but with the same mesogenic azobenzene groups and the same aliphatic terminal groups taken in the equimolar ratio (1:1). Such LC dendrimers are of considerable interest for understanding relationships between molecular structure and properties, and in the following sections we briefly discuss some of these.

10.6.1 Phase Behavior and Structure

The first four generations of photochromic nG-(Hex-Azo)homo homodendrimers displayed crystalline phase and enantiotropic SmA mesophases, with both transition temperatures from crystalline phase to SmA and from SmA to isotropic melt only slightly depending on the generation number (see Fig. 10.30). The SmA mesophase exhibited a typical lamellar structure consisting of alternate layers of mesogenic groups and flexible dendrimer interiors, as already described for other carbosilane LC dendrimers (see Section 10.4.1). Long spacings, d_{001}, calculated from X-ray diffraction data as a function of generation number, suggest a model for this lamellar SmA structure shown in Fig. 10.31. On passing from low generations to the fifth generation, the packing type drastically changes and the SmA phase is replaced by a columnar phase, the structure of which is close to the centro-symmetrical rectangular columnar mesophase with a rectangular two-dimensional cell (see for example Fig. 10.21).

The situation is significantly different for copolymers, where photochromic mesogenic groups are "diluted" by the long flexible aliphatic fragments (see copolymer series **11** of Fig. 10.9). Introduction of flexible fragments, statistically

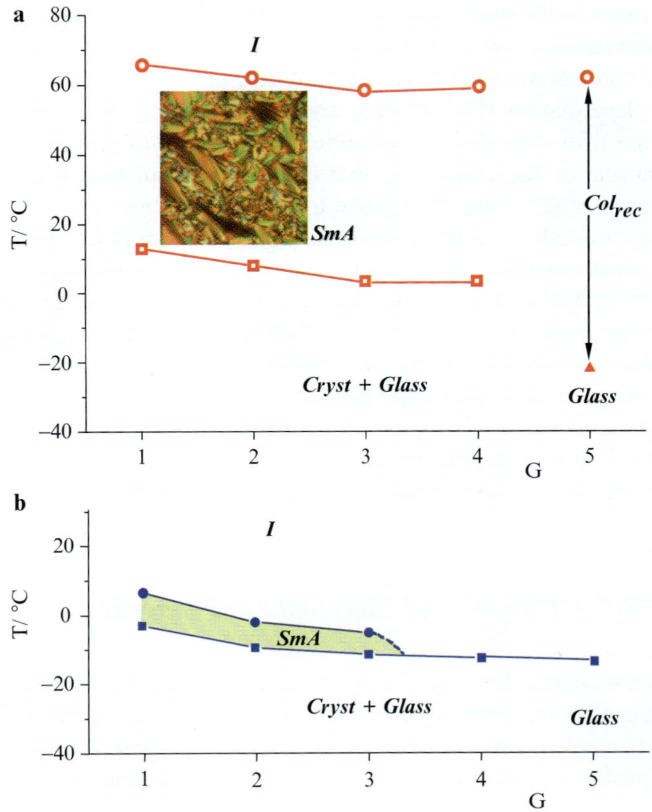

Fig. 10.30 Phase diagrams of a series photochromic (**a**) homodendrimers nG-(Hex-Azo)homo and (**b**) statistical co-dendrimers nG-(Hex-Azo)stat of the first–fifth generations. The inset shows a typical microphoto of SmA mesophase of the third generation homodendrimer.

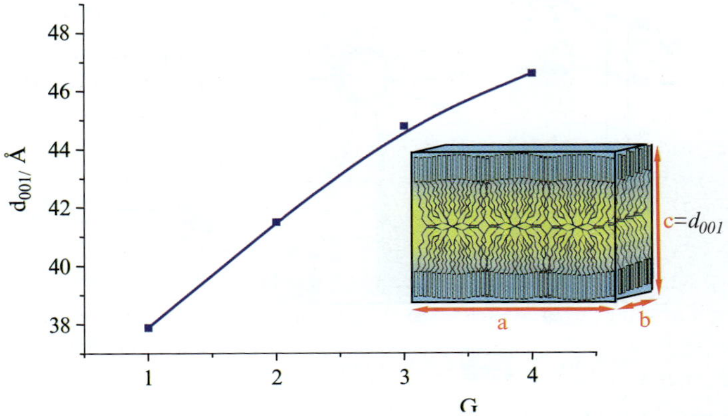

Fig. 10.31 Dependence of the long spacing d_{001} in SmA phase for a series of homodendrimers nG-(Hex-Azo)homo as a function of generation number and scheme of a dendrimer packing in SmA mesophase.

distributed between the rigid azobenzene mesogens, led to a substantial decrease in clearing temperatures (~60°C) and to depression of melting points. Phase diagrams of statistical codendrimers nG-(Hex-Azo)stat showed considerable reduction of the SmA mesophase region (Fig. 10.30b) and disappearance of LC phases for the fourth and the fifth codendrimer generation. The latter was probably caused by a relative decrease of the content of mesogenic groups in statistical copolymers which are responsible for the mesophase formation.

With respect to this, a comparison of the phase behavior of LC dendrimers with different molecular architectures but of the same generation is particularly interesting (see Fig. 10.32). Two factors can be observed. First, there is a steady reduction of the temperature range within which the LC phase appears from homodendrimers, to block- and to stat-codendrimers. Such behavior is associated with a decrease in the content of azobenzene mesogenic groups which cause the mesophase formation. Second, block-codendrimers occupy an intermediate position between homodendrimers and statistical codendrimers demonstrating a higher tendency for microphase separation with SmA phase formation than statistical codendrimers.

10.6.2 Photochemical and Photooptical Properties

UV-irradiation of all nG-(Hex-Azo) homodendrimers in dilute dichloroethane solutions, and the first generation of ethoxyazobenzene-functionalized dendrimer, **19** (see Fig. 10.28), led to marked spectral changes. At 360 nm, a significant decrease in the absorption peak associated with the π–π^* electron transition of the azobenzene

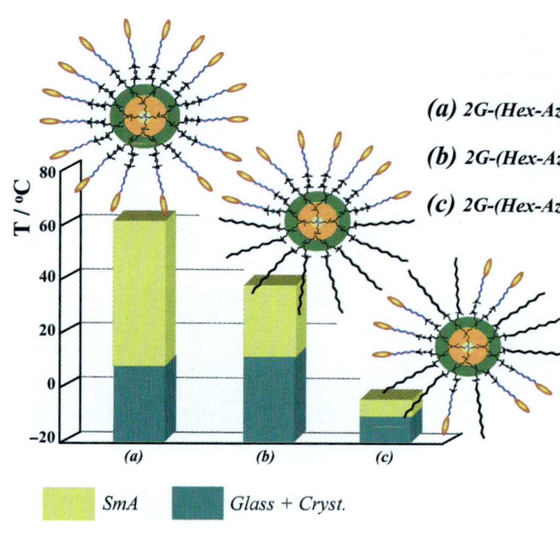

Fig. 10.32 Phase behaviour of the second generation (**a**) homodendrimers 2G-(Hex-Azo)homo, (**b**) block- and (**c**) statistical codendrimers, consisting of azobenzene and aliphatic terminal groups (see Fig. 10.9, codendrimers **11** and **13**).

groups was observed, as well as a slight absorbance increase in the region of the n–π* electron transition (450 nm) (see Fig. 10.33) [46, 71]. It is important, that the position of the absorbance maxima and the character of the spectral changes during UV-irradiation were practically the same for all dendrimer solutions, independent of the generation numbers and molecular architectures. Moreover, the process of *E–Z* isomerization was photochemically and thermally reversible; the back *Z–E* isomerization took place either by exposure to visible light or by annealing.

Figures 10.34a and b show the kinetics of the back reaction for solutions of homodendrimers of different generations (a) and of statistical and block-codendrimers (b) which were irradiated before measurements by UV-light for 200 s to achieve the photostationary state. It can be seen from this figure that the above-mentioned temperature dependences can be approximated by the same monoexponential function. It was found that the rate of isomerization depends only on the temperature and neither on generation number nor on molecular architecture. As an example, the rate constants of thermally-induced Z–E isomerization of all these dendrimers have the same value: $k = 1.7 \cdot 10^{-3}\ s^{-1}$ at 60°C. These data provide a convincing demonstration of the molecular dispersity of dendrimer solutions without any aggregation or micelle formation by the dendrimer molecules which behave as individual particles.

Another situation is observed for the thin (<1 μm) optically transparent dendrimer films prepared by the spin-coating technique. This technique prevents formation of LC or crystalline phases because of the fast solvent evaporation, and the resulting amorphous films are rather stable and retain their structures for several days at ambient temperature. It was found [46] that absorbance spectra of films from different dendrimer generations are a little different from each other and, in many respects, are similar to the absorbance spectra shown in Fig. 10.33. A strong difference, however, was observed when the back process, the *Z–E* isomerization of azobenzene fragments

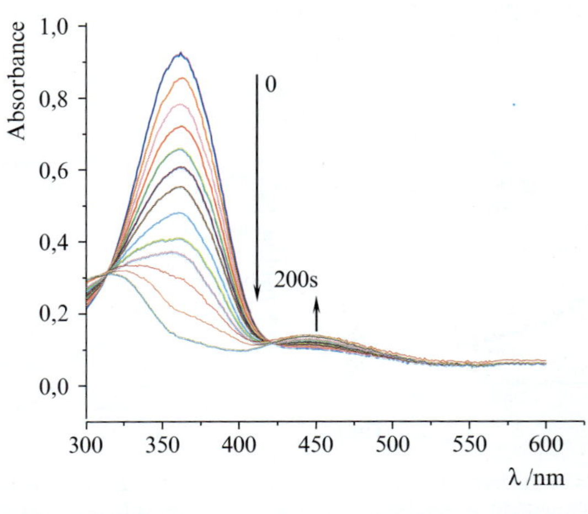

Fig. 10.33 Absorbance spectra changes of dendrimer 2G-(Hex-Azo)homo in dichloroethane solution during UV-irradiation (365 nm). Spectra were recorded each 15 s. (T = 52°C).

after UV-irradiation of the initial films to achieve the photostationary state, was studied. From the kinetic dependences of absorbance (shown in Fig. 10.34), the rate constants and activation energies were calculated for *Z–E* isomerization of azobenzene groups for different dendrimer generations and different topologies of the same generation.

Figure 10.35 shows temperature dependences of the rate constants for the *Z–E* isomerization of azobenzene groups of homodendrimers, demonstrating the influence of their generation numbers. The data indicate a more significant difference only between the fifth generation and the others; while the differences between the lower generations are much less pronounced. Activation energies for the *Z–E* isomerization process differed by a factor of 1.2 from the first to the fifth generations, and changed from 76 to 93 kJ/mol. These values are very close to the values of the corresponding low-molar-mass azobenzene derivatives. The small differences between these kinetic parameters for different homodendrimers generations can be explained by the close amorphous structure of the spin-coated samples that have undergone *E–Z* photo-isomerization by UV-irradiation. The difference in the rate constants is most clearly manifested at lower temperatures (40–50°C) and practically disappears at higher temperatures (~60°C) (see Fig. 10.36). Nevertheless, a small distinction between high and low dendrimer generations could be explained by pronounced steric hindrance imposed by the bent *Z*-isomers on dendrimer molecules, and faster decrease of bent isomers during the *Z–E* isomerization. In the case of statistical codendrimers, aliphatic terminal groups act as "diluents" providing sufficient free volume for *Z–E* isomerization. As a result of this, the values of the rate constants and activation energies of *Z–E* isomerization become independent of the generation number.

Comparison of LC homodendrimers, statistical and block-codendrimers of the same (second) generation shows, that photochemical Z–E isomerization of homodendrimers and block-codendrimers are characterized by similar kinetic parameters, while statistical codendrimers are isomerized at slightly lower rates (see Fig. 10.36a). Although the difference between dendrimers with different

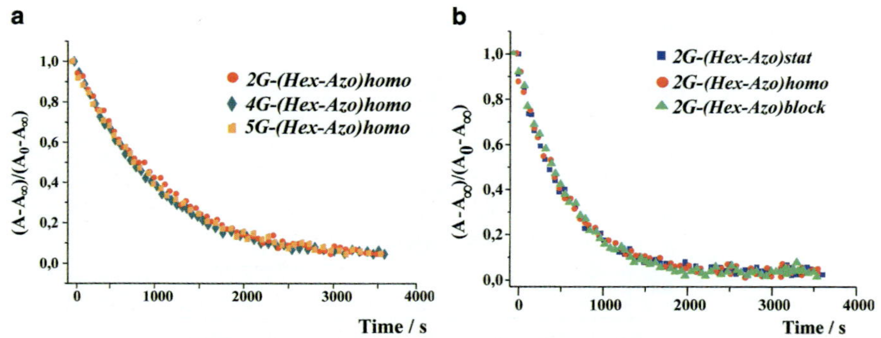

Fig. 10.34 Kinetics of the back thermally-induced process of *Z–E* isomerization of (**a**) homodendrimers at T = 52°C and (**b**) codendrimers at T = 60°C with different molecular architecture in dichloroethane solutions (A_0, A_t and A_∞ are the absorbances at 360 nm at the time equal to, 0, current time, t, and t→∞, respectively).

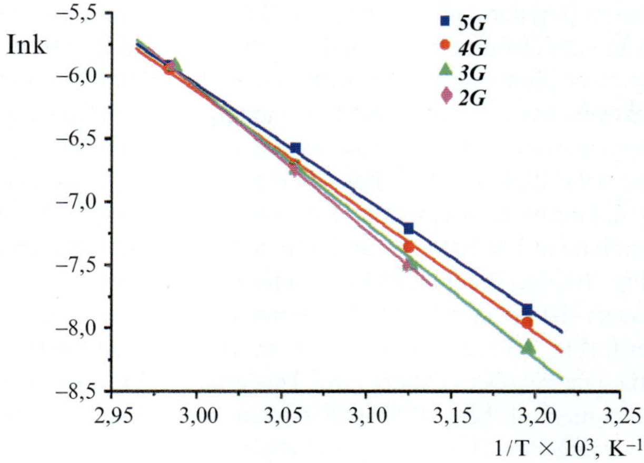

Fig. 10.35 Arrhenius plot for temperature dependence of the constant rates for *Z–E* isomerization for the amorphizable films of homodendrimers nG-(Hex-Azo)homo of the different generations.

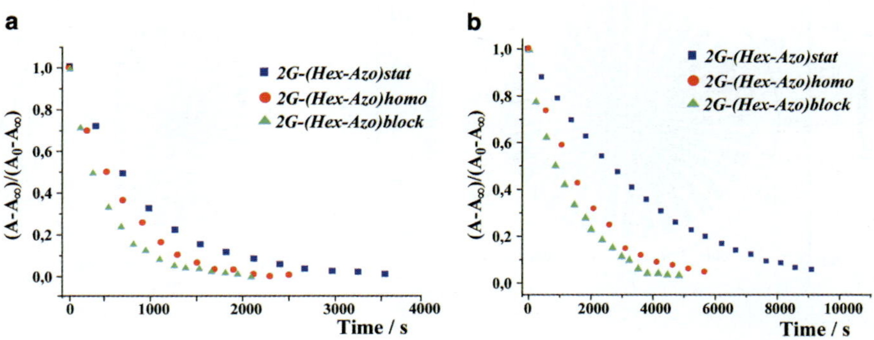

Fig. 10.36 Kinetics of *Z–E* isomerization of azobenzene groups for the films of homopolymers, statistical and block codendrimers of the second generation (**a**) at 62°C and (**b**) 47°C. Before the measurements the films were irradiated by UV-light (365 nm) during 200 s for achieving the photostationery state.

molecular architectures is clearly seen at low temperature (Fig. 10.36b), correlation between the molecular topology of LC dendrimers, including generation numbers, and their photochemical properties is not conclusively established.

The study of the photochemical behavior of photochromic LC dendrimers of different generations and various molecular topologies has again clearly demonstrated the dual nature of these dendrimer molecules. On one hand, photochromic azobenzene-containing groups undergo *E–Z* and *Z–E* isomerization seemingly independently on the dendrimer interior, while on the other hand, generation number influences kinetic parameters of the photoisomerization processes in a way which is not fully understood yet and needs additional investigations.

It is well-known [66] that cyclic photoinduced *E–Z–E* isomerization of azobenzene photochromes in side-chain polymers and dendrimers *irradiated by polarized light* leads to cooperative photoorientation of the photochromic and non-photochromic fragments. This process is accompanied by the appearance of photoinduced birefringence, which permits calculation of an order parameter (S) of photochromic fragments. The polar diagrams (see Fig. 10.37a and b) of the first generation dendrimer clearly demonstrate orientation of the mesogenic groups after irradiation of the film with polarized UV-light. It can be seen, that the isotropic character of the absorbance (Fig. 10.37a) is replaced by an anisotropic one (Fig. 10.37b), and that the order parameter sharply increases at the beginning of irradiation but then drastically decreases after it (Fig. 10.37c). It achieves its maximum value S = 0.2 very quickly (~100 s), but then decreases to almost zero. This effect is, most likely, related to the increase in the content of the bent-shaped *Z*-isomers having a very low anisometry of azobenzene fragments. This leads to full degeneration of the orientational order in the dendrimer during further irradiation. Thus, the highly branched dendritic

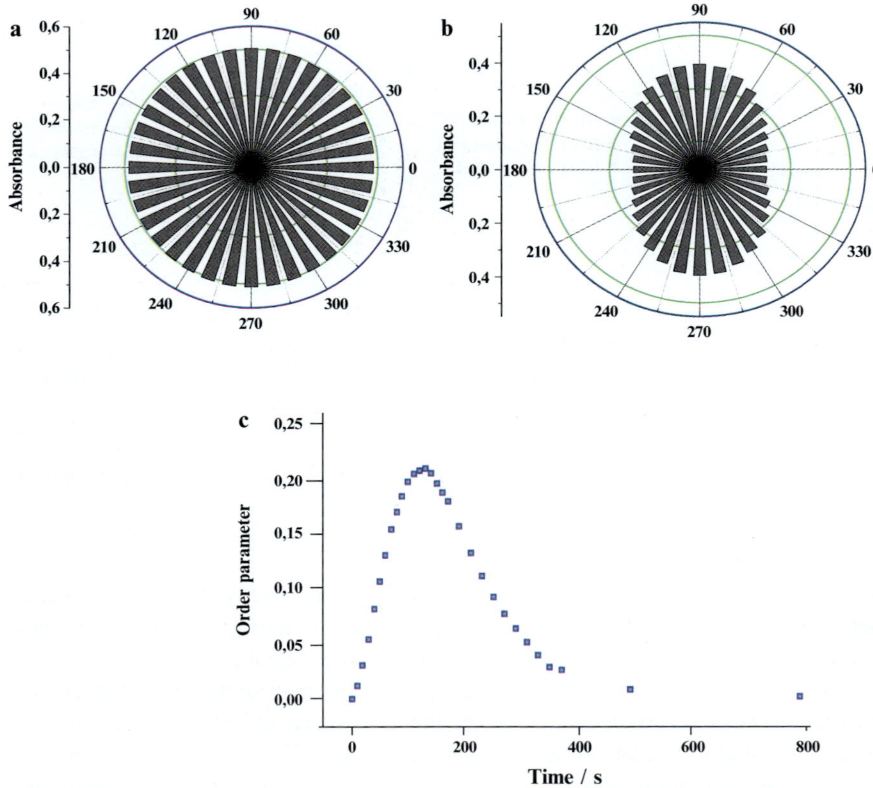

Fig. 10.37 Polar diagrams of photochromic dendrimer **19** (λ = 332 nm) (**a**) before and (**b**) after irradiation with polarized UV light (20°C); (c) change of order parameter for the same dendrimer (Reprinted from [70]. ©2002 Wiley-VCH Verlag GmbH & Co KG. With permission).

architecture exerts a destabilizing influence on the orientation of mesogenic groups, but generation number does not significantly influence the maximum values of the photoinduced order parameter.

10.7 Conclusions and Future Outlook

This chapter discusses the synthesis and development of liquid-crystalline dendrimers having silicon-containing ball-like interiors and peripheral rigid rod mesogenic fragments. Within the last few years, a number of papers describing research on LC dendrimers with other interiors, such as poly(propylene imine), poly(amidoamine), polyesters, polyurethanes etc. have been published [22, 72–76]. Moreover, in addition to the rigid rod mesogens the disk- and the bent-shaped fragments, as well as some more complicated structural units that lead to LC phase formation were also included into dendrimers. However, these, as well as the pioneering work by Percec et al. [72, 73] devoted to the synthesis and study of cone-shaped monodendrons capable of self-assembling by the mechanism of molecular recognition into LC columnar mesophases, are outside the scope of this chapter.

General relationships between the molecular structure and properties of LC dendrimers are governed by the closely related regularities regardless of their chemical composition. The combination of amorphous dendrimer interior and external mesogens with their pronounced tendency to anisotropic ordering predetermines all unusual structural and physico-chemical properties of these *nanoscale monodisperse organic-inorganic hybrids*. The main characteristic of LC dendrimers is the competition between their isotropic spherical interiors within and from which the branches tend to radiate isotropically (*entropic gain*) and the effective anisotropic interaction of mesogens, leading to mesophase formation (*enthalpic gain*). As a result of these contradictory tendencies, a molecule of any LC dendrimer tends to microphase separation because of its chemical incompatibility. It is very important to stress that in all LC dendritic systems, including those discussed in this chapter, primary *intramolecular micro-segregation, driven by the association of* like parts of dendritic molecules, directly leads to self-organization with formation of different nanostructured LC mesophases [71].

These micro-segregation processes are most clearly manifested in the codendrimers and block-codendrimers consisting of mesogenic and non-mesogenic fragments. In this respect, LC dendrimers and codendrimers can be regarded as linear or graft copolymers consisting of incompatible molecular fragments, the properties of which usually determine a dualism in their behavior. *This dual nature is one of the more intriguing and challenging features of LC dendrimers, and particularly so of the silicon-containing ones*. Their flexible dendritic interiors and methylene spacers are responsible for low values of glass transition temperatures which lie predominantly below 0°C. In fact, anisotropic mesophases are formed only in cases when the dendrimer interior is sufficiently flexible and may readily deform, accommodating to a certain type of mesogen packing and changing its spherical shape. This is true for dendrimers of lower generations (from the first to the fourth one) where microphase-separaion process leads to the formation of lamellar structures (SmA or/and SmC) due to the terminal mesogen packing. At the same time, conformation of the dendrimer interior, including the spacers, is governed by the ordering of mesogenic fragments.

A different situation is found in dendrimers of higher generations. The columnar structures formed by these dendrimers result from aggregation of their molecules into round-shaped disks which further self-organize into cylindrical columns. These columns self-assemble into hexagonal arrays to form a columnar hexagonal phase.

In this chapter, we discussed only the silicon-containing LC dendrimers which represent a unique class of nano-structurated and self-organizing systems. However, synthetic chemists continue to surprise by synthesizing new families of LC dendrimers with complex and unusual compositions and architectures, such as metal- and fullerene-containing LC dendrimers, banana- and other bent-shaped fragments, chiral nematic octasilsesquioxanes, ferroelectric and photochromic LC dendrimers, etc. Some examples of such systems, based on carbosilane dendrimer interiors, have also been considered in this chapter.

There is a little doubt that in parallel with homodendrimers, codendrimers and block-codendrimers consisting of mesogenic and other functionalized terminal fragments can provide valuable functions with a great tendency to self-assemble into a rich variety of different supramolecular structures with nanoscale dimensions. The remarkable, sophisticated architecture and multifunctional character of LC dendrimers should stimulate further research directed at understanding the structure-property relationships of end-functionalized LC dendrimers for the purpose of constructing novel families of tailored functional LC nanomaterials [77–79]. This class of materials can be widely used in such areas as photo- and optoelectronics, medicine (as drug carriers) and medical diagnostics, sensors and photonics. They can also find application in other directions of science and technology where molecular species of a few nanometers in size and capable of self-ordering and changing their properties under the action of external electromagnetic fields are required.

Acknowledgements The authors are greatly indebted to their colleagues and Ph.D. students and former Ph.D. students (E. Agina and A. Lysachkov) from the Chemistry Department of Moscow State University for joint experimental works relating to the synthesis and study of LC carbosilane dendrimers. We also gratefully acknowledge Prof. R. Richardson for X-ray measurements of LC dendrimers, Prof. A. Muzafarov, Dr. E. Rebrov and Dr. S. Ponomarenko for help with the synthesis of dendritic matrices. The authors are also very grateful to Ms. S. Amelekhina for her assistance in the manuscript and figures preparation.

This work was supported by the Russian Foundation for Basic Research (Grants № 07-03-01089 and № 08-03-00481)

References

1. Demus D, Goodby J, Gray G, Spiess, Vill V (eds) (1998) Handbook of liquid crystals, Wiley-VCH, Weinheim
2. Definitions of basic terms relating to low-molar-mass and polymer liquid crystals (IUPAC Recommendations 2001) (2001) Pure Appl Chem 73:845
3. Reinitzer F (1889) Monatsh 9:421
4. Fish M (2004) Liquid crystals, laptops and life, World Scientific, New York
5. Shibaev VP, Plate NA (1984) Adv Polym Sci 60/61:175
6. Finkelmann H, Rehage G (1984) Adv Polym Sci 60/61:100
7. Shibaev VP, Freidzon YaS, Plate NA (1975) Proceedings of the 2nd Mendeleev Congress on General and Applied Chemistry, Nauka, Moscow, vol 2, p164
8. Shibaev VP, Plate NA (1977) Vysokomol Soedin A19:923 (in Russian); Polym Sci USSR A19:1065 (1978) (English translation)
9. Finkelmann H, Ringsdorf H, Wendorff J (1978) Macromol Chem Phys 179:273
10. Plate NA, Shibaev VP (1980) Comb-shaped polymers and liquid crystals, Chemistry, Moscow (in Russian); The extended and additional version in English was published by Plenum, New York/London (1987)
11. Cifferri A, Krigbaum W, Meyer R (eds) (1982) Polymer liquid crystals, Academic, New York
12. Blumstein A (ed) (1985) Polymeric liquid crystals, Academic, New York
13. McArdle C (ed) (1989) Side chain liquid crystals polymers, Blackie, London
14. Plate NA (ed) (1993) Liquid crystalline polymers, Plenum, New York
15. Shibaev VP, Lui Lam (eds) (1994) Liquid crystalline and mesomorphic polymers, Springer, New York
16. Xin-Jiu Wang, Qi-Feng Zhou (eds) (2004) Liquid crystalline polymers, World Scientific, Singapore
17. Rebrov EA, Ponomarenko SA, Boiko NI, Muzafarov AM, Shibaev VP (1994) Proceedings of the IUPAC Conference on Liquid Crystal Polymers, Beijing, China, September 6–9, 1994, p100
18. Ponomarenko SA, Boiko NI, Rebrov EA, Muzafarov AM, Shibaev VP (1995) Proceedings of the IUPAC Symposium "Polymers for Advanced Technologies", Italy, Pisa, June11–15, 1995, p207
19. Shibaev VP, Ponomarenko SA, Boiko NI, Muzafarov AM (1996) Proceedings of the International Symposium "Liquid Crystals for Advanced Technologies of Materials Research Society", USA, San Fransisco, April 8–11, 1996, p158
20. Ponomarenko SA, Rebrov EA, Boiko NI, Vasilenko NV, Muzafarov AM, Freidzon YAS, Shibaev VP (1994) Polym Sci A36:896
21. Shibaev VP, Ponomarenko SA, Boiko NI, Rebrov EA, Muzafarov AM, Whitehouse IJ, Richardson RM (1998) Preprints of the 35th the IUPAC Symposium on Macromolecules, Australia, Brisbane, July 11–18, 1998, p771
22. Ponomarenko SA, Boiko NI, Shibaev VP (2001) Polym Sci C43:1
23. Mehl GH, Goodby JW (1996) Chem Ber 129:521
24. Mehl GH, Thomton AJ, Goodby JW (1998) Proceedings of the 17th International Liquid Crystalline Conference, Strasbourg, France, July 19–24, 1998, pp.0–3
25. Saez IM, Goodby JW (1999) Liq Cryst 26:1101
26. Elasser R, Mehl G, Goodby J, Weith M (2001) Angew Chem Int Ed 40:2688
27. Saez IM, Goodby JW, Richardson RM (2001) Chem Eur J 7:2758
28. Ponomarenko SA, Rebrov EA, Boiko NI, Muzafarov AM, Shiabev VP (1998) Polym Sci A40:763
29. Ponomarenko SA, Rebrov EA, Bobrovsky AYu, Boiko NI, Muzafarov AM, Shibaev VP (1996) Liq Cryst 21:1

30. Richardson RM, Ponomarenko SA, Boiko NI, Shibaev VP (1999) Liq Cryst 26:101
31. Ponomarenko SA, Boiko NI, Shibaev VP, Maganov SN (2000) Langmuir 16:5487
32. Ryumtsev EI, Evlampieva NP, Lezov AV, Ponomarenko SA, Boiko NI, Shibaev VP (1998) Liq Cryst 25:475
33. Ponomarenko SA, Boiko NI, Shibaev VP, Richardson RM, Whitehouses IJ, Rebrov EA, Muzafarov AM (2000) Macromolecules 33:5549
34. Lebedev BV, Smirnova NN, Ryabkov MV, Ponomarenko SA, Makeev EA, Boiko NI, Shibaev VP (2001) Polym Sci A43:323
35. Agina EV, Ponomarenko SA, Boiko NI, Rebrov EA, Muzafarov AM, Shibaev VP (2001) Polym Sci A43:1000
36. Frey H, Mulhaupt R, Lorenz K, Rapp V, Mayer-Pozher F (1995) Polym Mater Sci Eng 73:127
37. Frey H, Lorenz K, Mulhaupt R, Rapp V, Mayer-Rozner F (1996) Macromol Symp 102:19
38. Coen MC, Lorenz K, Kressler J, Frey H, Mulhaupt R (1996) Macromolecules 29:8069
39. Lorenz K, Holter D, Frey H, Stuhn B (1997) Polym Mater Sci Eng 77:168
40. Lorenz K, Holter D, Stuhn B, Mulhaupt R, Frey H (1996) Adv Mater 8:414
41. Lorenz K, Frey H, Stuhn B, Mulhaupt R (1997) Macromolecules 30:6860
42. Terunuma D, Kato T, Nishio R, Matsuoka R, Kuzuhara H, Aoki Y, Nohira H (1998) Chem Lett 27:59
43. Terunuma D, Nishio R, Aoki Y, Nohira H, Matsuoka K, Kuzuhara H (1999) Chem Lett 28:565
44. Lysachkov AI, Boiko NI, Shibaev VP, Rebrov EA, Muzafarov AM (2005) Synthesis of carbosilane dendrimers of different architecture with aliphatic and mesogenic terminal groups. Paper presented at the European Polymer Congress, Moscow, Russia, June 27–July 1, 2005 p123
45. Lysachkov AI, Boiko NI, Shibaev VP, Rebrov EA, Muzafarov AM (2007) Carbosilane liquid crystalline dendrimers containing mesogenic and aliphatic groups. Paper present at the 4th All-Russian Polymer Conference, Moscow, Russia, January 29–February 2, 2007, v.2 p179
46. Lysachkov AI, Boiko NI, Rebrov EA, Muzafarov AM, Shibaev VP (2007) Russian Chemical Bulletin 56:2407
47. Leshchiner ID, Agina EV, Boiko NI, Shibaev VP Richardson RM (2004) Liquid Crystals and their application 9/10:72 (in Russian), Ivanovo State University, Ivanovo, Russia
48. Gensob KL, Holzmuller J, Leshchiner ID, Agina EV, Boiko NI, Shibaev VP, Tsukruk VV (2005) Macromolecules 38:8028
49. Rebrov EA, Ignatieva GM, Lusachkov AI, Demchenko NV, Muzafarov AM (2007) Polym Sci A49:757
50. Agina E V, Boiko NI, Richardson RM, Ostrovskii BI, ShibaevVP, Rebrov EA, Muzafarov AM (2007) Polym Sci A49:412
51. Shibaev VP (2002) Mesophase formation in carbosilane based liquid crystalline dendrimers. Paper presented at the 19th International Liquid Crystalline Conference, Edinburgh, UK, June 30–July 5, 2002
52. Boiko NI, Agina EV, Ponomarenko SA, Shibaev VP, Richardson RM (2006) Liquid Crystals and Their Application 18:78 (in Russian), Ivanovo State University, Ivanovo, Russia
53. Shibaev VP, Kozlovsky MV, Beresnev LA, Blinov LM, Plate NA (1984) Polym Bull 12:299
54. Shibaev VP, Kozlovsky MV, Plate NA (1987) Polym Sci A29:1144
55. Zhu Xiaomin, Vinokur RA, Rebrov EA, Muzafarov AM, Boiko NI, Shibaev VP (2000) Polym Sci A42:1
56. Boiko NI, Zhu Xiaomin, Vinokur RA, Rebrov EA, Muzafarov AM, Shibaev VP (2000) Mol Cryst Liq Cryst 352:343
57. Lysachkov AI, Ponomarenko SA, Boiko NI, Shibaev VP (2002) Liquid crystals and their application 1:75 (in Russian), Ivanovo State University, Ivanovo, Russia
58. Boiko NI, Lysachkov AI, Ponomarenko SA, Shibaev VP, Richardson RM (2002) Colloid Polym Sci 283:1155
59. Meyer RB, Liebert L, Strzelecki L, Keller P (1975) J Phys 36:L-69
60. Niori T, Sekine F, Watanabe J, Furukawa T, Takezoe H (1996) J Mater Chem 6:1231

61. Pelzl G, Diele S, Weissflog W (1999) Adv Mater 11:707
62. Tschierske CJ (2001) J Mater Chem 11:2647
63. Dantlgrater G, Baumeister U, Dile S, Kresse H, Luhmann B, Lang H, Tschierske CJ (2002) J Am Chem Soc 124:14852
64. Hahr H, Keith C, Lang H, Reddy RA, Tschierske CJ (2006) Adv Mater 18:2629
65. Kosata B, Tamba G, Baumeister U, Pelz K, Diele S, Pelzl G, Galli G, Samaritani S, Agina EV, Boiko NI, Shibaev VP, Weissflog W (2006) Chem Mater 18:691
66. Shibaev VP, Bobrovsky AYu, Boiko NI (2003) Progr Polym Sci 28:729
67. Boiko NI, Zhu Xiaomin, Bobrovsky AYu, Shibaev VP (2001) Chem Mater 13:1447
68. Bobrovsky AYu, Pakhomov AA, Ponomaranko SA, Boiko MI, Shibaev VP (2002) Polym Prepr 43:93
69. Bobrovsky AYu, Pakhomov AA, Zhu Xiaomin, Boiko NI, Shibaev VP, Stumpe J (2002) J Phys Chem B106:540
70. Bobrovsky AYu, Ponomarenko SA, Boiko NI, Shibaev VP, Rebrov EA, Muzafarov AM, Stumpe J (2002) Macromol Chem Phys 203:1539
71. Boiko NI, Shibaev VP (2007) Comparative analysis of the phase behavior of liquid crystalline comb-shaped and dendritic polymers. Paper presented at the All-Russian Conference "Polymers for XXI century", Russia, Moscow, January 29–February 2, Book of abstracts, 2007, 3:43
72. Percec V, Chu P, Ungar G, Zhou J (1995) J Am Chem Soc 117:11441
73. Percec V, Johansson G, Ungar G, Zhou J (1996) J Am Chem Soc 118:9855
74. Tschierske C (2001) Annu Rep Prog Chem Sect C 97:191
75. Donnio B, Barbera J, Gimenez R, Guillon D, Marcos M, Serrano JL (2002) Macromolecules 35:370
76. Pastor L, Barbera J, McKenna M, Marcos M, Martiv0Rapun R, Serrano JL, Luckhurst G, Mainai A (2004) Macromolecules 37:9386
77. Pan Q, Gao L, Chen X, Fan X, Zhou Q (2007) Macromolecules 40:4887
78. Shibaev VP (2006) Liq Cryst 33:1497
79. Frechet JM (2003) J Polym Sci Part A Polym Chem 41:3713

Chapter 11
Silicon-Organic Dendrimers

Petar R. Dvornic, Michael J. Owen, and Rakesh Sachdeva

11.1 Introduction

Silicon can play a variety of different roles when incorporated in or combined with otherwise "purely" organic dendritic branch cells which in this chapter are meant to include those that contain carbon and some combination of hydrogen, nitrogen, oxygen and sulfur. As a result, a diversity of compositional and architectural hybrid dendrimers can be formed, containing one or more of the following types of building blocks (see Fig. 11.1):

(a) Silicon atoms (di-, tri- or tetra-functional) (see Chapters 2–5) or silicon-containing groups (e.g., cyclic siloxanes, see Chapter 6, or polyhedral oligosilsesquioxanes [POSS], see Chapter 7) as *dendrimer cores.*
(b) Silicon atoms in *dendrimer interiors* as (see Chapter 1)
 (b1) *Branch extenders* (i.e., constitutive elements of branches with 1→1 multiplicity between branch junctures).
 (b2) *Branch junctures* with di- (1→2) or tri- (1→3) branching multiplicity.
(c) Reactive or non-reactive (inert) silicon-containing *end-groups* providing organic dendrimers with properties characteristic for organo-silicon compounds.
(d) Copolymeric dendrimers having *silicon-containing and "purely" organic branch cells* organized either in a radially layered or segmented (i.e., individual dendrons of different compositions) architectural arrangement (see below and also Chapter 2).

Because of these structural differences, the properties of the resulting dendrimers may range from being quite similar to those of their purely organic counterparts to being dramatically different from them. Hence, the introduction of silicon into organic dendritic structures opens up vast new areas of molecular design and engineering aimed at unique new materials for specific targeted applications.

P.R. Dvornic, M.J. Owen, and R. Sachdeva
Michigan Molecular Institute, 1910 W. St. Andrews Rd., Midland, MI, 48640, USA
E-mails: dvornic@mmi.org; michaelowen01@chartermi.net; rakeshsac@gmail.com

P.R. Dvornic and M.J. Owen (eds.), *Silicon-Containing Dendritic Polymers,*
Advances in Silicon Science 2,

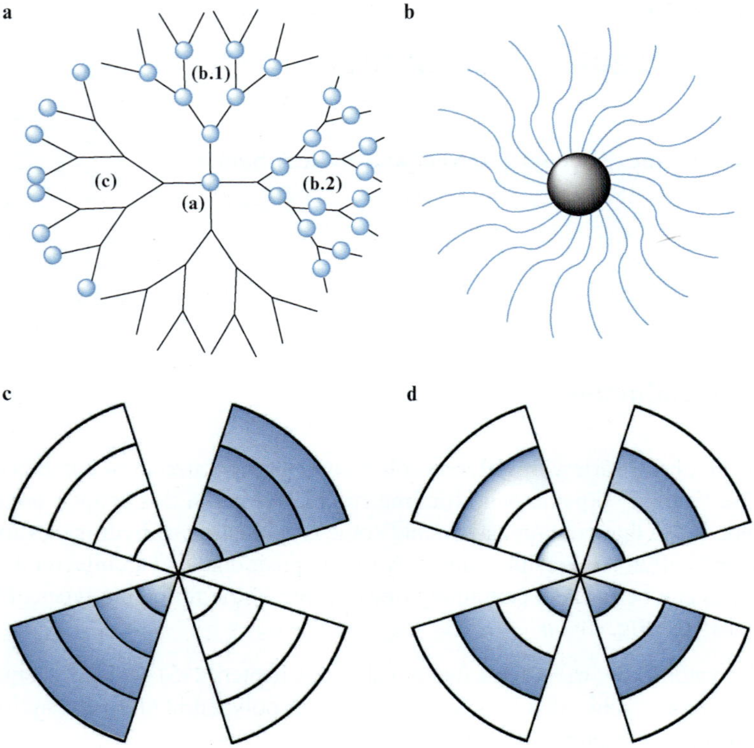

Fig. 11.1 Possible roles of silicon in silicon-organic dendrimers with silicon entities colored blue: (**a**) silicon atoms or silicon-containing groups as dendrimer cores (*a*), branch junctures (*b.1*), branch extenders (*b.2*) or end-groups (*c*); (**b**) silicon-containing arms in dendrimer-based multi-arm star polymers; (**c**) segmented copolymeric dendrimers with silicon-based segments (blue) and "purely" organic segments (non-colored); and (**d**) radially layered copolymeric dendrimers where blue-colored layers (tiers, generations) are silicon-based while the non-colored ones are "purely" organic.

By far the most investigated of all silicon-organic dendrimers have been the poly(amidoamine-organosilicon)s, PAMAMOS. These dendrimers were not only the first silicon-organic dendrimers to be reported [1–3], but also the first silicon-containing dendritic polymers to be commercialized (by Dendritech Inc, Midland, MI, in 2003 [4]). They can be classified into two main groups: (a) PAMAMOS with silicon in the end-groups, which can be considered as silicon-modified traditional polyamidoamines (PAMAMs) [3], and (b) radially copolymeric PAMAMOS, which are built of PAMAM interiors surrounded by at least two layers of organosilicon (OS) branch cells (Fig. 11.2) [5, 6]. To the first of these two groups also belong multi-arm star polymers (see also Chapters 3 and 10) where silicon is either used to simply connect a PAMAM dendrimer core to purely organic chains/arms, or in which the arms that are attached to PAMAM cores are essentially silicon-based, for example, polydimethylsiloxane (PDMS) arms [7, 8]. On the other hand, radially multilayered PAMAMOS are an example of truly copolymeric dendrimers with significantly different compositions of the branch cells (i.e., dendritic repeat units) and the resulting properties of combined "mers" [5, 6]. The only other known examples of radially layered copolymeric dendrimers are poly(carbosilane-siloxanes) of Chapter 2 and poly(benzylether-benzylester) dendrimers, described by Hawker and Frechet in 1992 [9, 10].

Other silicon-organic dendrimers that are reported in the literature include: (a) phthalocyanines with axial dendritic substituents [11, 12], (b) poly(alkyl-phenyl ethers) with silicon as an extending element in the dendrimer branches [13], (c) dendrimeric silatranes with silicon in the branch junctures [14], and (d) poly(propylene imine-organosilicon)s, PPIOS [2, 15, 16], which are close relatives of the corresponding PAMAMOS [1–3]. These dendrimers are described in the final section of this chapter.

11.2 PAMAMOS: PAMAM Dendrimers with Silicon-Containing End-Groups

PAMAMOS dendrimers with organosilicon end-groups can be represented as shown in Fig. 11.2a. They can be considered as inverted unimolecular micelles where the PAMAM interior is highly hydrophilic and nucleophilic, while the OS exterior is oleophilic and can contain a variety of OS functionalities of which, $Si–CH_3$, $Si–CH_2–CH_3$, $Si–O–CH_3$, $Si–O–CH_2–CH_3$, $Si–CH{=}CH_2$ and $Si–CH_2–CH{=}CH_2$

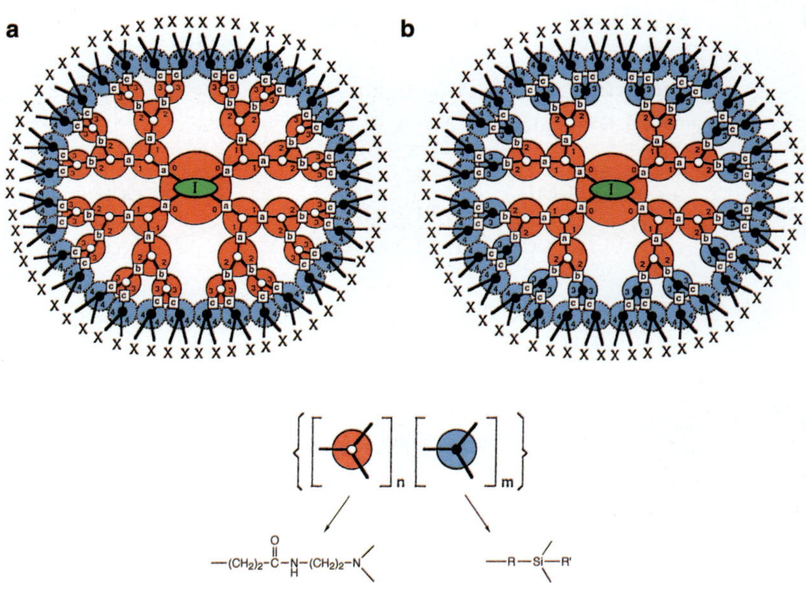

Fig. 11.2 Poly(amidoamine-organosilicon), PAMAMOS, dendrimers: (**a**) PAMAMOS dendrimers with silicon-containing end groups; (**b**) radially layered copolymeric dendrimers with two layers (tiers, generations) of OS (blue) branch cells built on top of a generation 2 PAMAM (red) core.

have been reported [1–3]. Depending on the generation of the PAMAM dendrimer used for their preparation, these PAMAMOS vary in their molecular sizes. In general, their diameters increase by about 1 nm per generation of the PAMAM interior and by about 1.5 nm per layer of the OS branch cells [17]. Currently, generation 0–7 PAMAMOS with Si–OCH_3 end-groups are commercially available [4, 18].

In principle, PAMAMOS dendrimers can be obtained from either amino, carbomethoxy, or hydroxy terminated PAMAMs. From the former, a variety of reactions have been evaluated, as shown in Reaction Scheme 11.1, including Michael addition (11.1a, b), isocyanate addition (11.1c), haloalkylation (11.1d) and epoxidation, but the former two were the easiest to perform at reasonable rates and in quantitative yields [2, 3]. Other reactions include amidation of carbomethoxy-terminated PAMAMs which can be performed with aminoalkylsilanes (or aminosilanes), and reaction of halosilanes with hydroxyl-terminated PAMAMs which give very interesting PAMAMOS dendrimers with hydrolizable C–O–Si "*Achilles' heel*" bonds built into the dendrimer structure. The latter enable controlled break-down of the outer OS shells and provide an effective "dendrimer box"-type opening mechanism.

The nature and the degree of substitution of the PAMAM end-groups cause the solubility of these dendrimers to change continuously during the course of their

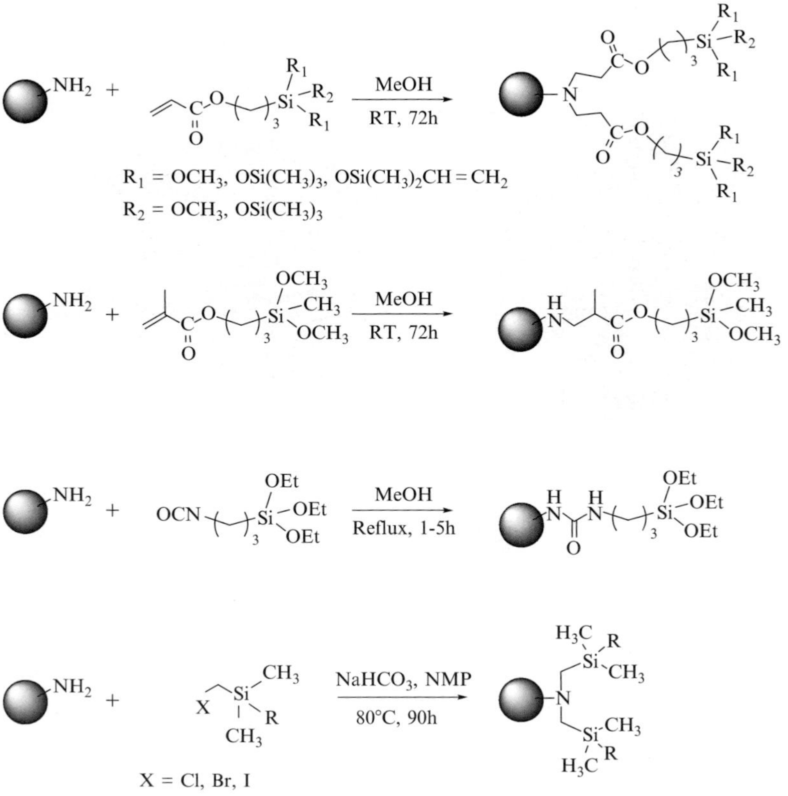

(11.1a-d)

formation. For example, in the case of PAMAMOS with trimethylsilyl end-groups (see Reaction Scheme 11.1d), the highly hydrophilic PAMAMs soluble in water and methanol but insoluble in hydrocarbons, turn oleophilic in the completely siliconized product that becomes soluble in toluene and insoluble in water [3]. Solubility in organic solvents is achieved at about 50% conversion of the PAMAM NH_2 groups, while the solubility in water is lost when two thirds of the end-groups become trimethylsilyl units. Surprisingly, the characteristic PAMAM solubility in methanol is retained by the PAMAMOS, in spite of the fact that methanol is a non-solvent for traditional silicone polymers. Computer modeling data indicate that only one layer of OS branch cells is not enough to completely block interactions of the PAMAM interior with the outside environment (see Fig. 11.3) [3, 19].

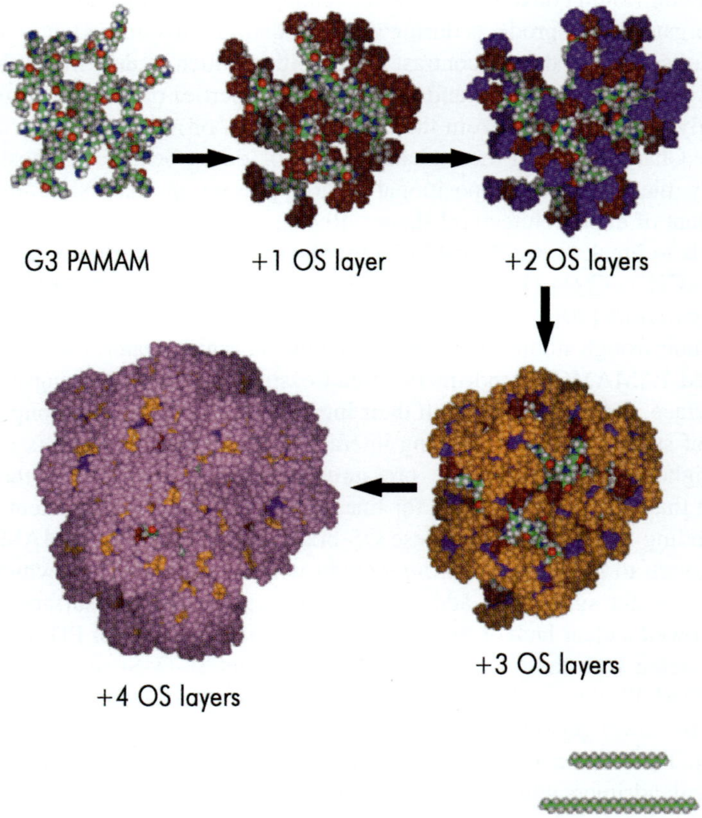

Fig. 11.3 A schematic representation, using CPK models, of the formation of higher generation radially copolymeric PAMAMOS dendrimers. Note the increasing encapsulation of the PAMAM interior with increasing number of OS branch cell layers (i.e., generations). Color code: green–red–white–blue – generation 3 PAMAM interior; brown – first OS layer; purple – second OS layer; yellow – third OS layer; magenta – fourth OS layer. *Bottom right*: two extended $-(CH_2)_n-$ chains, representing "*molecular rulers*" that are to be used to compare the dimensions of dendrimer species. The lengths of the rulers correspond to: *top* 2.30 nm; *bottom* 3.85 nm, respectively. Reprinted with permission from [35]. copyright 2006 Wiley Interscience.

The glass temperatures (T_g) of Si–CH_3 and Si–CH = CH_2 terminated PAMAMOS are practically identical to those of the corresponding generation PAMAMs and reach about 15°C for dendrimers having generation 4 and higher PAMAM interiors with at least half of the end-groups converted into OS units [3]. Among these, vinyl-PAMAMOS seem to have slightly lower T_gs (by about 2–5°C) probably because of increased intermolecular free volume created by the mutual repulsion of the rigid vinyl groups.

Thermal and thermo-oxidative degradation behavior of Si–CH_3 terminated PAMAMOS are almost identical with those of the corresponding PAMAMs with one major exception. While PAMAMs annihilate completely during dynamic thermal gravimetric analysis in air at about 650°C, the PAMAMOS leave about 5–10 wt% of the silica residue [1, 3]. The residue is, however, less than expected if all the silicon would convert into SiO_2, indicating that about one half of it is lost as volatile organosilicon products during the degradation. Most importantly, however, these results indicate that in contrast to solubility, which is determined by a fine interplay of the compositional and architectural properties of dendrimer molecules and clearly sets them apart from the traditional types of macromolecular architectures (see Chapter 1), their thermal and thermo-oxidative behavior is mostly determined by the weakest compositional link(s) present in their structure and is independent of their architectural shapes. Recent FTIR studies have shown that the first bonds to break in the PAMAMOS thermo-oxidative degradation are the ester groups, –$(CH_2)_2$–*C(O)–O*–$(CH_2)_3$–$Si(OR)_3$ which connect their PAMAM interiors with OS exteriors [20].

Langmuir trough studies have shown that the otherwise water insoluble Si–CH_3 terminated PAMAMOS dendrimers spread easily and interact strongly with the water surface, presumably through their interior amine and amide groups that are capable of strong hydrogen bonding interactions [21, 22]. This results in significantly higher maximum surface pressures of over 50 mN/m than the typical 10 mN/m that is usually observed for linear PDMS [8], and is consistent with the understanding that one layer of these OS branch cells around a PAMAM interior is not enough to prevent it from interacting with the environment, which in this case is the water surface. In addition, the surface pressure vs. surface area diagrams showed a clear lack of the plateau region characteristic for PDMS and usually associated with the pronounced flexibility of its Si–O–Si segments. Instead, the PAMAMOS always showed some degree of hysteresis which was more pronounced for lower generation PAMAM interiors with lesser content of OS exterior [2, 8, 22]. This can be associated with the relative efficiency of water penetration into these dendrimers which is more pronounced for more open and less congested molecules, along with a possible formation of a second dendrimer layer. The Langmuir trough data also showed that molecular surface areas of water insoluble PAMAMOS correlated well with the surface areas of the corresponding generation PAMAM dendrimers. This suggests that the increase in dendrimer size achieved by adding a layer of OS branch cells is similar to that obtained by adding another generation of PAMAM.

11.3 PAMAMOS Multi-arm Star Polymers

Two types of PAMAMOS multi-arm star polymers have been reported [7, 8]. One of these contained silicon atoms as connectors between the PAMAM dendrimer cores and $-(CH_2)_2-S-(CH_2)_x-CH_3$ polymethylene arms, and were obtained by a thiol addition of the corresponding mercaptans to the vinylsilyl-terminated PAMAMOS in the presence of azo*bis*isobutyronitrile (AIBN) (see Reaction Scheme 11.2a). Their arms ranged in length from C_3 (x = 2) to C_{18} (x = 17), comparable to the length of the arms of the other type that had significantly higher silicon content resulting from the addition of epoxypropyl-mono-terminated linear PDMS having a degree of polymerization of 11 to amine-terminated PAMAMs, as shown in Reaction Scheme 11.2b.

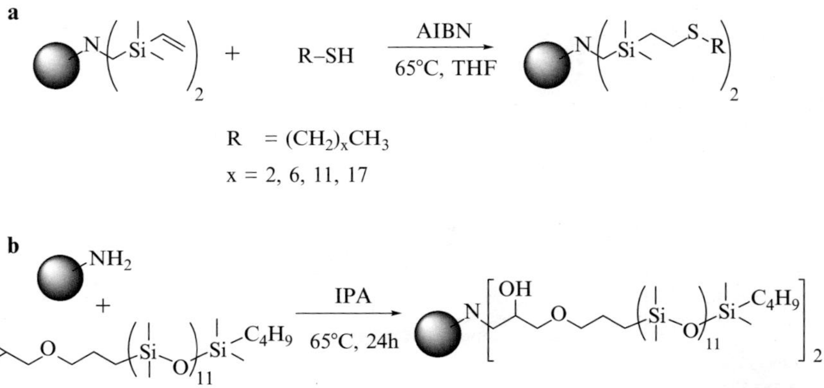

(11.2a, b)

The PDMS multi-armed PAMAMOS star polymers showed a Langmuir film behavior that is unique among the reported dendrimer-based amphiphilic multi-arm star polymers [8]. Their Langmuir isotherm characteristics were intermediate between the behavior of the corresponding linear polymer (i.e., PDMS) and that of the related PAMAMOS dendrimers, and consisted of (a) a low surface pressure pseudo-plateau over the range of larger and intermediate molecular surface areas, typical for linear PDMS, and (b) a fast upswing in surface pressure at lower molecular areas to relatively high surface pressures characteristic of PAMAMOS dendrimers with one or two layers of OS branch cells in the molecular exterior (see Fig. 11.4). Surprisingly, this upswing in surface pressure originated at unexpectedly large molecular surface area values (e.g., 8,700 Å^2/mol or 87 nm^2/mol for a multi-arm star polymers with generation 3 PAMAM core), which were several times larger than those of the corresponding PAMAMOS dendrimers from generations 3 and 4 PAMAM and one OS layer exteriors, respectively. This was also much larger than the limiting molecular areas of the related amphiphilic star polymers with PAMAM cores and various types of alkyl chain arms for which a model of a flattened or oblate core with hydrophobic chains extending away from the water surface was proposed to account for the

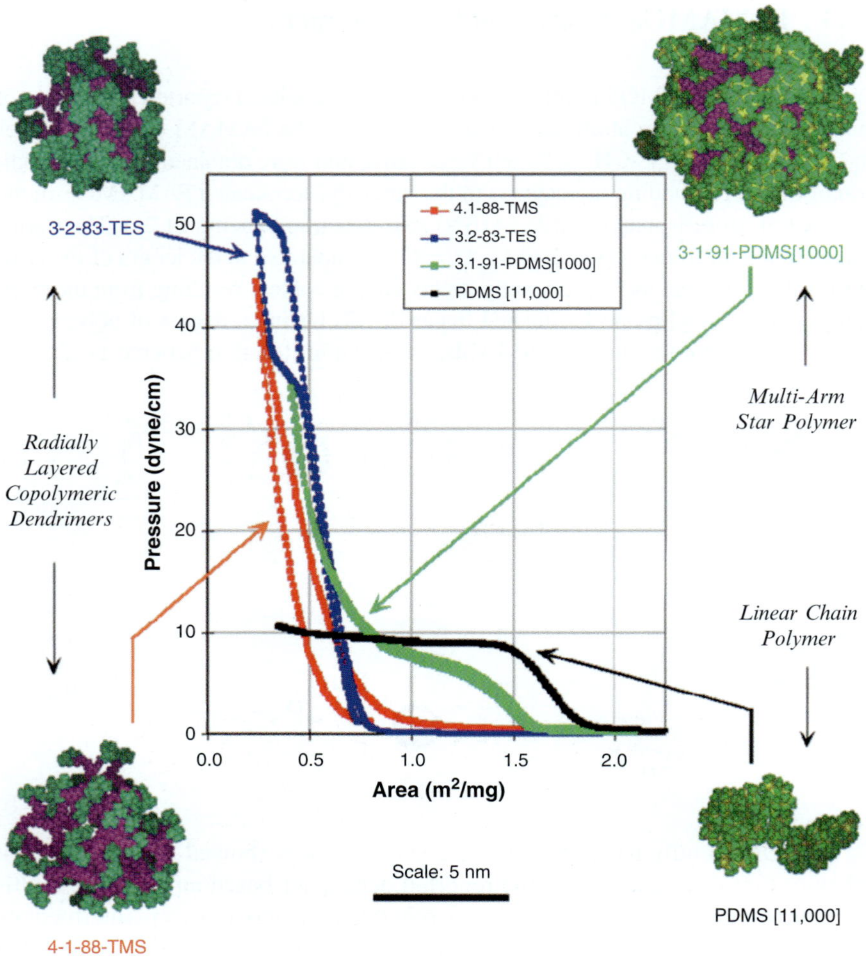

Fig. 11.4 Surface pressure versus area isotherms for: PAMAMOS dendrimer with one OS layer and a generation 4 PAMAM core (*bottom left*); copolymeric PAMAMOS dendrimer with two OS layers (generations) surrounding a generation 3 PAMAM core (*top left*); a PAMAMOS multi-arm star polymer having linear PDMS arms (M_w = 1,000) attached to a generation 3 PAMAM core (*top right*); linear PDMS (M_w = 11,000) (*bottom right*). In the molecular models, the violet color indicates the part of the PAMAM interior that is still "visible" (i.e., accessible) to the environment, the green color indicates methyl groups of the OS compositions and the yellow color represents the siloxane backbone.

observed behavior [23–25]. Hence, it seems that in contrast to these, the PDMS chains of the PAMAMOS stars are capable of spreading, at least to a certain extent, over the water surface, and thus occupy a greater area than the corresponding C–C counterparts. This behavior, also seems to be yet another consequence of the pronounced flexibility of the siloxane backbone (energy barrier for free rotation around the Si–O bonds is generally accepted to be less than 2.1 kJ/mol) which can orient itself so as to interact with its polar, hydrophilic Si–O–Si units with the water surface while pointing its hydrophobic CH_3 groups away from the backbone and into the air.

These multi-arm PAMAMOS stars are also very effective phase-transfer agents capable of "dissolving" metal cations, such as copper(II), in organic media where they are normally insoluble or only sparingly soluble (see Fig. 11.5a) [8]. To account for this ability, we proposed a multi-step pathway made possible by the amphiphilic character of these multi-arm stars and occurring in three main steps. First, the stars orient themselves at the water-organic solvent interphase with their PAMAM cores towards the aqueous phase and long arms extended into the organic phase (see Fig. 11.5b). Then, the PAMAM cores interact with the cations dissolved in water and absorb them into their highly nucleophilic interiors. There, the cations form complexes with the tertiary nitrogens helped by the neighboring carbonyl oxygens, and finally, the organic arms pull the entire host-guest assembly into the bulk of the organic phase rendering it the color of the metallic complex (e.g., blue in the case of Cu^{2+}, see Fig. 11.5a). Due to high solubility of these multi-arm stars in organic solvents, significant quantities of metals can be transferred from the

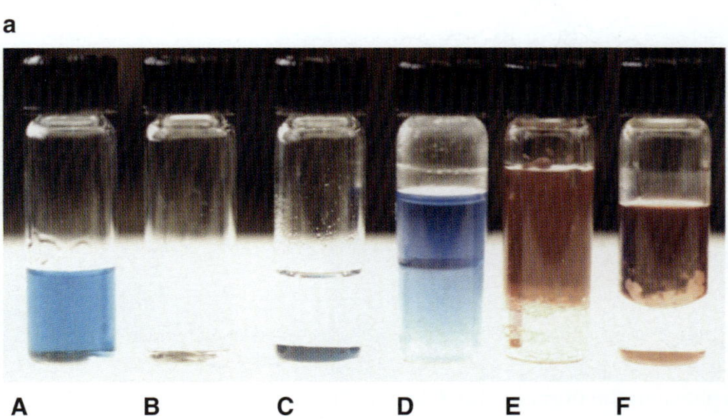

Fig. 11.5 (**a**) Dissolving the insolubles. Preparation of a hexane-soluble copper nanocomplex and nanocomposite using PAMAMOS-PDMS 128-arm star polymer. *A*: water solution of $Cu(Ac)_2$; *B*: hexanes solution of PAMAMOS-PDMS star polymer; *C*: $Cu(Ac)_2$ is insoluble in water; *D*: two-phase system obtained after thorough shaking of *A* and *B* and subsequent separation of immiscible layers (*upper layer*: organic; *lower layer*: aqueous); *E*: *D* after Cu^{2+} reduction with hydrazine to form "dissolved" Cu^0 metal; *F*: *E* after decantation. (**b**) Possible mechanism of metal dissolution in organic solvent: metal ion encapsulation by the PAMAM dendrimer interior in water (*top*), transfer of the organic solvent soluble dendrimer-metal complex through the water-organic solvent interface (*middle line*), "dissolution" of metal in organic solvent by PAMAMOS multi-arm star polymers (*bottom*). Three different polymer conformations at the interface (*B*) and in organic phase (*A* and *C*) are shown. Reprinted with permission from [19]. copyright 2002 American Chemical Society.

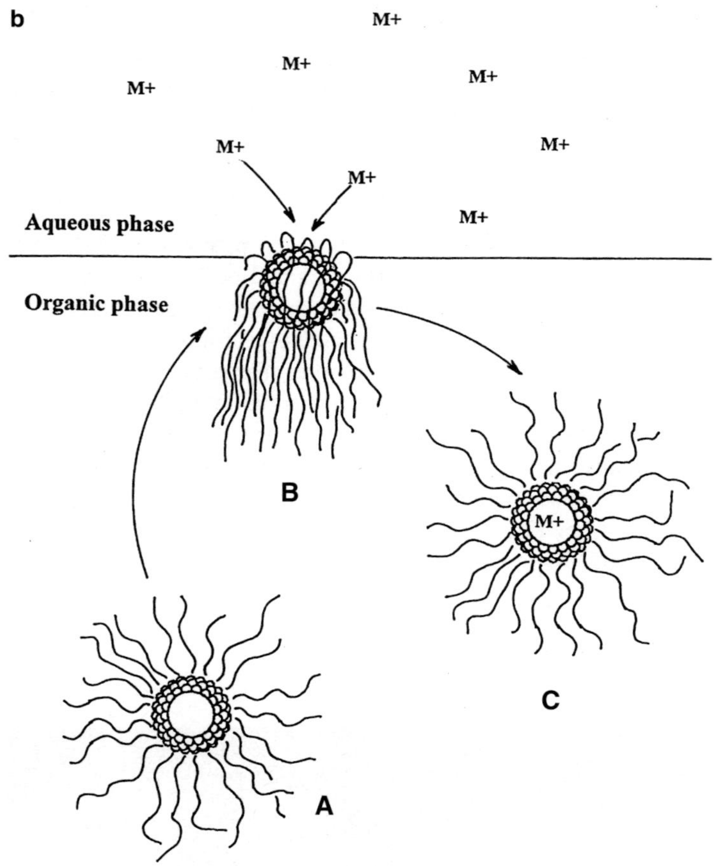

Fig. 11.5 (continued)

aqueous phase to the organic one. Clearly, the cations have stronger affinity towards the PAMAM than towards the water.

11.4 PAMAMOS Networks

Of particular interest among PAMAMOS dendrimers, especially for materials engineering applications, are their cross-linkable derivatives, including those with vinyl, allyl and alkoxysilyl end-groups [3]. While the former two can be crosslinked using classical free-radical polymerization initiators, such as AIBN, benzoyl peroxide, etc. (see Reaction Scheme 11.3a), the alkoxysilyl-functionalized dendrimers easily undergo hydrolysis and silanol condensation reactions, typical for sol-gel processes (see Reaction Scheme 11.3b) [26–30]. Note that in Reactions 11.3 snaky lines (∿∿∿∿∿∿∿∿) represent either other interdendrimer bridges, or intradendrimer loops of the same composition, or unreacted crosslinkable groups.

a

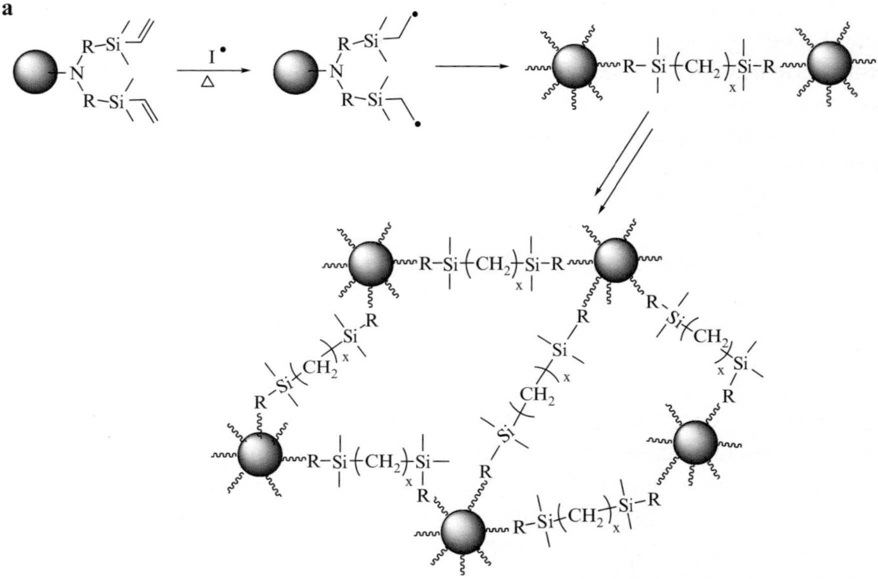

(11.3a)

b

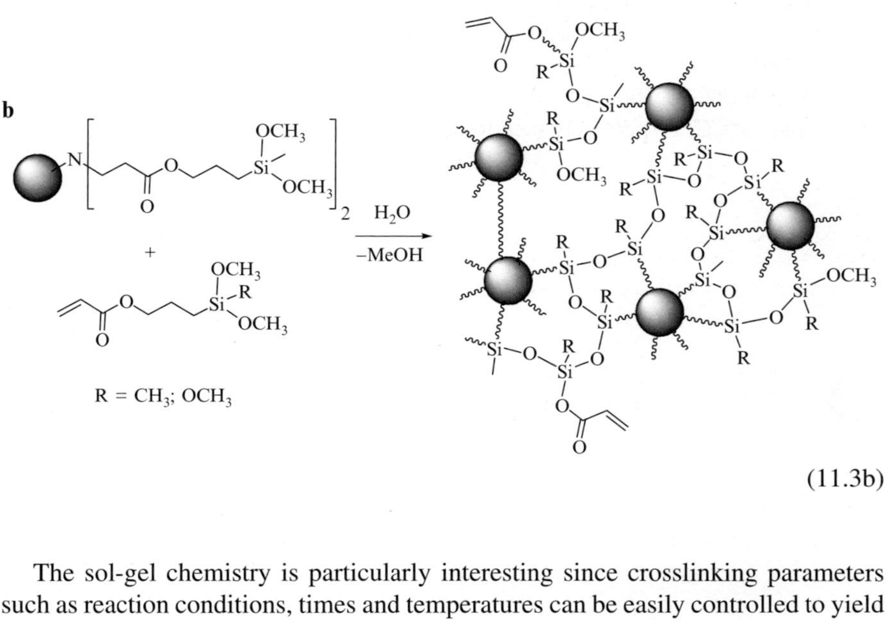

(11.3b)

The sol-gel chemistry is particularly interesting since crosslinking parameters such as reaction conditions, times and temperatures can be easily controlled to yield a variety of different products with controlled thermal and mechanical properties [30]. The reaction rates can be controlled in a variety of ways, for example (a) by selecting either the methoxysilyl or ethoxysilyl-functionalized PAMAMOS, (b) by carrying out the reaction with or without added catalyst (e.g., traditional organotin compounds), since PAMAM interiors are basic enough to self-catalyze the hydrolysis of alkoxysilyl groups in the presence of moisture, (c) by reacting either with liquid water or water vapor, and (d) by operating at temperatures which may range from normal ambient temperatures for extended periods of time to about 120–130°C for only several hours.

The resulting networks consist of spheroidal PAMAM nanodomains regularly distributed within the OS matrix (see Reaction Scheme 11.3), and they can be very easily shaped into self-supporting films, sheets or coatings on a variety of substrates, including glass, paper, metals, wood, ceramics, various plastics, etc. [19, 30–32]. X-ray studies showed that at 25°C the diameters of these PAMAM domains in the networks correspond very closely to the size of the same generation PAMAM dendrimers in a good solvent [33]. It was also found that they exhibit practically identical host behavior towards electrophilic guests as their soluble multi-arm star counterparts of Section 11.3, being able to complex over 20 different cations tested, including Ag^{+}, Cu^{+}, Cu^{2+}, Ni^{2+}, Zn^{2+}, Cr^{3+}, Cd^{2+}, Fe^{2+}, Fe^{3+}, Au^{3+}, Co^{2+}, Pd^{2+}, Rh^{3+}, Pt^{2+}, Pt^{4+}, and lanthanides, such as Eu^{3+}, Tb^{3+}, Sm^{3+}, Dy^{3+} [34–36].

Networks can be prepared from alkoxysilylated PAMAMOS either directly or from mixtures with other reagents, which enables significant influence on the mechanical properties of the resulting products [30]. For example, combination of methoxysilylated PAMAMOS with 10 to 40 wt% tetraethoxysilane (TEOS), yielded networks with higher crosslinking density having T_g values by about 15–25°C, respectively, higher than the networks obtained using the same dendrimers alone. Likewise, reduction of the crosslinking density by addition of α,ω-alkoxy telechelic linear PDMS (degree of polymerization of 12, in relative amounts of 10–40 wt%) resulted in more flexible, i.e., elastomeric, networks having T_g values between −15°C and −25°C, respectively (see Reaction Scheme 11.4) [30].

a

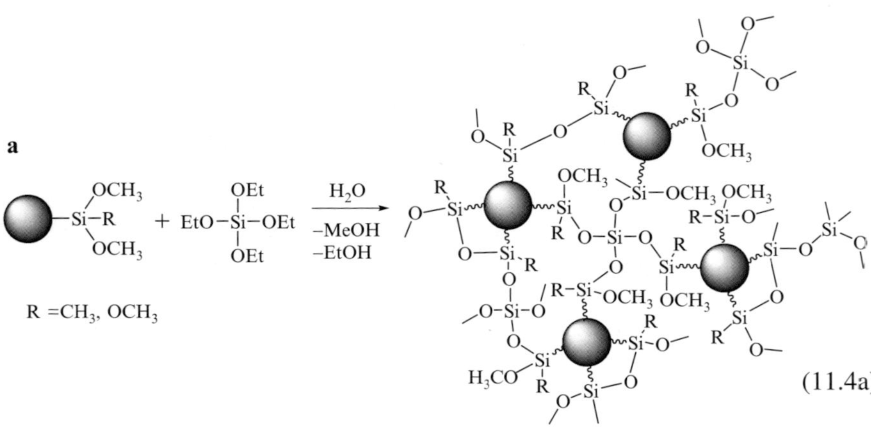

(11.4a)

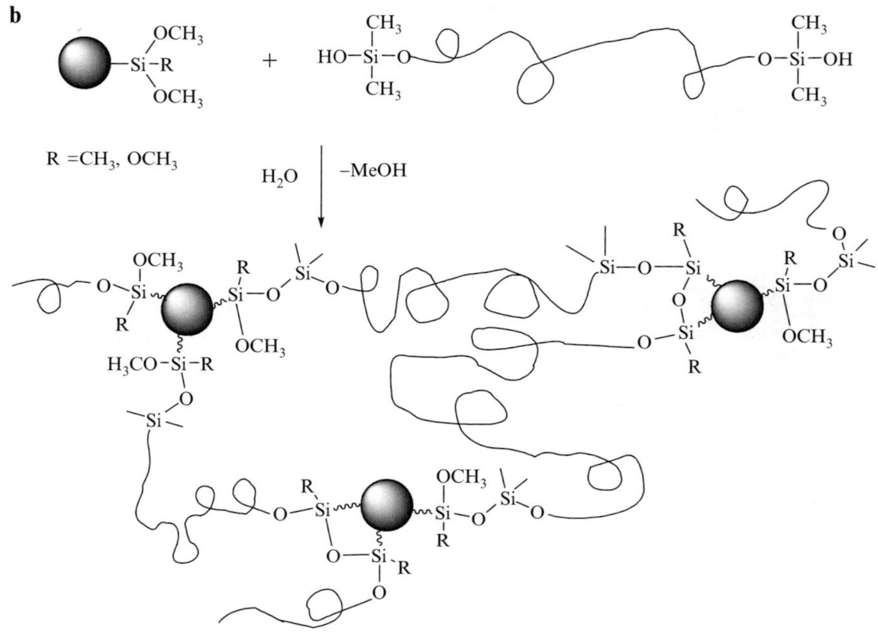

(11.4b)

For any given crosslinking composition, the hardness of the resulting networks can be fine-tuned by appropriate combination of the curing time and temperature, resulting in coatings with pencil hardness values ranging from HB to 5 H+.

This sol-gel crosslinking chemistry is by no means discriminative and it can occur with any available alkoxysilyl or silanol group present in the system. Thus, PAMAMOS coatings can be covalently bonded to various appropriately reactive substrates, including silanol-functionalized glasses or silicas (see Fig. 11.6) [19, 30]. When such reactive groups are not available, the coatings can still tenaciously attach to various substrates through a multitude of van der Waals and electrostatic interactions of dendrimer end-groups.

Methoxysilyl PAMAMOS dendrimers were also used for the preparation of interpenetrating polymer networks (IPNs) with poly(methyl methacrylate) (PMMA) and with cellulose acetobutyrate (CAB), respectively [37]. The PAMAMOS-PMMA IPNs were prepared by an in situ process involving sequential formation of two networks in a mixture of all the required components. First, the PAMAMOS network was formed by a dibutyltin dilaurate-catalyzed sol-gel reaction with atmospheric moisture at room temperature; following which an AIBN-initiated polymerization of methyl methacrylate (MMA) and subsequent crosslinking of the resulting polymer by ethylene glycol dimethacrylate was performed by heating to 60°C. In contrast to this, the PAMAMOS-CAB IPNs were prepared starting from a preformed crosslinkable CAB polymer instead of the polymerizable monomer and the consecutive crosslinkings were effected at

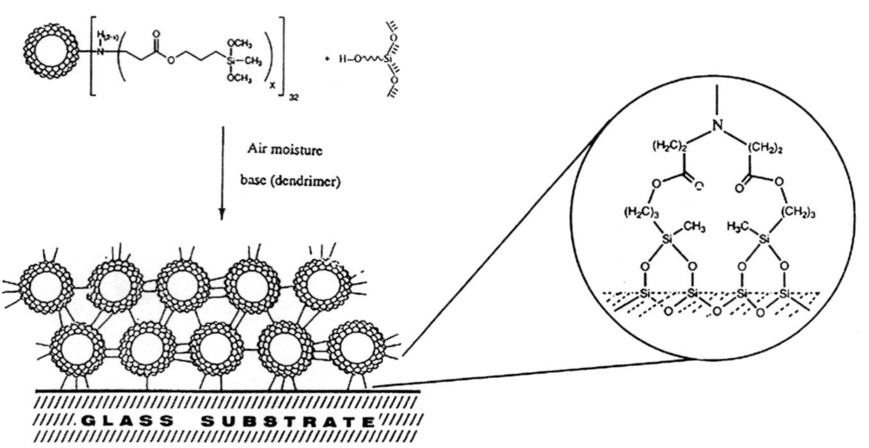

Fig. 11.6 Schematic representation of the formation of covalently bonded coating from methoxysilylated-PAMAMOS dendrimers on a glass substrate having silanol surface groups. Reprinted with permission from [19]. copyright 2002 American Chemical Society.

both room temperature and at 55°C. The obtained IPNs were not affected by solvents, indicating that they were indeed truly covalently crosslinked. Further, they had single T_gs, were optically clear and transparent, and capable of complexing metal cations, such as Cu^{2+}, just like pure PAMAMOS networks (see the following section). These examples represent the first reported true IPNs containing a dendritic and a linear polymer network (i.e., both covalently crosslinked components) intimately and permanently combined at the supramolecular level.

11.5 PAMAMOS Networks Nanocomplexes and Nanocomposites

One of the most interesting and potentially useful properties of PAMAMOS networks, either in the form of films, sheets or coatings, is their pronounced ability to interact with electrophilic guests, such as metal cations, organic or organometallic compounds, and create various types of solid, inorganic/organic host-guest nanocomplexes, much like their precursor dendrimers or multi-arm star polymers do in solution (compare Figs. 11.5 and 11.7) [34–39]. These network complexes are illustrated in Reaction Schemes 11.5, where 11.5a assumes a tetracoordinate complexation of metal cation M^+ and 11.5b is a schematic representation of the large complexing capacity of individual PAMAM dendrimer molecules

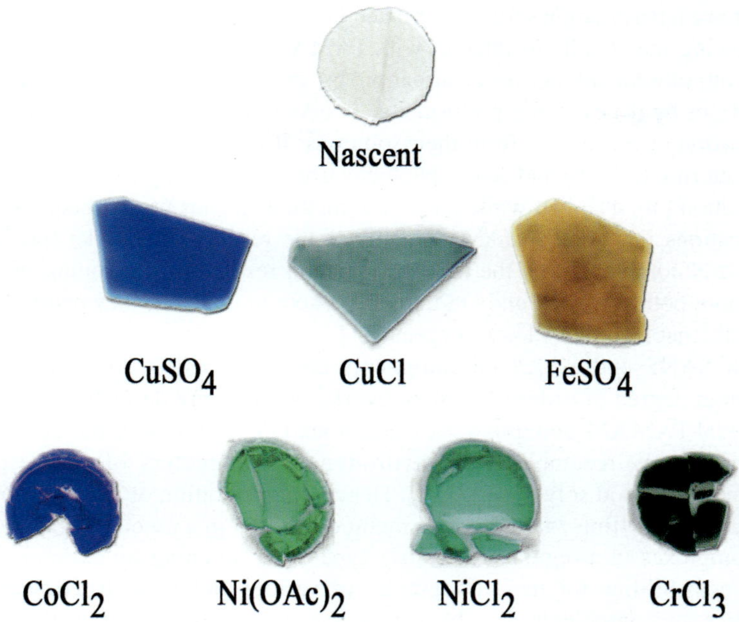

Fig. 11.7 The colorful world of PAMAMOS nanocomplexes. Note optical clarity and transparency of the nascent network/film (*top*). Same film after complexation of various salts indicated in rows 2 and 3 from the top. Reprinted with permission from [35]. copyright 2006 Wiley Interscience.

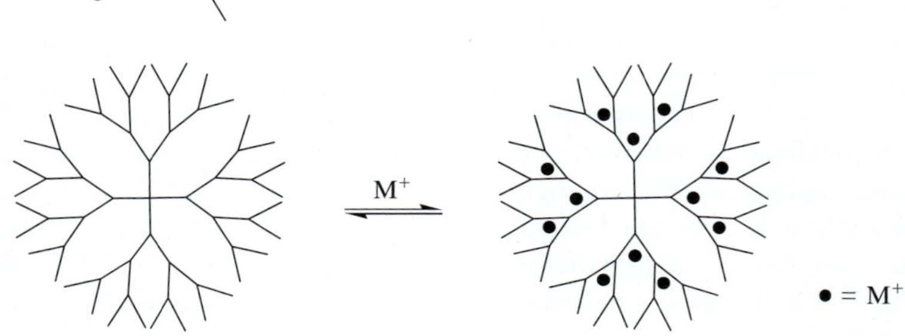

(11.5a, b)

which have tertiary amines for 1→2 branch junctures. These can be made either by: (a) allowing metal salts to interact with PAMAMOS dendrimers in solution followed by subsequent network formation by crosslinking (the so-called *in situ* method), or by (b) exposing preformed networks (i.e., films, sheets or coatings) to the *diffusion* of metal salts from their solutions. It was shown by small angle neutron scattering (SANS) and X-ray photoelectron spectroscopy (XPS) for Cu^{2+} as a metal cation [40, 41], that while the *in situ* method enables precise sequestering of metal cations into the PAMAM domains of the resulting networks (preferential Cu^{2+}/*tert*-N complexation), the diffusion method results in partitioning of cations throughout both PAMAM and OS domains where in the latter they reside presumably at the nucleophilic siloxy oxygens [41].

Both SANS and SAXS measurements also provided strong evidence of a fairly high degree of order in these networks, which were described as arrays of spheroidal PAMAM domains in a continuous OS matrix with the core-to-core spacing closely resembling the hydrodynamic diameters of the respective dendrimers in good solvents [33, 41]. Hence, incorporation of metal cations into the PAMAMOS films via the *in situ* method resulted in a precise organization of nanocomplexes in a regular dot-matrix-type order, opening up extremely interesting possibilities for nanolithography in various fields, such as electronics, photonics and optoelectronics [42]. It was also demonstrated that the core-to-core distances between the neighboring nanocomplex domains increased with increasing metal salt concentration, indicating that PAMAM domains expanded until the limiting metal content was reached. Above this limiting level, the cations "spilled over" from the PAMAM cells into the surrounding OS matrix [41]. In the case of Cu^{2+} and generation 4 PAMAM domains, for example, the limiting amount of metal was reached between 5 wt% (by SANS) and 8 wt% (by XPS), in good agreement with the 9 wt% value calculated for the idealized perfect dendrimer structure and tetra-coordinate copper binding [41].

The diffusion method of metal-nanocomplex formation followed Type II diffusion kinetics and the depth of the metal penetration could be easily controlled by optically following the progress of penetration of the colored front [42]. This enables preparation of very precise, thin, metal-containing layers and multi-metal, multi-layered structures in which metal cations are distributed throughout both compositional domains because their local concentration in the network layers immediately following the diffusion front is always higher than the limiting metal content.

It was also shown that encapsulated cations inside the PAMAM domains of PAMAMOS dendrimer networks retain their chemical reactivity towards both liquid and gaseous reagents [35]. For example, nanocomposites with zero-valent metal particles can be obtained by reduction of complexed cations with various reagents, such as aqueous sodium borohydrate, hydrazine, or even ascorbic acid, leading to reflective materials in the case of silver, or conductive materials in the case of copper (see Fig. 11.8). On the other hand, exposure of nanocomplexes of $CuSO_4$ or $Cd(OAc)_2$ to H_2S or H_2Se gases at room temperature yielded the corresponding CuS, CuSe, CdS and CdSe composites (see Fig. 11.9) [35]. Since the sizes of the

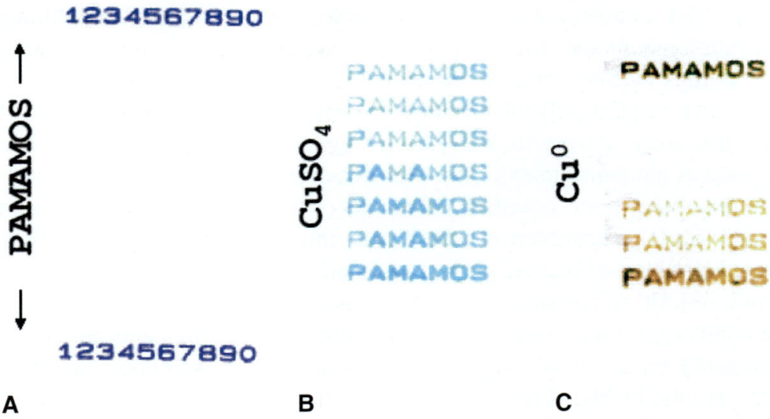

Fig. 11.8 An overture to dendrimer-based nanolithography. From *left* to *right*: (**a**) a section of white paper on which word "PAMAMOS" was printed seven times (from *top* to *bottom*) using Si–OCH_3 functionalized PAMAMOS in isopropanol (IPA) solution as "ink" in an ink-jet printer; (**b**) the same section of paper after exposure to (i.e., "development" in) aqueous $CuSO_4$; (**c**): paper from B after exposure to aqueous $NaBH_4$. Lines in yellow/coppery letters formed after reduction to Cu^0 nanocomposite showed conductivity of 10^2–10^3 S/cm. Reprinted with permission from [35]. copyright 2006 Wiley Interscience.

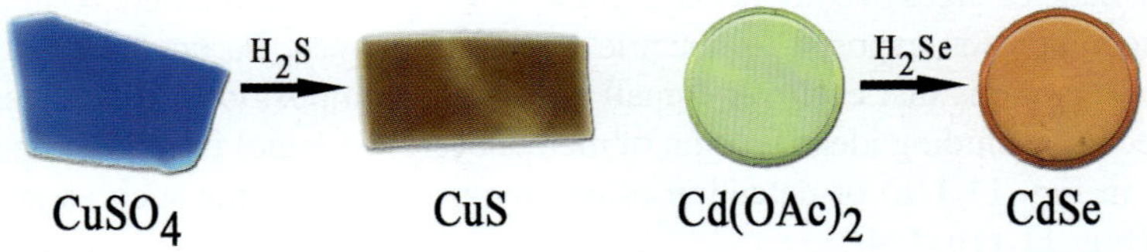

Fig. 11.9 Preparation of CuS and CdSe quantum dots templated in PAMAMOS films by exposure of $CuSO_4$ and $Cd(OAc)_2$ nanocomplexes to gaseous H_2S and H_2Se, respectively. Reprinted with permission from [35]. copyright 2006 Wiley Interscience.

resulting particles are controlled by the dendrimer generation, this provides an extremely easy and practical way of making encapsulated quantum dot-containing materials without the need for added chemical protection. In addition, these quantum dots can be very precisely placed within the well defined size domains of the PAMAM cells within the networks (within the 1–10nm range), separated by the 1–3nm thin OS walls, with the placement precision of ±1nm and with great potential for dot-matrix-type printing and templating in various electronic, photonic, photovoltaic and biomedical applications.

In addition to nano-lithography and quantum dots preparation, some other suggested applications [43] of PAMAMOS networks and their nanocomposites and nanocomplexes include: (a) environmentally benign antifouling coatings that are effective against zebra mussels, quagga mussels and barnacles, and useful for protection of man-made objects submerged in fresh water and marine environments

[35, 44]; (b) fouling-resistant semi-permeable reverse osmosis, nanofiltration or ultrafiltration membranes for water purification [35, 44]; (c) coatings for activation of mesoporous carriers for absorption of heavy metals, arsenic, selenium, pesticides, aromatic and/or polychlorinated hydrocarbons from polluted waters [45]; (d) colorimetric array sensors for detection of organic as well as inorganic contaminants, such as alcohols, arsenic, etc. [35]; (e) decontamination filters for detection and elimination of toxic industrial chemicals or threat agents, such as cyanide gas from air [46]; (f) preparation of nanoporous interlayer dielectrics for 22 nm node integrated circuits and beyond [20]; (g) printed wiring boards for electronic equipment [47, 48], (h) micropatterning [49] and microcontact printing [50].

For example, a unique new class of environmentally benign antifouling coatings was prepared by encapsulating selected biocides, such as copper or zinc compounds, into the PAMAM domains of the PAMAMOS networks, from which they were prevented from leaching into the surrounding water by dual action of strong complexation and high degree of steric hindrance imposed by the highly branched dendritic structure, while retaining antifouling activity by being placed within nanometers from the coating-water interface [35, 44]. The results showed that these coatings were quite effective against barnacles and zebra mussels, and either did not leach biocides at all or leached orders of magnitude less than the traditional antifouling paints (see Fig. 11.10). In another example, arrays of different cationic nanocomplexes were tested for colorimetric sensors (Fig. 11.11), and it was found that dramatic changes in color patterns could be obtained from the right combinations of nanocomplexes exposed to specific targeted analytes. Particularly interesting were observations that even very small changes in compositions of analytes could be detected, including identification of methanol from ethanol from isopropanol (as shown in Fig. 11.11a) or detection of dangerous contaminants such as arsenic in water (Fig. 11.11b) [54].

11.6 Copolymeric PAMAMOS Dendrimers

In analogy with traditional copolymers [5, 6, 51], copolymeric dendrimers consist of at least two different types of branch cells (see Fig. 11.2 and also Chapter 1). In principle, intramolecular arrangement of these cells can be random or ordered, but only examples of the latter have been demonstrated so far. Ordered copolymeric dendrimers resemble traditional block copolymers and can be classified into two main groups: (a) segmental, where at least two dendrons of a dendrimer are composed of compositionally different branch cells, and (b) radially layered, where different generational layers of all dendrons are composed of different branch cells (see Fig. 11.1).

Until now, only radially layered copolymeric dendrimers, including PAMAMOS [5, 6] have been described. They were prepared by a combination of divergent and convergent methodologies where in the first step polycarbosilane (PCS) dendrons of various generations and with haloalkyl focal groups were synthesized via the

a

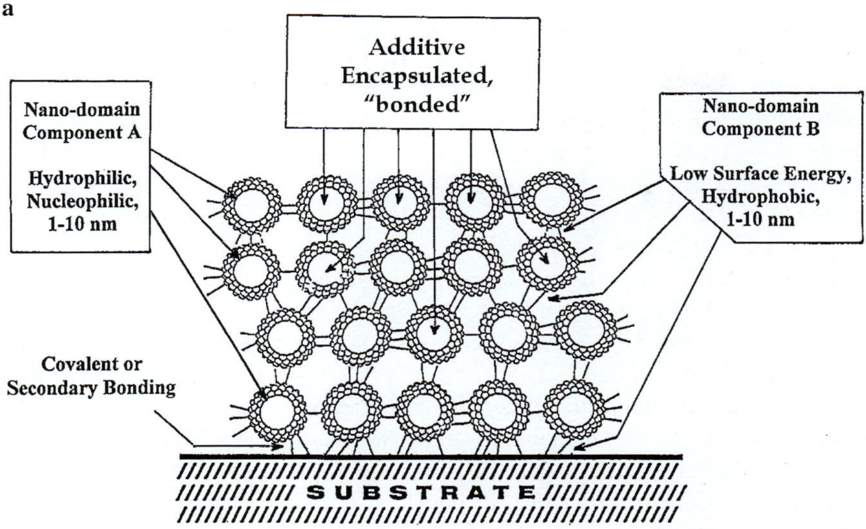

b

Fig. 11.10 (**a**) Schematic representation of a cross-section of the honeycomb-like structure of the nano-domained antifouling coatings. The PAMAM domains (circular sections) act as nano-sized containers capable of effectively encapsulating and retaining additives (i.e., biocides such as organocopper or organozinc compounds) yet positioning them immediately under the surface (at 1–2 nm depth) to ensure unobstructed effectiveness. The coating surface has methyl-silicone composition and associated "non-stick" characteristics. Depending on the type of the substrate the coatings can bond either covalently (permanently) or with a multitude of very strong secondary bonds. Component A: PAMAM, component B: OS. (**b**) PAMAMOS-coated steel test-plates after exposure to zebra mussel infested waters of Sanford Lake, MI. Top row: plates coated with Cu^{2+}–containing nanocomplex; bottom row: non-coated control plates. Small specks on the bottom row plates are settled zebra mussels; note their absence on the coated plates. Reprinted with permission from [35]. copyright 2006 Wiley Interscience.

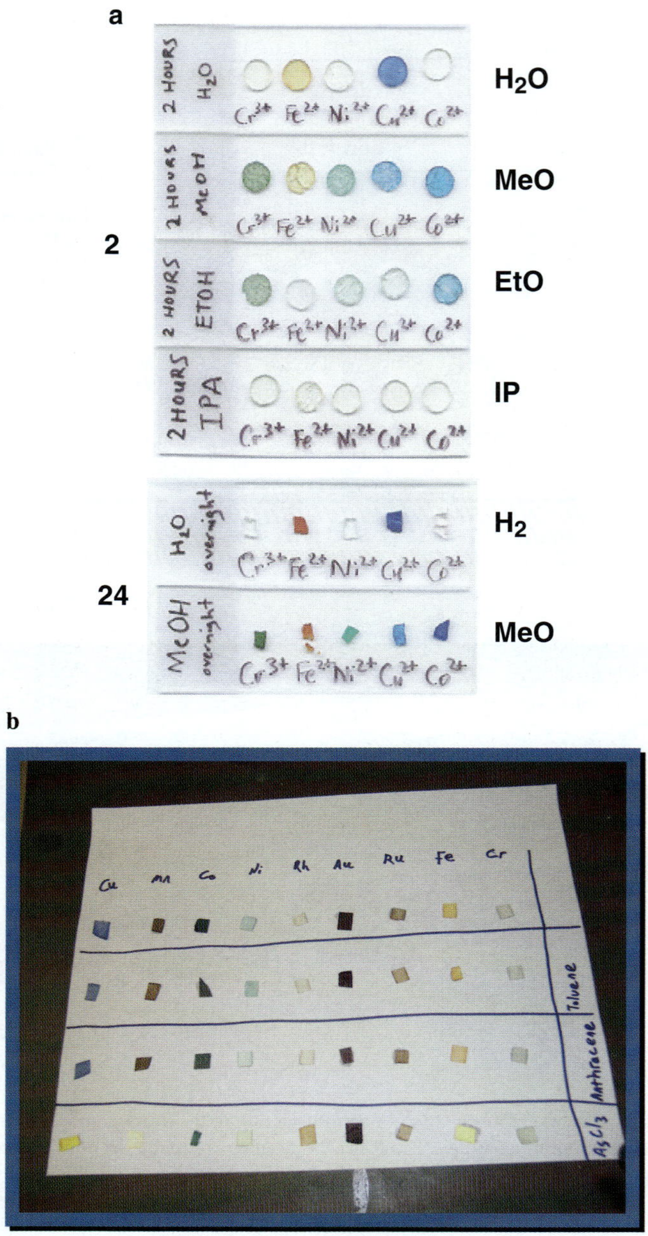

Fig. 11.11 PAMAMOS-based colorimetric array sensors. (**a**) An array of (from *left* to *right*) Cr^{3+}, Fe^{2+}, Ni^{2+}, Cu^{2+} and Co^{2+} PAMAMOS nanocomplexes after 2-h exposure to (*top* to *bottom*): water, methanol, ethanol and isopropanol. Note dramatic color changes resulting from very small changes in chemical composition of alcohols. *Bottom*: two arrays after extended 24-h exposure to water and methanol. Note increased color intensity with exposure time. (**b**) An array of (from *left* to *right*) Cu^{2+}, Mn^{4+}, Co^{2+}, Ni^{2+}, Rh^{3+}, Au^{3+}, Ru^{3+}, Fe^{3+}, Cr^{3+} PAMAMOS nanocomplexes after 30 min exposure to (*top* to *bottom*) water, toluene, anthracene and $AsCl_3$. Reprinted with permission from [35]. copyright 2006 Wiley Interscience.

well known Roovers–van Luewen hydrosilylation–Grignard reaction scheme (see Chapter 3), as shown in Reaction Scheme 11.6.

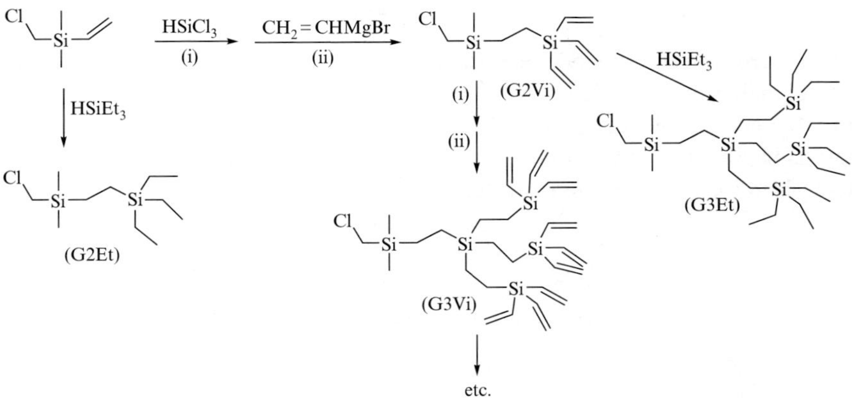

(11.6)

In the second step, these dendrons were attached to the amine-terminated PAMAM cores by a chloroalkylation reaction to yield radially layered copolymeric PAMAMOS containing from generation 0 to generation 4 PAMAM interiors surrounded by two to three PCS layers (i.e., generations), as shown in Reaction Scheme 11.7 (compare with Fig. 11.2).

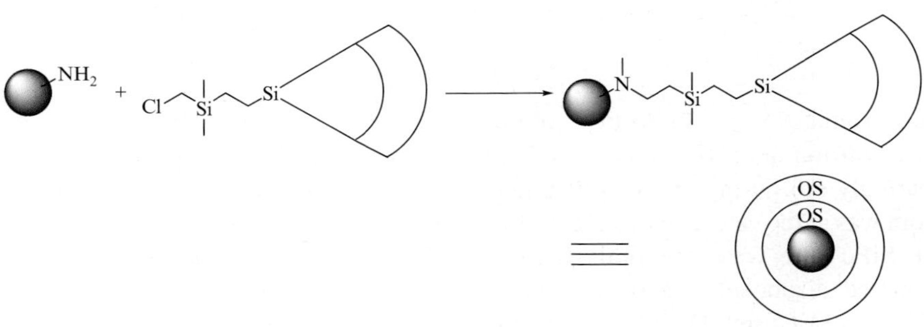

(11.7)

Numerous attempts to grow such dendrimers divergently, from vinyl or allylsilyl-terminated PAMAMOS via the hydrosilylation-Grignard reaction scheme were unsuccessful due to the complexation of the platinum catalyst with PAMAM interiors and its consequent effective withdrawal from the desired reaction location at Si–CH=CH_2 or Si–CH_2–CH=CH_2 end-groups. Only when PAMAM tertiary amines were quantitatively neutralized with either platinum or other appropriate electrophile prior to reaction, was Karstedt's catalyst able to catalyze hydrosilylation in the usual manner [6].

11.7 Other Silicon-Organic Dendrimers

11.7.1 Dendrimers with Silicon in the Core

In 1998, McKeown and co-workers prepared a series of didendron dendrimers having silicon phthalocyanine, SiPc, cores. This was achieved by substitution of the axial chloro ligands of dichloro(phthalocyaninato)silicon with two poly(aryl ether) dendrons having benzylic alcohol focal groups [11, 12], as shown in the Reaction Scheme 11.8 for the generation 2 dendrons:

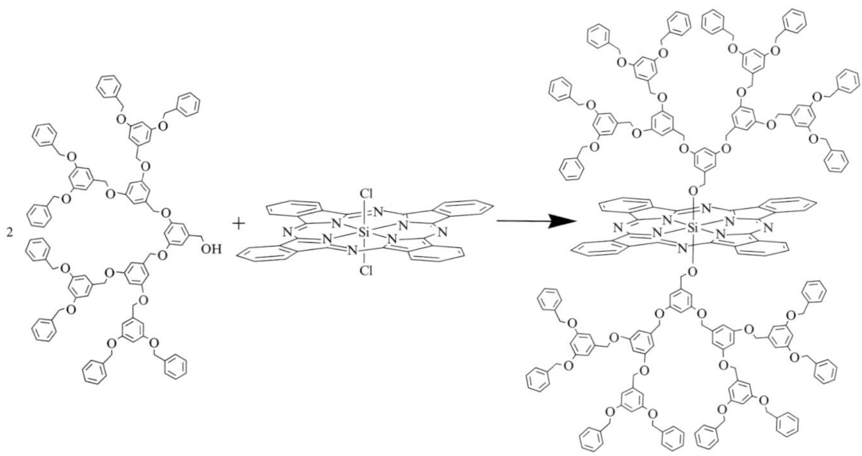

(11.8)

The products were all solids, soluble in common organic solvents, such as tetrahydrofuran (THF), toluene and chloroform (interestingly: generation 1 only sparingly so). X-ray analysis of a generation 2 sample showed that the silicon atom was at the center of symmetry, the O–Si–O bond angle was almost 180°, and the SiPc core was essentially planar [11, 12]. Each of the three dendrimers prepared displayed a distinct glass temperature on cooling from the melt (at 139°C, 124°C and 110°C for generations 1, 2 and 3, respectively; while melt temperatures for generations 1 and 2 were 253°C and 168°C, respectively), enabling facile fabrication of good quality solid films by melt processing [12]. The UV–Vis spectra of films from generation 2 and 3 dendrimers prepared either by cooling from the melt or by spin-coating were very similar and showed relatively weak edge-to-edge exciton interactions, indicating effectively suppressed cofacial intermolecular excitonic interactions by large axial dendron substituents separating the SiPc cores by as much as 14 Å in the case of generation 2, as determined by X-ray spectroscopy. Polarized optical microscopy showed that these films were optically clear and defect-free, providing quite uniform, isotropic and glassy, true solid solutions. The author concluded that the molecular arrangement within these glasses was controlled by the size, number and position of the dendrons attached to the SiPc macrocycle and suggested that they may have interesting optical properties [12].

More recently, Kobayashi and co-workers described a very interesting self assembly of amphiphilic constructs from silicon phthalocyanines (SiPc) which had a hydrophilic generation 3 poly(aryl ether) dendron with terminal carboxyl groups for one axial ligand and a hydrophobic C_8H_{17} alkyl chain for another [52]. In contrast to the McKeown-type didendron dendrimers, these amphiphiles self-assembled in water into spherical supramolecular micelles with hydrophobic stems of their mushroom-resembling molecules organizing inside the micelles and hydrophilic carboxylated "dendritic caps" pointing outward, as shown in the Reaction Scheme 11.9:

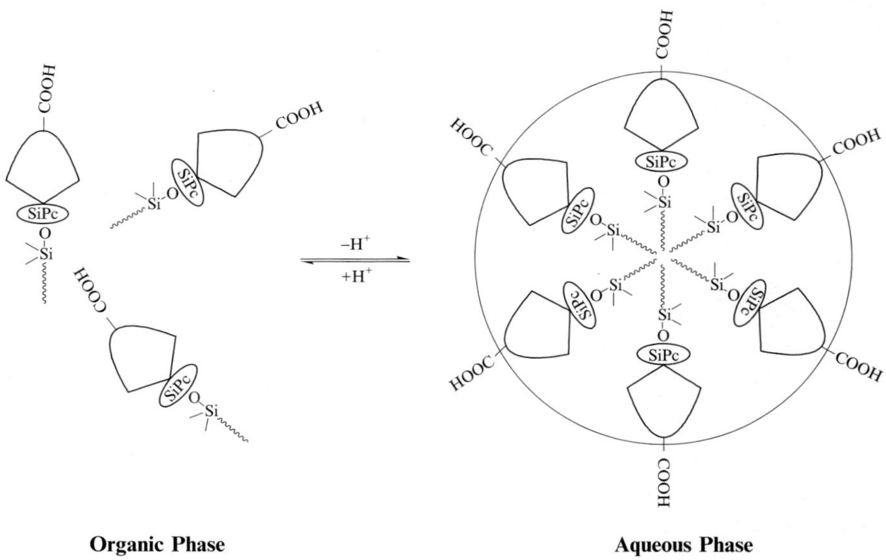

(11.9)

The "mushrooms" from generation 3 dendrons had a diameter of about 7 nm, as determined by low temperature transmission electron microscopy (cryo-TEM) [52]. Compared to traditional surfactants that usually agglomerate from several tens to over a hundred molecules per micelle, only about ten "mushrooms" were enough to make up a micelle, and their critical micelle concentration (10^{-7}–10^{-6} M) was significantly lower than that typical for conventional anionic surfactants (10^{-1}–10^{-3} M). On contact with the organic phase, the supramolecular micelles opened up and broke into individual "mushrooms", which enabled an effective transfer of organic guests, such as perylene, from organic into aqueous environment [52] by a process opposite of the transfer of inorganic species, such as metal cations, from water into organic phases by PAMAMOS dendrimers shown in Fig. 11.5.

11.7.2 Dendrimers with Silicon in Their Interiors

There are two types of dendrimers with silicon in their interiors: those containing silicon atoms in the linear segments of their branch arms, namely extenders with 1→1 multiplicity, and those containing silicon as the branch junctures with either 1→2 or 1→3 multiplicity (see Fig.11.1 and also Chapter 1). An example of the first mentioned type was reported by Astruc and his co-workers of the CNRS, France, in their quest for the largest dendrimer ever prepared [13]. The authors utilized the reiterative sequence of (a) hydrosilylation of the polyallyls with dimethylchloromethylsilane catalyzed by Karstedt's catalyst in ether, and (b) nucleophilic substitution of the resulting chloride by the phenolate group of triallylphenol catalyzed by sodium iodide in dimethylformamide (DMF) to prepare nine generations of the dendrimer shown in structure **1**.

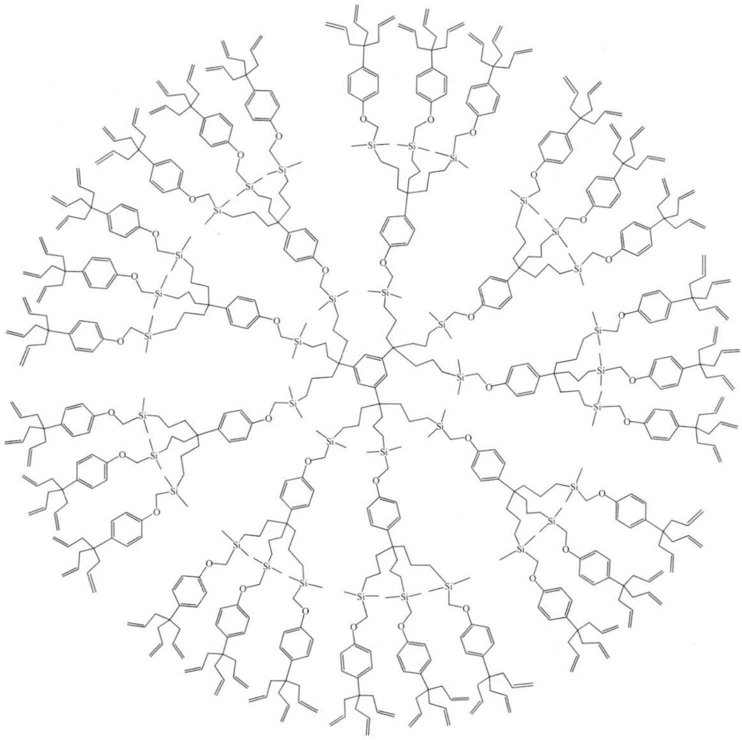

1

The strategy was based on a 1→3 (carbon) branch connectivity aimed at a faster increase of the number of end-groups per generation during the synthesis and the role of silicon was to enable extension of the flexible linear branches. It was reported that, although in a very low yield of only about 8% of purified structure, a generation 9 dendrimer with 177,147 allyl end-groups was obtained, which was far beyond the calculated de Gennes dense-packed stage for this composition (see Chapter 1). The authors suggested that this was possible because of the small size of the end-groups and high conformational flexibility of the branches that enabled pronounced bending-back of the termini, which spent only about 6% of their estimated time on the dendrimer periphery and 94% in its interior. As a consequence, they concluded that this dendrimer construction was limited by the density of the interior rather than by the bulk of peripheral termini [13].

Some of the best examples of dendritic polymers with silicon in the branch junctures are, of course, those found in the polysiloxane, polycarbosilane, polycarbosiloxane, polysilane and polysilazane dendrimers described in Chapters 2–5 and the polycarbosilane and polycarbosiloxane hyperbranched polymers of Chapters 12, 13 and 16. However, it is rather surprising that the ability of silicon to introduce different and easily controlled 1→2 and 1→3 branching functionality into dendritic structures has not been significantly utilized in otherwise "organic" derivatives as well. A notable exception to this was the work of Kemmitt and Henderson from the New Zealand Institute for Industrial Research and Development, and the University of Waikato in Hamilton reported in 1997 (see also Chapter 5) [14].

These authors described the synthesis of two types of silatrane-containing dendrons, representing the first examples of dendritic structures containing pentacoordinate silicon (**2** and **3**). Their synthetic strategy used (a) the capping of the glycidoxy group of trimethoxy(glycidoxypropyl)silane (TMGS) with an alkanolamine to form a higher alkanolamine with an additional hydroxyl group, followed by (b) reaction of the resulting alkanolamine with the trimethoxysilyl group of another TMGS molecule to form a silatrane [14]:

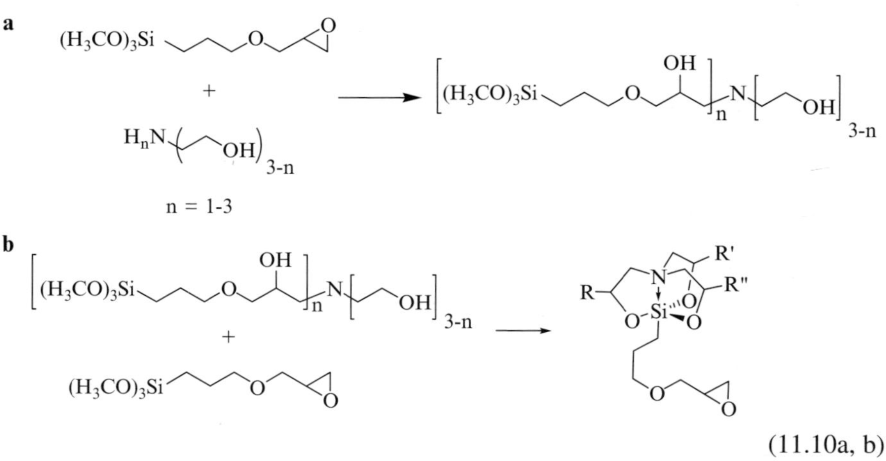

(11.10a, b)

This enabled sequential addition of reagents to build dendrons in a convenient and controlled quasi-one-pot way where the reaction stopped after each successive step because each functionality was alternately protected. It also allowed preparation of two different types of dendrons: one in which all branch junctures were of the same functionality, if the same alkanolamine was used throughout the synthesis (i.e., structure **2** with all 1→2 branching multiplicity throughout), and another (i.e., structure **3**) in which different generational tiers could have different branching multiplicity by selecting different alkanolamines (i.e., 1→1, 1→2 and 1→3 for mono-, di- or triethanolamine, respectively). The resulting polymers were all viscous oils or amorphous solids in which the large number of diastereomers prevented crystallization. A detailed characterization of these structures was performed by ^{1}H, ^{13}C and ^{29}Si NMR, and electrospray mass spectroscopy.

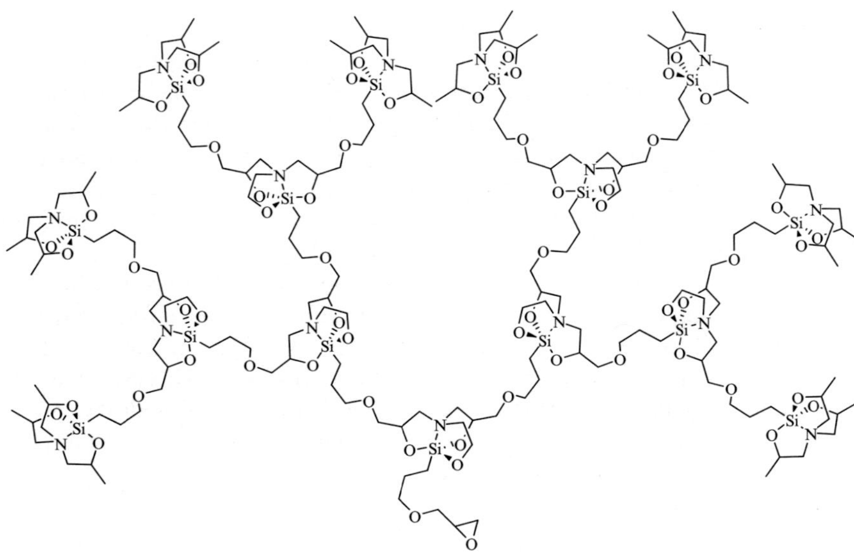

2

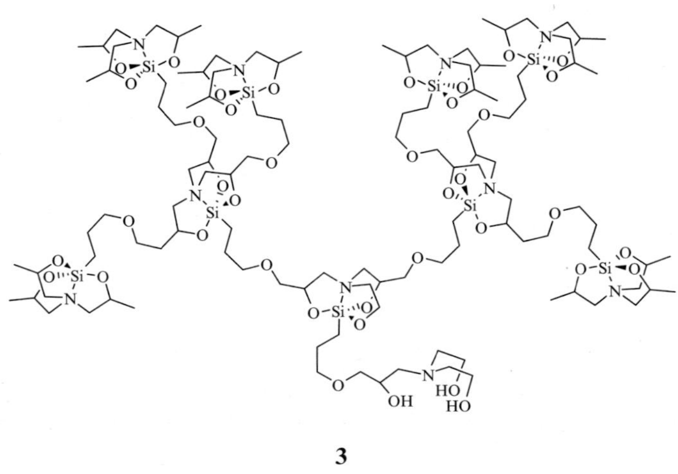

3

11.7.3 Dendrimers (Other than PAMAMOS) with Silicon in the End-Groups

A special type of PAMAMOS are dendrimers with PAMAM interiors and polyhedral oligomeric silsesquioxane (POSS) cages appended as their end-groups, **4** [53]. These dendrimers represent a unique case of organo-inorganic hybrids made from two of the main building blocks presently available for bottom-up soft nanotechnology (see Fig. 1.3) and their synthesis and properties are treated in more detail in Section 7.4.5.

4

In addition to PAMAMOS, another example of dendrimers with silicon-containing end groups was reported by Newkome and co-workers from the University of South Florida [15, 16]. In their quest for combinatorial construction of dendrimers, these authors developed a series of stable isocyanate tri-branched monomers of the general structure OCN–CR_3 where in one example R was the –$(CH_2)_3$–O–$Si(CH_3)_2C(CH_3)_3$ group. In a reaction analogous to Reaction Scheme 11.1c, these isocyanates were successfully reacted with a fourth generation amine-functionalized poly(propylene imine), –$[(CH_2)_3$–$N]_n$ <, PPI, dendrimer to yield a silicon-modified product having 96 end-groups containing hydrolytically sensitive C–O–Si units.

PPI dendrimers were also claimed for the preparation of PPIOS [2] which can be viewed as more thermally stable analogues of the PAMAMOS. These dendrimers can be used for the preparation of a variety of more complex nano-structured products [38], similar to those described for PAMAMOS in the preceding sections. For example, Paleos and co-workers from the Institute of Physical Chemistry, NCSR "Demokritos", Attiki, Greece, reported on the utilization of PPIOS-impregnated ceramic filters for water purification. They showed that just like the PAMAMOS (see Fig. 11.6), these dendrimers easily crosslinked and covalently bonded to M-OH functionalized ceramic surfaces, and could effectively attract and encapsulate toxic pollutants, such as polycyclic aromatics, to reduce their concentration in contaminated water to only a few ppb by continuous filtration [45]. In addition to this, the filters loaded with pollutants could be effectively regenerated by treatment with acetonitrile. Based on these results, and the fact that PPI nuclei should have affinity to electrophiles similar to that of PAMAMs, it is to be expected that either PAMAMOS or PPIOS should also offer high promise for water purification from heavy metals and other electrophilic contaminants.

Acknowledgements The authors gratefully acknowledge financial support for a part of the herein described work on PAMAMOS dendrimer networks by the National Science Foundation (grants no. DMI-0419193 and DMI-0522183).

References

1. de Leuze-Jallouli AM, Swanson DR, Perz SV, Owen MJ, Dvornic PR (1997) Polym Mater Sci Eng 77: 67.
2. Dvornic PR, de Leuze-Jallouli AM, Swanson DR, Owen MJ, Perz SV (1998) US Patent 5,739,218.
3. Dvornic PR, de Leuze-Jallouli AM, Owen MJ, Perz SV (2000) Macromolecules 33: 5366.
4. www.Dendritech.com
5. Dvornic PR, de Leuze-Jallouli AM, Owen MJ, Perz SV (2000) Radially layered poly(amidoamine-organosilicon) copolymeric dendrimers and their networks containing controlled hydrophilic and hydrophobic nanoscopic domains, in Silicones and Silicone-Modified Materials, Clarson SJ, Fitzgerald JJ, Owen MJ, Smith SD, Eds., ACS Symposium Series 729, American Chemical Society, Washington, DC, pp. 241–269.
6. Dvornic PR, de Leuze-Jallouli AM, Owen MJ, Perz SV (2000) US Patent 6,077,500.
7. Dvornic PR, Hu J, de Leuze-Jallouli AM, Owen MJ, Parham PL, Perz SV, Reeves SD (2002) US Patent 6,350,384 B1.
8. Dvornic PR, Hu J, Reeves SD, Owen MJ (2002) Silicon Chem 1: 177.
9. Hawker CJ, Fréchet JMJ (1992) J Am Chem Soc 114: 8405.
10. Hawker CJ, Fréchet JMJ (1995) Three-dimensional dendritic macromolecules: Design, synthesis, and properties, in New Methods of Polymer Synthesis, Ebdon JR, Eastmond GC, Eds., Blackie Academic & Professional, London, Volume 2, pp. 290–330.

11. Brewis M, Clarkson GJ, Goddard V, Helliwell M, Holder AM, McKeown NB (1998) Angew Chem Int Ed 37: 1092.
12. McKeown NB (1999) Adv Mater 11: 67.
13. Ruiz J, Lafuente G, Marcen S, Ornelas C, Lazare S, Cloutet E, Blais J-C, Astruc D (2003) J Am Chem Soc 125: 7250.
14. Kemmitt T, Henderson W (1997) J Chem Soc Perkin Trans 1: 729.
15. Newkome GR, Baker GR, Moorefield CN, He E, Epperson J, Weis CD (1997) Polym Mater Sci Eng 77: 65.
16. Newkome GR, Weis CD, Moorefield CN, Baker GR, Childs BJ, Epperson J (1998) Angew Chem Int Ed 37: 307.
17. Uppuluri S, Keinath SE, Tomalia DA, Dvornic PR (1998) Macromolecules 31: 4498.
18. Aldrich handbook of fine chemicals (2007–2008) p. 823.
19. Dvornic PR, Owen MJ (2002) Poly(amidoamine-organosilicon) (PAMAMOS) dendrimers and their derivatives of higher degree of structural complexity, in Synthesis and Properties of Silicones and Silicone-Modified Materials, Clarson SJ, Fitzgerald JJ, Owen MJ, Smith SD, Van Dyke ME, Eds., ACS Symposium Series 838, American Chemical Society, Washington, DC, pp. 236–259.
20. Sachdeva R, Dvornic PR (2008).
21. de Leuze-Jallouli AM, Swanson D, Dvornic PR, Perz SV, Owen MJ (1997) Polym Mater Sci Eng 77: 93.
22. Dvornic PR, de Leuze-Jallouli AM, Perz SV, Owen MJ (2000) Mol Cryst Liq Cryst 353: 223.
23. Sayed-Sweet Y, Hedstrand DM, Spindler R, Tomalia DA (1997) J Mater Chem 7: 1199.
24. Iyer J, Hammond PT (1999) Langmuir 15: 1299.
25. Zhang T, Dvornic PR, Kaganove SN (2007) Langmuir 23: 10589.
26. Dvornic PR, de Leuze-Jallouli AM, Owen MJ, Perz SV (1998) The first dendrimer-based networks containing hydrophobic organosilicon and hydrophilic amidoamine nanoscopic domains, in Silicones in Coatings II, Paint Research Association, London, Paper 5.
27. Dvornic PR, de Leuze-Jallouli AM, Owen MJ, Perz SV (1998) Polym Preprints 39(1): 473.
28. de Leuze-Jallouli AM, Dvornic PR, Perz SV, Owen MJ (1998) Polym Preprints 39(1): 475.
29. Dvornic PR, de Leuze-Jallouli AM, Owen MJ, Perz SV (1999) US Patent 5,902,863
30. Dvornic PR, Li J, de Leuze-Jallouli AM, Reeves SD, Owen MJ (2002) Macromolecules 35: 9323.
31. Dvornic PR, de Leuze-Jallouli AM, Owen MJ, Perz SV (1999) Polym Preprints 40(1): 408.
32. Dvornic PR, de Leuze-Jallouli AM, Owen MJ, Dalman DA, Parham P, Pickelman D, Perz SV (1999) Polym Mater Sci Eng 81: 187.
33. Bubeck RA, Bauer BJ, Dvornic PR, Owen MJ, Reeves SD, Parham PL, Hoffman LW (2001) Polym Mater Sci Eng 84: 866.
34. Balogh L, de Leuze-Jallouli AM, Dvornic PR, Owen MJ, Perz SV, Spindler R (1999) US Patent 5,938,934.
35. Dvornic PR (2006) J Polym Sci Part A: Polym Chem 44: 2755.
36. Vohs JK, Fahlman BD (2007) New J Chem 31: 1041.
37. Vidal F, Hémonic I, Teyssié D, Boileau S, Reeves SD, Dvornic PR (2001) Polym Preprints 42(1): 128.
38. Dvornic PR, Owen MJ, Keinath SE, Hu J, Hoffman LW, Parham PL (2001) Polym Preprints 42(1): 126.
39. Ruckenstein E, Yin W (2000) J Polym Sci Part A: Polym Chem 36: 1443.
40. Owen MJ, Dvornic PR, Bubeck RA (2002) Nanostructured coatings from radially layered poly(amidoamine-organosilicon) (PAMAMOS) dendrimers, in Silicones in Coatings IV, Paint Research Association, London, Paper 3.
41. Bubeck RA, Dvornic PR, Hu J, Hexemer A, Li X, Keinath SE, Fischer DA (2005) Macromol Chem Phys 206: 1146.
42. Dvornic PR, Bubeck RA, Reeves SD, Li J, Hoffman LW (2003) Silicon Chem 2: 207.
43. Dvornic PR, Hartmann-Thompson C, Kaganove SN, Rousseau J, Sarkar A, Zhang T (2006) Polym Preprints 47(2): 1166.

44. Dvornic PR, Sarkar A, Rousseau J, Hartmann-Thompson C, Merrington A, Carver P, Zhang T, Keinath S (2008) Dendritic polymer networks: A new class of nano-structured antifouling coatings, 235 ACS National meeting, New Orleans, LA.
45. Arkas M, Tsiourvas D, Paleos CM (2005) Chem Mater 17: 3439.
46. Kaganove SN, Zhang T, Dvornic PR (2005) Presentations at IDS-4, Mount Pleasant, MI; ACS-Midland Section Fall Scientific Meeting, Midland, MI; Pacifichem 2005, Honolulu, HI.
47. Dalman DA, Dvornic PR, Hoover MF, Lentz T (2002) PAMAMOS copper nanocomposite coatings for the fabrication of printed wiring boards, in Microelectronics Gold Rush, Proceedings 35th International Symposium on Microelectronics, IMPAS, pp. 33–38.
48. Dalman DA, Dvornic PR (2005) US Patent 6,866,764 B2.
49. Kohli N, Dvornic PR, Kaganove SN, Worden RM, Lee I (2004) Macromol Rapid Commun 25: 935.
50. Kohli N, Worden RM, Lee I (2005) Chem Commun 3: 316.
51. Dvornic PR, Owen MJ (2009) Radially layered poly(amidoamine-organosilicon) dendrimers, submitted to Advances in Silicones and Silicone-Modified Materials, Clarson SJ, Owen MJ, Smith SD, Van Dyke ME, Eds., ACS Symposium Series, American Chemical Society, Washington, DC.
52. Uchiyama T, Ishii K, Nonomura T, Kobayashi N, Isoda S (2003) Chem Eur J 9: 5757.
53. Dvornic PR, Hartmann-Thompson C, Keinath SE, Hill EJ (2004) Macromolecules 37: 7818.
54. Sarkar A, Dvornic PR, (2007) Unpublished results, Michigan Molecular Institute.

Chapter 12
Hyperbranched Polycarbosilanes via Nucleophilic Substitution Reactions

Leonard Interrante and Qionghua Shen

12.1 Introduction

Nucleophilic substitution reactions involving organomagnesium (Grignard) [1] and organolithium reagents have been used extensively for many years to form Si–C bonds (see Reaction Scheme 12.1). However, their use for the construction of hyperbranched polymers whose backbone contains, as a major structural component, silicon–carbon bonds, i.e., polycarbosilanes [2] is relatively more recent.

$$R_3SiX + MR' \rightarrow R_3SiR' + MX$$
$$(R, R' = \text{alkyl or aryl; } M = Mg(X), Li, Na; X = \text{halogen, OR''}) \qquad (12.1)$$

This chapter focuses on the application of such nucleophilic substitution reactions toward the synthesis of hyperbranched polycarbosilanes, with particular emphasis on those preparations that have resulted in relatively well characterized products. These syntheses are organized by the type of AB_n monomer unit used (see Section 1.2), where A and B refer to the (C)X and $(Si)X_n$, respectively, functional ends of the monomer unit and where the nature of the coupling reaction leads to entirely or primarily Si–C bond formation. In most cases, these are "*one-pot*" reactions that employ monomers that bear halogen or alkoxy groups on the C and Si ends of the unit. Indeed, hyperbranched polycarbosilanes have been described, in general, as "*obtained in one synthetic step via a random, one-pot polymerization of multifunctional monomers of AB_n type*" [2]. Treatment of the AB_n monomer with either elemental Mg or an organolithium reagent, ideally (but not always) forms a complexed carbanion (the nucleophile) by reaction with the C–X end of the monomer unit, resulting in an intermediate of the type, $(X_xM)CSiX_n$, where M = Mg or Li, X = halogen or alkoxy,

L. Interrante
Department of Chemistry and Chemical Biology, Rensselaer Polytechnic Institute, Troy, NY 12180, USA
E-mail: interl@rpi.edu

Q. Shen
Starfire Systems, Inc., 10 Hermes Rd., Malta, NY 12020, USA
E-mail: ShenQ@starfiresystems.com

and x = 1 (Mg) or 0 (Li). Self-coupling of this reagent via reactions of the type shown in Reaction Scheme 12.1 leads to oligomeric and polymeric products that are connected primarily through Si–C bonds and yield an inorganic MX_x by-product.

Due to the greater reactivity and lower selectivity, in general, of organolithium reagents towards C–X and Si–X bonds, C–C and Si–Si bond formation (Wurtz coupling) is more likely to compete effectively with Si–C bond formation when M = Li. This often leads to a mixture of these three types of bonds in the corresponding products and, consequently, cross-linking occurs to an appreciable extent, forming a substantial proportion of insoluble products. Moreover, formation of the desired $(M)CSiX_n$ reagent in the case of Li is much more problematic than with Mg, where elemental Mg can discriminate effectively between the Si–X and C–X bonds in a $XCSiX_n$-type monomer, forming largely or entirely the desired $(XM)CSiX_n$ intermediate, thereby avoiding appreciable Si–Si coupling and subsequent cross-linking.

Thus, the most frequently used nucleophilic substitution reaction for building up the backbone structure of hyperbranched polycarbosilanes has been the Grignard reaction. Here, chain growth occurs via a stepwise coupling of the Grignard reagent with a SiCl or SiOR group and mostly gives rise to head-to-tail connections, i.e., to Si–C bonding. The Wurtz coupling reaction, which employs alkali metals, can also be used, but as noted above, it generally leads to appreciable Si–Si and C–C bonding, in addition to Si–C bonding, and extensive cross-linking. The electrosynthesis of hyperbranched polycarbosilanes has been reported recently [30]. It is essentially a kind of nucleophilic substitution reaction where an organometallic reagent is formed via electrolysis of a C–X group. In all of these cases the resultant hyperbranched polycarbosilane contains uncoupled Si–X groups and can be further functionalized with groups such as allyl, alkoxy, H, phenyl, vinyl, etc., via reactions with the corresponding nucleophilic reagents. Thus, nucleophilic reactions can be used both for the formation and for the subsequent modification of hyperbranched polycarbosilanes. Other synthetic approaches, such as hydrosilylation, can also be used to prepare and to modify hyperbranched polycarbosilanes. However, these methods are described in Chapters 3, 7 and 13, and will not be discussed further here.

As polymers that contain both Si and C in their backbone structure, polycarbosilanes have been of particular interest as precursors to silicon carbide, an important high temperature structural ceramic and semiconductor material. Early work by Yajima and coworkers in the 1970s led to a polycarbosilane with a nominal "$SiH(CH_3)CH_2$"composition, which subsequently found commercial application as a source of "SiC" fibers. This "polymer", which was prepared from polydimethylsilane, $[Si(CH_3)_2]_n$, by thermal treatment in an autoclave, was subsequently shown to have a complex structure with extensive branching, along with cyclization and intermolecular Si–Si bonding [3]. Owing to its largely insoluble character and lack of functionality appropriate for cross-linking after fiber spinning, processing of "SiC" fibers typically involved melt spinning followed by thermal treatment in air to "fix" the fiber structure. The resultant fibers have a "SiC_xO_y" composition with excess carbon and substantial oxygen content, making them less desirable than "stoichiometric SiC" for many high temperature applications [4]. Later work in the 1990s, which was driven, in part, by the search for a stoichiometric SiC precursor, yielded a better-defined, soluble, liquid, hyperbranched polycarbosilane having a "SiH_2CH_2" composition. This polymer, "HPCS" (for hydridopolycarbosilane) or "HBPCS" (for hyperbranched polycarbosilane) and its partially allyl-substituted derivative, "AHPCS", sold under the trade name SMP-10, have also found commercial interest as SiC precursors [5, 6]. A wide range of applications has been found for SMP-10, particularly as a SiC matrix source for making ceramic composites. Further details regarding the synthesis, structures and properties of this initial example of a hyperbranched polycarbosilane obtained by nucleophilic substitution reactions, are given in the following section.

However, hyperbranched polycarbosilanes with specific functional groups have applications beyond their use as ceramic precursors. For example, hyperbranched polycarbosilanes bearing pendant oligoethyleneoxo groups [7, 8], dissolve various LiX salts and exhibit appreciable Li^+ ion conductivity. Their potential use as electrolytes in Li batteries has been suggested. Also, a thin film made by sol-gel processing of an alkoxy-substituted hyperbranched polycarbosilane was found to have a dielectric constant (κ) of 2.5 to 2.8 and excellent mechanical properties [9]. This film has been patented for use as an interlayer dielectric (ILD) in semiconductor processing and its hyperbranched alkoxycarbosilane precursor is now also commercially available [6]. Related applications as low κ dielectric materials from branched polycarbosilanes have been reported in several patents [10–12]. A chemical sensor for explosives made from hyperbranched polycarbosilanes with hydroxyl groups has also been claimed [13].

12.2 Hyperbranched Polycarbosilanes from AB_3 Monomers

Several different AB_3-type monomers have been reported to form hyperbranched polycarbosilanes via nucleophilic substitution reactions. In general, compounds meeting the requirement of the AB_3-type monomer include: $XRSiCl_3$ (X = Br or Cl) and $XRSi(OR')_3$ (R = $-CH_2-$, $-CH_2CH_2-$, $-C_6H_4-$; R′ = $-CH_3$ or $-C_2H_5$ etc.). Specific examples are: chloromethyltrichlorosilane, bromoethyltrichlorosilane, chloropropyltrichlorosilane, 2-bromo-5-trimethoxysilylthiophene, chloromethyltrimethoxysilane, *p*-chloromethylphenylenetrichlorosilane, 3- or 4-bromophenyltriethoxysilane, etc. Depending on the reaction conditions and the monomer structure, some of these monomers can form cyclic molecules, in addition to hyperbranched polymers.

12.2.1 From Chloromethyltrichlorosilane

The first hyperbranched polycarbosilane derived from the Grignard reaction of chloromethyltrichlorosilane $ClCH_2SiCl_3$ in ether was reported in 1991 [5]. The purpose of this study was to obtain a pre-ceramic polycarbosilane with an average $[SiH_2CH_2]_n$ formula, as shown in Reaction Scheme 12.2. The Grignard reaction initially generated a Cl-substituted polymer $[SiCl_2CH_2]_n$, which was reduced subsequently by $LiAlH_4$ to form its hydrido derivative. Thus, in this case, both reactions can be considered as nucleophilic substitution processes.

$$ClCH_2SiCl_3 \xrightarrow{\text{Mg/ether}} [SiCl_2CH_2]_n \xrightarrow{LiAlH_4} [SiH_2CH_2]_n \qquad (12.2)$$

These two reactions were carried out in a one-pot process without the need to separate the polymers from the solvent and by-products. As was deduced by multinuclear NMR studies carried out after quenching the reaction at various stages with $LiAlH_4$ and methanol, as well as on the final reaction product, the real structure of both the chloro and hydrido polymers are much more complicated than their average formula [5]. The evolution of their branched structures can be considered to occur via the sequence of nucleophilic substitution reactions shown in Scheme 12.3:

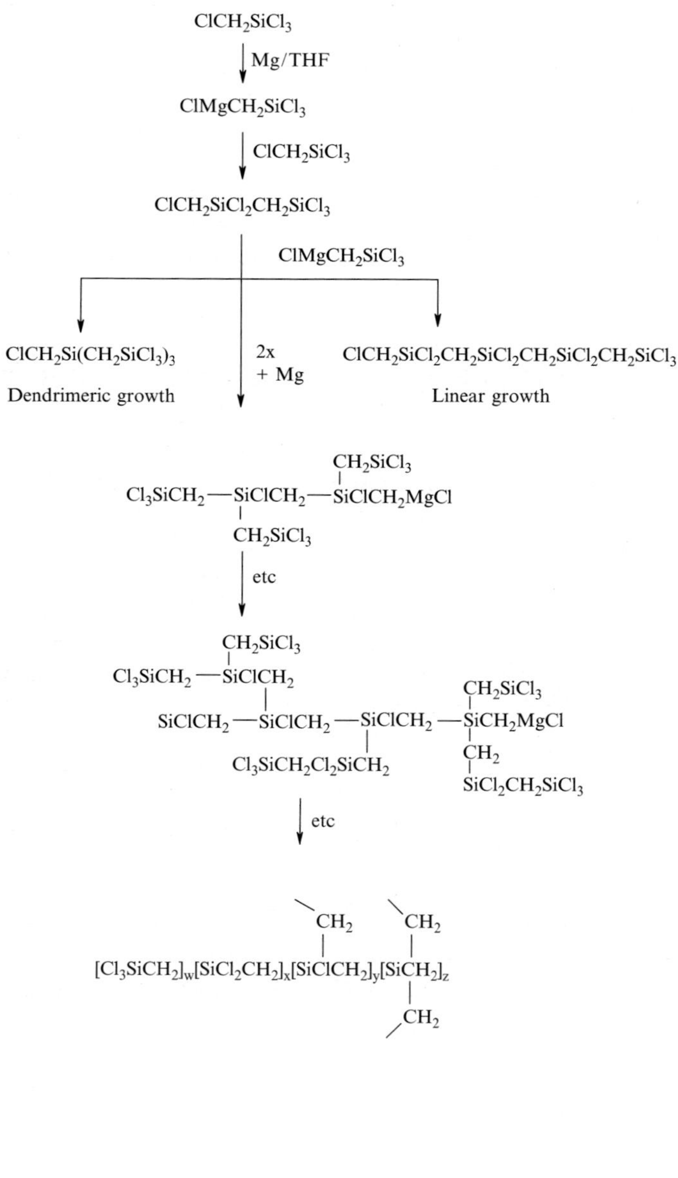

(12.3)

The basic chain-growth mechanism from $ClCH_2SiCl_3$ involves the initial formation of the Grignard intermediate $ClMgCH_2SiCl_3$. This reactive intermediate has a trifunctional $–SiCl_3$ "tail", and a monofunctional $–CH_2MgCl$ "head". It will replace one of the Si–Cl terminal groups in the starting $ClCH_2SiCl_3$ monomer and form a dimer, $ClCH_2SiCl_2CH_2SiCl_3$. The dimer now has five Si–Cl groups, thus its further coupling with the Grignard intermediate $ClMgCH_2SiCl_3$ has more options than the original monomer. If the subsequent substitution reactions by the Grignard intermediate occur sequentially on the remaining two terminal Cl groups of the initial monomer, the resultant products will be a trimer $ClCH_2SiCl(CH_2SiCl_3)_2$ and tetramer $ClCH_2Si(CH_2SiCl_3)_3$, as shown in the left side of the Reaction Scheme 12.3. The tetramer is an ideal generation-one dendrimer with a $ClCH_2–$ group in its "core". If the following reactions continuously replace the nine Cl groups from the tetramer, a dendrimer with a well defined second layer would be obtained. As can be expected, however, the reactions do not continue in such an organized manner, in part, because there is no reason why the $ClCH_2–$ group in the "core" will not also form a Grignard reagent, which will then couple with a Si–Cl group from a monomer, dimer, or trimer, etc.

Moreover, there are other, more likely, pathways for the reaction of the Grignard intermediate with the initially formed dimer in this Scheme. For example, if the propagation reaction always adds monomer at the newly formed $–SiCl_3$ tail, it will end up with an entirely linear $–[SiCl_2CH_2]_n–$ structure, as shown in the right side of Reaction Scheme 12.3. Such an exclusive linear growth was also not observed, because there is no strong driving force for the Grignard reagent to couple with a Si–Cl group of a $–SiCl_3$ unit. If instead the choice of available Si–Cl reaction sites is more or less equivalent, chain growth will occur at both $=SiCl_2$ and $–SiCl_3$ locations in the dimer, leading to a randomly branched structure. These randomized reactions of the initial Grignard intermediate with $Si–Cl_n$ groups on the growing chain, along with chain growth at the $–CH_2Cl$ end of the intermediates through reaction with excess Mg, are illustrated in the middle of the Reaction Scheme 12.3. As the chain grows, its sub-structure can form dendrimeric or linear units. The NMR studies provided support for this more general reaction pathway, where all possible structural units, such as $–CH_2SiCl_3$, $–CH_2SiCl_2–$, $(–CH_2)_2SiCl–$, and $(–CH_2)_3Si–$ were found in the final mixture of products. A generalized structure which reflects the branched units can be represented as $[Cl_3SiCH_2]_w[SiCl_2CH_2]_x[SiCl(CH_2–)CH_2]_y[Si(CH_2–)_2CH_2]_z$, although the average compositional formula of the polymer can be represented as $[SiCl_2CH_2]_n$.

Reduction by $LiAlH_4$ converts this chloro-polymer to the corresponding silane, $[H_3SiCH_2]_w[SiH_2CH_2]_x[SiH(CH_2–)CH_2]_y[Si(CH_2–)_2CH_2]_z$ (see Reaction Scheme 12.4). A quantitative analysis of the ratio of the units: $–CH_2SiH_3$, $–CH_2SiH_2–$, $(–CH_2)_2SiH$, and $(–CH_2)_3Si–$ in the reduced polymer, as obtained from ^{29}Si NMR spectroscopy, gave 11/20/8/2 [8]. The lower content of branched units than that which would be predicted by statistics for random coupling to the available $Si–Cl_n$ units is understandable based on steric considerations. The reduction by $LiAlH_4$ should not change the degree of branching; it only replaces Cl groups by H.

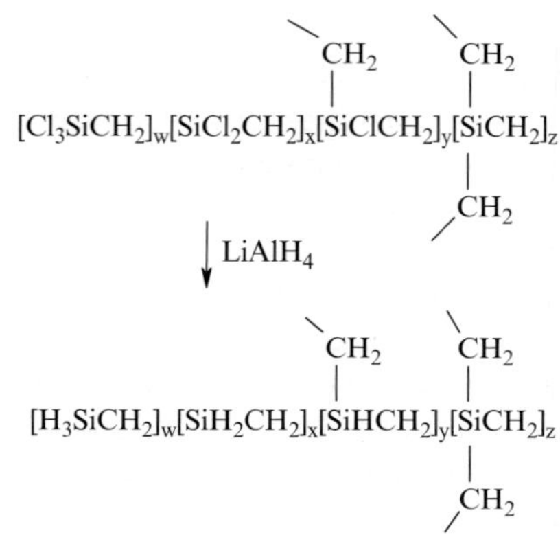

(12.4)

Although the reaction pathways shown in Reaction Scheme 12.3 can account for the observed NMR results, they are clearly not the only pathways possible for this complex polymerization process, even if the assumptions of exclusive Si–C coupling and no side reactions are completely valid. As is discussed below, at least one possible side reaction was evidenced in this early study, which, if not inhibited, does lead to other, undesired, products. However, even if we ignore this (and other) possible side reactions, intramolecular reactions involving dimer, trimer and higher linear oligomers shown in Reaction Scheme 12.5, are also possible.

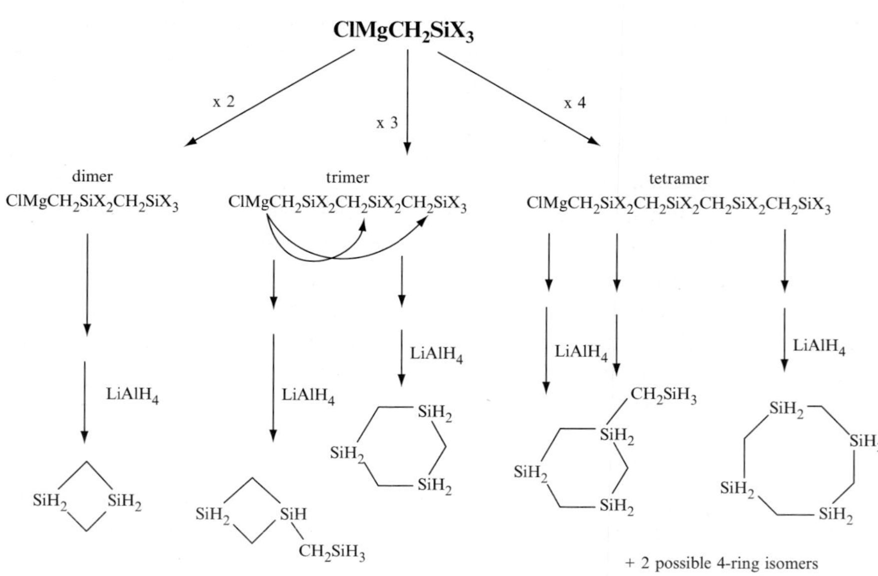

(12.5)

These reactions involve the formation of cyclic products, by attack of the reactive organomagnesium "tail" on a Si–Cl group in the same molecule. A later study, reported thus far only in the form of a Ph.D. thesis [14], provides support for the existence of this alternative series of cyclization reactions, depending on the reaction conditions used (including, among others, the concentration of the chlorosilane in the initial reaction mixture). In fact, by using alternative, less reactive leaving groups in these Grignard coupling reactions in place of Si–Cl, such as alkoxy, the formation of low molecular weight cyclic products can be favored relative to the polymer formation. This has been used, along with variations in the concentration of the silane and the order of addition of Mg and silane, to prepare specific members of this series of cyclic carbosilanes (such as 1,3-disilacyclobutane [15] and 1,3,5-trisilacyclohexane) [16], as well as a mixture of relatively low molecular weight oligomeric carbosilanes for use in the chemical vapor deposition (CVD) of silicon carbide films [6]. Although the formation of such low molecular weight cyclic and other oligomeric products can be suppressed to a considerable extent when the goal is to prepare a polymeric precursor for silicon carbide ceramics, cyclization reactions of this type can, and probably do, occur to a significant extent even in this case, leading to a further level of complexity, and potential variability, in the structure of these polymers.

In the initial study of the $ClCH_2SiCl_3$/Mg system by Whitmarsh and Interrante, the occurrence of a side reaction involving cleavage of the diethyl ether solvent during the Grignard reaction of $ClCH_2SiCl_3$ was evidenced, leading to the incorporation of a small amount of ethyl and ethoxy groups into the polymer structure [5]. In the final reduction steps, all chloro and ethoxy groups are eliminated, yielding a liquid product with an approximate compositional formula $[SiEtH_{1.85}CH_2]_n$. Gel permeation chromatography (GPC) of this hydridopolycarbosilane (HPCS) indicated a wide molecular weight distribution from 300 to about 50,000 amu (M_n = 750, M_w = 5,200, relative to polystyrene standards), with the majority of the polymer falling between 300 and 3,000 amu.

An improved preparation of HPCS was subsequently developed by Starfire Systems, Inc. [6], which yields a hyperbranched polycarbosilane having an overall composition $[SiH_2CH_2]_n$ and a M_n of 790 (determined by Vapor Pressure Osmometry, VPO), corresponding to an average degree of polymerization of ca. 13 and a polydispersity of 1.5–1.8 [8]. On pyrolysis to 1,000°C, this liquid polycarbosilane was found to yield an amorphous, stoichiometric, SiC. However, due to its relatively high cross-linking temperature (ca. 300°C), loss of volatile oligomeric components occurs on unconfined pyrolysis in a flowing nitrogen atmosphere, leading to a ceramic yield of ca. 55%. A significant improvement in the ceramic yield was observed when olefinic side chain groups, such as vinyl or allyl, were added to the backbone of this hyperbranched polymer by a Grignard reaction prior to reduction with $LiAlH_4$ (see Reaction Scheme 12.6). This enhancement of the ceramic yield, and lowering of the effective cross-linking temperature (to ca. 200°C), was attributed to the contribution of a thermally-induced, intramolecular, hydrosilylation reaction involving Si–H and Si–allyl (or –vinyl) groups. A detailed study of the linear polymer relative of HPCS, $[SiH_2CH_2]_n$, polysilaethylene and its deuterated derivative, $[SiD_2CH_2]_n$, indicated that, in the absence of such olefinic groups, the main cross-linking mechanism involves the 1,1-elimination of molecular hydrogen from SiH_n (n > 2) groups, a reaction whose rate becomes significant only above ca. 300°C [17]. The resultant polymers, known as VHPCS or AHPCS, respectively, have an average formula $[SiRHCH_2]_m[SiH_2CH_2]_{n\text{-}m}$ (m = up to ca. 0.9 n; R = vinyl or allyl) but contain units such as $–CH_2(R)SiH_2$, and $CH_2(R)SiH–$, etc. in addition to the original chain units: $–CH_2SiH_3$, $–CH_2SiH_2–$, $(–CH_2)_2SiH–$, and $(–CH_2)_3Si–$ [18].

$$ClCH_2SiCl_3 \xrightarrow{Mg/ether} [SiCl_2CH_2]_n \xrightarrow{RMgX} [SiClRCH_2]_m[SiCl_2CH_2]_{n\text{-}m}$$

R=allyl or vinyl

$$\downarrow LiAlH_4$$

$$[SiHRCH_2]_m[SiH_2CH_2]_{n\text{-}m} \quad (12.6)$$

About 2.5–10% substitution of such groups has been found sufficient to promote thermal cross-linking without incorporation of substantial excess carbon on pyrolysis, yielding products which on heating at 300–400°C convert to a glassy solid and then at 1,000°C to a ceramic in an overall yield of 70–80%. Due to this high ceramic yield, its apparent (kinetic) stability for extended periods at room temperature in air, and its fluidity, AHPCS has found applications as a matrix source for SiC-fiber-reinforced/SiC matrix composites as well as for joining SiC ceramics [19]. A repair kit, based on AHPCS, has been developed to repair damage to the U.S. space shuttle's heat shield in orbit [20].

In general, the reactivity of the intermediate chloro-polymers obtained after the Grignard coupling of $ClCH_2SiCl_3$ by nucleophilic substitution provides a convenient means of preparation of a wide range of derivatives having a hyperbranched polycarbosilane structure. In particular, substitution of the Cl by alkoxy was reported to generate an interesting silicon oxycarbide precursor [21]. The substitution by ethoxy groups was carried out by heating the chloropolymer with an excess of ethanol which yielded a polymer with an average formula $[Si(OEt)_2CH_2]_n$, and average structure illustrated in Reaction Scheme 12.7. Acid-catalyzed sol-gel processing of the alkoxy substituted polymers in ethanol was found to yield, after drying and heating to ca. 600°C, a solid with an average formula $[Si(O)CH_2]_n$ having an inorganic/organic network structure with both $Si–CH_2–Si$ and Si–O–Si bridging groups. This polymer was further converted into silicon oxycarbide glass in high yield upon heating under nitrogen to 1,000°C. Both the $[Si(O)CH_2]_n$ "gel" and the silicon oxycarbide ceramic obtained at 1,000°C were found to possess a high degree of microporosity, with a BET surface area of 700–900 m^2/g for the $[Si(O)CH_2]_n$ "gel" and 250 m^2/g for the SiO_xC_y ceramic, and a peak in the pore size distribution at ca. 20 Å in both cases.

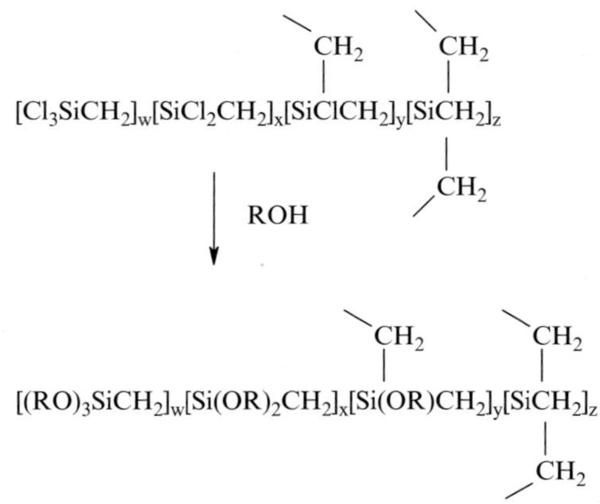

(12.7)

Using this hyperbranched alkoxy-substituted polycarbosilane, a sol-gel based spin-coating process was developed by Lu to prepare dielectric films on Si wafers [9, 14]. A dielectric constant (κ) of ca. 2.5 was reported for these films, which could be lowered to $\kappa < 2.0$ by using a tri-block copolymer as a porogen. Recent studies indicate improved mechanical properties (substantially higher modulus) for these films, both in the porous and in the dense forms, as compared to those obtained by using methylsilsesquioxane (MSQ), the most common alternative organosilicate material currently used in electronic processing [11]. A commercial product based on the methoxy-substituted polycarbosilane has been developed for use as insulating films for semiconductor devices. This hyperbranched polymer, which is sold under the trade name of DMPCS, has an average formula $[Si(OMe)_2CH_2]_n$ [6].

The reactivity of Si–Cl groups can also be used to attach other useful functional groups to the hyperbranched PCS skeleton by nucleophilic substitution. For example, the three hyperbranched polymers shown in Reaction Scheme 12.8 were prepared from the chloropolycarbosilane by using the corresponding Grignard reagents. Unlike the partial substitution by vinyl or allyl in Reaction Scheme 12.6, here the substitution was controlled so as to achieve a maximum conversion. These substituted branched polymers were reported to further react with hexafluoroacetone to form $-CF_3$ and hydroxyl functional groups, which can be used to produce hydrogen bond acidic coatings for chemical sensor applications; in particular for detecting the presence of explosives [13]. It was found that the branched structure from these polymers not only improves the sensitivity to organophosphorous species, but also shows high selectivity and sensitivity toward nitrogen-substituted chemicals, such as those present in certain explosives.

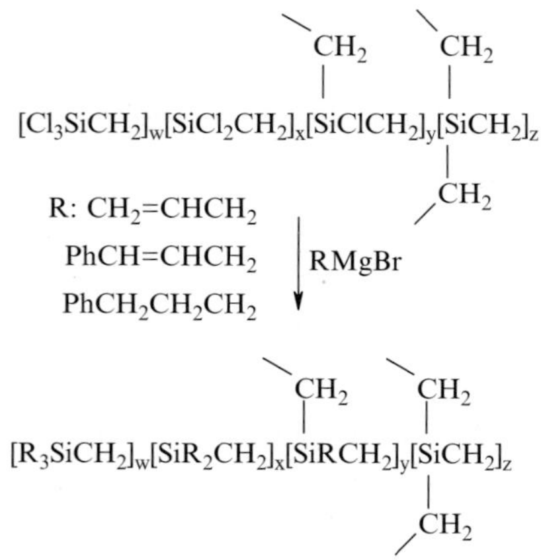

(12.8)

In addition to the above described polymers which have commercial interest, other novel polymers were also prepared using a branched chloro (or bromo) polymer as the starting material as shown in Reaction Scheme 12.9:

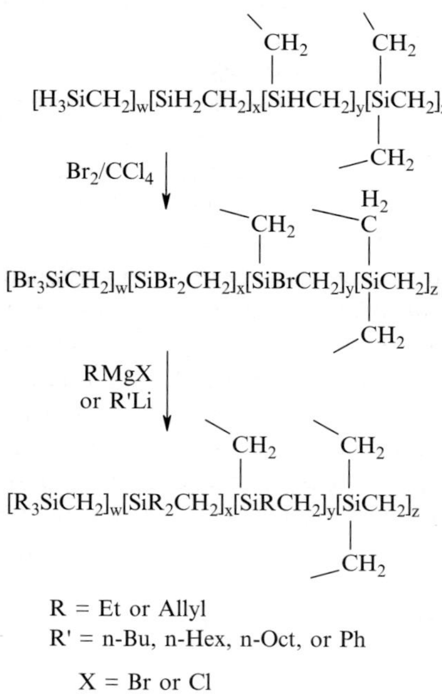

(12.9)

For example, the bromo-substituted polycarbosilane obtained from the branched hydridopolymer $[SiH_2CH_2]_n$ by reaction with Br_2, was reported to react with organolithium and Grignard reagents to form various substituted polymers, including a polyethyleneoxo (PEO)-derivative that was prepared by using the following Reaction Scheme 12.10 [8].

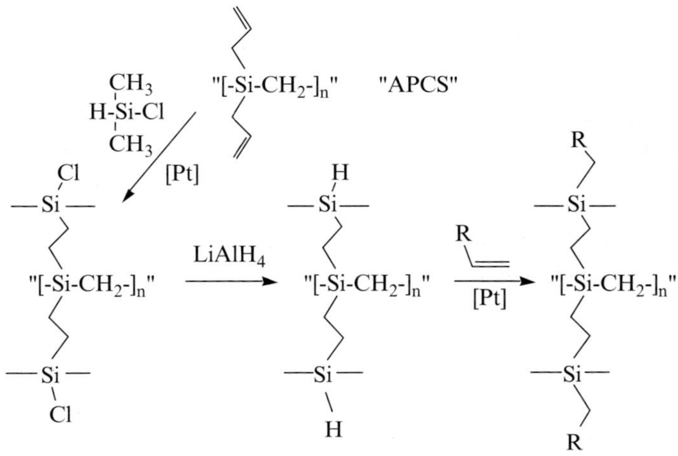

Where R = $CH_2O(CH_2CH_2O)_mCH_3$; m = 2, 7, 12, *ca.* 15 (12.10)

The substitution of Si–H by Br was quite efficient (>97%, based on the residual SiH peak intensity in the ^{1}H NMR spectrum), and the overall degree of substitution, based on the original Si–H content, was 75–80% for the two Grignard reagents (R = allyl or Et) and 85–90% for most of the lithium reagents (R = Bu, Hex, Oct, or Ph). Interestingly, the use of HBPCS for the hydrosilylation of terminal olefins, which would have represented a more direct route to the HBPCS-PEO derivatives, was precluded by inability to effect hydrosilylation using this hydridopolymer, presumably because of the known low reactivity of the predominant SiH_2 and SiH_3 groups towards hydrosilylation of olefins.

Subsequent efforts have led to a series of polyethyleneoxo-HBPCS derivatives of this type having different PEO chain lengths (see Reaction Scheme 12.10). The "$[–Si–CH_2–]_n$" formulas used in this scheme to designate the polycarbosilane main chain represent the average composition of the highly branched backbone structure of these polymers, as is discussed in Section 12.2.1). These viscous liquid polymers were found to dissolve various LiX (X = PF_6, etc.) salts and to exhibit relatively high Li^+ ion conductivities (comparable to existing poly(phosphazine)-PEO (MEEP) side chain polymers) while resisting attack by strong bases [7].

12.2.2 From 2-Bromo-5-trimethoxysilylthiophene

The synthesis of hyperbranched poly(2,5-disilylthiophenes) was pursued in order to investigate the possibility of σ-π conjugation in three dimensions [22]. As shown in Reaction Scheme 12.11, this polymer was prepared from 2-bromo-5-(trimethoxysilyl) thiophene via a Grignard reaction. Since 2-bromo-5-(trimethoxysilyl)thiophene is also an AB_3 type monomer, its chain growth and branching behavior are very similar to those of the polymers made from $ClCH_2SiCl_3$. It has a C–Br bond to form the Grignard center, and a $(MeO)_3Si–$ tail to be coupled with the Grignard reagent. The resulting methoxy-substituted polymer contains four types of branched units: $(OMe)_3$Si(thiophenyl),$(OMe)_2$Si$(thiophenyl)_2$,(OMe)Si$(thiophenyl)_3$andSi$(thiophenyl)_4$ (see Reaction Scheme 12.11).

The methoxy substituted polymer was converted into the methyl-, vinyl-, thiophenyl-, and phenyl-substituted polymers by reaction with the corresponding RMgBr reagents, and the methyl substituted polymer was used to determine the ratio of the different branched units by ^{1}H NMR spectroscopy. It was found that the ratio w/x/y/z for the above four branched units was 27:36:25:1, indicating that the low content of the Si$(thiophenyl)_4$ branched units was due to the high steric hindrance for substitution of the last methoxy group on the silicon atom. The molecular weight (M_n) of these polymers was in the range of 4,460–6,600 by GPC. VPO gave the corresponding M_n in the range of 4,200–10,500. The methyl substituted polymer had melting point of 43°C, while the vinyl, phenyl and thiophenyl polymers had glass transition temperatures at 14°C, 120°C and 135°C, respectively.

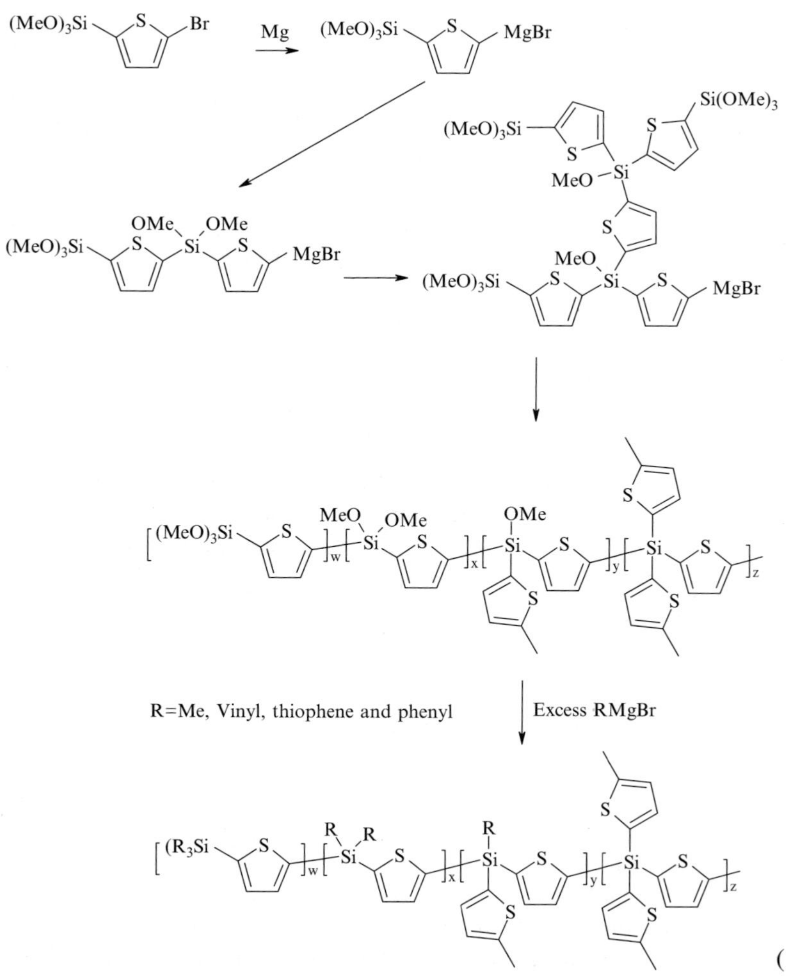

(12.11)

The extent of σ–π conjugation was examined for the methyl, vinyl, phenyl and thiophenyl substituted polymers by UV–visible spectroscopy, and a significant red shift was observed for the π–π^* transition, indicating a significant degree of conjugation through the silicon atoms. The UV absorptions appeared to be relatively independent of the polymer substituents. The effects of the three-dimensional polymer structure versus the more typical one-dimensional structure were not entirely clear, but at least for the π–π^* absorptions they appear to be relatively minor.

12.2.3 From 2-Bromoethyltrichlorosilane

Together with chloromethyltrichlorosilane, 2-bromoethyltrichlorosilane was also briefly mentioned as a source of a branched polycarbosilane in the patent by Houser and McGill [13]. Based on the experimental example described in this patent, the reaction can be summarized as shown in Reaction Scheme 12.12. The bromoethyltrichlorosilane initially formed a Cl-substituted branched polymer by reaction with Mg. This chloropolymer was then further reacted with chloromethyltriallylsilane via a second Grignard reaction. The allyl groups in the final branched polymer were used to form $-CF_3$ and hydroxyl-functionalized product with hexafluoroacetone for chemical sensor applications.

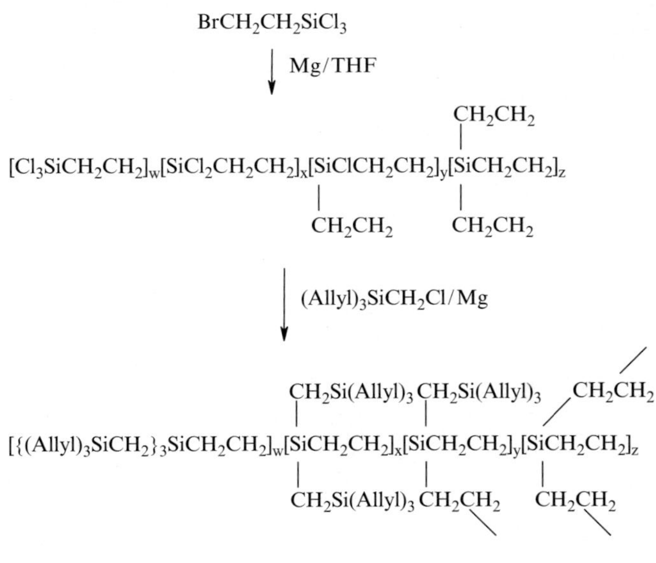

(12.12)

12.2.4 From 3- or 4-Bromophenyltriethoxysilane

Ohshita et al. [23] studied the synthesis and thermal properties of various substituted poly(silylenephenylenes) prepared by Grignard reactions of 3- or

4-bromophenyltriethoxysilane monomers (see Reaction Scheme 12.13). The Grignard reactions were conducted at low temperature, 0°C, and the polymers were separated by adding hexane to the resulting reaction mixture and re-precipitating twice from chloroform-ethanol solution. It is interesting to note that, as indicated in Reaction Scheme 12.13, both polymers were found to form a linear structure, although the starting monomers, 3- or 4-bromophenyltriethoxysilane are AB_3-type monomers.

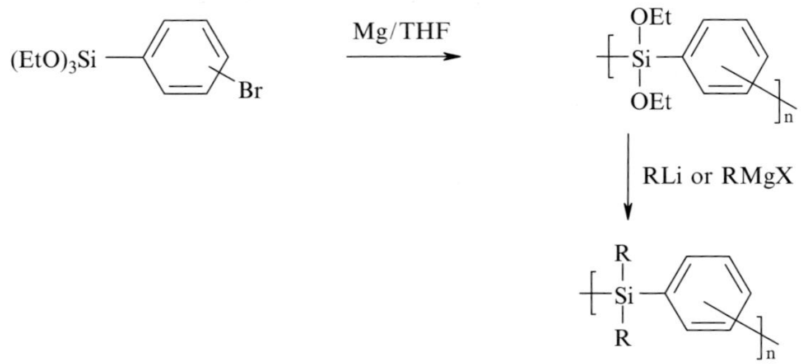

(12.13)

The apparent absence of branching in the above polymers may be due to the use of a very low reaction temperature or to the loss of branched fraction during reprecipitation of the crude products. Alternatively, the reduced reactivity of the Si–OEt leaving group (relative to Si–X (X = Cl or Br)) toward the Grignard reagent, along with the relatively high steric inhibition associated with the alternative branching reactions (see Reaction Scheme 12.14), could have been responsible for the apparent exclusive formation of the linear polymer in this case.

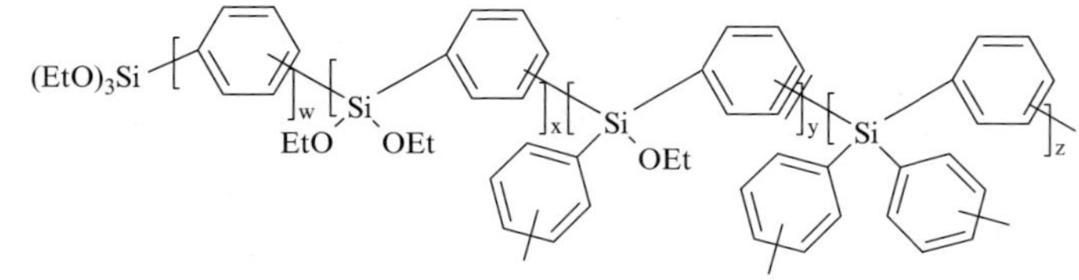

(12.14)

The ethoxy groups on these polymers could be subsequently converted into Cl, F, H and aromatic substituents.

12.2.5 *From Other Monomers*

As shown in Section 12.2.4, not all AB_3 type monomers generate hyperbranched polycarbosilanes exclusively. For example, chloropropyltrichlorosilane,

$ClCH_2CH_2CH_2SiCl_3$, 2-chloromethyl-3-trichlorosilylpropene, $CH_2{=}C(CH_2Cl)CH_2SiCl_3$, and chloromethyltrimethoxysilane, $ClCH_2Si(OMe)_3$, constitute three examples of AB_3-type monomers that do not form predominantly hyperbranched polymers on Grignard coupling. Chloropropyltrichlorosilane usually forms a small ring compound, 1,1-dichlorosilacyclobutane, in over 50% yield via Grignard reaction in diethylether [24]. The minor by-product, although not reported in the paper, should be a branched oligomeric material shown in Reaction Scheme 12.15:

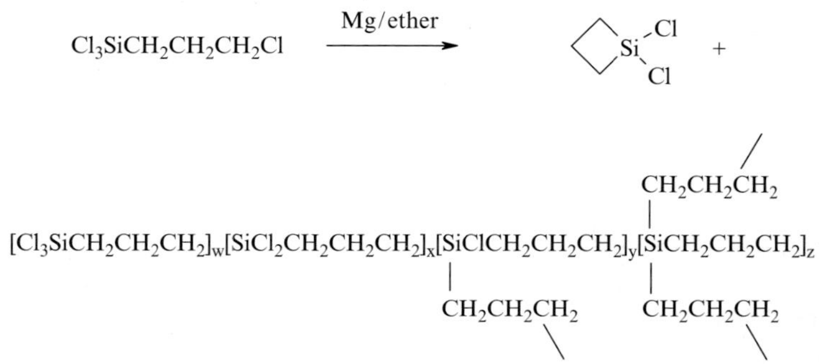

(12.15)

A hyperbranched polycarbosilane having overall composition $[SiH_2{-}C_6H_4{-}CH_2]_n$, was prepared by Lu from a mixture of mono-, di- and tri-methoxysilanes with the average composition, $ClCH_2{-}C_6H_4{-}SiCl(OMe)_2$, by using a Grignard coupling reaction and reduction with $LiAlH_4$ as follows [14]:

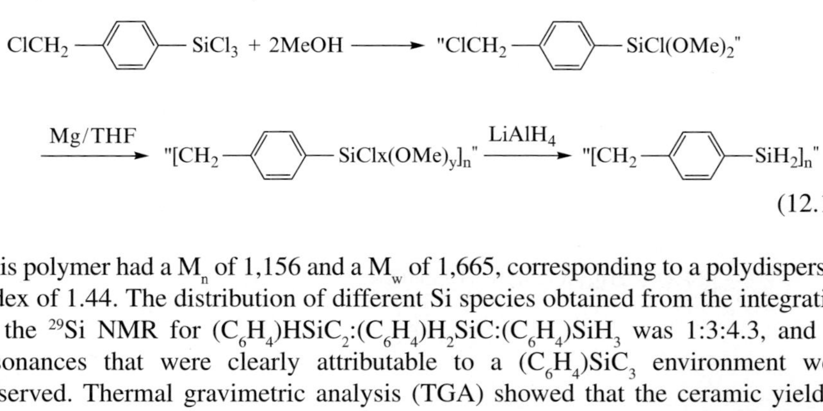

(12.16)

This polymer had a M_n of 1,156 and a M_w of 1,665, corresponding to a polydispersity index of 1.44. The distribution of different Si species obtained from the integration of the ^{29}Si NMR for $(C_6H_4)HSiC_2$:$(C_6H_4)H_2SiC$:$(C_6H_4)SiH_3$ was 1:3:4.3, and no resonances that were clearly attributable to a $(C_6H_4)SiC_3$ environment were observed. Thermal gravimetric analysis (TGA) showed that the ceramic yield at 1,000°C was 70.1%. Given that the rate of nucleophilic substitution of Si–OMe by a Grignard reagent is generally slower than that of Si–Cl, it seems likely that the mixture of Si–X units (X = Cl and OMe, with an average of one Cl per monomer unit) that was used in this case, led to a more linear type of chain growth than would be the case for typical AB_3 (B = Si–OMe, Si–Cl) monomers.

2-Chloromethyl-3-trichlorosilylpropene, $CH_2{=}C(CH_2Cl)CH_2SiCl_3$, was used to form a four-membered ring in 10% yield by a Grignard reaction [25], as shown in Reaction Scheme 12.17. The major products were not discussed, but are presumably oligomers or polymers with head-to-tail and possibly head-to-head (C–C) chain linkages. The branched polymer shown in Reaction Scheme 12.17 contains only the head-to-tail structure. A similar example of a monomer with allylic chlorine groups is described in Section 12.3.2.

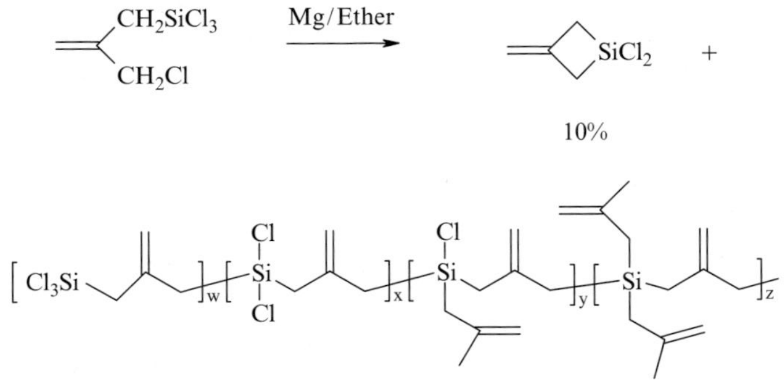

(12.17)

Chloromethyltrimethoxysilane, $ClCH_2Si(OMe)_3$, was reported to form 1,1,3,3, 5,5-hexamethyl-1,3,5-trisilacyclohexane in 45% yield and another six-membered ring with an attached $CH_2Si(OMe)_3$ group, as shown in Reaction Scheme 12.18 [26]. The remaining products should be oligomers with a branched structure. The Grignard reaction was conducted in tetrahydrofuran (THF) at low temperature (–20°C to 10°C).

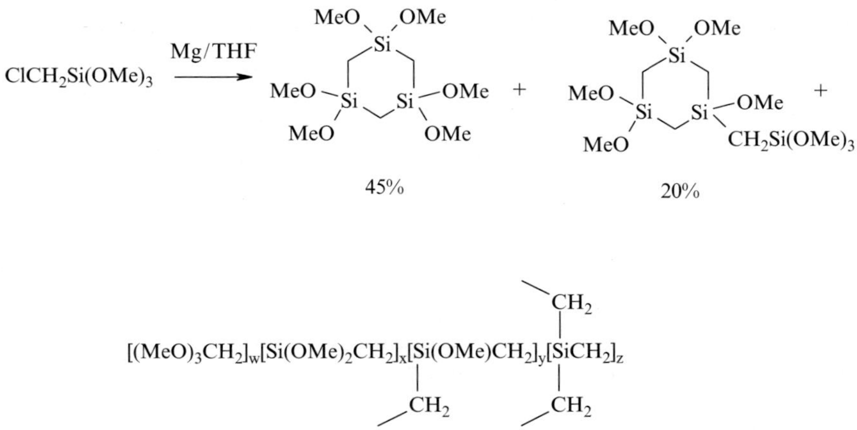

(12.18)

It is noteworthy that even the six-membered ring with a $-CH_2Si(OMe)_3$ group that was obtained in 20% yield already contains three different types of Si units: $-CH_2Si(OMe)_3$, $-(CH_2)_2Si(OMe)_2$, and $-(CH_2)_3SiOMe$. This information, along with the results of Lu described in Section 12.2.1 for the $ClCH_2SiCl_3$ system (see Reaction Scheme 12.5) [14], suggests that some of the branched polycarbosilanes discussed above should also contain cyclic structures, in addition to acyclic branched units.

12.3 Hyperbranched Polycarbosilanes from AB_2 Monomers

Hyperbranched polycarbosilanes made from AB_2-type monomers through nucleophilic substitution reactions are quite limited. Thus far, only $ClCH_2SiMeCl_2$ and $ClCH_2CH{=}CHSiMeCl_2$ were reported to yield distinct branched polymers.

12.3.1 From Chloromethylmethyldichlorosilane

The Grignard reaction of chloromethylmethyldichlorosilane, $ClCH_2SiMeCl_2$, was reported by Kriner in the 1960s [27]. The reaction was conducted in a reverse manner (powdered Mg was added slowly to the chlorosilane) in dilute solution in order to improve the yield of 1,3-dichloro-1,3-dimethyl-1,3-disilacyclobutane. It was reported, however, that the desired cyclic compound was obtained in less than 10% yield, but there was no information provided concerning the identity of the other reaction products. In 1993, Froehling employed a similar synthetic approach to that used by Whitmarsh and Interrante for $ClCH_2SiCl_3$ (see Section 12.2.1, Reaction Scheme 12.6) to obtain a branched polyhydridopolycarbosilane from $ClCH_2SiMeCl_2$ [28]. As shown in Reaction Scheme 12.19, the Grignard reaction generated a chloro-substituted polycarbosilane with three types of units: $–CH_2SiMeCl_2$, $(–CH_2)_2SiMeCl$, and $(–CH_2)_3SiMe$. The branched chloropolymer should have an average formula $[SiMe(Cl)CH_2]_n$. Subsequent reduction by $LiAlH_4$ converted it into a H-substituted polymer, which has an average formula $[SiMe(H)CH_2]$ and three structural units: $–CH_2SiMeH_2$, $(–CH_2)_2SiMeH$, and $(–CH_2)_3SiMe$ with a weight average molecular weight (M_w) of 3,200. This polymer was reported to give 19% SiC residue after heating beyond 700°C under argon. If the polymer was pre-cross-linked at 150°C and then subjected to pyrolysis, the ceramic yield was enhanced to 51%.

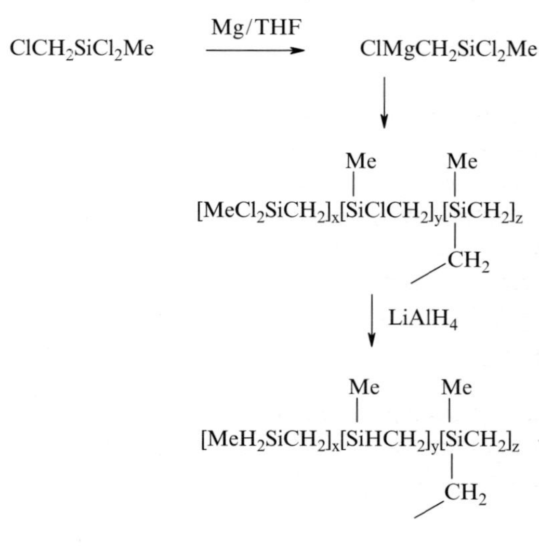

(12.19)

Fry et al. further characterized the product from the Grignard and reduction reactions of $ClCH_2SiMeCl_2$ [29]. The reduced polymer was found, by GC/mass spectrometry, to contain low molecular weight oligomers, such as trimers and tetramers. The following isomers were identified among these trimers and tetramers:

Trimers:

$CH_3SiH_2CH_2SiH(Me)CH_2SiH(Me)CH_3$

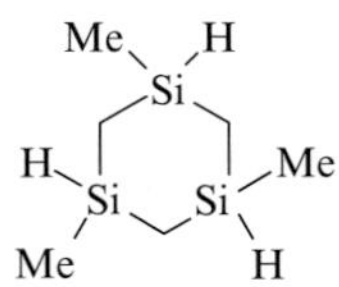

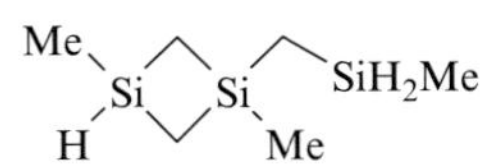

Tetramers:

$CH_3SiH_2CH_2SiH(Me)CH_2SiH(Me)CH_2SiH(Me)CH_3$

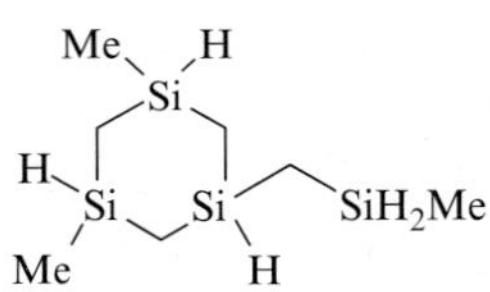

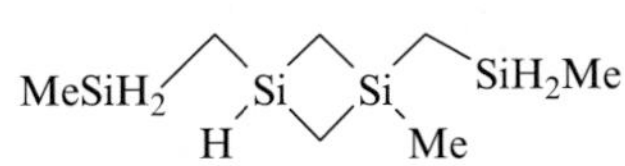

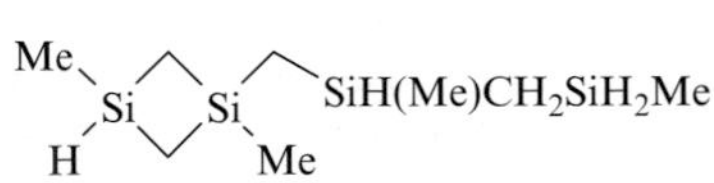

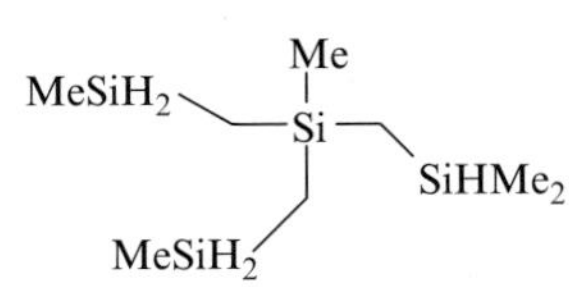

The three possible Si structural units, $-CH_2SiMeH_2$, $(-CH_2)_2SiMeH$, and $(-CH_2)_3SiMe$, also occur in these isomers. Their overall ratio, $-CH_2SiMeH_2/(-CH_2)_2SiMeH/(-CH_2)_3SiMe$, in the polymer, as determined by ^{29}Si NMR spectroscopy, was 1.17:1.02:0.81. A photo-activated hydrosilylation curing of the branched polymer from $ClCH_2SiMeCl_2$ was achieved by using a bis(acetylacetonato)platinum complex as a catalyst and tetravinylsilane as a cross-linking agent.

In contrast to the HBPCS polymer, where hydrosilylation of terminal olefins does not represent an effective route to side chain modification of the hyperbranched polycarbosilane backbone, the hyperbranched "[$SiMe(H)CH_2$]" polymer (HB-MePCS) participates readily in Pt complex (Karsted's catalyst)-catalyzed hydrosilylation of terminal olefins to give the corresponding Si–C bonded side chain derivatives in good yield. In his Ph.D. thesis, Zheng reported the use of this polymer along with various $CH_2{=}CHCH_2{-}(OCH_2CH_2)_n{-}OCH_3$ compounds (n = 2, 7, 12, 15) in hydrosilylation reactions analogous to those employed in Scheme 12.10 of Section 12.2.1 to prepare a series of HB-MePCS-PEO derivatives, for use as Li^+ electrolyte materials [7]. In fact, these are much simpler to prepare than the corresponding hydrosilylation-PEO derivatives of Reaction Scheme 12.10 and give very similar conductivity properties when doped with LiX salts, despite a lower degree of branching.

Electropolymerization of $ClCH_2SiMeCl_2$ was investigated by using three types of anodes: Zn, Mg and Al [30]. When Zn was used, the major products were $H–[CH_2SiHMe]_n–H$ oligomers (n = 2, 3, 4) of undetermined structure. Electropolymerization of $ClCH_2SiMeCl_2$ using Mg as the anode resembles the conventional Grignard coupling reaction, and gives rise to branched polycarbosilanes having a broad molecular weight distribution and a wide range of linear and cyclic oligomers. The reaction was strongly exothermic, and was accompanied by rapid coagulation of the electrolyte solution to the extent of disrupting the electropolymerization. A soluble polycarbosilane with M_n around 1,060 and a monomodal distribution ($M_w/M_n = 4.9$) was isolated in 39% yield. A mild electropolymerization in which coagulation of the electrolyte solution was avoided was achieved by using an Al anode. A high yield of polycarbosilane oligomers that were partially soluble and partially insoluble was typically obtained. The soluble polycarbosilanes exhibited a bimodal molecular weight distribution with a high molecular weight fraction having $M_n = 33{,}000$ and $M_w/M_n = 1.8$ and a low molecular weight fraction having $M_n = 270$ and $M_w/M_n = 1.6$. Preliminary studies indicated that the mechanism of electropolymerization of $ClCH_2SiMeCl_2$ consists of an iterative stepwise reaction sequence, which involves electroreduction of the CH_2–Cl bond to a carbanion that is stabilized by the α-silyl hyperconjugation effect, followed by S_N2 nucleophilic displacement of a carbanion with a Cl–Si group, where Cl serves as a good leaving group.

12.3.2 From $ClCH_2CH{=}CHSiMeCl_2$ and $CH_2{=}C(CH_2Cl)SiMeCl_2$

A hyperbranched polycarbosilane with both Si–H and alkene groups was reported by using a mixture of $ClCH_2CH{=}CHSiMeCl_2$ and $CH_2{=}C(CH_2Cl)SiMeCl_2$ [31]. Both monomers were obtained as isomers from the reaction of propargyl chloride and methyldichlorosilane as shown in Reaction Scheme 12.20; however, it proved difficult to separate them completely by distillation due to their close boiling point.

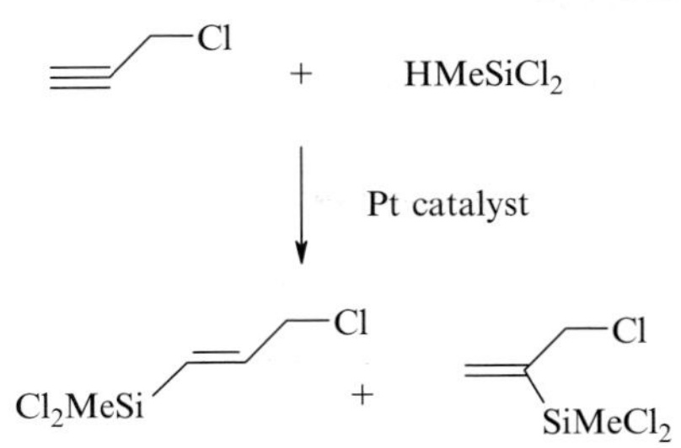

(12.20)

The chlorines at the allylic position in both monomers should have similar reactivity in the subsequent Grignard reaction. Hence, the polymerization of this monomer mixture was suggested to follow the reaction mechanism shown in Reaction Scheme

12.21. Both reactions (I) and (II) lead to a head-to-tail connection, while the side reaction (III) leads to the formation of a head-to-head structure. Reaction III also contributes chain termination, which decreases the polymer's molecular weight. The reaction shown in equation III is a well known side reaction that occurs during the preparation of allyl Grignard reagents from allyl halides. The labeled protons, H_a, H_b, H_c and H_d, were all observed by ^{1}H NMR spectroscopy in the polymer product.

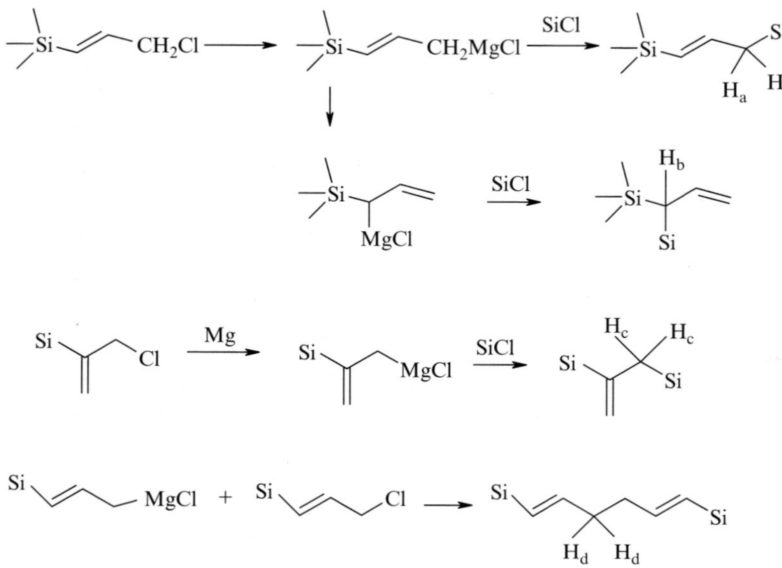

(I)

(II)

(III)

(12.21)

Like the other chlorocarbosilanes of the AB_2 and AB_3 types, the coupling of the Grignard reagents, $Cl_2MeSiCH{=}CHCH_2MgCl$ and $CH_2{=}C(CH_2MgCl)SiMeCl_2$, with multiple Si–Cl groups was found to form a random branched structure. The ^{29}Si NMR spectrum of the polymer after reduction with $LiAlH_4$ showed SiH_2, SiH, and Si atoms with four alkyl substituents. However, it is very difficult to draw a general formula for the final polymer product due to the use of mixed monomers and side reactions. Nevertheless, the polymer should have the following six units connected in head-to-tail or head-to-head fashion.

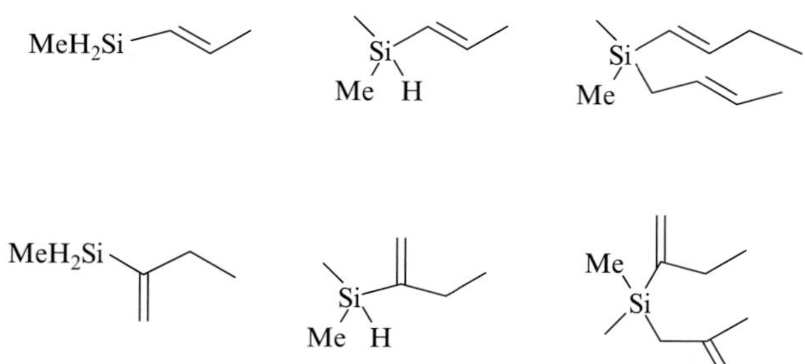

An absolute molecular weight (M_n) of 670 for the above polymer was measured by VPO in chloroform, which corresponds to an average of eight monomer units per oligomer molecule. Due to the existence of unsaturated C=C double bonds and Si–H groups, this branched polymer can be efficiently cross-linked by using a Pt catalyst, generating an infusible gray solid. Upon heating under nitrogen to 975°C, the cross-linked polymer gave a black ceramic residue in 69% yield.

12.3.3 Other Monomers

Certain other organochlorosilanes which have been employed as reagents in Grignard-type reactions, such as 3-chloropropylmethyldichlorosilane, $ClCH_2CH_2CH_2SiMeCl_2$, chloromethylmethyldiethoxysilane, $ClCH_2SiMe(OEt)_2$, chloromethylmethylchloroethoxysilane, $ClCH_2SiMe(OEt)Cl$, and 3 (or 4)-bromophenylmethyldiethoxysilane can also be considered as AB_2-type monomers, although the intended products of these reactions were small cyclic oligomers or thermally stable materials. In particular, $ClCH_2CH_2CH_2SiMeCl_2$ [32] has been used to prepare 1-chloro-1-methyl-silacyclobutane in 75% yield by means of a Grignard reaction (see Reaction Scheme 12.22). The by-products, although not reported, should be oligomers with the following branched structure:

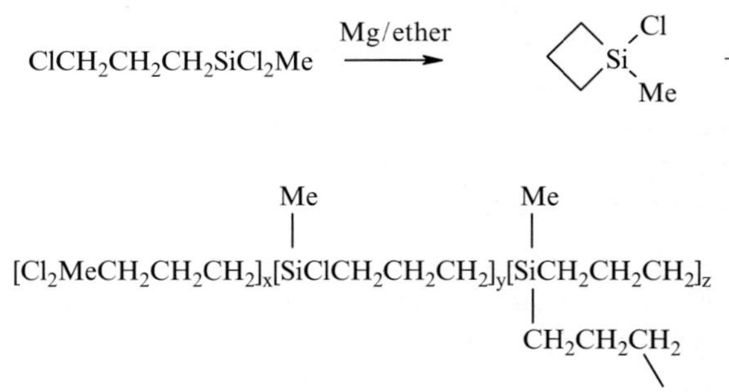

(12.22)

Similarly, both $ClCH_2SiMe(OEt)_2$ and $ClCH_2SiMe(OEt)Cl$ were reported to form 1,3-diethoxy-1,3-dimethyl-1,3-disilacyclobutane in less than 35% yield by a Grignard reaction [27]. The by-products from this Grignard reaction are also likely to be branched oligomers as shown in Reaction Scheme 12.23:

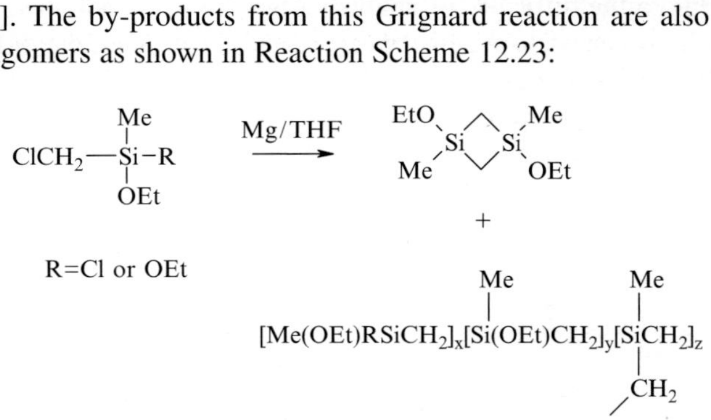

(12.23)

3 (or 4)-Bromophenylmethyldiethoxysilane, like 3 (or 4)-bromophenyltriethoxysilane, also has functional groups to form a branched polycarbosilane by a Grignard reaction [23]. However, it was claimed that the major products from the Grignard reaction of 3

(or 4)-bromophenylmethyldiethoxysilane were linear, instead of branched polymers. As was discussed previously in Section 12.2.4, the apparent lack of an appreciable yield of branched products in these cases can be understood on the basis of the relatively high steric inhibition associated with the formation of the third Si–Ph bond.

12.4 Hyperbranched Polycarbosilanes from A_2B_2-Type Monomers

Compounds such as (dichloromethyl)methyldichlorosilane, $Cl_2CHSiMeCl_2$, and bis(chloromethyl)dichlorosilane, $(ClCH_2)_2SiCl_2$, have functional groups corresponding to an A_2B_2-type monomer. Both of these compounds are by-products of the synthesis of chloromethylmethyldichlorosilane, $ClCH_2SiMeCl_2$, by chlorination of dimethyldichlorosilane, $(CH_3)_2SiCl_2$, and there has been strong interest in finding applications for them. In 1993, a patent from DuPont disclosed the synthesis of polycarbosilanes from $Cl_2CHSiMeCl_2$ by using both Grignard and Wurtz coupling reactions [33]. The polymers described in the patent were originally called "cross-linked polycarbosilanes", instead of hyperbranched polymers. The patent authors found that Wurtz coupling reaction of $Cl_2CHSiMeCl_2$ by metallic sodium formed an insoluble solid of a formula $[MeSiCH]_n$. It is well known that in addition to Si–C bonds the Wurtz coupling reaction can also generate both C–C and Si–Si units. Thus, the polymer made from $Cl_2CHSiMeCl_2$ by the Wurtz reaction is likely to have a polycarbosilane/polysilane network structure, as shown in Reaction Scheme 12.24:

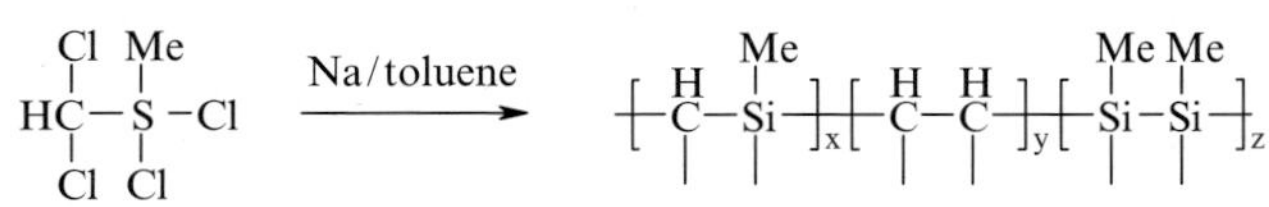

(12.24)

No analysis of the ratio of the three kinds of bonding in this product was reported, presumably due to the difficulty of analyzing the insoluble polymer. Note that in this case, branching can occur at both the carbon and silicon sites, due to the existence of two chlorine atoms on both ends of the monomer molecules. As shown by the following structures, both carbon and silicon atoms can form three possible branching units from three types of bondings: C–Si, C–C, and Si–Si, respectively:

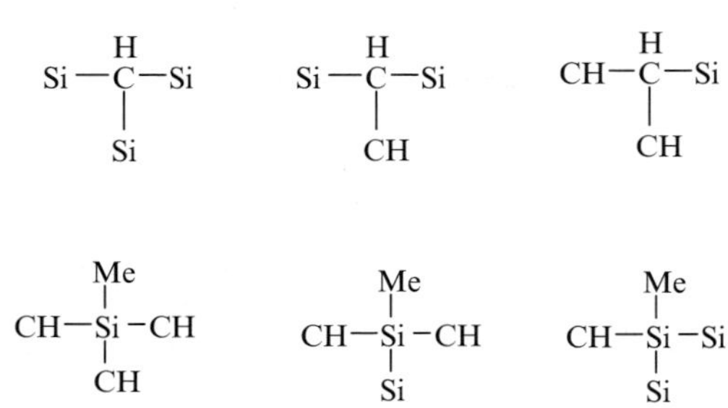

On pyrolysis, the solid polymer made from $Cl_2CHSiMeCl_2$ by sodium coupling produced a black ceramic in 79% yield. Despite this high ceramic yield, however, there is limited application for this solid polymer due to its insolubility.

In contrast to the Wurtz coupling reaction, a soluble polymer was obtained from $Cl_2CHSiMeCl_2$ by using a Grignard reaction conducted in THF solution (see Reaction Scheme 12.25). In this case, little or no C–C or Si–Si coupling occurred and the product was a polycarbosilane with Si–C bonding. However, elemental analysis indicated that the Grignard reaction could not convert all of the chlorine atoms on carbon into an active CHMgCl Grignard reagent, presumably due to steric hindrance, and all of the Cl left on C was unlikely to be replaced by H during the reduction step. Therefore, the final polymer contained un-reacted Cl on some of the C atoms:

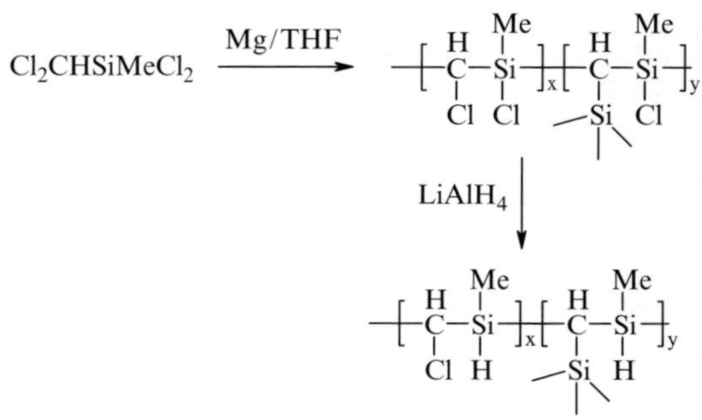

(12.25)

As both carbon and silicon atoms in this polycarbosilane can form branching sites, both –CHCl– and –CH< structural units are likely present in the backbone. The –CH– links to a Cl atom as shown in the x unit, while –CH< links to two Si atoms to form a branching site, as shown in the y unit. The silicon site is much more complicated. Each Si atom in its average x and y composition can form three types of units from the Grignard reaction: $–CHSiMeCl_2$, $(–CH)_2SiMeCl$, and $(–CH)_3SiMe$, based on the Si units observed for the polycarbosilane made from $ClCH_2SiMeCl_2$ (see Reaction Scheme 12.19). By reduction with $LiAlH_4$, these Si sites yield $–CHSiMeH_2$, $(–CH)_2SiMeH$, and $(–CH)_3SiMe$, respectively. Due to the subtle structural difference, it is difficult to resolve the Si units between the x and y composition in the ^{29}Si NMR spectra. In fact, the patent [33] reported three major units: Si(no H), SiH, and SiH_2, and the ratio of the three was given as 81:14:5. Elemental analysis suggested that after reduction the final polymer had a $[C_2H_{4.5}SiCl_{0.11}]$ composition. At room temperature, this hyperbranched polycarbosilane is a solid that is soluble in most common organic solvents. It had a molecular weight (M_w) of 9,500, which was much higher than the M_w (3,200) of the polymer made from the $ClCH_2SiMeCl_2$ monomer, presumably due to the additional chain growth through the C end of the monomer unit that can occur in this case. The ceramic yield from heating this polymer under nitrogen to 1,000°C was in the range of 55–63% without a pre-cross-linking treatment, which was much higher than the ceramic yield of the polymer, $[SiMeHCH_2]_n$, made from $ClCH_2SiMeCl_2$. Relative to other precursors to silicon carbide, the excess carbon (over the stoichiometric 1:1 SiC ratio) and residual Cl present in this polymer represent potential disadvantages for certain applications.

The same patent also reported a hydrolyzed product that was made from the chloro polymer, as shown in Reaction Scheme 12.26. The hydrolysis was conducted with excess distilled water to quench the residual Si–Cl groups. It is possible that some of the silanol groups were condensed to siloxane bonds during the process.

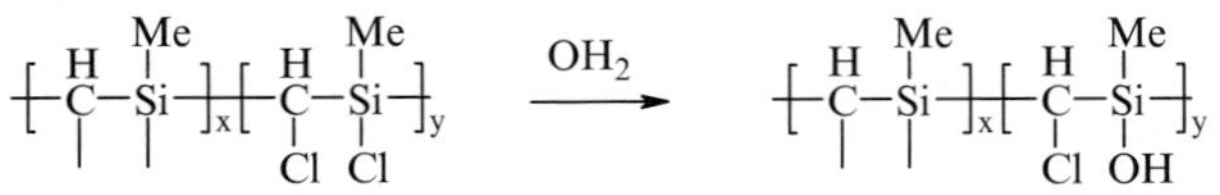

(12.26)

The final polymer was a viscous liquid with formula $[MeSiCHCl_{0.056}O_{0.44}]_n$, derived from elemental analysis data. This formula suggests that more than 50% of the monomer was converted into a [>CHSiMe<] structure during the Grignard reaction. The hydrolyzed polymer gave a 69% ceramic yield based on TGA.

The patent further disclosed a copolymer obtained by using $Cl_2CHSiMeCl_2$ and vinylmethyldichlorosilane $(CH_2{=}CH)MeSiCl_2$ in a Grignard reaction, followed by reduction with $LiAlH_4$. Vinylmethyldichlorosilane does not provide any functional groups that can form a Grignard regent, but does provide Si–Cl groups for coupling with the Grignard reagents generated from $(Cl_2CH)SiMeCl_2$. The addition of vinylmethyldichlorosilane will certainly change the ratio of branching units, although no details were reported. It was mentioned, however, that the residual Cl content in the copolymer after reduction was decreased. For instance, a 5 mol % of vinylmethyldichlorosilane used for the copolymerization decreased the Cl content from 6.20% in the reduced homo-polymer to 2.21% in the copolymer. This was probably due to less steric hindrance in the copolymer system, which allowed more dichloromethyl groups to form a Grignard reagent and couple with SiCl groups. The reduced copolymer was a liquid at room temperature, and afforded a 46% ceramic yield based on TGA.

12.5 Hyperbranched Co-Polycarbosilanes

Hyperbranched co-polycarbosilanes made by nucleophilic substitution reactions have also been reported in the literature. However, it is difficult to classify these reactions based on the monomer structure. Therefore, a few examples will be summarized in order to illustrate their nature.

12.5.1 Co-Polycarbosilanes from Phenyltrichlorosilane, Diphenyldichlorosilane and Dibromomethane

Habel and co-workers investigated the Wurtz coupling of a mixture of phenyltrichlorosilane, $PhSiCl_3$, diphenyldichlorosilane, Ph_2SiCl_2, and CH_2Br_2, as shown in Reaction Scheme 12.27 [34]. The coupling reaction was found to form predominately

C–Si bonds, although the Wurtz reaction also results in C–C and Si–Si bonding. It is possible that the more reactive C–Br group may have helped to drive the formation of C–Na, which coupled with the Si–Cl groups to generate mainly Si–C bonds. An average formula $[(Ph_2Si)_nSiPh(CH_2)_{n+1.5}]_x$ was reported for the resultant product:

$$nxPh_2SiCl_2 \quad + \quad (n+1.5)\ x\ CH_2Br_2 \quad + \quad xPhSiCl_3$$

$$\downarrow \text{Na/xylene}$$

$$[(Ph_2Si)_nSiPh(CH_2)_{n+1.5}]_x$$

or

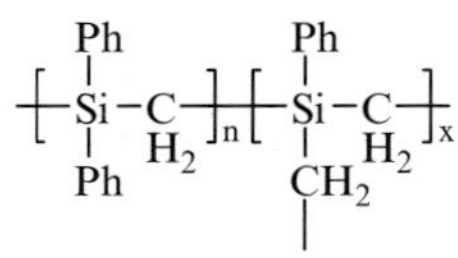

(12.27)

To better understand the structure of this copolymer, a structural formula is suggested at the bottom of Reaction Scheme 12.27. The copolymer contained mainly $-SiPh_2CH_2-$ and $>SiPhCH_2-$ units, which came from diphenyldichlorosilane and phenyltrichlorosilane, correspondingly. The branching units were obviously derived from $PhSiCl_3$; hence, by simply changing the content of $PhSiCl_3$ in the starting material mixture, the branching sites in the polymer could be controlled quantitatively. This is different from the branched polycarbosilanes described in the previous sections, where the branching units were formed randomly from a single monomer unit. The molecular weight of the polymers also depended on the content of $PhSiCl_3$ in the starting monomer mixture. A higher content of $PhSiCl_3$ (more branching sites) gave a higher molecular weight. The replacement of Si–phenyl in these polymers by Si–Cl was achieved by $AlCl_3$-catalyzed reaction with HCl (see Reaction Scheme 12.28). Further reaction of chloro-substituted polymer with $LiAlH_4$ yielded the hydridopolycarbosilane. Both the chlorination and reduction reaction were carried out with high yield. However, on pyrolysis the final hydrido "poly"carbosilane gave a relatively low ceramic yield (30.9 wt%), presumably because of its largely oligomeric character.

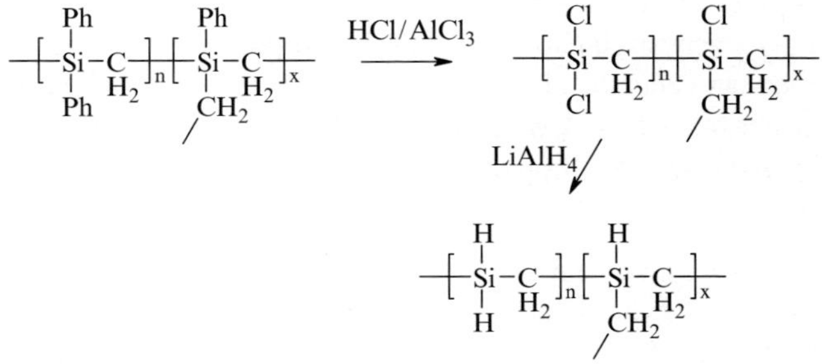

(12.28)

12.5.2 Copolymer from $ClCH_2Si(OMe)_{1.75}Cl_{1.25}$ and $ClCH_2SiMeCl_2$

Branched homo-polycarbosilanes from $ClCH_2SiCl_3$ and $ClCH_2SiMeCl_2$ have been reported as summarized in the preceding sections. Although the hydridopolycarbosilane made from $ClCH_2SiCl_3$ is a high yield SiC ceramic precursor, its production using a large amount of $LiAlH_4$ is a cost concern. Therefore, solutions to eliminate or reduce the use of $LiAlH_4$ have been explored by one of the authors of this chapter (QS). As a result, a branched copolymer made from $ClCH_2Si(OMe)_{1.75}Cl_{1.25}$, $ClCH_2SiMeCl_2$ and allylchloride was disclosed in a patent recently [35]. As shown in Reaction Scheme 12.29, the procedure comprised a Grignard reaction in THF and a reduction reaction in the same solvent.

$$ClCH_2Si(OMe)_{1.75}Cl_{1.25} \quad + \quad ClCH_2SiMeC_{l2} \quad + \quad CH_2{=}CHCH_2Cl$$

$$\big\downarrow \; \text{Mg/THF}$$

$$[Si(OMe)_2CH_2]_x[\underset{\displaystyle CH_2CH{=}CH_2}{\underset{|}{Si}}(OMe)CH_2]_y[SiMe(Cl)CH_2]_z$$

$$\big\downarrow \; LiAlH_4$$

$$[SiH_2CH_2]_x[\underset{\displaystyle CH_2CH{=}CH_2}{\underset{|}{Si}}HCH_2]_y[SiMe(H)CH_2]_z \qquad (12.29)$$

The partially methoxylated silane is a mixture of three compounds: $ClCH_2Si(OMe)Cl_2$, $ClCH_2Si(OMe)_2Cl$ and $ClCH_2Si(OMe)_3$. The Grignard reagents formed from the CH_2Cl groups and allychloride can couple with both the methoxy and Cl groups on the Si atoms. Reaction Scheme 12.29 only gives the average formula for both the methoxy- and hydrido-polycarbosilanes. The branched units discussed for the homo-polymers (see Sections 12.2.1 and 12.3.1) made from $ClCH_2SiCl_3$ and $ClCH_2SiMeCl_2$, respectively, should all exist in this copolymer. Physical properties of the copolymer with an average formula $[SiH_2CH_2]_{0.8n}[SiH(Allyl)CH_2]_{0.1n}[SiMe(H)CH_2]_{0.1n}$ were similar to those of the homo-polymer made from $ClCH_2SiCl_3$. This copolymer can generate a black ceramic in 70% yield, which is comparable to the homopolymer made from $ClCH_2SiCl_3$. By using a co-monomer $ClCH_2SiMeCl_2$, the required $LiAlH_4$ for reduction can be decreased, since less OMe and Cl groups in the copolymer need to be reduced after the Grignard reaction.

12.5.3 Copolymer from $Cl_2CHSiMeCl_2$ and $ClCH_2Si(OMe)_{1.75}Cl_{1.25}$

Other types of branched co-polycarbosilanes were disclosed in another patent [36] by one of us (QS). One specific copolymer was made from $Cl_2CHSiMeCl_2$ and $ClCH_2Si(OMe)_{1.75}Cl_{1.25}$ by using a Grignard reaction and a reduction by $LiAlH_4$. The reaction steps are shown in Reaction Scheme 12.30:

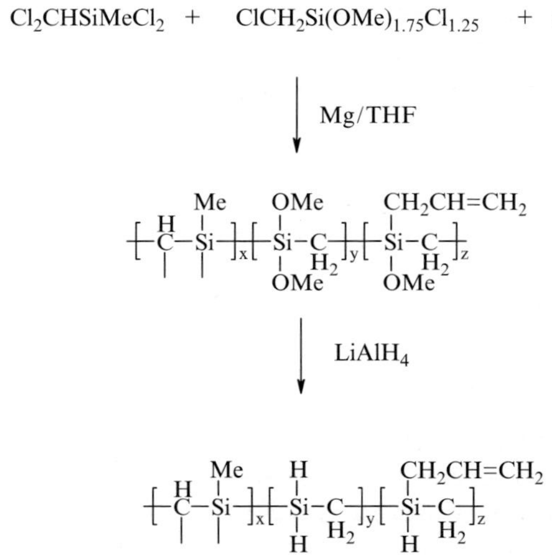

(12.30)

The fraction of $Cl_2CHSiMeCl_2$ in the monomer mixture can be as high as 75% and the resulting copolymer is a solid at room temperature. When the content of $Cl_2CHSiMeCl_2$ is less than 50%, the copolymer is a liquid. The physical state of these polymers presumably relates to the proportion of >SiMe– branching units. The use of co-monomers can allow all dichloromethyl groups in $Cl_2CHSiMeCl_2$ to form Grignard reagents and couple with Si–Cl or Si–OMe groups completely, due to less steric hindrance in the copolymer system. The copolymers typically gave a ceramic yield in the range of 65–75%. When the content of $Cl_2CHSiMeCl_2$ was 75%, the required $LiAlH_4$ for reduction could be cut down to 20%, compared to when no $Cl_2CHSiMeCl_2$ was used in the polymerization of $ClCH_2SiCl_3$. An 80% saving of $LiAlH_4$ can substantially decrease the production cost for the pre-ceramic polycarbosilanes. Hence, although the black ceramics obtained from this copolymer contain excess carbon, this is still an attractive alternative for certain applications, due to its lower cost and capability of structure tailoring. A commercial product based on this co-polymer has been developed for making C/SiC composites by a hot-melting process [6].

12.6 Conclusion and Future Outlook

The application of nucleophilic substitution reactions in the preparation of hyperbranched polycarbosilanes represents a logical extension of the organomagnesium (Grignard) and organolithium coupling reactions with halo- and alkoxysilanes that have been used in organosilane chemistry for over a century for Si–C bond formation. In the past several decades, this methodology has been used successfully, particularly in its organomagnesium manifestation, to obtain a wide variety of hyperbranched polycarbosilanes. These polymers have been of primary interest as potential precursors to silicon carbide, a ceramic material which is of considerable technological importance due to its high thermal, chemical and oxidative stability, as well as for its electrical (semiconductor) and mechanical (high hardness, shock resistance, strength, etc.) properties. At least one of the polymers made in this manner, a hyperbranched polycarbosilane prepared from $ClCH_2SiCl_3$ by reaction with magnesium, followed by reduction with $LiAH_4$, is now being produced commercially for use in the fabrication of ceramic composites. Various other applications for these polymers, including the fabrication of low-κ dielectric films, Li ion battery electrolytes, and chemical sensors have been suggested, primarily in the patent literature, which constitutes a major source of reports on these systems.

In addition to $ClCH_2SiCl_3$, a variety of other AB_3, as well as AB_2 and A_2B_2, monomers has been used in reactions of this type, alone and as mixtures, to prepare a range of different hyperbranched carbosilanes. The products of these reactions generally consist of a polydisperse mixture of oligomers and polymers that are coupled primarily though Si–C bonds, but which can also contain appreciable Si–Si bonding, especially when the more reactive Li reagents are employed. Depending on the conditions used, in addition to branching, these reactions can also lead to cyclization, adding a further level of complexity to the structures and molecular composition of the resulting products. The use of polynuclear NMR and mass spectrometry, along with various molecular separation techniques, has provided considerable information regarding the nature of these products and the course of the coupling reactions in a few cases. However, a full characterization of most of these systems is still far from complete and many of them have received little or no attention.

In addition to the use of nucleophilic reactions to form a hyperbranched polycarbosilane backbone, this methodology has also been used extensively to add side chains and otherwise modify the structure of the initially obtained polycarbosilane through reaction with Si–Cl or Si–OR groups on the polymer backbone. Thus, this methodology has been, and is likely to continue to be, an important component of the chemical tool kit that is available for the preparation and effective utilization of this important class of organosilicon polymers.

References

1. Kipping F S (1904) Proc Chem Soc 20:15
2. Interrante L V, Shen Q (2000) Polycarbosilanes. In: Jones R et al. (eds) Silicon-Containing Polymers, Kluwer, Dordrecht/Boston/London, p. 312

3. (a) Yajima S et al. (1976) Nature 260:683 (b) Yajima S et al. (1978) J Mater Sci 13:2569
4. Jones R H et al. (2002) J Nucl Mater Part 2 307–311:10157
5. Whitmarsh C K, Interrante L V (1991) Organometallics 10:1336
6. Starfire Systems, Inc. 10 Hermes Rd, Suite 100 Malta, NY12020
7. Zheng Chang-Feng (2006) Synthesis and characterization of polycarbosilane with grafted polyethylene oxide as polymer electrolytes, Ph.D. dissertation, Rensselaer Polytechnic Institute, Troy, NY
8. Rushkin I L et al. (1997) Macromolecules 30:3141
9. Interrante L V, Lu N (2004) US Patent 6,809,041, 1 Jan 2004; Rathore JS, Interrante LV, Dabois G, (2008) Adv. Funct. Mater., 18:1
10. Apen P, Wu H-J (2005) US Patent 6,841,256, 11 Jan 2005
11. Babich K et al. (2007) US Patent 7,172,849, 6 Feb 2007
12. Nakagawa H et al. (2006) US Patent 20,060,134,336, 22 June 2006
13. Houser E J, McGill R A (2006) US Patent 7,078,548, 18 July 2006
14. Lu Ning (2004) Hyperbranched polycarbosilanes: Synthesis, characterization and application, Ph.D. dissertation, Rensselaer Polytechnic Institute, Troy, NY
15. (a) Shen Q, Interrante L V (1997) J Polym Sci Part A Polym Chem 35:3193 (b) Shen Q, Interrante L V (1996) Macromolecules 29:5788
16. Shen Qionghua (1995) Preparation and characterization of new polycarbosilanes: Part 1, synthesis, characterization and thermal properties of substituted poly(silylenemethylenes); part 2, stoichiometric polycarbosilane precursors to SiC, Ph.D. dissertation, Rensselaer Polytechnic Institute, Troy, NY
17. Liu Q et al. (1999) Chem Mater 11:2038
18. Whitmarsh C K, Interrante L V (1992) US Patent 5,153,295, 6 Oct 1992
19. (a) Interrante L V et al. (1998) Low cost, near net shape ceramic composites by polymer infiltration and pyrolysis. In: Newaz G M, Gibson R F (eds) Proceedings of the 8th Japan–US Conference on Composite Materials, Technomic, Lancaster, PA, pp. 506–515 (b) Interrante L V et al. (1997) Fabrication and properties of fiber- and particulate-reinforced SiC matrix composites obtained with (A)HPCS as the matrix source. In: Key Engineering Materials, Transtec, Switzerland, Vols. 127–131, pp. 271–278 (c) Interrante L V et al. (1994) MRS Symp Proc 346:593
20. Anderson E (2007) Malta product a NASA option. In: Albany Times Union, Albany, New York, Aug 14, 2007
21. Babonneau F et al. (1994) Chem Mater 6:51
22. Yao J, Son D Y (1999) Organometallics 18:1736
23. Ohshita J et al. (1997) Macromolecules 30:1540
24. Lanne L (1967) J Am Chem Soc 89:1144
25. Damrauer R et al. (1990) J Organomet Chem 391:7
26. Brodani D J et al. (1993) Tetrahedron Lett 34:2111
27. Kriner W A (1964) J Org Chem. 29:1601
28. Froehling P E (1993) J Inorg Organomet Polym 3:251
29. Fry B E et al. (1997) J Organomet Chem 538:151
30. Wang X et al. (2007) Macromolecules 40:3939
31. Yao J, Son D Y (1999) J Polym Sci Part A Polym Chem 37:3778
32. (a) Vdovin V M et al. (1961) Dokl Akad Nauk SSSR 141:843 (b) Vdovin V M et al. (1963) Dokl Akad Nauk SSSR 150:799
33. Michalczyk M J (1993) US Patent 5,270,429, 14 Dec 1993
34. Habel W et al. (1994) J Organomet Chem 467:13
35. Shen Q, Sherwood W (2007) US Patent 20,070,093,587, 26 Apr 2007
36. Shen Q (2007) US Patent 20070167599, 19 July 2007

Chapter 13
Hyperbranched Polycarbosilanes and Polycarbosiloxanes via Hydrosilylation Polymerization

Hanna Schüle and Holger Frey

13.1 Introduction

As pointed out in Chapter 1, silicon chemistry offers a variety of quantitative, high yielding reactions, i.e. hydrosilylation, Grignard reactions and controlled condensation of silanols that are suitable for the synthesis of organic-inorganic hybrid materials. Thus, silicon-based chemistry played a prominent role in the evolution of dendrimer chemistry [1–4], and it did not take long until the first examples of silicon-containing hyperbranched polymers were reported. Hyperbranched polymers are generally prepared by one-pot polymerization of AB_x ($x \geq 2$) (see also Section 1.2) monomers and are characterized by polydispersity as well as a randomly branched structure due to the multifunctional polycondensation or polyaddition process. The statistical treatment of such polyfunctional polycondensations was achieved in the early 1950s by Flory, who calculated both molecular weights and polydispersity in such systems, as is discussed in Section 13.3 of this chapter [5, 6]. The properties of hyperbranched polymers are significantly different from their linear analogs and are characterized by good solubility, low viscosity and a large number of end-groups that can be used for further functionalization. Despite imperfections in branching and structure of hyperbranched polymers compared to monodisperse dendrimers, these properties render them easily accessible competitors for dendrimers, particularly in applications where structural perfection is not a mandatory prerequisite.

As for dendrimers, the research interest for silicon-containing hyperbranched polymers has focused on the preparation of carbosilanes, carbosiloxanes and siloxanes (see Chapters 2 and 3). These compositions generally give highly flexible materials with low glass transition temperatures and are characterized by good

H. Schüle
Institute of Organic Chemistry, Organic and Macromolecular Chemistry, Duesbergweg 10-14, Mainz, Germany
E-mail: schuelh@uni-mainz.de

H. Frey
Johannes Gutenberg-University Mainz, D-55099 Mainz, Germany
E-mail: hfrey@uni-mainz.de

P.R. Dvornic and M.J. Owen (eds.), *Silicon-Containing Dendritic Polymers*,
Advances in Silicon Science 2, © Springer Science+Business Media B.V. 2009

thermal and chemical stability. The majority of silicon-containing hyperbranched polymers has been synthesized by polymerization of suitable AB_x monomers either via hydrosilylation (this chapter), via nucleophilic substitution (see Chapter 12) or condensation (Chapter 14). However, the preparation of hyperbranched polymers based on $A_2 + B_3$ approaches, i.e., by copolymerization of multifunctional monomers, is also known and summarized in Chapter 16. To date, the development of silicon-containing hyperbranched polymers has only been summarized as a subchapter in otherwise dendrimer-oriented reviews [7, 8]. This chapter exclusively covers the field of hyperbranched silicon-containing polymers prepared via hydrosilylation polymerization of AB_x monomers. Independent of the basic structure, monomers suitable for the preparation of hyperbranched silicon-containing polymers via hydrosilylation have to comprise a Si–H bond as well as a π-bond as units suitable for polyaddition. Either the Si–H or the π-bond may represent the A-unit. The resulting structures are characterized by a large number of functional end-groups, i.e., silanes or alkenes, respectively, which can be used for further functionalization. In what follows, the synthesis and characterization of the different classes of polymers are discussed separately. However, first, the basic principles and reaction mechanism of transition metal-catalyzed hydrosilylations are summarized in a brief manner.

13.2 Hydrosilylation

Catalytic hydrosilylation (see also Chapters 3 and 7) is one of the most important industrial techniques for the formation of Si–C bonds, only outdone by the Direct Process, the transition metal catalyzed oxidative addition of alkylhalides to silicon. However, hydrosilylation is more generally applicable, especially in laboratory setups, and can be broadly used for the introduction of functional silanes into organic molecules in high yield. The reaction leads to the formation of Si–C bonds by addition of the relatively weak Si–H across π-bonds, normally alkenes and alkynes (see Reaction Scheme 13.1).

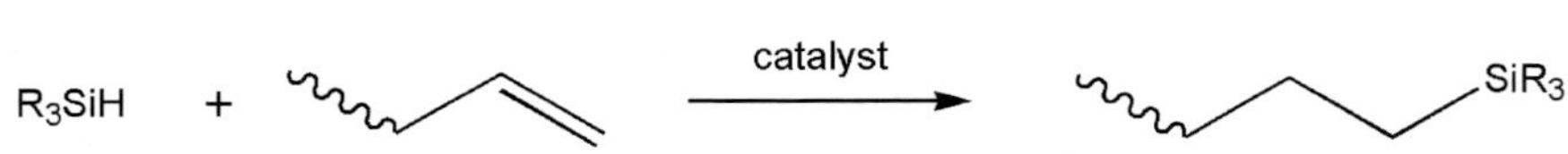

(13.1)

Regarding the synthesis of hyperbranched structures, hydrosilylation can be used in a polyaddition-type reaction to produce polymeric products. Hence, a mechanistic analysis is of crucial importance for understanding of the polymerization pathway and of possible side-reactions that can have a significant influence on the molecular weight and molecular weight distributions of the polymers. The reaction can also be performed under free-radical conditions (i.e., initiated with ultraviolet light or peroxides) [9–11], but metal-assisted hydrosilylation is presently more popular and has been proven to be more efficient [12]. Several transition metals show catalytic activity in hydrosilylation reactions, however platinum complexes like Speier's catalyst (H_2PtCl_6-*i*PrOH) [13] or Karstedt's catalyst [14, 15] are most commonly employed.

Solid-supported catalysts, like Pt/C, are less reactive, but they can be removed by filtration after completion of the reaction, increasing the shelf-life of the materials. Platinum catalysts tolerate a variety of polar functional groups (e.g., halogens, NO_2, CN, COOR, amines), as long as they are poor ligands. Strongly coordinating groups can compete with the alkene to be hydrosilylated for the platinum center, inhibiting the hydrosilylation (see for example Chapter 11, Section 11.6). Hydrosilylation polymerization has been successfully used for the synthesis of linear polycarbosilanes and polycarbosiloxanes. High molecular weight linear polymers with up to 75,000 g/mol were obtained applying Karstedt's catalyst [16, 17].

The most commonly accepted mechanism for platinum-catalyzed hydrosilylation has been proposed by Chalk and Harrod and is illustrated in Reaction Scheme 13.2 [18] in its classical [19, 20] and modified [21–23] form. The first step involves oxidative addition of the silane to the platinum-alkene complex I and subsequent formation of the platinum-alkene silyl hydride complex II. This is followed by migratory insertion of alkene into either the [Pt]-hydride (classical Chalk-Harrod) or [Pt]-silyl bond (modified Chalk-Harrod) of complex II to give the resulting complexes III or IV, respectively. The product V is obtained after Si–C or C–H reductive elimination and the metal-alkene complex is regenerated in excess of alkene.

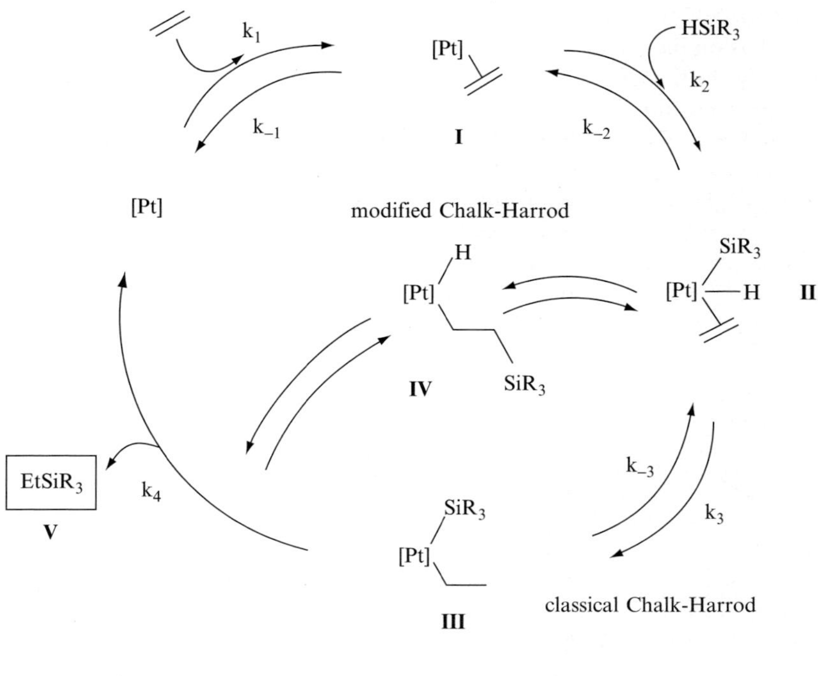

(13.2)

Despite the general acceptance of the Chalk-Harrod mechanism, further approaches have been made to explain some phenomena observed in experiments. Colloidal platinum, which is responsible for the yellow to brown color of hydrosilylation reactions, was first proposed to be the catalytically active species by Lewis and coworkers [24]. However, they reported later that colloidal platinum is only an unreactive end-product and instead a mononuclear two-coordinate platinum complex is the active species formed during the induction period in a series of ligand exchanges [25]. For platinum complexes in higher oxidation states, e.g., Speier's catalyst, the induction period was assigned to the initial reduction of platinum. They also stated that reaction of the silane with platinum precedes the coordination of the olefin. For several reaction systems, small amounts of oxygen are necessary for the hydrosilylation to occur. According to the Lewis mechanism, oxygen impedes the formation of catalytically inactive multinuclear platinum species by coordination to platinum. Recently, the kinetics of Pt(0) catalyzed hydrosilylations targeting linear polymers was treated. In these studies the presence of a mononuclear platinum complex as the active species was confirmed [26, 27].

Electron withdrawing substituents on the silane and electron donating groups on the olefin generally accelerate the addition, resulting in higher reaction rates for monomers with –OR or $–OSiR_3$ substituents compared to mere alkyl groups [24, 28, 29]. However, steric effects are also of great importance in hydrosilylation and may eventually counteract this influence. Two possibilities exist for the addition of a silane R_3SiH to a substituted alkene: namely, Markovnikov or anti-Markovnikov addition, which results in the formation of branched (α-adduct) or linear (β-adduct) products, respectively. The distribution of both products depends on the catalyst used as well as on the nature of the substituents on both the alkene and the silane [12]. Generally, the β-adduct tends to predominate, since the silane favors the addition to the terminal position to minimize steric interactions. The formation of side-products can have a significant influence on the outcome of polymerization. Thus, in the following reaction schemes (Reaction Scheme 13.3) possible side-reactions known for platinum catalysts are briefly summarized:

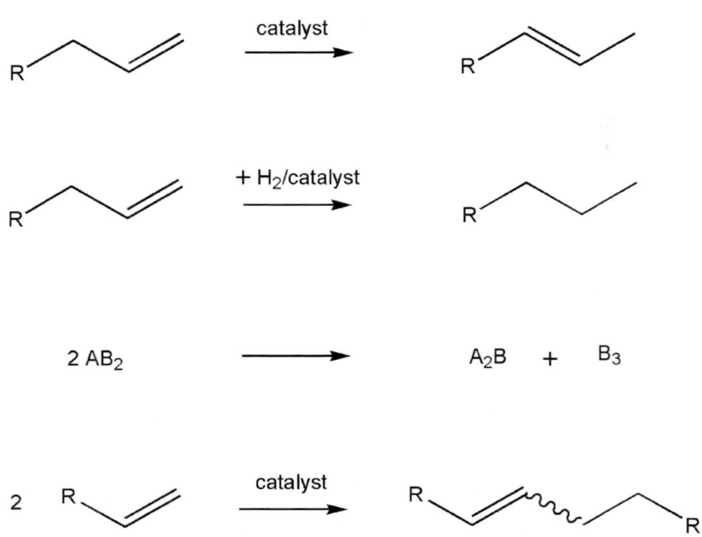

(13.3)

Alkene isomerization of terminal double bonds leads to the formation of internal double bonds [19, 20, 29–31]. These are unreactive in hydrosilylations and thus limit the number of functionalities available for polymer growth. The same holds true for hydrogenated double bonds resulting from platinum-catalyzed addition of hydrogen, which can be evolved in minor amounts during the side-reactions [24]. Dyotropic rearrangements at silicon comprise simultaneous exchange of substituents between two silanes. During the synthesis of hyperbranched polymers the exchange of functional groups of AB_2 monomers can lead to the formation of B_3 and A_2B species with B_3 representing a possible crosslinker, which can cause undesired gelation. If the metal complex is capable of coordinating more than one alkene, the possibility of alkene coupling arises, resulting in dimerized products [32]. Additionally, dry reaction conditions are necessary during hydrosilylation, since water present in the reaction mixture can hydrolyze silanes to give silanols and hydrogen. Although the aforementioned side-reactions are normally present only in minor amounts, they can be the reason for limited molecular weights and gelation observed in certain systems. In summary, hydrosilylation is a high yielding reaction working best for activated silanes containing electron withdrawing groups like halogens or alkoxy groups and electron-rich alkenes. Further details on catalytic hydrosilylation may be found in other reviews on the subject [33–36].

13.3 Synthesis and Characterization of Silicon-Containing Hyperbranched Polymers

It is sometimes overlooked that the synthesis of hyperbranched polymers is based on an old concept, already described in basic work of Flory in the early 1950s, where he provided the theoretical foundation as well as experimental evidence for the formation of branched macromolecules in the polycondensation of AB_x-type monomers (see also Chapter 1) [5, 6]. In this type of polycondensation, the A and B functional groups of these monomers can react with each other to form new linkages leading to randomly branched structures as the polymerization progresses. The most important expressions derived by Flory for the number average degree of polymerization ($\overline{DP_n}$) and polydispersity (PDI) of the step growth of AB_x monomers are given in Equations (E.13.1) and (E.13.2) where p = conversion:

$$\overline{DP_n} = \frac{1}{1-p} \qquad \text{(E.13.1)}$$

$$PDI = \frac{\overline{M_w}}{\overline{M_n}} = \frac{1-\frac{p^2}{x}}{1-p} \approx \overline{DP_n}/2 \qquad \text{(E.13.2)}$$

The deduction of these equations was based on the assumption that all functional groups of a given type (A or B) show equal reactivity at any stage of the reaction, and the occurrence of intramolecular cyclization reactions was intentionally neglected (compare Chapter 15). In essence, these equations demonstrate that conversion p of the reaction of A- and B-groups is the key parameter to obtain high molecular weights, i.e., only high-yielding transformations are suitable for the preparation of extended hyperbranched structures. On the other hand, the expressions show that the resulting structures will always exhibit broad molecular weight distributions, characterized by large polydispersities. For instance, for a conversion of 99% ($p = 0.99$) a degree of polymerization of 100 and a polydispersity $\bar{M}_w / \bar{M}_n$ of 51 would be expected, which is, of course, dramatically large. On the other hand, cyclization of the single, "focal" A-group of the macromolecules with one of the multiple B-groups limits molecular weights, but also keeps polydispersity lower than the values calculated by using (E.13.2).

As a consequence of interesting properties found for dendrimers, the synthesis of hyperbranched polymers attracted attention as an alternative leading to materials with similar properties, but facile preparation. Thus, the simple one-pot polymerization of AB_x ($x \geq 2$) monomers results in branched structures that are less defined than dendrimers, but display similar physical and structural properties, like low viscosity and a large number of end-groups. However, as a result of the synthetic pathway, hyperbranched polymers possess a lower degree of branching (DB) than the monodisperse dendrimers. If side-reactions are excluded, the AB_x stoichiometry in these reactions does not permit gelation, unless an unlikely 100% intermolecular reaction of A-groups is achieved.

In the last decade, theoretical concepts have been developed for the control of molecular weight, degree of branching and polydispersity of hyperbranched polymers [37–40]. Copolymerization with core molecules of the B_f-type as well as slow monomer addition techniques can be employed to reduce polydispersity and to enhance control in the polymerization. These approaches permit the synthesis of rather well-defined hyperbranched structures with moderate, sometimes low polydispersity. For instance, a highly reactive monomer can be added slowly to the growing hyperbranched structure to avoid homopolymerization and favor attachment to the growing species. In the case of such an ideal slow monomer addition polymerization of an AB_2 monomer with a B_f-core it was found that polydispersity can be lowered to PDI = $1 + 1/f$, emphasizing the need for a high number of core functionalities f [38, 40]. An example of this strategy is shown in Section 13.4.3.

Hyperbranched silicon-containing structures usually exhibit excellent solubility and can, therefore, be well characterized. Theoretical considerations for the description of the degree of branching (DB) of hyperbranched polymers and its relationship to the number of end-groups enabled the determination of this key parameter by NMR-spectroscopy [41]. This is especially interesting for silicon chemistry, since well-separated signals can generally be observed for the different branching units in ^{29}Si-NMR spectra. Commonly, the analysis is based on low molecular weight model compounds possessing structures analogous to the respective branching units. Hyperbranched polymers obtained by polymerization of an AB_2 monomer possess dendritic, linear and terminal (D, L, T) units. Dendritic units are perfectly branched as found in dendrimers, i.e., all pendant functional groups are converted. In the case of AB_3 systems, additionally, imperfectly branched units, called semidendritic (sD), with one unreacted group are found. For an assignment of different branching units in an AB_3 system see Fig. 13.1 in Section 13.4.2.

The DB for AB_2 and AB_3 systems, as derived by Hölter and Frey is given in Equations (E.13.3) and (E.13.4), respectively (where D = dendritic units, sD = semidendritic units, L = linear units) [41]:

$$DB(AB_2) = \frac{2D}{2D+L} \tag{E.13.3}$$

$$DB(AB_3) = \frac{2D+sD}{\frac{2}{3}(3D+2sD+L)} \tag{E.13.4}$$

In a random polycondensation the maximum value for DB is 0.5 for AB_2 monomers and 0.44 for AB_3 monomers. IR-spectroscopy can be a helpful tool to monitor the conversion during polymerization. Particularly for monomers with a Si–H functionality as the A-group, the disappearance of the Si–H stretching mode around $\tilde{\nu} = 2{,}100\,\mathrm{L/cm}$ is a significant indication of the occurrence of the reaction. On the other hand, characterization of hyperbranched polymers via size exclusion chromatography (SEC) based on linear polystyrene standards is problematic and molecular weights are often significantly underestimated due to the more compact nature of hyperbranched structures. Thus, unless molecular weights are determined by absolute methods, the results can only be taken as an indication of the actual degree of polymerization.

Among various silicon-based hyperbranched polymers reported, polycarbosilanes and polycarbosiloxanes have received the strongest attention. Polycarbosilanes are kinetically and thermodynamically very stable compounds, resulting from the low polarity of the Si–C bond and a dissociation energy of 306kJ/mol, roughly comparable to that of the C–C bonds (345kJ/mol). On the other hand, polycarbosiloxanes contain Si–$(CH_2)_x$–Si as well as Si–O–Si units in their polymer backbone, but the Si–O bond in siloxanes is also characterized by excellent chemical resistance and a dissociation energy of 465kJ/mol. Additionally, this structural element leads to further flexibility of the polymer scaffold. In contrast, under mildly acidic conditions, alkoxysilane linkages of the Si–O–CH_2-type can undergo hydrolysis, which makes them suitable for the synthesis of degradable polymers (see also Chapters 6 and 14). The synthesis of hyperbranched polymers via hydrosilylation results in materials that bear a large number of either π-bonds or Si–H groups, which are interesting for further functionalization. In the case of Si–H end-groups a transformation is generally necessary in order to obtain air- and moisture-stable structures, since these moieties are easily hydrolyzed by water.

13.4 Polycarbosilanes

13.4.1 General Synthetic Strategy

The preparation of monomers for the synthesis of hyperbranched polycarbosilanes (HB-PCS) via hydrosilylation polymerization reaction is generally based on the nucleophilic substitution of halides in commercially available silanes, e.g., trichlorosilane or dichloromethylsilane, by metal alkyls (Reaction Scheme 13.4).

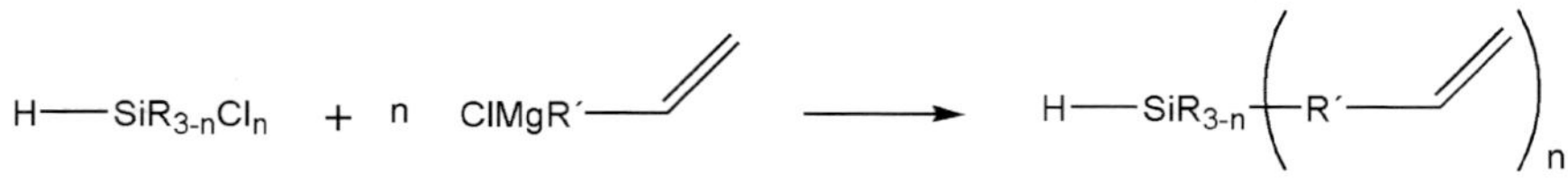

(13.4)

To this end, Grignard and alkyl lithium reagents are used, which can conveniently be prepared by the reaction of alkyl halides with the respective metal. The branching multiplicity can be varied by using either di- or trihalide silanes. Additionally, the choice of the organometallic compound determines the chemical structure of the polymer backbone and the spacer length between the functionalities, which can be important for the polymerization result.

The first hyperbranched polycarbosilanes were reported in 1993 by Muzafarov et al. who polymerized methyldiallylsilane, **1**, triallylsilane, **2**, methyldivinylsilane, **3**, and methyldiundecenylsilane, **4**, using platinum catalysis [42].

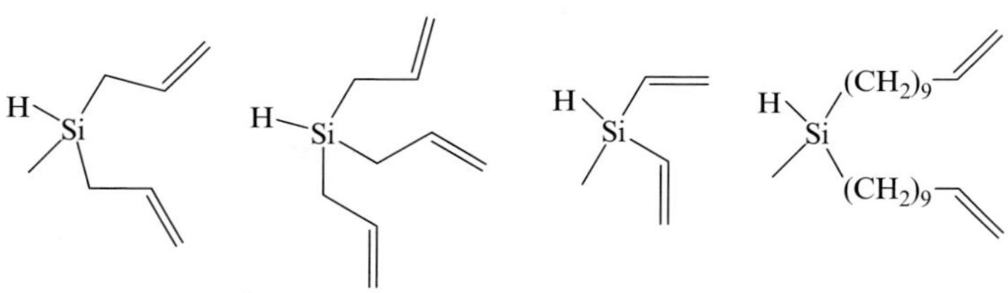

1 **2** **3** **4**

Molecular weights of the resulting polymers depended on the polymerization conditions and decreased with the decrease of monomer concentration as well as with increase of the catalyst concentration. Interestingly, upon addition of new monomer to the system, no increase in molecular weight of the polymers could be observed, which is unexpected on the basis of an idealized polyaddition scheme. On the contrary, the reaction of the polymers with dichloromethylsilane proceeded quantitatively, which ruled out steric hindrance as the cause for this behavior. Therefore, the authors explained the limited growth by kinetic factors but, in agreement with more recent studies, this limitation is most likely due to competing cyclization that renders further polyaddition of the single A-group of the polymers impossible (see also Chapter 15).

The hyperbranched polymers of this study were used as effective modifiers of polysiloxane rubber blends to increase the tensile strength as well as tear strength of the blend. Modification of the double bonds allowed adjustment of the compatibility with the polymer matrix. However, no detailed experimental data were presented in this work. An elaborated investigation of the branching structure of hyperbranched polycarbosilanes was reported by Frey et al., who determined the degree of branching of polytriallylsilane by ^{29}Si-NMR spectroscopy [43]. A value of 0.48 was found for the polymerization of this AB_3 monomer, which is in good agreement with 0.44 expected for a random polymerization. Copolymerization of triallylsilane with B_1-core 2-(10-decen-1-yl)-1,3-oxazoline resulted in the first example of hyperbranched polycarbosilane macromonomers, **5**, with an oxazoline group at the focal point.

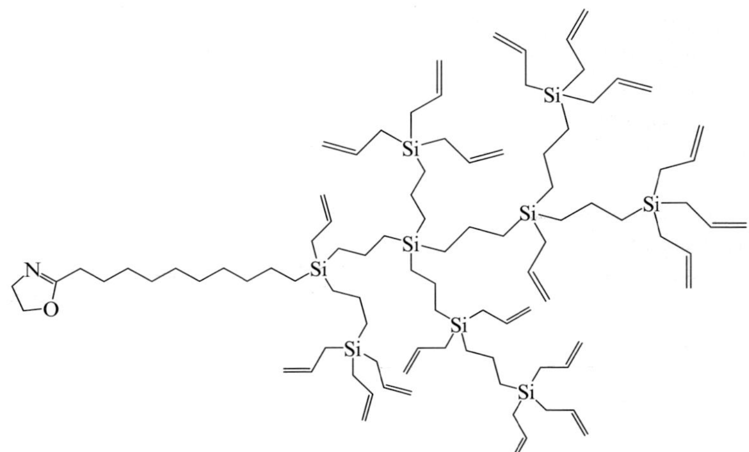

5

As expected, SEC measurements revealed a decrease of molecular weight and polydispersity with increasing amounts of the core molecule added. In subsequent work by Drohmann et al., the investigation was extended to study the effect of variation of the monomer structure on the kinetics of the addition reaction, the branching structure and the occurrence of side reactions [44, 45]. A strong influence of reaction conditions and the spacer length on the polymerization result was found. Besides the decrease of the molecular weight of the resulting polymers upon dilution of the monomer, in some cases multimodal distributions and eventually crosslinking were observed. Since the monomodal molar mass distributions could be obtained by keeping the reaction temperature and the catalyst concentration low, the occurrence of multimodal distributions was assigned to rearrangement reactions at elevated temperatures for vinyl and allyl monomers. It was confirmed that for diallylmethylsilane and methyldivinylsilane subsequent addition of monomer did not increase the molecular weight. In contrast, for methyldiundecenylsilane a gradual increase in molecular weight was observed upon further monomer addition. The authors suggested that the formation of less sterically crowded polymers in the case of monomers with a longer alkenyl spacer could be the reason for this, compared to a high structural density for the polymers from monomers with short alkenyl spacers. However, intramolecular cyclization consuming Si–H functionalities is most likely the reason for the limited molecular weights of polymers from short alkenyl-containing monomers, as found in the works by Fréchet et al. [46] and Frey et al. [47]. For large monomers cyclization is kinetically disfavored, enabling further polymer growth. These observations represent the first step towards molecular weight control of hyperbranched polycarbosilanes.

Son and Yoon reported the first examples of polycarbosilanes containing aromatic moieties [48, 49]. A series of AB_3 carbosilarylene monomers **6**, **7** and **8** was synthesized from dibromobenzenes using lithium-halogen exchange and subsequent treatment with chlorosilanes.

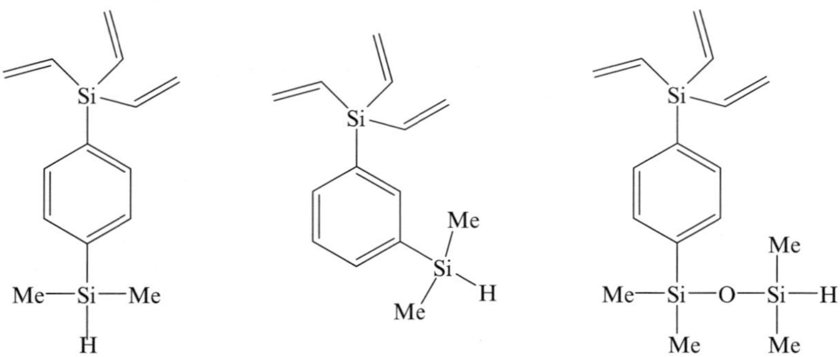

6 **7** **8**

The polymer obtained from 1-dimethylsilyl-4-trivinylsilylbenzene, **6**, exhibited improved thermal stability and better solubility than its linear analog. Despite the rigid aromatic moieties in the backbone, the polymer shows a relatively low glass transition temperature (T_g) of 12°C due to the hyperbranched structure. The influence of *meta*-bonding and the incorporation of siloxane linkages into the monomers was also studied. Polymerization of a mixture of the monomers allowed tailoring of the materials properties. The degree of branching was determined by ^{29}Si-NMR to be 0.42, close to the expected value, indicating similar reactivity for all B-groups. The authors suggested the application of these materials in advanced elastomers, using the unreacted vinyl groups for crosslinking. In continuing work the same group reported the introduction of furan and thiophene moieties into a series of different AB_2 (**11a/b**) and AB_3 monomers (**9a/b, 10a/b**) [50, 51].

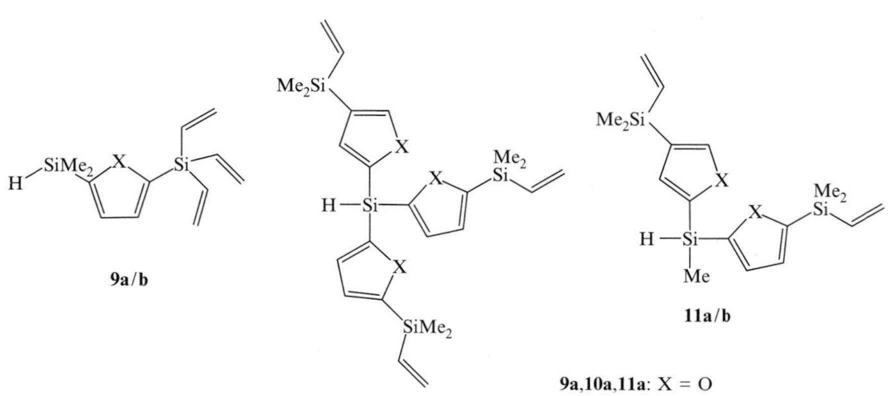

9a/b **10a/b** **11a/b**

9a,10a,11a: X = O
9b, 10b, 11b: X= S

The presence of oxygen and sulfur in the polymers was meant to enable their application in host-guest or metal-stabilization chemistry. The functional groups can be easily incorporated into the monomers via replacement of chlorosilanes with the respective organometallic compounds. The resulting polymers were air-stable, highly soluble and exhibited low T_gs between −61°C and −7°C. Molecular weights were rather low, between 1,400 and 3,200 g/mol, as determined by vapor pressure osmometry (VPO), and the branched structure was confirmed using ^{29}Si-NMR.

McGill et al. reported on the synthesis of naphthalene-containing polycarbosilanes [52]. The concept was to functionalize free vinyl groups as well as the naphthyl moieties with hexafluoroacetone to impart a response to nitroaromatics and organophosphonates, permitting application as a chemical sensor. First results revealed a rapid response time for the polymer-coated sensors; however, no further experimental data were given.

Hydrosilylation addition of Si–H to π-bonds is not only applicable to alkenes, but also to alkynes. The polymerization of ethynyl-containing monomers leads to polymers with interesting properties, since unsaturations are incorporated into the resulting polymer backbone. In the first example Son and coworkers polymerized methyldiethynylsilane, **12**, using Karstedt's catalyst to polymer **13** [53], as shown in Reaction Scheme 13.5.

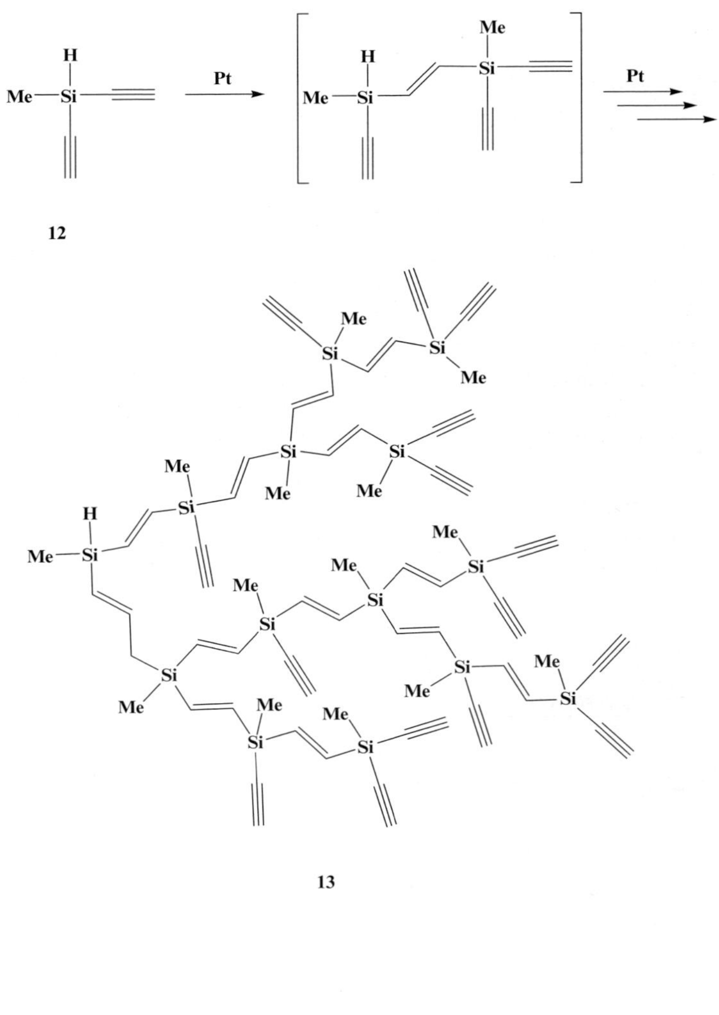

(13.5)

As expected, no additional hydrosilylation was observed for internal double bonds formed in the reaction due to their low reactivity. The resulting polymer was air- and moisture-stable and exhibited excellent thermal stability in nitrogen with only 14% mass loss on heating to 1,300°C. The remaining ethynyl groups can be used for crosslinking and as attachment sites for transition metals. An interesting study targeting the simple preparation of σ–π-conjugated polymers has been reported by Masuda et al. [54, 55]. They polymerized bis(4-ethynylphenyl)methylsilane, **14**, and *m*-bis(dimethylsilyl)phenyl-acetylene, **15**, using Rh-catalyzed hydrosilylation.

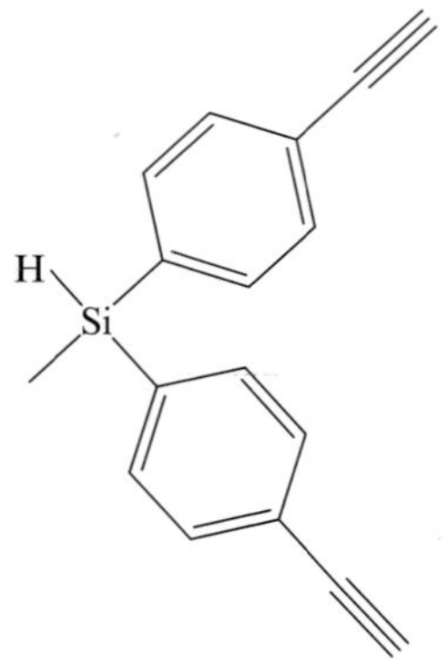

14

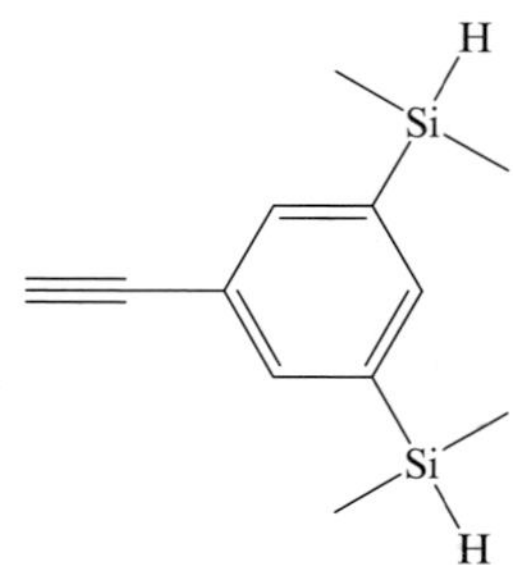

15

Performing the polymerization at elevated temperatures, i.e., 80°C, directed the stereoselectivity towards *trans*-rich polymers. The soluble products were obtained by polymerization in dilute solution or subsequent conversion of the Si–H groups with phenylacetylene, respectively. The branched, *trans*-vinylene structure was confirmed by ^{1}H-NMR. Optical properties of the ethynyl-functional polymer were analyzed and compared to the linear analog. A weak absorption around 330 nm, hardly seen in the linear polymer, was assigned to σ-to-π charge-transfer [55]. The authors concluded that this was a result of the more compact structure of the hyperbranched polymer in solution, facilitating the required through-space interactions. The excess of Si–H groups in the polymer, based on *m*-bis(dimethylsilyl)phenylacetylene, could be used for aerial oxidation or hydrolysis to give a crosslinked transparent film with siloxane linkages (see Reaction Scheme 13.6).

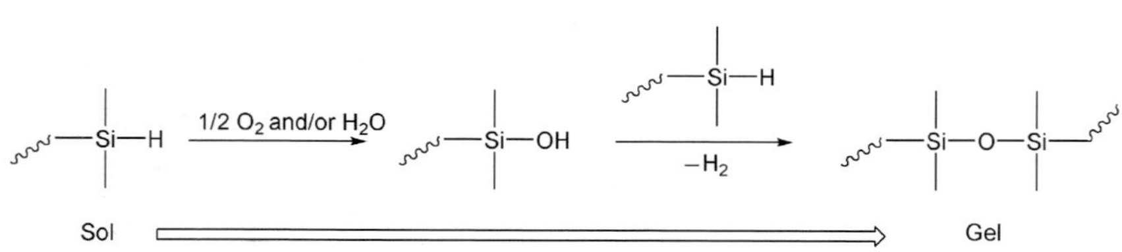

(13.6)

The homogenity of the surface as well as the uniform distribution of siloxane linkages in the film was proven using scanning electron microscopy (SEM) and energy dispersive X-ray spectroscopy (EDX). The charge-transfer phenomenon was accountable for the emission of an intense blue light, when the film was excited at the charge transfer absorption wavelength (>300 nm), and excellent solvent resistance of the film was confirmed in extraction experiments [54].

13.4.2 Functionalization

The robustness and inertness of polycarbosilane structures towards most reaction conditions provides the basis for a variety of modification reactions. Several of the approaches described for carbosilane dendrimers (see Chapter 3) have been adopted for the functionalization of the analogous hyperbranched polymers. Surprisingly, some of the most promising research areas for dendrimers, such as their application as supports for metal complexes (see Chapter 8) and catalysts (see Chapter 9) or as liquid crystalline materials (see Chapter 10) have so far only been rarely studied. Frey et al. have shown that the combination of concepts known from dendrimer chemistry, namely the hydrosilylation of terminal double bonds with chlorosilanes and subsequent substitution with Grignard reagents can be applied for the enhancement of the degree of branching of hyperbranched polymers [56]. The strategy of post-synthetic conversion of the double bonds of a hyperbranched polytriallylsilane led to a structure without linear units that exhibited a formal degree of branching close to 1 that may be viewed as a "*pseudodendrimer*". The conversion of linear and semidendritic units to dendritic units was clearly confirmed by the decrease of the respective signals in the ^{29}Si-NMR, as shown in Fig. 13.1. This approach is generally applicable for hyperbranched polymers; however, the resulting structures do not possess the symmetry and regularity known for dendrimers, regardless of the high degree of branching.

The first report on the supramolecular properties of hyperbranched polymers was published by Frey et al. in 1997. Oxazoline-functional polycarbosilanes were homopolymerized via cationic polymerization as well as copolymerized with phenyloxazoline to give side chain hyperbranched polymers. Coupling of the oxazoline functionality with trimesic acid resulted in trimers of the hyperbranched fragments. Interestingly, the trimers formed relatively well-defined aggregates in solution, despite the randomly branched structure of the building blocks. Aggregation into columnar structures was studied by wide angle X-ray scattering (WAXS), dynamic and static light scattering as well as atomic force microscopy (AFM) and explained by hydrogen bonding between the polar cores, which form the centers of the stacks surrounded by an apolar, disordered corona [57, 58].

Carbosilane dendrimers are also studied for their application as scaffolds for catalytically active metal complexes. The soluble complexes permit good access to the reactive sites, and in addition the catalyst system can be removed using ultracentrifugation or ultrafiltration and reused in analogy to a heterogenous

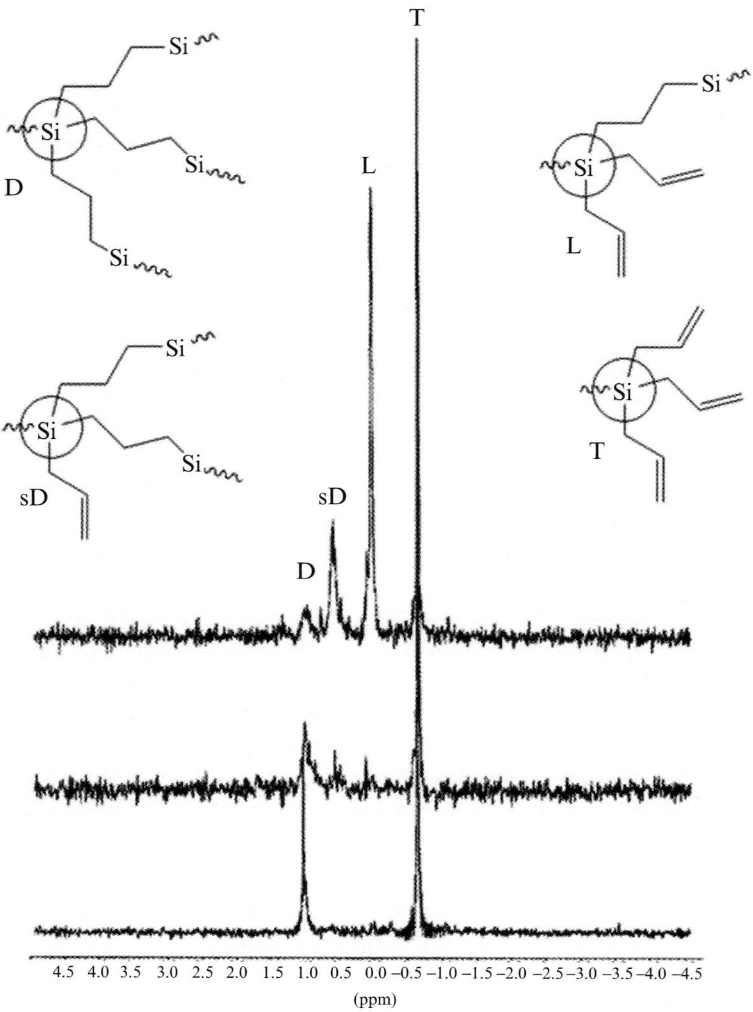

Fig. 13.1 Comparison of ^{29}Si-NMR spectra for hyperbranched polymer, hyperbranched polymer after modification and dendrimer (from *top* to *bottom*) [56].

catalyst (compare with Chapter 9). The only example where this approach has been transferred to hyperbranched carbosilanes to date was reported by Schlenk et al. [59, 60]. Here, polytriallylsilane was used as a soluble macromolecular support for catalyst **16**, obtained by introduction of dimethylaminoaryl palladium(II) "pincer" complexes as catalytically active sites at the free double bonds by a sequence of lithiation/transmetalation reactions.

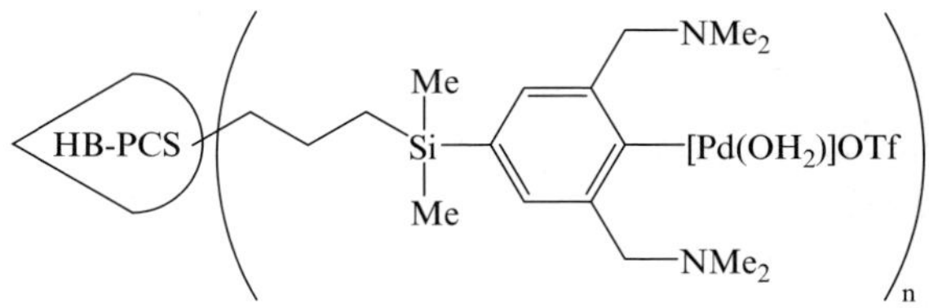

16

The obtained soluble catalyst was employed in the aldol condensation of benzaldehyde and methyl isocyanoacetate where it showed similar reactivity and conversion as the molecular catalyst (see Table 13.1). However, it was impermeable for dialysis membranes, which permits its application in continuous membrane reactors.

Kim and Kim gave further examples for possible modifications of terminal double bonds of polytriallylsilane and polydiallylmethylsilane [61]. After hydrosilylation addition of dichloromethylsilane or chlorodimethylsilane, the resulting chlorosilanes were substituted with allyl magnesium bromide, lithium amide, cholesterol, lithium bis(trimethylsilyl)amide or lithium phenylacetylenide to give functional polymers. The obtained polymers were characterized by NMR and matrix assisted laser desorption ionization-time of flight (MALDI-TOF), however no examination of the resulting properties or possible applications was reported. Hydrophilization of polyallylsilanes with hydroxyl end-groups was reported by Getmanova et al. [62] who synthesized 1-(3-dimethylsilyl)propyloxy-2-trimethylsilyloxyethane, **17**, and coupled it to the polymer double bonds. Hydrolysis of the trimethylsiloxy end-groups afforded the desired hydroxyl derivative **18** (see Reaction Scheme 13.7). The formation of a hydrogen bond network was studied using IR spectroscopy and it revealed high sensitivity towards temperature and dilution.

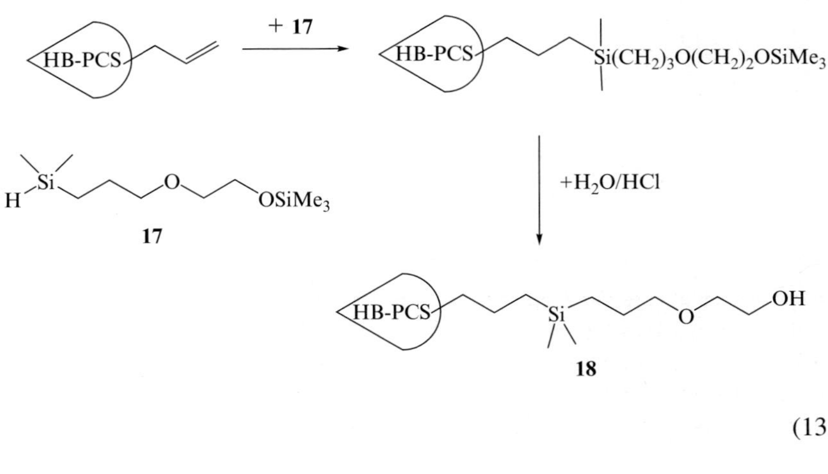

(13.7)

Table 13.1 Aldol condensation catalysis using benzaldehyde and methyl isocyanoacetate as substrates

Catalyst	Mol %	Yield (%)	TOF[a]	TTN/Pd site[b]
None	–	<5	–	–
Molecular	0.99	99	37	100
Polymer supported	0.79	>99	19	78

[a]Turnover frequency per Pd center per hour; during the first hour
[b]After 24 h obtained by specific signal integration using ^{1}H-NMR and mesitylene as internal standard (duplo experiment); TTN = total turnover number

An interesting approach to create nanometer-size cavities in a polymer matrix was reported by Möller et al. [63]. They modified polydiallylmethylsilane with methacrylate end-groups via a degradable Si–O linkage **19** (see Reaction Scheme 13.8). The resulting structures were then copolymerized with different amounts of methacrylate monomers to form homogeneous copolymer networks. Finally, the sensitivity of Si–O linkages towards alcoholysis was utilized for degradation, resulting in small functional pores within the polymer matrix.

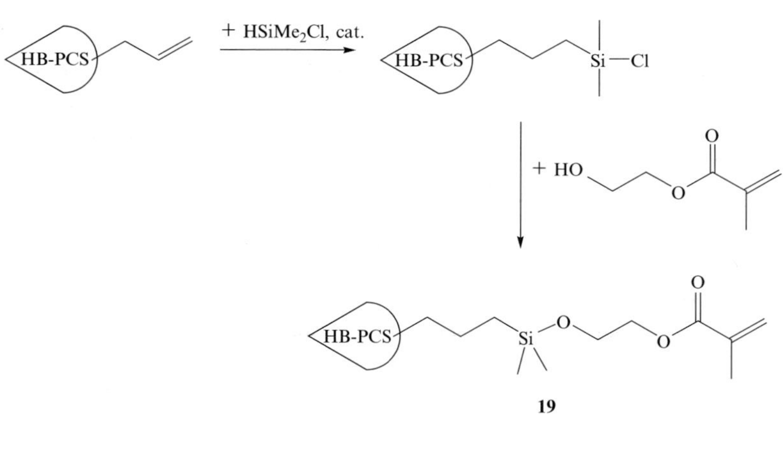

19

(13.8)

Bystrova et al. prepared ethoxysilyl derivatives of polydiallylmethylsilane and studied the network formation by hydrolysis with acetic acid [64]. The formation of siloxane bonds was accompanied by the evolution of ethyl acetate, alcohol and water, which allowed determination of the conversion of ethoxysilyl groups (70%). After extraction of the soluble part of the network, the resulting yield was >90% (see also

Chapter 11, Section 11.4). The networks were stable in a wide temperature range. Unfortunately, no analysis of the mechanical properties of the materials was given. The spatial structure of polydiallylmethylsilane was studied by Ozerin and coworkers using X-ray scattering and molecular modelling [65, 66]. The primary finding was that the hyperbranched structure was less dense than a dendrimer of the same molecular weight, but considerably more dense than a comparable Gaussian coil of similar molecular weight. A further detailed structural analysis of polydiallylmethylsilane was performed by Muzafarov et al., who studied fractionated samples using absolute molecular weight detection [67]. As expected, it was found that the Mark-Houwink parameter *a* (0.26–0.28) was low and intermediate between that for dendrimers and linear polymers. Additionally, intrinsic viscosities were found to be low and comparable to those of dendrimers. An improved synthesis for core-functional polydiallylmethylsilane using the B_3-core (11-trivinylsilanyl-undecyl)-isoindol-1,3-dion was reported by Frey et al. [68]. Here, the vinylsilane groups of the core showed higher reactivity than the double bonds of the monomer diallylmethylsilane, resulting in an efficient functionalization and control of polymerization. Removal of the phthalimide core with hydrazine and bisglycidolization of the resulting amine afforded a macroinitiator for the attachment of a hyperbranched polyglycerol block via ring opening polymerization of glycidol to yield hyperbranched diblock copolymers.

13.4.3 Block Copolymers

Recently, the first incorporation of hyperbranched polymer structures into block copolymers has been reported. In an innovative approach, García-Marcos et al. employed a polystyrene-*b*-1,2-polybutadiene (PS-*b*-PB) block copolymer **20** as a multifunctional core for the hypergrafting of a hyperbranched polycarbosilane block **21** (see Reaction Scheme 13.9) [69, 70]:

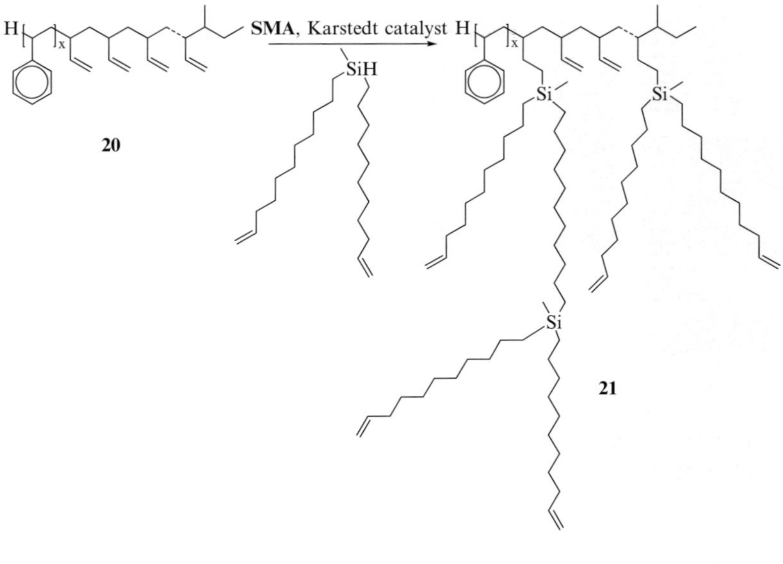

20

21

(13.9)

In this strategy, the short polybutadiene block represents a multifunctional core with a number of double bonds that are capable of reacting with carbosilane monomers. In order to avoid polyaddition of the monomer, the slow monomer addition (SMA) technique was applied. It utilizes slow and dropwise addition of a dilute monomer solution, which favors reaction with the core and permits control of the size of the hyperbranched polycarbosilane block, while keeping the molecular weight distributions extremely narrow (PDI <1.1). Morphological studies by transmission electron microscopy (TEM), AFM and small angle X-ray scattering (SAXS) of these systems demonstrated that various strongly separated microdomain structures were obtained, despite the additional structural isomerism of the branched block adding to the polydispersity of molecular weights. Manipulation of materials composition allowed morphological control analogous to that in linear diblock copolymers with observation of similar microphase-separated states. However, the branched structure appeared to contribute a topological factor to incompatibility resulting in ordered cylindrical structures for the block copolymer PS_{520}-b-[PB_{47}-HB-PCS_{142}] with 49 wt % of the hyperbranched component (see Fig. 13.2). Since an irreversible transformation to the expected lamellar phase took place after annealing, the authors suggested that cylindrical morphology represented a kinetically controlled or metastable state rather than an equilibrium morphology. All of the block copolymers with hyperbranched polycarbosilane-blocks showed extremely rapid phase segregation due to pronounced incompatibility of the linear polystyrene blocks and hyperbranched polycarbosilane structure.

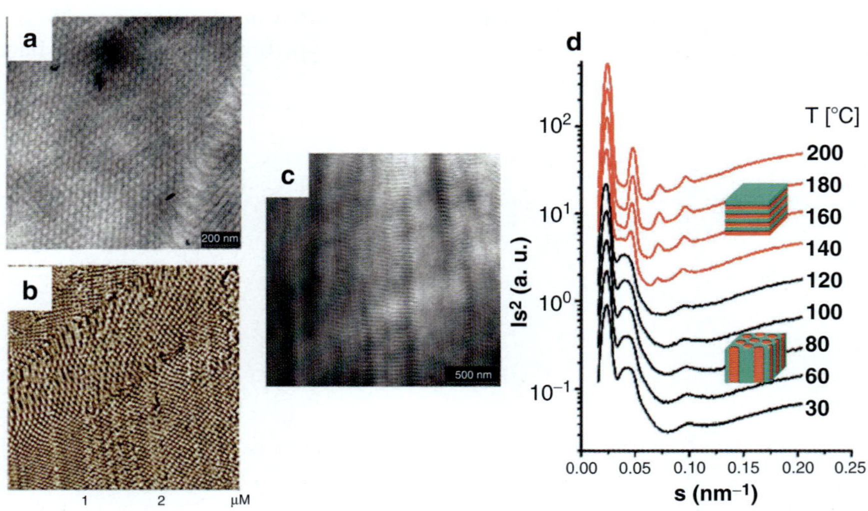

Fig. 13.2 Morphology of PS_{520}-*b*-[PB_{47}-HB-PCS_{142}]: (**a**) and (**b**) unstained TEM and AFM phase images of the melt-pressed sample at room temperature, (**c**) TEM at room temperature of the solution-cast sample, (**d**) temperature-dependent one-dimensional SAXS pattern.

13.5 Polycarbosiloxanes

13.5.1 General Synthetic Strategy

Most of the basic synthetic routes for the preparation of Si–O containing structures have already been demonstrated in dendrimer research [71]. Monomers for the preparation of hyperbranched polycarbosiloxanes or polysiloxysilanes are mainly synthesized by condensation of silanols and silyl chlorides. In order to minimize the number of possible condensation products, e.g., self-condensation of silanols, bases have to be added to keep the reaction medium neutral. Particularly, the system triethylamine and dimethylaminopyridine (DMAP) is known to favor SiCl/SiOH condensation.

Polycarbosiloxanes (PCSO) were the first examples of silicon containing hyperbranched structures reported as early as 1991 in a seminal work by Mathias and Carothers [72–74]. The AB_3 monomer allyltris(dimethylsiloxy)silane, **22**, was prepared by condensation of allyltrichlorosilane and dimethylchlorosilane, and its polymerization afforded a hyperbranched structure **23** with a molecular weight of 19,000 g/mol (Reaction Scheme 13.10).

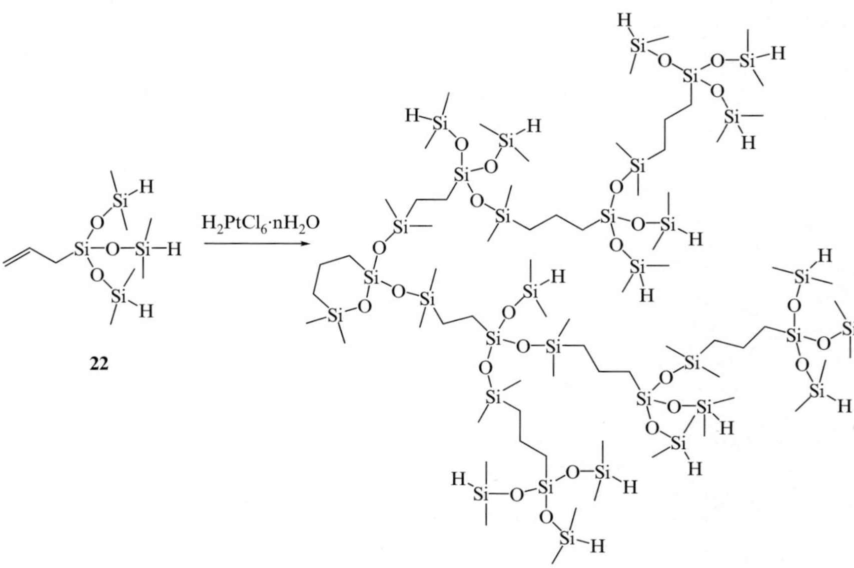

22 **23** (13.10)

The large number of Si–H end-groups led to a decreased stability of the polymer and resulted in gradually decreasing solubility of the material due to the Si–H oxidation followed by Si–O–Si formation. End-capping with allylphenyl ether or allyl-terminated PEO afforded long-term stable polymers. In subsequent work the authors showed that intramolecular cyclization into a six-membered ring constituted a significant side reaction [75]. This cyclic acted as a bifunctional core during the polymerization, broadening the molecular weight distribution, but its formation was disfavored when longer spacer groups between the Si–H moiety and the double bond were employed [76].

Following this, Rubinsztajn confirmed formation of six-membered cyclics during the polymerization of allyltris(dimethylsiloxy)silane, **22** [77], and reported polymerization of vinylsilyl derivatives vinyltris(dimethylsiloxy)silane **24** and tris(dimethylvinylsiloxy)silane **25**. In the latter cases, possible cyclization products were five-membered rings, which have a higher ring strain than the six-membered rings and are thus less prone to intramolecular reaction. As expected, higher polymer yields and molecular weights were achieved, and this result was most pronounced in bulk polymerization. Slow monomer addition technique resulted in polymers of considerably higher molecular weights and broad molecular weight distributions. Later, the same group reported on the synthesis of an aromatic AB_3 monomer (4-vinylphenyl)tris(dimethylsiloxy)silane, **26** [78]. This monomer can be polymerized to give either the hyperbranched polycarbosiloxane or linear tris(dimethylsiloxy)silyl-substituted polystyrene. As expected, an increase in T_g of the aromatic hyperbranched polymer (T_g = −58°C) was observed relative to its aliphatic analog prepared from tris(dimethylsiloxy)vinylsilane (T_g = −100°C). All polymers reported by this group were capable of further functionalization, e.g. Si–H end-groups were functionalized with trimethylvinylsilane and vinyl-terminated polymers were reacted with pentamethyldisiloxane.

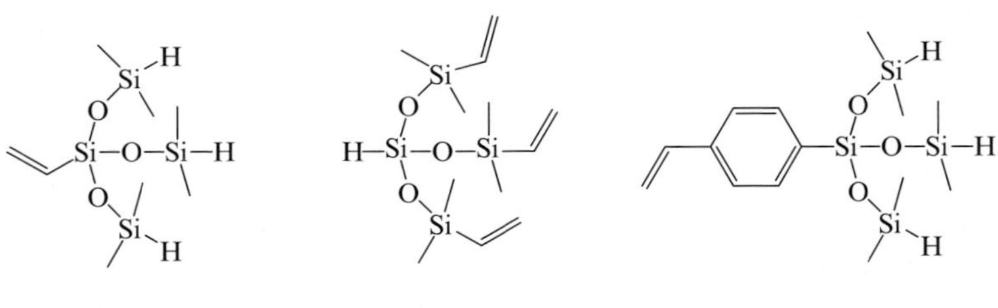

24 **25** **26**

In a detailed study of polycarbosiloxanes, Miravet and Fréchet examined the effects of the branching multiplicity of monomers on the polymerization of AB_2 **27**, AB_4 **28** and AB_6 **29** in the presence of solid-supported Pt/C catalyst [79, 80]:

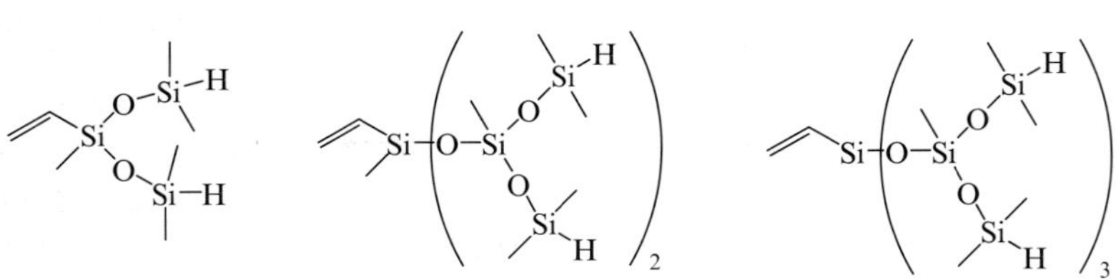

27 **28** **29**

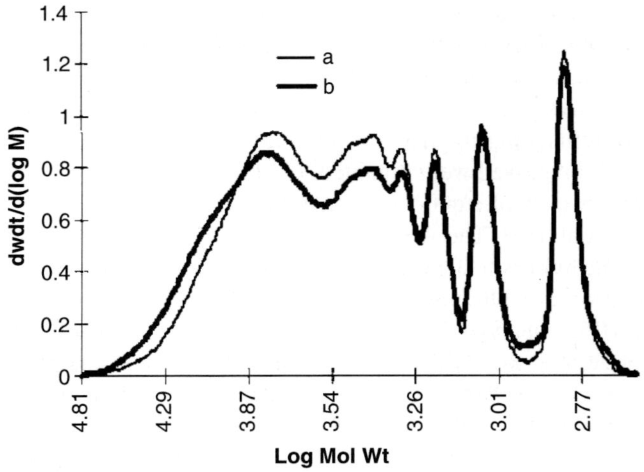

Fig. 13.3 SEC traces for the polymer obtained from vinyltris[methylbis(dimethylsiloxy)siloxy] silane; (**a**) directly after polymerization; (**b**) after a second addition of monomer and catalyst [80].

In all cases silicon hydride terminated polymers with molecular weights of up to 9,000 $gmol^{-1}$ were obtained. SEC traces showed multiple resolved peaks with elution volumes corresponding to low molecular weight compounds (i.e., monomer, dimer, trimer, etc.). These peaks were most probably due to cyclized products since no remaining vinyl groups could be detected in the materials. In all cases, only a moderate increase of molecular weights was observed when extra monomer was added to yield materials that, like initial polymers, contained low molecular weight oligomers (see Fig. 13.3).

The accessibility of terminal Si–H groups was proven by the quantitative reaction with different olefins and aldehydes. Polymers were modified with antioxidant groups, triethylene glycol or benzylchloride moieties via hydrosilylation addition of double bond-containing reagents. The T_gs of unmodified polymers revealed the high flexibility of the structures (T_g below −100°C), but were strongly affected by the type of the terminal groups. In order to extend the range of available molecular weights, the same group reported on slow monomer addition approaches, which should reduce the occurrence of cyclization [46]. The slow addition of monomer allowed preparation of polymers with molecular weights in the range of 6,000–86,000 g/mol and polydispersities between 2 and 15. The authors stated that the slower the addition or the larger the amount of the monomer added, the higher the molecular weight and the polydispersity of the resulting polymers. The same correlation was found when the monomer was slowly added to a preformed and purified polymer core.

A slow addition approach was also reported by Deubzer and Herzig [81] who added vinyltris(dimethylsiloxy)-silane to different cores with multiple Si–H groups. Molecular weights of the polymers depended on the amount of monomer added and were established by viscosity measurements. However, the formation of small cyclic compounds could not be avoided and it constituted up to 10% of the material.

The hydrosilylation polymerization of poly(dimethylsiloxane) macromonomers to afford long-chain hyperbranched polycarbosiloxanes was reported by Muzafarov et al. [82]. The macromonomers employed had degrees of polymerization of 10, 50 and 100, respectively, and contained one vinyl and two silicon-hydride groups. Molecular weights ranging from 15,000 to 800,000 g/mol depending on the macromonomer employed were obtained.

One of the few examples of optically active hyperbranched polymers was reported by Kawakami et al. [83]. The AB_2 monomer (S,S)-1,5-di(1-naphthyl)-1,5-diphenyl-1,5-divinyl-3-methyl-trisiloxane, **30**, was synthesized by coupling of an enantiomeric pure silanolate with dichloromethylsilane. Optical purity of the monomer was confirmed using HPLC analysis on a chiral stationary phase. Hydrosilylation polymerization was performed using different feed ratios of the AB_2 monomer and optically inactive A′B′ monomer 1,1,3,3-tetramethyl-1-vinylsiloxane, **31**, to give optically active hyperbranched polymers **32** ($[a]_D^{25}$ = +14.1–+4.5) as shown in Reaction Scheme 13.11:

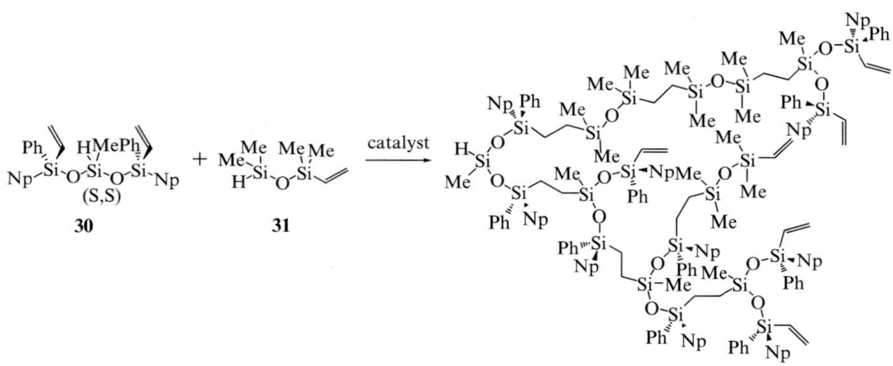

(13.11)

A detailed investigation of the degree of branching was given by comparison with model compounds using ^{29}Si-NMR. The DB was found to be 0.43 for the homopolymer, slightly lower than expected from theory, and, as expected, a decrease of the DB was observed for larger feed ratios of A′B′ monomer. The optical rotation power for the homopolymer was found to be higher than for the monomer and for the linear analog. The authors assigned this unusual finding to distinct local environments of each assymmetric center in the branched structure. End-functionalization was demonstrated using hydrosilylation or hydroboration to obtain optically active hyperbranched polycarbosiloxanes with silyl chloride or hydroxyl end-groups. Recently, a slightly improved synthesis of tris(dimethylvinylsiloxy)silane, **25** (previously reported by Rubinsztajn [77]) was published by Kakimoto et al. [84]. Detailed spectroscopic characterization of the monomer and the resulting polymer was given. β-addition was found in 74% of the linkages, compared to 80–90% reported by Rubinsztajn. The vinyl end-groups were fully converted to epoxides using 3-chloroperoxybenzoic acid. Radical addition of 2-mercaptoacetic acid and

2-mercaptoethanol resulted in carboxylated or hydroxylated polymers with a slight increase of polydispersity.

13.5.2 Polymer Modification and Application

In recent years, the functionalization and utilization of polycarbosiloxanes has become the most intensely studied area in the field of hyperbranched silicon-containing polymers. Gong and Fréchet dedicated further research to the transformation of multiple Si–H functionalities of poly(methylvinylbis(dimethylsiloxy) silane) [85]. Targeting applications in the field of crosslinking or curing materials the polymers were modified with epoxy, **33**, amine, **34**, or hydroxyl groups, **35** (see Reaction Scheme 13.12):

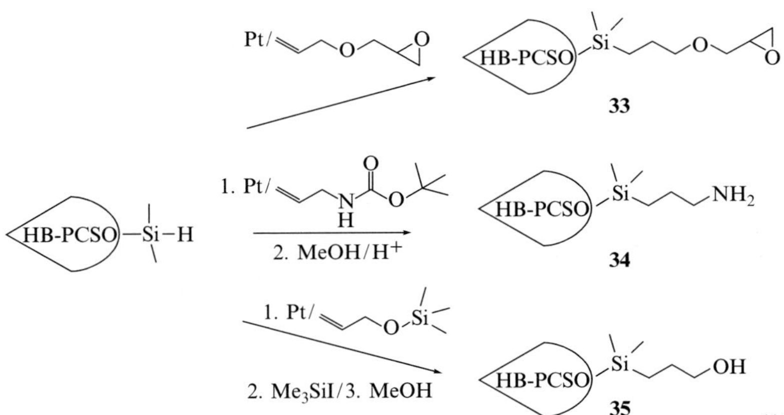

(13.12)

The high efficiency of polymer modification was expanded to the preparation of hyperbranched–linear star polymers. The attachment of vinyl-terminated polyisobutylene chains resulted in enhanced hydrophobicity of the polymers, as determined by contact angle measurements with water. On the other hand, amphiphilic star polymers were prepared by reaction with allyl-terminated PEO resulting in water-soluble materials.

Using a modified reaction procedure, Wang et al. prepared several AB^2 siloxysilane monomers, including methylbis(methylethylvinylsiloxy)silane, methylvinylbis (methylethylsiloxy)silane and methylbis(dimethylallylsiloxy)silane [86]. Their hydrosilylation polymerizations afforded vinyl, silyl and allyl terminal polymers, respectively. In order to obtain UV-curable products the silicon-hydride functional polymer was further reacted with glycidylmethacrylate to yield an epoxy-functional hyperbranched structure. The UV curing behaviors of all polymers was investigated in air and under nitrogen using different photoaccelerators. In a follow-up paper the same group reported modification of silicon-hydride polymers with alkyl phenone acrylates via hydrosilylation addition [87]. The resulting UV-sensitive materials were used as polymeric photoinitiators for the curing of epoxy acrylate resins. Thermal stability of the curing systems was higher for the polymeric photoinitiators than for the low molecular ones.

Hyperbranched polymers and especially silicon-containing structures are usually amorphous materials characterized by high chain flexibility resulting in low T_g values. However, crystalline hyperbranched polymers were reported by Kakimoto et al. [88]. Hydro-heptacyclopentyl-substituted polyhedral oligomeric silsesquioxane (POSS-H) **36** was attached to preformed hyperbranched PCSO via hydrosilylation with controllable degrees of substitution **37** (see Reaction Scheme 13.13 and compare with Section 7.4.5). The solid materials displayed a T_g at 16–18°C as well as a melting point, T_m. The latter was assumed to be due to the POSS regions and decreased with decreasing POSS content. Wide-angle X-ray scattering revealed a similar diffraction pattern for POSS-H and the fully substituted polymer, suggesting a hexagonally packed structure in both cases. Again, the degree of crystallization was found to be proportional to the content of POSS groups.

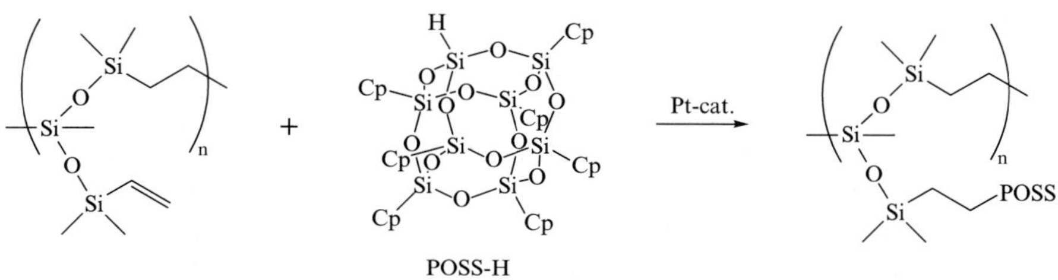

36 **37**

(13.13)

The performance of conductive polymer aluminum solid electrolytic capacitors is strongly influenced by the affinity between the aluminum oxide and the conductive polymer (see Fig. 13.4). Therefore, Nogami et al. modified the aluminum oxide film with hyperbranched poly(1,5-divinyl-1,1,3,5,5-pentamethyltrisiloxane) (PDVS) before the deposition of the electro active film (here poly(3,4-ethylenedioxythiophene), PEDOT) [89, 90]. Compared to the untreated device, an increase in capacitance and a decreased equivalent series resistance was observed (ESR). The effect was even more pronounced when a slightly crosslinked PDVS of higher molecular weight was used. In contrast, no positive effects were observed when linear vinyl-terminated polydimethylsiloxane was applied. This result was attributed to an increased hydrophobic-hydrophilic ratio for hyperbranched PDVS compared to that of the linear polymer, resulting in improved affinity to aluminium oxide as well as to the conducting polymer. In a subsequent analysis it was shown that HB-PDVS inhibits the damage to the oxide dielectric layer during the chemical polymerization of PEDOT [90]. Thus, an improved adhesion between the dielectric oxide film and the PEDOT resulting in an increased contact area is achieved, constituting the main factor for the improved performance of the coated devices.

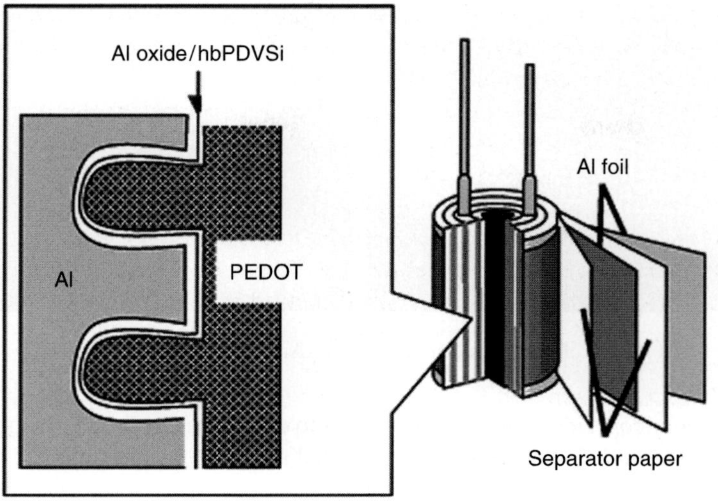

Fig. 13.4 Aluminum solid electrolytic capacitor (wound type) [90].

The same polymer was used in an interesting application of polysiloxysilane-based star polymers targeting column materials for temperature responsive HPLC reported by Kakimoto et al. [91]. Polymerization of 1,5-divinyl-1,1,3,5,5-pentamethyltrisiloxane afforded hyperbranched structures that were hydroborated to give the hydroxyl modified polymers. Reaction with 2-bromoisobutyryl bromide resulted in ATRP macroinitiators **38**. Grafting of poly(*N*-isopropylacrylamide) (PIPAAm) arms in solution led to star polymers **39** with Mn = 23,500 g/mol and PDI of 1.31 that showed a lower critical solution temperature at 34°C in water (see Reaction Scheme 13.14). Additionally, the macroinitiator was immobilized on silica beads and likewise used as initiator for the ATRP of *N*-isopropylacrylamide (see Fig. 13.5). PIPAAm grafted silica beads were successfully applied as column material for the separation of hydrophobic and hydrophilic steroids.

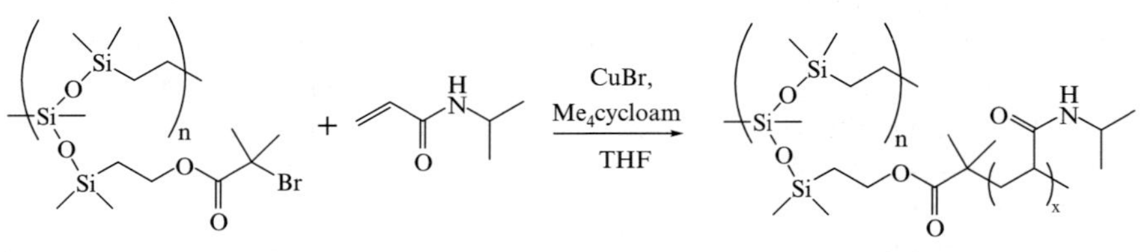

38 **39**

(13.14)

Kakimoto et al. used branched poly(1,5-divinyl-1,1,3,5,5-pentamethyl-trisiloxane) as a scaffold to inhibit fluorescence quenching of rhodamine B dye [92]. The polymer was converted into the hydroxyl-terminated derivative via radical addition of mercaptoethanol. Esterification with N,N′-dicyclohexylcarbodiimide (DCC) and

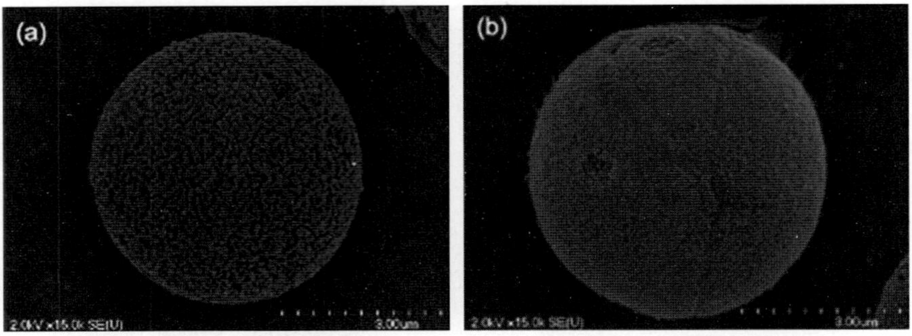

Fig. 13.5 SEM images of original silica (**a**) and PIPAAm-hbPCSO-silica (**b**) [91].

4-dimethylaminopyridine (DMAP) was used to attach the dye to the polymer scaffold and NMR and UV measurements indicated a degree of substitution of 32–34%. As a result of the dye separation in the branched structure, no fluorescence quenching was observed for the polymer-supported dye even at high concentrations.

13.6 Polyalkoxysilanes

A general procedure for the synthesis of monomers suitable for hyperbranched polyalkoxysilanes is based on the hydrolysis of chlorosilanes with unsaturated alcohols. These monomers are easily accessible from commercially available materials, however the potential moisture sensitivity of the resulting polymer products has to be kept in mind (compare with Chapters 6 and 14). A few selected examples of such monomers are shown in structures **40–43**.

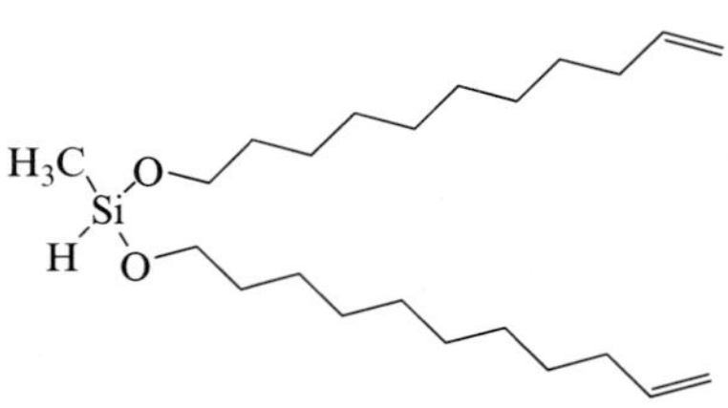

40

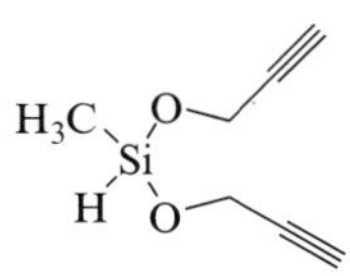

41

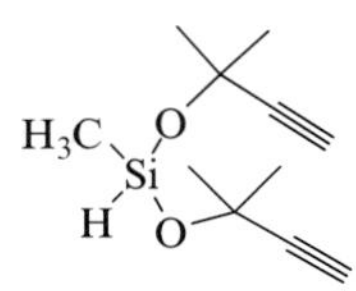

42

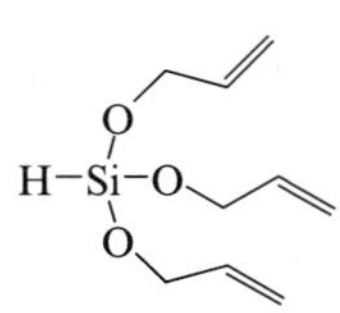

43

Hyperbranched poly(bis(undecenyloxy)methylsilane), from monomer **40**, was reported by Möller and coworkers in 1995 [93]. The authors capitalized on the labile nature of the silicon-oxygen-carbon bonds for deliberate degradation of the polymer. Low molecular weight compounds were obtained by reaction with refluxing methanol or by reaction with aqueous HCl. This enabled the use of this polymer as a degradable template for creation of nanometer-size cavities in polymer resins [63]. For example, photoreaction with methacrylate monomers resulted in a phase-segregated structure with agglomerates of hyperbranched polymer, presumably due to incompatibility of the hyperbranched polymer functional groups with the resin material, from which the removal of hyperbranched structure by hydrolysis resulted in the targeted pores in the matrix.

Xiao and Son prepared AB_2 and AB_3 monomers by reaction of dichloromethylsilane or trichlorosilane with propargylalcohols [94]. Polymerization with Pt/C led to polymers containing pendant acetylenic groups that underwent thermally induced crosslinking between 220°C and 260°C. Heating the polymers to 1,400°C in a nitrogen flow resulted in weight losses ranging from 33% to 66%. Their thermal stability was found to be lower than for polymers based on methyldiethynylsilane. Also, reaction of terminal groups with phenylethynyldimethylsilane further decreased this thermal stability.

A hyperbranched polycarbosiloxane was used as a core for dendrimer-like growth by Kim and Kim [95]. Triallyloxysilane was polymerized and modified by subsequent addition of trichlorosilane and substitution with allyl alcohol. Two generations were added to the core with an almost unchanged polydispersity index. An attempt to increase the control of the polymerization of hyperbranched polyalkoxysilanes was reported by Wang et al. [96]. A boron core with three functional allyloxysilane groups was added to the bulk polymerization of diallyoxymethylsilane. With increasing amount of core added the polydispersity of the samples decreased from 3.33 for the unfunctionalized polymer to 2.30 for the system with 2.5 mol % core.

13.7 Polycarbosilazanes

Although carbosilazane dendrimers are known and show promise as preceramic polymers, little effort has so far been dedicated to the preparation of hyperbranched polycarbosilazane structures (see Chapter 5). Carbosilazanes contain $N(Si)_x$ as branching points and are characterized by the basicity of nitrogens and relative sensitivity of the Si–N bond, which provides a pathway for directed degradation. Such materials were also considered as precursors for Si_3N_4 ceramics. In 1999 Yoon and Son reported that an attempt at the hydrosilylation polymerization of silazane AB_2 monomers failed and instead the intramolecularly cyclized product was observed. Platinum catalysis in dilute solution resulted in intramolecular α-addition to give 1-aza-2,4-disilacyclobutanes in nearly quantitative yields [97]. Only recently, Qi et al. reported the first synthesis of hyperbranched polysilazanes **45** from AB_2 **44** and AB_4 type monomers using H_2PtCl_6 as a catalyst, as shown in Reaction Scheme 13.15 [98]. NMR analysis revealed that, likewise, α-addition prevailed, with molecular weights in the range of 3,900–14,500 g/mol as determined by multi-angle laser light scattering (MALLS). At this time, this class of hyperbranched materials is still in its infancy.

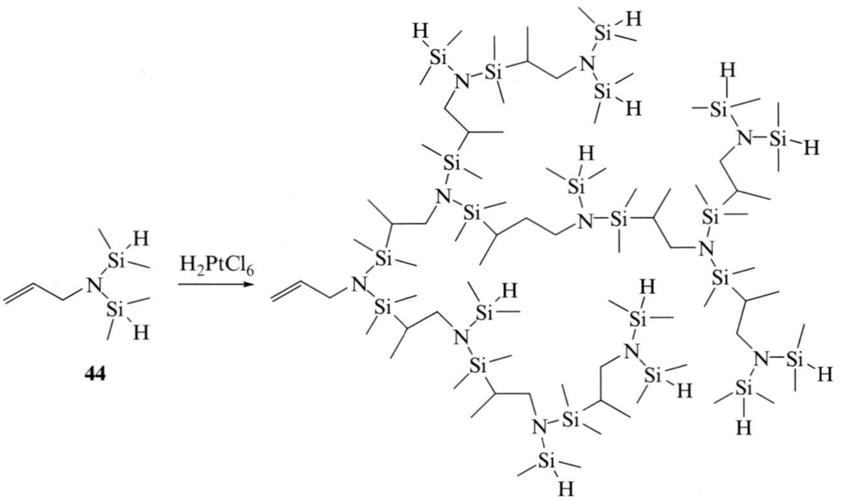

44

45

(13.15)

13.8 Summary and Perspectives

Although numerous basic synthetic pathways have been established for the preparation of hyperbranched materials based on hydrosilylation chemistry, it appears that the field is by no means fully exploited or mature. For example, the progress reported for this field to date has only little capitalized on the facile combination with sol-gel chemistry based on alkoxysilanes, which may be quite promising for toughening ceramics as well as for preparation of new, unusual hybrid materials. Furthermore, the flexibility of carbosilanes may be used to create liquid crystalline architectures in analogy to perfectly branched LC-dendrimers (see Chapter 10). The synthesis of "tailor-made" hyperbranched silicon-containing architectures including star or linear-hyperbranched block copolymers will certainly receive more systematic future attention. First promising approaches in this direction have already been reported, as summarized in this review.

Further studies of structure formation in these unusual topologies will also increase the general understanding of structure-property relationships in block copolymers. Branched or hyperbranched polysiloxanes possessing multiple end-groups bear potential for a large variety of applications and it is safe to expect that more work in this area will be carried out in the near future. Hydrosilylation chemistry has also played a very important role in the preparation of hyperbranched silicon-containing polymers by bimolecular nonlinear polymerization (BMNLP) reactions involving $A_x + B_y$ monomer systems (see Chapter 16). In summary, this area of hyperbranched polymer chemistry offers interesting and manifold synthetic and structural challenges for the future.

References

1. Mathias LJ, Carothers TW (1995) Advances in Dendritic Macromolecules. CAI, Greenwich, p 101
2. Frey H, Lach C, Lorenz K (1998) Adv Mater 10: 279
3. Majoral JP, Caminade AM (1999) Chem Rev 99: 845
4. van der Made AW, van Leuwen PWNM (1992) J Chem Soc Chem Comm 19:1400–1401
5. Flory PJ (1952) J Am Chem Soc 74: 2718
6. Flory PJ (1953) Principles of Polymer Chemistry. Cornell University Press, Ithaca, NY
7. Son DY (2001) In: Rappoport Z, Apeloig Y (eds) The Chemistry of Silicon Compounds, vol 3. Wiley, New York, p 745
8. Frey H, Schlenk C (2000) Top Curr Chem, vol 210. Springer, Heidelberg, Berlin, p 69
9. Chatgilialoglu C (1992) Accounts Chem Res 25: 188
10. Kopping B, Chatgilialoglu C, Zehnder M, Giese B (1992) J Org Chem 57: 3994
11. Sommer LH, Pietrusza EW, Whitmore FC (1947) J Am Chem Soc 69: 188
12. Speier JL (1979) Adv Organomet Chem 17: 407
13. Speier JL, Webster JA, Barnes GH (1956) J Am Chem Soc 79: 974
14. Ashby BA, Modic FJ (1981) US Patent 4,288,345
15. Karstedt BD (1973) US Patent 3,715,334
16. Dvornic PR, Gerov VV (1994) Macromolecules 27: 1068
17. Dvornic PR, Gerov VV, Govedarica MN (1994) Macromolecules 27: 7575
18. Marciniec B, Maciejewski H, Duczmal W, Fiedorow R, Kitynski D (2003) Appl Organomet Chem 17: 127
19. Harrod JF, Chalk AJ (1965) J Am Chem Soc 87: 16
20. Harrod JF, Chalk AJ (1977) In: Wender I, Pino P (eds) Organic Synthesis via Metal Carbonyls. Wiley, New York, p 673
21. Randolph CL, Wrighton MS (1986) J Am Chem Soc 108: 3366
22. Reichel CL, Wrighton MS (1980) Inorg Chem 19: 3858
23. Schroeder MA, Wrighton MS (1977) J Organomet Chem 128: 345
24. Lewis LN (1990) J Am Chem Soc 112: 5998
25. Stein J, Lewis LN, Gao Y, Scott RA (1999) J Am Chem Soc 121: 3693
26. Antic VV, Antic MP, Govedarica MN, Dvornic PR (2007) J Polym Sci Polym Chem 45: 2246
27. Antic VV, Antic MP, Govedarica MN, Dvornic PR (2007) Mater Sci Forum 555: 485
28. Stein J, Lewis LN, Smith KA, Lettko KX (1991) J Inorg Organomet Polym 1: 325
29. Uriarte RJ, Lewis LN (1990) Organomet 9: 621
30. Chalk AJ, Harrod JF (1964) J Am Chem Soc 86: 1776
31. Haszeldine RN, Parish RV, Taylor RJ (1974) J Chem Soc A: 2311
32. Hill AF (2002) Organotransition Metal Chemistry. Royal Society of Chemistry, Cambridge
33. Hiyama T, Kusumoto T (1991) In: Trost BM, Fleming I (eds) Comprehensive Organic Synthesis, vol 8. Pergamon, Oxford, p 763
34. Ojima I (1989) In: Rappoport Z, Apeloig Y (eds) The Chemistry of Organic Silicon Compounds, vol 1. Wiley, Chichester, UK, p 1479
35. Ojima I, Zi Z, Zhu J (1998) In: Rappoport Z, Apeloig Y (eds) The Chemistry of Organic Silicon Compounds, vol 2. Wiley, Chichester, UK, p 1687
36. Brook MA (2000) Silicon in Organic, Organometallic and Polymer Chemistry. Wiley, Canada
37. Burgath A, Sunder A, Frey H (2000) Macromol Chem Physics 201: 782
38. Hanselmann R, Hölter D, Frey H (1998) Macromolecules 31: 3790
39. Hölter D, Frey H (1997) Acta Polymer 48: 298
40. Radke W, Litvinenko G, Müller AHE (1998) Macromolecules 31: 239
41. Hölter D, Burgath A, Frey H (1997) Acta Polymer 48: 30
42. Muzafarov AM, Gorbacevich OB, Rebrov EA, Ignat'eva GM, Chenskaya TB, Myakushev VD, Bulkin AF, Papkov VS (1993) Polym Sci 35: 1575

43. Lach C, Müller P, Frey H, Mülhaupt R (1997) Macromol Rapid Commun 18: 253
44. Drohmann C, Gorbatsevich OB, Muzafarov AM, Möller M (1998) Polym Prepr Am Chem Soc 39: 471
45. Drohmann C, Möller M, Gorbatsevich OB, Muzafarov AM (2000) J Polym Sci Pol Chem 38: 741
46. Gong CG, Miravet J, Fréchet JMJ (1999) J Polym Sci Polym Chem 37: 3193
47. Burgath A, Sunder A, Frey H (2000) Macromol Chem Phys 201: 782
48. Yoon K, Son DY (1999) Macromolecules 32: 5210
49. Son DY, Yoon K (1999) Polym Mater Sci Eng 80: 200
50. Rim C, Son DY (2003) Macromolecules 36: 5580
51. Son DY, Rim C (2002) Polym Prepr Am Chem Soc 43: 1180
52. Simonson DL, Houser EJ, Stepnowski JL, Pu L, McGill RA (2003) Polym Mater Sci Eng 89: 866
53. Wong RA, Xiao Y, Son DY (2000) Polym Prepr Am Chem Soc 41: 608
54. Kwak G, Takagi A, Fujiki M, Masuda T (2004) Chem Mater 16: 781
55. Kwak G, Masuda T (2002) Macromol Rapid Commun 23: 68
56. Lach C, Frey H (1998) Macromolecules 31: 2381
57. Lach C, Hanselmann R, Frey H, Mülhaupt R (1998) Macromol Rapid Commun 19: 461
58. Frey H, Schlenk C, Pusel T, Lach C (2000) Polym Prepr Am Chem Soc 41: 568
59. Schlenk C, Kleij AW, Frey H, van Koten G (2000) Angew Chem Int Edit 39: 3445
60. Schlenk C, Kleij AW, Frey H, van Koten G (2000) Angew Chem Int Edit 39: 3736
61. Kim C, Kim H (2001) J Polym Sci Pol Chem 39: 3287
62. Getmanova EV, Chenskaya TB, Gorbatsevich OB, Rebrov EA, Vasilenko NG, Muzafarov AM (1997) React Funct Polym 33: 289
63. Muzafarov AM, Rebrov EA, Gorbacevich OB, Golly M, Gankema H, Moller M (1996) Macromol Symp 102: 35
64. Bystrova AV, Tatarinova EA, Buzin MI, Muzafarov AM (2005) Polym Sci Ser A 47: 820
65. Fadeev MA, Rebrov AV, Ozerina LA, Gorbatsevich OB, Ozerin AN (1999) Polym Sci Ser A 41: 189
66. Ozerin A (2001) Macromol Symp 174: 93
67. Tarabukina EB, Shpyrkov AA, Potapova DV, Filippov AP, Shumilkina NA, Muzafarov AM (2006) Polym Sci Ser A 48: 974
68. Schüle H, Nieberle J, Frey H (2007) Polym Mater Sci Eng 96: 252
69. Garcia-Marcos A, Pusel TM, Thomann R, Pakula T, Okrasa L, Geppert S, Gronski W, Frey H (2006) Macromolecules 39: 971
70. Frey H, Garcia-Marcos A, Pusel T, de Castro BD, Geppert S, Thomann R, Gronski W (2003) Polym Prepr Am Chem Soc 44: 534
71. Lang H, Lühmann B (2001) Adv Mater 14: 1523
72. Mathias LJ, Carothers TW (1991) Polym Prepr Am Chem Soc 32: 633
73. Mathias LJ, Carothers TW (1991) J Am Chem Soc 114: 4043
74. Mathias LJ, Carothers TW, Bozen RM (1991) Polym Prepr Am Chem Soc 32: 82
75. Carothers TW, Mathias LJ (1993) Polym Prepr Am Chem Soc 34: 538
76. Carothers TW, Mathias LJ (1993) Polym Prepr Am Chem Soc 34: 503
77. Rubinsztajn S (1994) J Inorg Organomet Polym 4: 61
78. Rubinsztajn S, Stein J (1995) J Inorg Organomet Polym 5: 43
79. Miravet JF, Fréchet JMJ (1997) Polym Mater Sci Eng 77: 141
80. Miravet JF, Fréchet JMJ (1998) Macromolecules 31: 3461
81. Herzig C, Deubzer B (1998) Polym Prepr Am Chem Soc 39: 477
82. Vasilenko NG, Rebrov EA, Myakushev VD, Muzafarov AM (1998) Polym Prepr Am Chem Soc 39: 603
83. Oishi M, Minakawa M, Imae I, Kawakami Y (2002) Macromolecules 35: 4938
84. Ishida Y, Yokomachi K, Seino M, Hayakawa T, Kakimoto M (2007) Macromol Res 15: 147
85. Gong C, Fréchet JMJ (2000) J Polym Sci Polym Chem 38: 2970
86. Si QF, Wang X, Fan X-D, Wang SJ (2005) J Polym Sci Pol Chem 43: 1883

87. Si QF, Fan XD, Liu YY, Kong J, Wang SJ, Qiao WQ (2006) J Polym Sci Pol Chem 44: 3261
88. Seino M, Hayakawa T, Ishida Y, Kakimoto MA (2006) Macromolecules 39: 8892
89. Nogami K, Kakimoto M, Hayakawa T, Yokomatchi K, Seino M, Sakamoto K (2006) Chem Lett 35: 144
90. Nogami K, Sakamoto K, Hayakawa T, Kakimoto M (2007) J Power Sources 166: 584
91. Seino M, Yokomachi K, Hayakawa T, Kikuchi R, Kakimoto M, Horiuchi S (2006) Polymer 47: 1946
92. Ding LJ, Hayakawa T, Kakimoto MA (2007) Polym J 39: 551
93. Muzafarov AM, Golly M, Möller M (1995) Macromolecules 28: 8444
94. Xiao YX, Son DY (2001) J Polym Sci Pol Chem 39: 3383
95. Kim C, Kim H (2004) C R Chimie 7: 503
96. Kong J, Fan XD, Si QF, Zhang GB, Wang SJ, Wang X (2006) J Polym Sci Pol Chem 44: 3930
97. Yoon K, Son DY (1999) Org Lett 1: 423
98. Zhang GB, Fan XD, Kong J, Liu YY, Wang MC, Qi ZC (2007) Macromol Chem Phys 208: 541

Chapter 14
Rearranging Hyperbranched Silyl Ether Polymers

Daniel Graiver

14.1 Introduction

Recent developments in the control of macromolecular architecture have led to important progress in dendritic structures, such as dendrimers and hyper-branched polymers, which exhibit fundamentally different properties from their linear counterparts, including impeded crystallization, minimized chain entanglements, unusual viscosity profiles and solubility behavior (see also Chapter 1). Furthermore, in contrast to linear polymers where the number of functional end-groups quickly diminishes as the molecular weight is increased, in these highly branched macromolecules the end-group functionality increases directly proportional to the molecular weight. This combination of high molecular weight, large number of terminal functional groups, overall globular shape of their molecules, a lower than expected viscosity and low degree of entanglements provides potential advantages that can be utilized in various applications.

Most common synthetic approaches to hyperbranched polymers have been based on a divergent growth method, where the monomer contains two types of functional groups that can react with each other but cannot react with themselves, and the overall functionality is greater than two (i.e., reacting monomer molecules are of the type AB_x, where $x \geq 2$) (see also Section 1.2). The simplest suitable monomers of this type contain a single functional group A and two functional groups B (i.e., AB_2 type), and polymerize as shown schematically in Reaction Scheme 14.1:

D. Graiver
Department of Chemical Engineering and Material Science, Michigan State University, East Lansing, MI, 48824, USA
E-mail: graiverd@egr.msu.edu

P.R. Dvornic and M.J. Owen (eds.), *Silicon-Containing Dendritic Polymers*, Advances in Silicon Science 2,

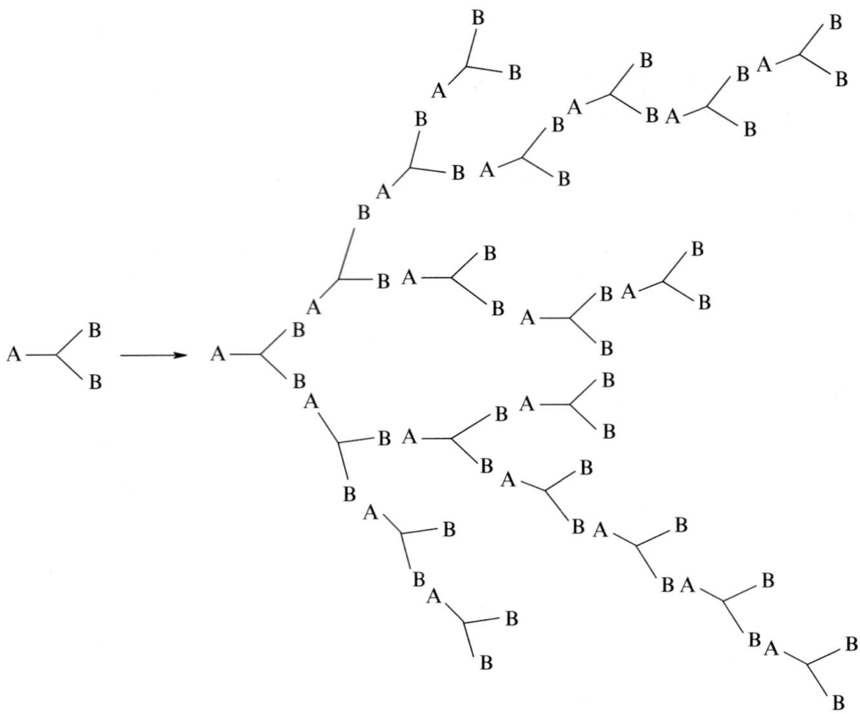

(14.1)

This one-step polymerization leads to uncontrolled statistical growth and molecular weight distributions with wide polydispersity indices of the resulting polymer products, since the statistical nature of the coupling steps and steric hindrance associated with the growing chains cause different polymer segments (or branches) to have different chain lengths [1]. Note, however, that the general picture shown in Reaction Scheme 14.1 does not include possible cyclization reactions at any reaction mixture concentration or at any stage of the polymerization. The effect of these cyclization reactions is limited chain-growth of the polymer and decreased polydispersity to some finite value that is less than infinity, as predicted by Flory's calculation [2]. However, since the focus of this chapter is on the concept of structural rearrangement, the omission of these cyclization reactions is not of critical importance. For more details about the importance of cyclization reactions in non-linear hyperbranched polymerizations see Chapter 15.

Monomers with a higher number of functional groups result in polymers with denser structures. Thus, a monomer containing a single functional group A and three functional groups B (i.e., an AB_3 type) will produce higher branching density polymers and a higher molecular concentration of B functional groups on their chain-ends. In principle, AB_x type polymers can be prepared where x is any integer with a value of 2 or greater. Another key property of hyperbranched polymers is that their degree of branching can be controlled by increasing the chain length between the functional groups. This approach minimizes the end-group crowding and tends to yield higher molecular weight polymers compared with identical reactions where steric effects hinder the polymerization. Due to the large number of branches in hyperbranched polymers, their hydrodynamic volumes are smaller than those of the linear polymers of the corresponding molar masses and chemical compositions. These different relationships between the volume and the molecular weight can be directly correlated with differences in viscosity, solubility, and some other physical properties. In silicon-containing hyperbranched polymers the presence of organosilicon groups affects the surface properties (e.g., adhesion to substrates, polymer–polymer interface composites, hydrophobicity, etc.), thermal and oxidative stability, glass temperature, compatibility with other materials, solubility, reactivity, etc. (see also Chapters 12, 13 and 16).

It should be noted that there are cases where silanes were used as protective groups during the preparation of the monomers for hyperbranched polymers but were then removed prior to polymerization. Obviously, the resulting hyperbranched polymers did not include silicon atoms in their structures but the use of organosilicon chemistry in the preparation of these monomers was an important step in their preparation. One such example [3] is a family of A_2B type hyperbranched aromatic polycarbonates which were prepared from 1,1,1-tris(4′-hydroxyphenyl)ethane and in which one of the three phenol groups was protected by a trimethylsiloxy group during the preparation of the monomer.

14.2 Monomer Synthesis

Most common hyperbranched polymers containing silicon atoms in their structures are produced by hydrosilylation reactions of silanes and siloxanes containing vinyl and Si–H functionalities (either in the same or in the different molecules, as described in Chapters 13 and 16, respectively). Consequently, all these hyperbranched polymers contain thermally and chemically stable carbosilane linkages. In contrast to these, in this chapter, hyperbranched polymers containing hydrolytically unstable Si–O–C linkages are described together with the effects of specific rearrangements that lead to their hydrolysis and condensation and ultimate formation of permanent thermoset resin structures (see also Chapter 6 and Section 13.6).

One such example is based on the exchange reaction of hydroxyl (A) and alkoxysilane (B) groups. For example, an AB_2 monomer can be prepared by reacting equimolar amounts of 3-aminopropylmethyldiethoxysilane with either cyclic 1,2-propylene carbonate, ethylene carbonate or γ-butyrolactone, as shown in Reaction Scheme 14.2. These ring opening reactions proceed smoothly under an inert atmosphere even at moderate temperatures with excellent yields [4–6]. They obey second-order kinetics with respect to the concentrations of the cyclic carbonates and the amine, whereby the rate of the reaction increases with increasing electron-withdrawing character of the substituents. Hence, ethylene carbonate has been reported to be more reactive than other carbonates.

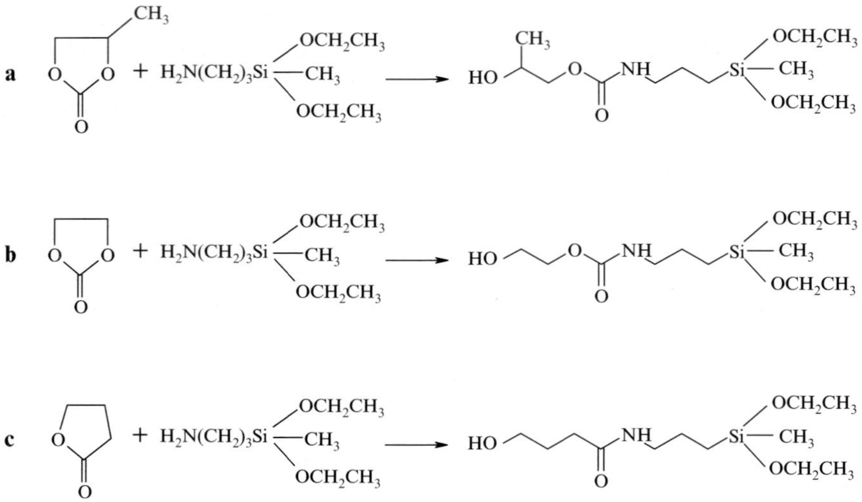

(14.2)

As shown in Reaction Scheme 14.3, this ring opening reaction actually yields two urethane adducts, one with a primary hydroxyl and the other with a secondary hydroxyl group. In general, the adduct containing a secondary hydroxyl group is predominantly formed when a stronger electron-withdrawing group is present. Although we were aware of the presence of these two different adducts, which would produce slightly different hyperbranched polymers, we have not tried to separate the monomer mixture.

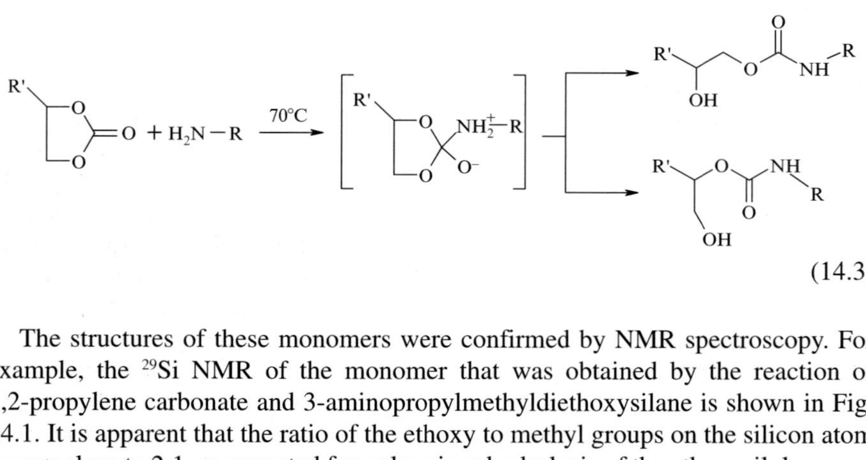

(14.3)

The structures of these monomers were confirmed by NMR spectroscopy. For example, the ^{29}Si NMR of the monomer that was obtained by the reaction of 1,2-propylene carbonate and 3-aminopropylmethyldiethoxysilane is shown in Fig. 14.1. It is apparent that the ratio of the ethoxy to methyl groups on the silicon atom is very close to 2:1, as expected for only minor hydrolysis of the ethoxysilyl groups. Similarly, the ^{29}Si NMR of the monomer that was obtained by the reaction of ethylene carbonate and 3-aminopropylmethyldiethoxysilane (Fig. 14.2) indicates three ethoxy groups attached to the silicon atom. Further confirmation of this AB_3 monomer structure was obtained from the ^{13}C NMR (Fig. 14.3). It is apparent that the cyclic carbonate opened and reacted with the amine. The extra peaks at 55.8 and 18.3 ppm are due to residual ethanol that was not removed completely from the reaction mixture, while the multiple peak around 7.5 ppm indicates the presence of residual, unreacted 3-aminopropylmethyldiethoxysilane.

14.3 Polymerization Reactions

Typical polymerizations of these AB_2 and AB_3 monomers to yield hyperbranched structures **1** were accomplished by heating the monomers at 90°C under gentle nitrogen flow [7] while collecting the by-product ethanol in a Dean-Stark trap:

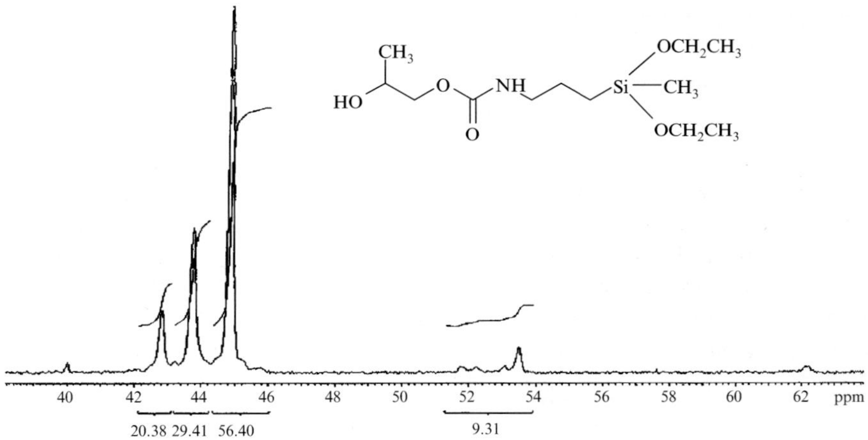

Fig. 14.1 ^{29}Si NMR of AB_2 monomer.

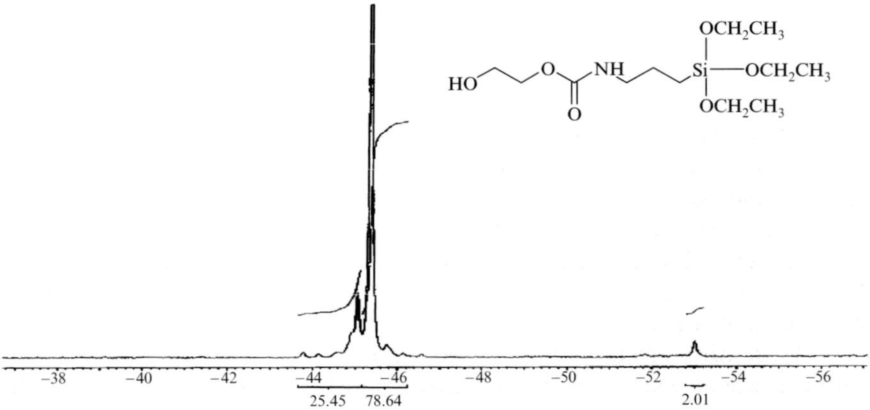

Fig. 14.2 ^{29}Si NMR of AB_3 monomer.

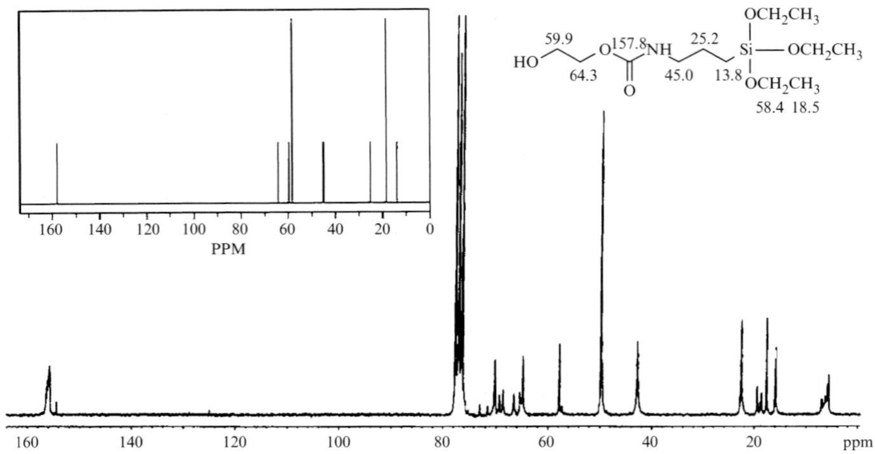

Fig. 14.3 ^{13}C NMR of AB_3 monomer.

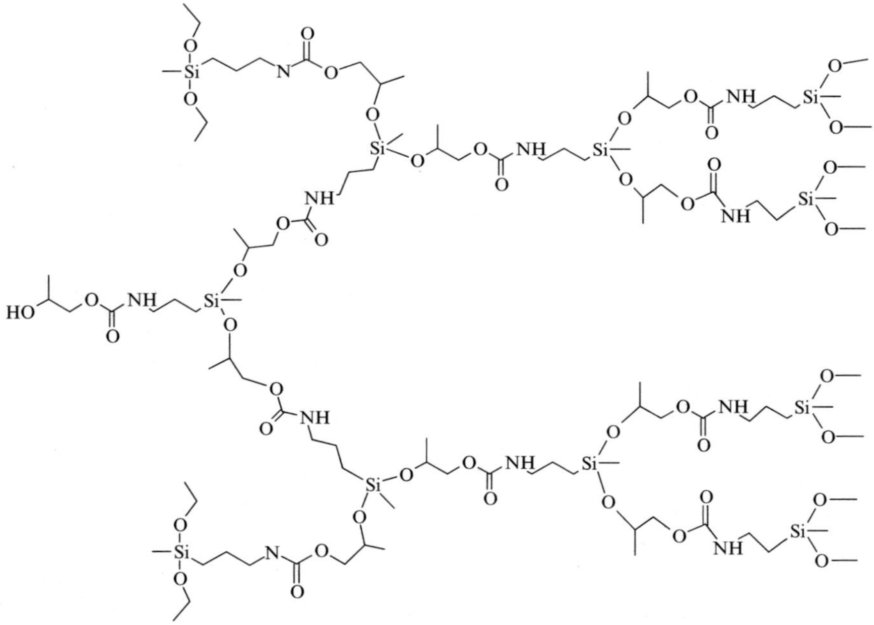

1

The change from methyldiethoxysilane to trimethoxysilane does not require any change in the experimental polymerization procedure, other than a somewhat longer reaction time to compensate for the additional crowding due to the higher

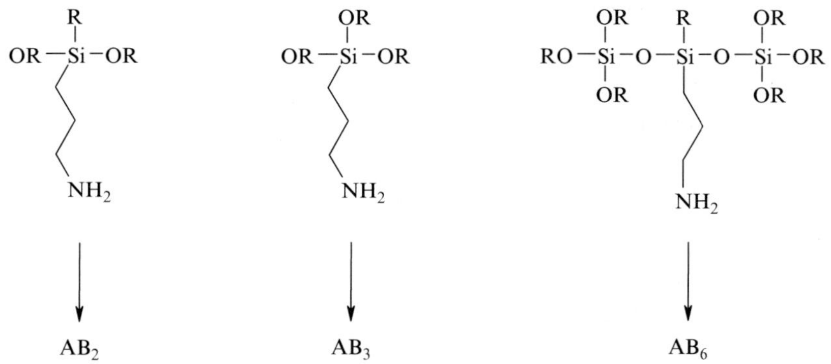

Fig. 14.4 Possible AB_x hyperbranched architectures derived from the silicone component.

branch density of the polymers resulting from AB_3 vs. AB_2 monomers. In fact, even higher branch density can be obtained by using higher alkoxy terminated siloxanes, such as the one shown in Fig. 14.4. Unfortunately, such high branch density polymers are more difficult to prepare due to the excessive crowding, which results in high structural defects and a large number of short branches. Nevertheless, this excessive crowding is easily alleviated by substituting the silane (left and middle in Fig. 14.4) with a higher molecular weight siloxane that is terminated with multiple alkoxy groups [8–10], such as AB_6 shown on the right of the same figure.

On polymerization, all these AB_2 and AB_3 monomers yield viscous liquids that were further characterized by gel permeation chromatography (GPC), NMR and viscosity. Although the collected amount of ethanol indicated a high degree of polymerization, this was not sufficiently accurate due to some loss during the process. GPC of the product indicated M_n of only $1.3–2.0 \times 10^3$. However, since this value was obtained relative to conventional linear polystyrene (PS) standards, it was also inaccurate and undoubtedly much lower than the actual molecular weight obtained due to the globular nature of the hyperbranched polymer molecules, which cannot be compared with linear PS standards. Although the precise shape of a hyperbranched molecule in solution is still under discussion, it is generally agreed that it will adopt a more globular shape (and, hence, smaller size) than a linear polymer of the comparable degree of polymerization, identical repeat unit composition and functionality. Unfortunately, it is not always easy to determine the shape of hyperbranched polymer molecules in solution since their interaction with solvents is governed not only by the nature of their backbones, but also by the nature and the number of the end-groups.

Clear evidence for hyperbranched polymer structure can be obtained from ^{29}Si NMR spectra (see Fig. 14.5) which showed four resonance peaks between –42 and –47 ppm corresponding to unbranched, single branched, double branched and triple branched silicon centers, respectively. The small peak around –50.6 ppm indicates that a very small fraction (estimated to be less than 1%) of siloxane linkages is present in the polymer as well. These Si–O–Si linkages could

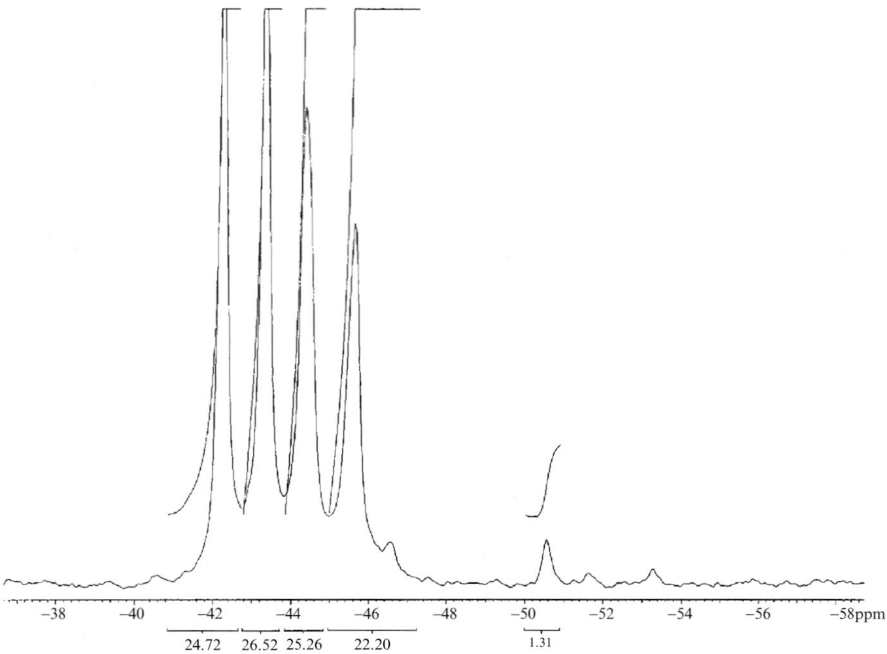

Fig. 14.5 ^{29}Si NMR of AB_3 hyperbranched polymer.

have resulted from the hydrolysis of ethoxysilyl groups followed by the silanol condensation. Unfortunately, the ^{13}C NMR spectrum of the polymer is identical to that of the monomer and although the ethoxy groups attached to the silicon atoms are removed during the polymerization, the new silyl ether linkages resonate essentially at the same place.

The increase in the molecular weight during the polymerization is conveniently observed by monitoring the change in the reaction mixture viscosity (see Fig. 14.6). Initially, this viscosity increases gradually as the alcohol is collected in the Dean-Stark trap and then much more rapidly toward the end of the polymerization. This behavior was characteristic for polymerizations of all of these AB_x silane monomers and consistent with the general trend that is typically observed in step-growth polycondensation reactions.

14.4 Rearrangement Reactions

One of the key attributes of these silicon-containing hyperbranched polymers is their ability to rearrange due to the hydrolytically unstable silyl ether, Si–O–C, linkages. Thus, upon exposure to moisture in the presence of an acid or base catalyst, these bonds are cleaved and the silanols that are formed condense to yield stable

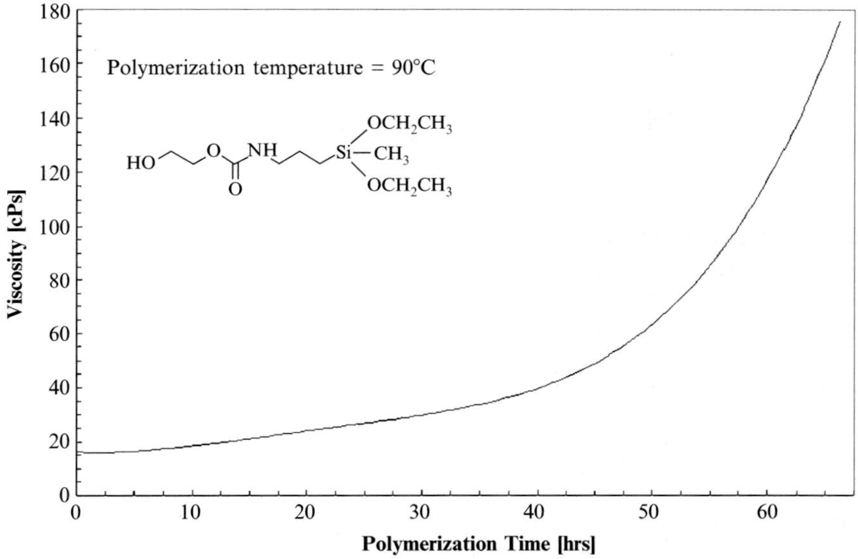

Fig. 14.6 Viscosity build-up during polymerization of AB_2 monomer.

siloxane Si–O–Si bonds. Through such a process, hyperbranched polymers from AB_2 monomers are converted into linear polysiloxanes with regularly grafted hydroxyl side groups (see Reaction Scheme 14.4), while hyperbranched polymers from AB_3 monomers are hydrolyzed and rearranged into 3D resinous networks.

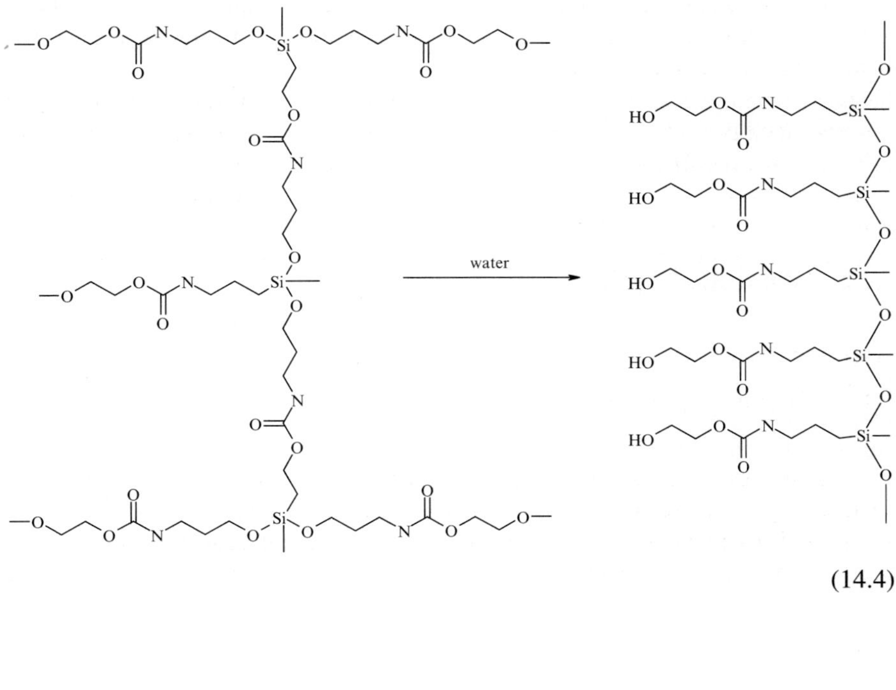

(14.4)

These unique rearrangement reactions are important for various applications, most notably as sealants against moisture absorption. For example, hyperbranched polymers from AB_3 monomers can be added to porous substrates such as wood, wool, concrete, brick, stone or leather to protect and preserve them following an appropriate condensation [11]. In contrast to sealants based on moisture activated silanes and linear polysiloxanes, in this case practically no volatile alcohols are produced on crosslinking since the carbinol functionality remains attached to the silicone matrix and the large number of these grafted carbinols contributes to adhesion to substrates. The lower viscosity of the hyperbranched precursors, relative to the corresponding linear polymers of the same molecular weights, and their high functionality further contribute to better wetting and penetration of these polymers into the porous structures. Finally, the resin networks, resulting from rearrangement reactions, are insoluble, locked into the pores of the substrates, and since the polymers are transparent, the surface appearance of the treated substrates is preserved.

Since in the case of silyl esters [12] the rearrangement is initiated by an S_N2 attack at the silicon atom with competing attack by the nucleophile at the carbonyl carbon, the relative prominence of the latter increases with increasing steric hindrance or with electron-donating character of the substituents on silicon. Hence, this variable mechanism of cleavage for the silyl ester bond may allow for a much broader range of stabilities to poly(silyl esters), than for the structures composed of carbon–oxygen bonds only, such as esters, anhydrides, etc.

Although a catalyst is not essential in order to initiate this hydrolysis and the subsequent silanol condensation reactions, common condensation catalysts such as tetrabutyl titanate will accelerate the reaction sequence, while raising the temperature will accelerate the condensation reaction. The non-crystalline nature of the hyperbranched polymers and the presence of amide and carbonyl functional groups provide a further advantage of enabling common condensation catalysts to dissolve in the polymer matrix. Conversely, the shelf-life is greatly improved when insoluble catalysts such as imidazole or its derivatives are dispersed in the hyperbranched matrix [11], as evidenced by no gel formation even after prolonged storage. Upon heating to a temperature above its melting point, the catalyst starts to flow, becomes miscible with the polymer and capable of initiating the desired rearrangement to a permanently crosslinked siloxane matrix.

14.5 Possible Applications

As was briefly mentioned above, the degree of branching, the viscosity of the hyperbranched polymer, and the rigidity of the resin that is formed after the rearrangement can be controlled by selecting the chain length between the reactive functional groups of the monomer or by using AB_x monomers with larger values of x. These parameters also control the mechanical properties of the resulting siloxane networks, yielding rigid, resin-like products when highly branched structures with larger values for x are used, or more ductile, softer structures when monomers with linear siloxane segments between the end-groups that result in branching points and/or smaller values of x are employed.

It should be further mentioned that since these hyperbranched polymers are soluble in organic solvents, they can be applied from solutions when low viscosity is required, as is the case, for example, in spraying. Many studies [13] of solution viscosities have shown that hyperbranched polymers have a significantly lower intrinsic viscosity, Mark–Houwink exponent, hydrodynamic volume, and ratio of the radius of gyration to hydrodynamic radius than their linear analogues of the corresponding molar weights. This results from the fact that branching decreases the second virial coefficient in "good" solvents which, therefore, becomes always lower than for the homologous linear polymer [14, 15]. Furthermore, in addition to increasing hydrophobicity and protecting the substrate from moisture, these rearranging hyperbranched polymers can also improve abrasion resistance, enhance the surface lubricity, and improve chemical resistance. Bio-active additives can also be incorporated to improve the fungal resistance of the substrates, and since the polymers are transparent, the natural beauty of the substrates (such as wood, for example) is not compromised upon coating.

The rearrangement of hyperbranched structures of these silicone-containing polymers could also be viewed as "degradation". However, unlike most degradation processes, the products of the process are not monomers or low molecular weight species but high molecular weight linear polymers and crosslinked resins. Furthermore, degradation rates of most polymers at ambient temperatures (and certainly, those of silicone-containing polymers) are generally low. They depend on steric and electronic effects of the substituents as well as on the physical properties of polymer backbones, hydrophobicity, solubility, crystallinity, crosslinking, additives, etc., and are measured on a timescale of weeks and months whereas the rearrangements/degradation rates of the hyperbranched polymers described in this chapter are of the order of minutes to hours. Somewhat similar hyperbranched poly(silyl ester)s have been recently prepared by a cross dehydro-coupling [16] and their degradation behavior was studied and compared with linear polymers. It was found by GPC that unlike the degradation mode of similar linear poly(silyl ester)s, which showed a rapid initial degradation followed by continuously decreasing decomposition rates, these hyperbranched silyl ether polymers exhibited a slow initial molecular weight loss, followed by rapid hydrolytic cleavage. This unexpected observation was determined to be an artifact of the GPC method used to monitor the molecular weight, and it was concluded that cleavage occurred in a random fashion to rapidly break down the structure and reduce the molecular weight in smaller increments than observed in similar linear poly(silyl ester)s.

Controlled degradation of hyperbranched polymers containing Si–O–C linkages (e.g. poly[bis(undeceny1oxy)methylsilane]s) was also studied with an aim to obtain well defined cavities with particular chemical groups on the walls [17]. Such porous materials would be of great interest for low dielectric constant insulators in electronics, heat resistant insulation, low refractive index transparent coatings, enzyme mimicry, high efficiency catalysis, and as new membranes for molecular separation. It was hoped that if the end-groups would remain attached to the matrix, the hyperbranched structure could be used as a template to create well-defined nanometer-sized cavities. This concept was discussed in further detail at a later date [18] and it was argued that if substituents are removed from the hyperbranched polymer network, ideally, a microporous material would be formed. The monomers for these hyperbranched polymers were prepared by alkoxylating methyldichlorosilane followed by hydrosilylation with 10-undecen-1-ol. Some gelation problems were encountered during the hydrosilylation step but they were resolved by diluting with hexane and using a more selective hydrosilylation catalyst. Several degradation routes were then examined; each yielding a different product, but no evidence was presented for the formation of porous structures.

A somewhat different approach to controlled porosity with the Si–O–C linkage was to use it as an organic–inorganic coupling agent. In this approach, the end-groups of hyperbranched aliphatic polyesters were reacted with chlorotrimethoxysilanes, followed by blending with methylsilsesquioxane, MSQ, prior to inducing thermal degradation [19]. It was argued that the presence of large number of siloxy groups at the chain-ends of hyperbranched structure will aid in compatibilizing the organic polyester with MSQ, and thus impact the morphology during the thermal degradation. Indeed, thermal gravimetric analysis (TGA) data showed that quantitative degradation of this template hyperbranched hybrid started at about 300°C and porous structures were observed and measured by transmission electron microscopy (TEM) and atomic force microscopy (AFM). It was also found that finer pore sizes were obtained as the concentration of alkoxysilane was increased. For example, the nominal pore size was reduced from 200 nm for an 80:20 wt% mixture of hyperbranched polyester and MSQ to about 50 nm when the end-groups of hyperbranched polymer were terminated with trimethylsiloxy groups. Furthermore, good control of pore sizes was achieved over a relatively wide range of compositions. It is important to note, however, that the pore size was not determined by the size of the hyperbranched molecule. Instead, it appears that a decrease in molecular mobility associated with increasing molecular weight of the hyperbranched polymer led to finer microstructures.

References

1. Inoue K (2000) Prog Polym Sci 25: 453–571.
2. Flory PJ (1953) Principles in Polymer Chemistry, Cornell University Press, Ithaca, New York, Chapters III and VIII.
3. Bolton DH, Karen KI (1997) Macromolecules 30(7): 1890.
4. Tomita H, Sanda F, Endo T (2001) J Polym Sci Part A Polym Chem 39: 3678.
5. Shivarkar AB, Gupte SP, Chaudhari RV (2006) Synlett 9: 1374.
6. Fujita S-I, Bhanage BM, Arai M (2004) Chem Lett 33(6): 742.
7. Decker GT, Graiver D, Tselepis AJ (1999) US Patent 6,001,945.
8. White MA (1983) US Patent 4,395,526.
9. Chung RH (1985) US Patent 4,528,324.
10. Lucas RM (1986) US Patent 4,539,085.
11. Decker GT, Graiver D, Tselepis AJ, Williams D (1999) US Patent 5,997,954; 6,103,848.

12. Sommer LH (1965) Stereochemistry, Mechanism and Silicon, McGraw-Hill Series in Advanced Chemistry; McGraw-Hill, New York.
13. Mourey TH, Turner SR, Rubinstein M, Frechet JMJ, Hawker CJ, Wooley KL (1992) Macromolecules 25: 2401.
14. Lue L (2000) Macromolecules 33(6): 2266.
15. Striolo A, Prausnitz JM, Bertucco A, Kee RA, Gauthier M (2001) Polymer 42(6): 2579.
16. Wang M, Gan D, Wooley KL (2001) Macromolecules 34: 3215.
17. Muzafarov AM, Monika Golly M, Misller M (1995) Macromolecules 28: 8444.
18. Kobayashi N, Kijima M (2007) J Mater Chem 17: 4289.
19. Plummer CJG, Garamszegi L, Nguyen T-Q, Rodlert M, Manson J-AE (2002) J Mater Sci 37: 4819.

Chapter 15
Cyclization Issues in Silicon-Containing Hyperbranched Polymers

David Y. Son

15.1 Introduction

Theoretical descriptions of AB_n ($n \geq 2$) hyperbranched polymerization systems have been known for some time [1], but in them, cyclization is a factor that is generally and largely ignored. However, it is now understood that cyclization is prevalent in polymerizations of this type, and that it can often affect to a significant extent both polydispersity and molecular weights of the polymer products. Since research in hyperbranched polymers has increased dramatically in recent years [2–4], a number of experimental and theoretical studies have focused on the presence and effects of cyclization in these systems (see, for example [5–10] and references cited therein). In essence, intramolecular cyclization of an oligomer in an AB_n polymerization results in the consumption of the focal A group (see Section 15.3), which converts the oligomer into a B_x core. Although the newly formed core can continue to grow through the reaction of other A groups with the B groups, this growth is limited, especially if other A groups in the polymerization system are also consumed through similar intramolecular cyclization reactions. Thus, for the control and optimization of the resulting polymer molecular weight, it is necessary to understand these issues and the methods that can be used to avoid excessive amounts of cyclization.

This chapter describes cyclization in organosilicon hyperbranched polymer synthesis, and techniques that have been used to minimize its occurrence. The focus is primarily on AB_n ($n \geq 2$) systems, although many of the principles apply to the similar $A_2 + B_3$ bimolecular systems that are now gaining more research attention (see Chapter 16) [11].

D.Y. Son
Department of Chemistry, Southern Methodist University, Dallas, TX 75275-0314, USA
E-mail: dson@smu.edu

P.R. Dvornic and M.J. Owen (eds.), *Silicon-Containing Dendritic Polymers*,
Advances in Silicon Science 2, © Springer Science+Business Media B.V. 2009

15.2 Intramolecular Cyclization of the Monomer

Before examining cyclization of oligomers or polymer molecules, it is advisable to consider the potential cyclization of the starting AB_n monomer itself. Clearly, one of the objectives in hyperbranched polymer synthesis is to use monomers that will not intramolecularly cyclize. If intramolecular reaction is favored over the intermolecular one, little if any polymer will be obtained.

For example, in an attempt to prepare hyperbranched carbosilazanes (see Chapter 5), in this group we synthesized AB_2 monomers **1** and **2** (see Reaction Scheme 15.1) [12]. However, on treatment with a Pt catalyst, the only products obtained in both cases were the four-membered azadisilacyclobutanes and no polymer was formed. This result was perhaps not totally unexpected, as analogous allylamines readily cyclize in high yields [13, 14].

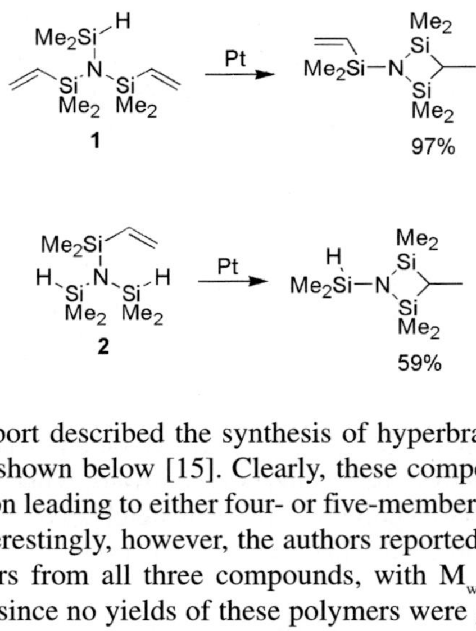

(15.1)

A more recent report described the synthesis of hyperbranched carbosilazanes from the monomers shown below [15]. Clearly, these compounds are expected to be prone to cyclization leading to either four- or five-membered rings in the fashion described above. Interestingly, however, the authors reported isolation and characterization of polymers from all three compounds, with M_w values ranging from 3,900 to 13,500, but since no yields of these polymers were given, it is impossible to judge the extent of monomer cyclization that may have occurred.

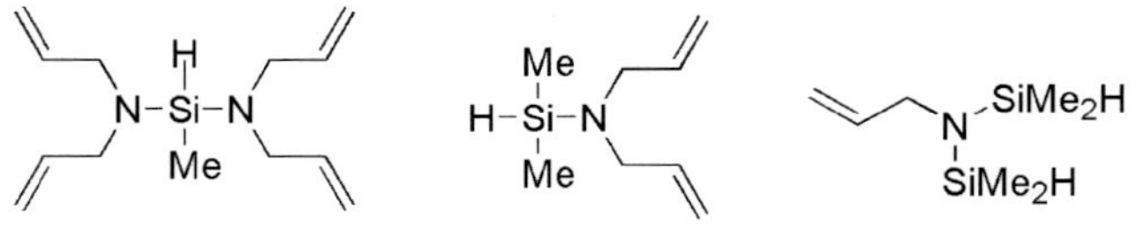

In their seminal report on the first preparation of a hyperbranched polymer by hydrosilylation, Mathias and Carothers described the synthesis of hyperbranched polcarbosiloxane from monomer **3** (see Reaction Scheme 15.2) [16] which had a molecular weight of 19,000 as determined by size exclusion chromatography (SEC) relative to polystyrene standards. However, since no yield of the polymer was reported in this case either, it is again difficult to judge the extent of the monomer cyclization that occurred. Nevertheless, in a later paper, the same authors [17] mentioned that cyclization did occur in this system, specifying that the six-membered cyclic compound was obtained in 88% yield in a rather dilute reaction mixture (0.5% w/w). This observation was later confirmed by Rubinsztajn [18], who repeated this polymerization and obtained a low yield of polymer accompanied with a significant amount of the six-membered cyclized product.

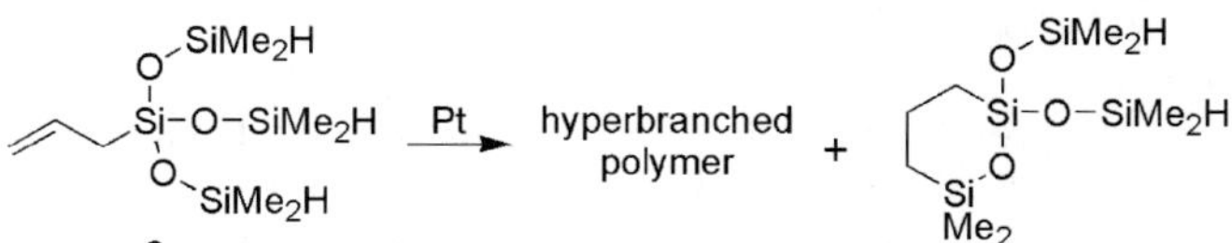

3

(15.2)

In 1998–1999, Fréchet and coworkers reported the polymerization of similar siloxane monomers **4–6** [19, 20]. In each case, gel permeation chromatography (GPC) showed a large peak corresponding to the molecular weight of the starting monomer. Since spectroscopy of the particular fractions indicated loss of the vinyl group, it was suggested that the large peak corresponded to the cyclized monomer. However, since the yields of polymers were 55–70%, this monomer cyclization process was apparently not the dominant reaction.

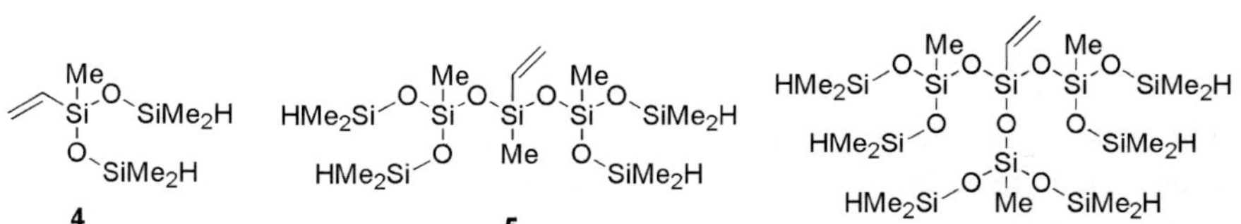

4 5 6

In a related study, Herzig [21] reported the co-polymerization of related monomer **7** [18] with various core compounds (see Reaction Scheme 15.3). In this case, only 8–10% of **7** cyclized to give the five-membered ring compound, as determined by ^{29}Si NMR spectroscopy.

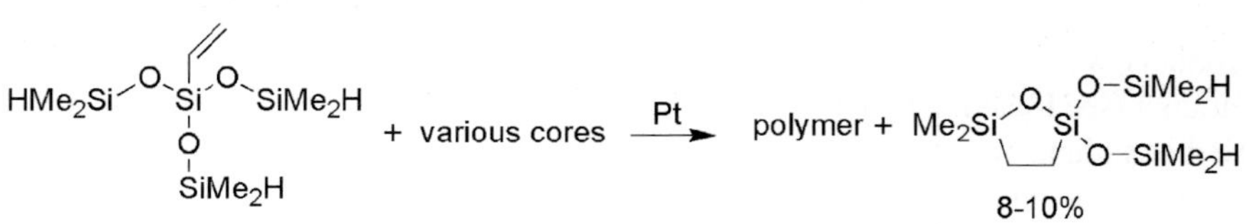

7

(15.3)

15.3 Intramolecular Cyclization of Oligomers

If intramolecular cyclization of an AB_n monomer is not a highly favored reaction, then intermolecular reactions can lead to polymer growth. However, cyclization can occur at any time during that growth through reaction of the focal A group with any

B group in the same molecule. In a simple case, two AB_3 monomers can react with each other to give an "AB_5" product, which can then cyclize if ring formation is favorable (see Reaction Scheme 15.4), giving a "B_4" core. Certainly, as mentioned above, polymer growth can proceed further through reaction of this B_4 core with an A group from another molecule, but as A groups are consumed through cyclization of other oligomers, their actual concentration in the reaction system will steadily decrease together with the likelihood of polymer growth.

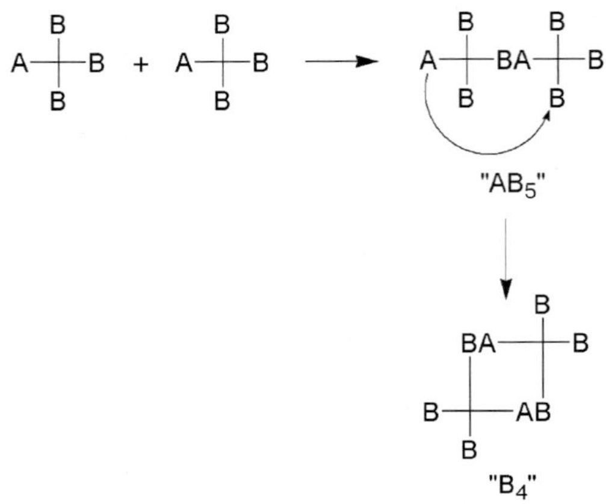

(15.4)

In certain cases, the presence of these larger cyclics can be observed or inferred. Organosilicon hyperbranched polymer synthesis through nucleophilic substitution reactions is a good example. The first such system was described by Whitmarsh and Interrante [22] (see Chapter 12), who polymerized (chloromethyl)trichlorosilane by first converting it to the Grignard reagent and then allowing head-to-tail reactions to occur. While it is clear that intramolecular cyclization of the monomer cannot occur, subsequent studies on similar systems revealed that cyclization of dimers, trimers, and oligomers does occur. For example, Corriu and coworkers converted (chloromethyl)trimethoxysilane to the Grignard reagent and found that the main products were the six-membered rings resulting from cyclization of a trimer or tetramer (see Reaction Scheme 15.5) [23]. The Grignard reagent from (chloromethyl) triethoxysilane was found to be more stable, but with mild heating it also gave primarily the analogous six-membered ring products.

(15.5)

Neckers and coworkers [24] synthesized a hyperbranched carbosilane from (chloromethyl)dichloromethylsilane using the general method of Interrante. Gas chromatography–mass spectrometry (GC–MS) data revealed the presence of oligomers, many of which consisted of three to seven monomer units, in agreement with the molecular weight data which suggested that many of the trimers and tetramers were cyclic oligomers (see Reaction Scheme 15.6). Oligomers containing four- and six-membered rings were observed as significant products in the electropolymerization of the same monomer, particularly when Al was used as the anode [25]. These oligomers were also identified by GC–MS.

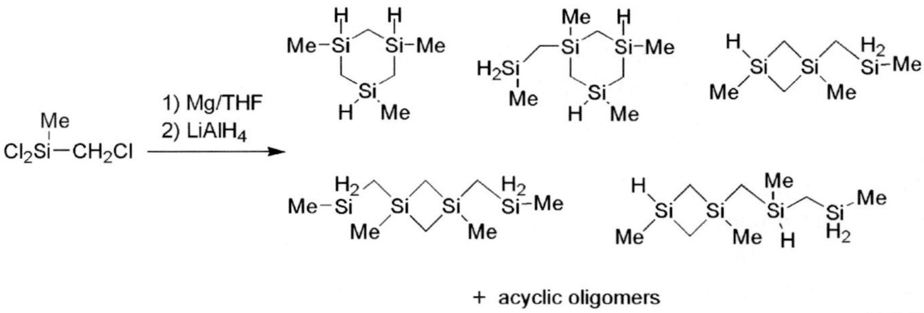

(15.6)

Another example of cyclization during a hyperbranched polymerization by nucleophilic substitution is the condensation polymerization of triethoxysilanol and acetoxytriethoxysilane to give hyperbranched polysiloxanes [26]. ^{29}Si NMR spectroscopy revealed the presence of cyclic units in the polymer structure, but the sizes of the rings were not determined. It was hypothesized that the formation of cyclics occurred via side reactions such as transesterification and head-to-head coupling reactions. However, cyclization via conventional intramolecular A–B coupling should not be ruled out.

Cyclization was also observed in organosilicon hyperbranched polymers prepared via hydrosilylation reactions (see Chapter 13). Monomers with siloxane linkages are especially good candidates for cyclization due to the flexibility of the Si–O–Si groups. As mentioned above, Fréchet and coworkers observed intramolecular cyclization of monomers **4–6** to a small extent [19, 20]. The isolated polymers from these reactions were not of high molecular weight (M_w 5,800–8,900) as suggested by GPC, and these limited molecular weights were attributed to the presence of cyclized products in the system. A more detailed study of the polymerization of **4** [20] indicated the formation of cyclic dimer, trimer, tetramer, pentamer, and hexamer, as determined by GPC, ^{1}H NMR spectroscopy, and matrix assisted laser desorption ionization-time of flight (MALDI-TOF) mass spectrometry. The amount of cyclized monomer through tetramer represented approximately 20% of the total mass of crude polymer. The low molecular weight of the polymer obtained (M_w ~ 5,000) was attributed to the rapid loss of vinyl groups due to intramolecular cyclization reactions. In contrast, the polymerization of AB_2 siloxane monomer **8** apparently gave a minimal amount of cyclics, as the relative amount of Si–H groups detected in the ^{1}H NMR spectrum corroborated well with the molecular weights determined by vapor pressure osmometry (VPO) [27].

Polymerization conditions included 12 h of stirring at 60–80°C, and the polymer was isolated in 87% yield with $M_n = 4{,}320$.

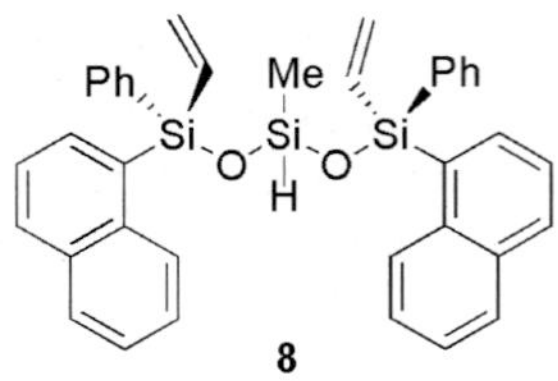

8

Oligomer cyclization also occurs in the hydrosilylation of organosilicon monomers without flexible Si–O–Si linkages. For example, Möller and coworkers [28] found that the hydrosilylation polymerization of methyldivinylsilane gave a product with significant amounts of low molecular weight components. GC–MS analysis of these components indicated the formation of trimers, tetramers, and pentamers. Since spectroscopy proved the near total loss of Si–H, these low molecular weight products were likely cyclic rather than acyclic oligomers, which would have the focal Si–H group intact.

15.4 Controlling Cyclization

From the discussion in Sections 15.2 and 15.3, it is evident that in order to obtain reasonably high molecular weights of hyperbranched polymers, intramolecular cyclization should be controlled.

15.4.1 Controlling Monomer Cyclization

The first consideration should be preventing intramolecular cyclization of the monomer itself. As described in Section 15.2, this cyclization can occur if a monomer can form a stable (typically six-membered) ring. To avoid this, monomers should be employed in which cyclization: (1) would give a highly strained ring, (2) is difficult due to the presence of an aryl or hetaryl unit between the reactive A and B groups, and (3) is entropically disfavored due to a long distance separating A and B groups.

A few selected examples of monomers that fall into the first category are shown below [28–30]:

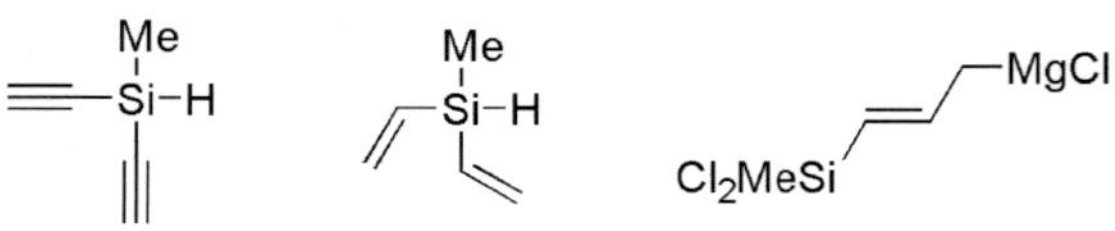

Another example of ring strain affecting polymer yields was reported by Rubinsztajn, who synthesized monomers **7** (see Reaction Scheme 15.3) and **9** [18]. Since cyclization of **7** and **9** would lead to five-membered rings that are more strained compared to the analogous six-membered ring from **3** [17] (by about 30 kJ/mol), it was expected that more polymer would be obtained, which indeed turned out to be the case. However, as mentioned above, monomer **7** nevertheless cyclized intramolecularly, although to a small extent [21].

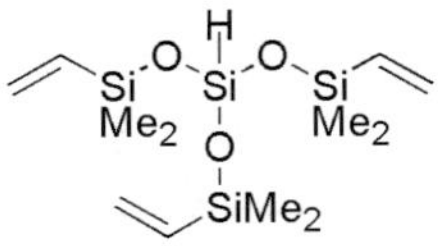

9

Separating A and B groups in the monomer structure by aryl or hetaryl units is another effective way of preventing monomer cyclization reactions (see structures below) [31–34]. In none of these cases was monomer cyclization cited as a significant side-reaction.

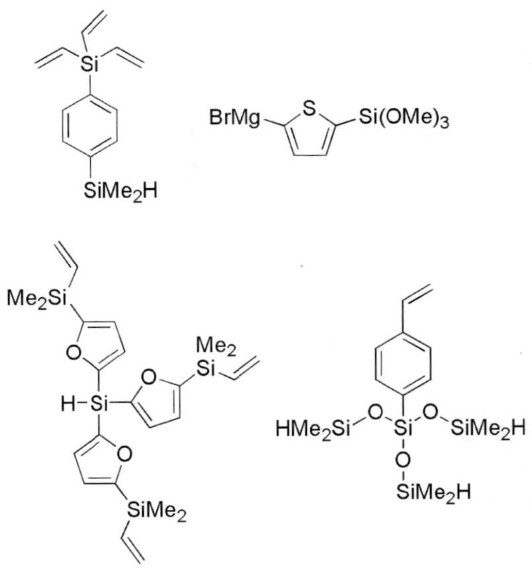

Finally, separating A and B groups by long spacers also discourages monomer cyclization due to entropy considerations. A good example of this is methyldiundec-10-enylsilane, shown below [28]. Polymer yields from this monomer were quantitative and molecular weights were higher than for polymers obtained from monomers with shorter spacers. This suggests that monomer (and oligomer) cyclization was not a significant side reaction in these polymerizations.

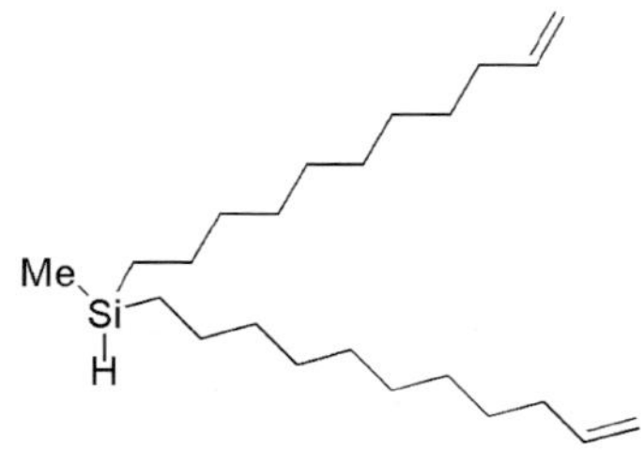

15.4.2 Controlling Cyclization of Oligomers

The factors that prevent monomer cyclization do not necessarily also apply for oligomers. For example, intramolecular cyclization of a monomer may be prevented by the ring strain of the resulting cyclic, but such ring strain may not apply to intramolecular cyclization of a dimer or tetramer. A good example of this is the above described polymerization of methyldivinylsilane [28], which does not cyclize itself, but yields significant amounts of cyclized trimers, tetramers, and pentamers.

One method that can reduce oligomer cyclization is to increase the concentration of the monomer solution, even to the point of conducting polymerization in the bulk (see also Section 16.3). An increased concentration of monomer/oligomer increases the probability of intermolecular reaction, while dilution favors intramolecular cyclization reactions. A number of groups have reported increased formation of cyclics as the polymerization reaction mixture concentration was diluted [17, 18, 25]. However, although bulk polymerizations favor polymer growth, cyclic formation is often still prevalent, as was observed in the bulk polymerizations of certain siloxanes [19, 20]. In addition, the use of bulk polymerizations is limited to monomers with sufficiently low viscosity.

An interesting method for reducing cyclization and increasing polymer molecular weight is slow addition of AB_n monomer to the polymerization medium or to a B_n core (see also Chapter 14) [35]. The idea here is that incoming monomer molecules will preferentially react with the existing polymer or core molecules (containing a large amount of available B groups) rather than with themselves. The overall result will then be a progressive increase in polymer molecular weight and a decreased amount of oligomeric cyclics formed. This was indeed demonstrated in the polymerization of monomer **4** [20]. While M_w values of a single-batch bulk process were approximately 5,800, slow addition experiments resulted in M_w values of up to 61,000 depending on the rate of addition. Slow addition of monomer to a core also resulted in higher molecular weights. Thus, the use of a M_w = 41,000 core resulted in polymer with M_w = 280,000 when monomer was added slowly. Similar increases in molecular weight by slow addition of monomer were also observed in the polymerizations of methyldiethynylsilane [36] and methyldiundec-10-enylsilane [28], but not in the polymerizations of methyldivinylsilane and methyldiallylsilane [28]. In another study, slow addition of methyldiundec-10-enylsilane to a core consisting of a polybutadiene-polystyrene block copolymer resulted in a systematic increase in molecular weight with the increase depending on the amount of monomer added [37]. A low molecular weight fraction, that was also formed, was attributed to cyclized oligomers. In contrast to this, conducting this same polymerization as a single-batch process resulted in a lower overall molecular weight of the polymer obtained, and formation of a much larger amount of the cyclic oligomeric by-product.

15.5 Concluding Remarks

This chapter emphasizes that intramolecular cyclization is a significant issue in the preparation of hyperbranched polymers in general, including the silicon-containing ones regardless of whether their synthesis proceeds via nucleophilic substitution or hydrosilylation polymerization reactions. It is also clearly pointed out that for control of the resulting polymer molecular weight, cyclization reactions must be taken into account and if at all possible placed under synthetic control. Fortunately, there are methods that allow some degree of such control, and therefore also enable control of the polymer molecular weight. In spite of this, however, cyclization remains an important factor which, particularly when higher molecular weight hyperbranched products are desired, must always be taken into serious consideration.

Acknowledgements The author thanks the Robert A. Welch Foundation (grant no. N-1375) and the National Science Foundation (grant no. 0092495), who supported his own work in this area, as described in the text.

References

1. Flory PJ (1953) Principles of Polymer Chemistry. Chapter 9, Cornell University Press, Ithaca, NY
2. Yates CR and Hayes W (2004) Eur Polym J 40:1257
3. Gao C and Yan D (2004) Prog Polym Sci 29:183
4. Voit B (2005) J Polym Sci Part A Polym Chem 43:2679
5. Burgath A, Sunder A, and Frey H (2000) Macromol Chem Phys 201:782
6. Dusek K, Somvarsky J, Smrckova M, Simonsick Jr. WJ, and Wilczek L (1999) Polym Bull 42:489
7. Gooden JK, Gross ML, Mueller A, Stefanescu AD, and Wooley KL (1998) J Am Chem Soc 120:10180
8. Drohmann C and Möller M (2000) Polym Preprints 41:140
9. Cameron C, Fawcett AH, Hetherington CR, Mee RAW, and McBride FV (1998) J Chem Phys 108:8235
10. Fadeev MA, Rebrov AV, Ozerina LA, Gorbatsevich OB, and Ozerin AN (1999) Polym Sci Ser A 41:189

11. Dvornic PR, Hu J, Meier DJ, and Nowak RM (2004) Polym Preprints 45:585
12. Yoon K and Son DY (1999) Org Lett 1:423
13. Tamao K, Nakagawa Y, and Ito Y (1990) J Org Chem 55:3438
14. Tamao K, Nakagawa Y, and Ito Y (1993) Organometallics 12:2297
15. Zhang G-B, Fan X-D, Kong J, Liu Y-Y, Wang M-C, and Qi Z-C (2007) Macromol Chem Phys 208:541
16. Mathias LJ and Carothers TW (1991) J Am Chem Soc 113:4043
17. Carothers TW and Mathias LJ (1993) Polym Preprints 34(2):538
18. Rubinsztajn S (1994) J Inorg Organomet Polym 4:61
19. Miravet JF and Fréchet JMJ (1998) Macromolecules 31:3461
20. Gong C, Miravet J, and Fréchet JMJ (1999) J Polym Sci Part A Polym Chem 37:3193
21. Herzig C and Deubzer B (1998) Polym Preprints 39(1):477
22. Whitmarsh CK and Interrante LV (1991) Organometallics 10:1336
23. Brondani DJ, Corriu RJP, El Ayoubi S, Moreau JJE, and Man MWC (1993) Tetrahedron Lett 34:2111
24. Fry BE, Guo A, and Neckers DC (1997) J Organomet Chem 538:151
25. Wang X, Yuan Y, Graiver D, and Cabasso I (2007) Macromolecules 40:3939
26. Jaumann M, Rebrov EA, Kazakova VV, Muzafarov AM, Goedel WA, and Moller M (2003) Macromol Chem Phys 204:1014
27. Oishi M, Minakawa M, Imae I, and Kawakami Y (2002) Macromolecules 35:4938
28. Drohmann C, Möller M, Gorbatsevich OB, and Muzafarov AM (2000) J Polym Sci Part A Polym Chem 38:741
29. Xiao Y, Wong RA, and Son DY (2000) Macromolecules 33:7232
30. Yao J and Son DY (1999) J Polym Sci Part A Polym Chem 37:3778
31. Rubinsztajn S and Stein J (1995) J Inorg Organomet Polym 5:43
32. Yoon K and Son DY (1999) Macromolecules 32:5210
33. Rim C and Son DY (2003) Macromolecules 36:5580
34. Yao J and Son DY (1999) Organometallics 18:1736
35. Hanselmann R, Hölter D, and Frey H (1998) Macromolecules 31:3790
36. Son DY and Xiao Y (2001) Polym Mat Sci Eng 84:301
37. Marcos AG, Pusel TM, Thomann R, Pakula T, Okrasa L, Geppert S, Gronski W, and Frey H (2006) Macromolecules 39:971

Chapter 16
Hyperbranched Silicon-Containing Polymers via Bimolecular Non-linear Polymerization

Petar R. Dvornic and Dale J. Meier

16.1 Introduction

Bimolecular non-linear polymerization, BMNLP (see Reaction Scheme 16.1), represents "*the other method*" for preparation of hyperbranched polymers by the step-growth reaction mechanism. In contrast to the monomolecular polymerizations of AB_x monomers, discussed for the two most prominent groups of silicon-containing hyperbranched polymers in Chapters 12 and 13, this polymerization type involves, as its name implies, two reactive monomers A_x and B_y, where A and B denote two types of mutually reactive functional groups while x and y are integers which must both be equal to or larger than 2, while one of them (either x or y) must be equal to or larger than 3 (see Section 16.3). Thus, the most common BMNLP systems include $A_2 + B_3$, $A_2 + B_4$, and $A_3 + B_4$ monomer combinations. General representation of the simplest of these, the $A_2 + B_3$ system in which the minor component has completely reacted, is shown in Reaction Scheme 16.1.

The essential and common feature of all BMNLP systems is the need for careful control of the polymerization process, by appropriate selection of the relative concentrations of the reacting A and B groups and the extent of the reaction, in order to prevent their natural tendency to crosslink to a gel. Ideally, very stringent requirements must be satisfied in order to meet these conditions (see Section 16.3), but in reality reaction parameters can be selected so as to shift the polymerization equilibria to the formation of desired soluble hyperbranched polymer products. When such selections and adjustments are appropriately made, BMNLP reactions can offer some important advantages over "traditional" AB_x polymerization systems, as is listed in Table 16.1. Of these, of particular interest are: (i) the extreme versatility of the process to yield a practically unlimited variety of polymer compositions from a vast number of commercially available monomers, (ii) the unique ability to produce compositionally identical polymers, $-[AB]_n <$, with different (A or B)

P.R. Dvornic and D.J. Meier
Michigan Molecular Institute, 1910 W. St. Andrews Rd., Midland, MI 48640, USA
E-mails: dvornic@mmi.org; meier@mmi.org

P.R. Dvornic and M.J. Owen (eds.), *Silicon-Containing Dendritic Polymers*,
Advances in Silicon Science 2,

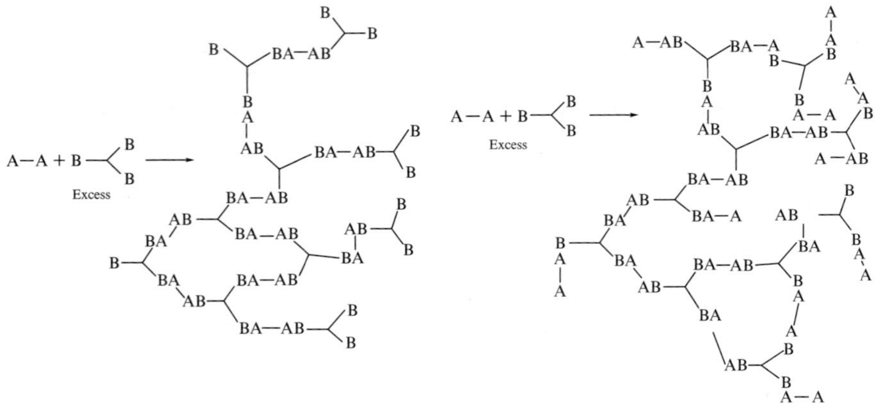

(16.1)

Table 16.1 Some general characteristics of non-linear polymerization systems $A_x + B_y$ and AB_x

	$A_x + B_y$ sytems	AB_x systems
Monomers	Commercially available in many variants	Rare, often difficult to synthesize and/or store
Polymers	Great variety of intramolecular compositions	Only one composition of polymer from each monomer is possible
Polymer end-groups	Can be either A or B	Only B possible
Polymer MW and DB[a]	Adjustable by tuning molar ratio of the reacting monomers and reaction conditions	Adjustable by tuning reaction conditions
Gelation	Can be avoided	Cannot occur

[a]MW = molecular weight; DB = degree of branching

end-groups, and (iii) the complete elimination of the shelf-life problems that are often encountered with many AB_x monomers which can polymerize either without the need for reaction catalyst or under less stringent storage conditions. Recently, a series of papers on modeling and experimental studies of the influence of polymerization procedure on polymer topology and other structural properties of hyperbranched polymers obtained from $A_2 + B_3$ BMNLP systems was published by Gallivan and her co-workers [1–4].

16.2 Historical Development of Bimolecular Non-linear Polymerization

The earliest suggestion that transient branched molecules are formed during the polymerization of a difunctional and a trifunctional monomer (such as phthalic anhydride and glycerol) were made by Kienle and co-workers from Rutgers

University in 1929–1939 [5, 6]. Although the authors did not call these structures hyperbranched (this name caught on only in the late 1980s and early 1990s), nor polymers (in those days Staudinger was still fighting for his "macromolecular hypothesis"), but instead "complex three-dimensional molecules", they clearly understood their branched character and described their fundamental building blocks as: (a) mono-connected (what we call today end-groups), (b) di-connected (linear segments), and (c) tri-connected (intramolecularly branched and/or looped) repeat units [6]. They even determined molecular weights of these transient species and showed that once they reached about 1,100 the viscosity of the reaction mixture containing stoichiometric amounts of the reacting functionalities rapidly increased toward gelation. Following this work, in 1941–1953, Flory and Stockmayer developed their statistical gelation theories explaining the formation of infinite cross-linked networks in such reaction systems and underlying again the fact that these networks develop through substantially branched, growing intermediates [7–11] However, the real interest in hyperbranched polymers has developed only in the last 15–20 years with the greatest attention being focused on the monomolecular polymerization of AB_x monomers [12–14], while until very recently the BMNLP approach has been almost completely neglected.

The first report on application of BMNLP for the synthesis of hyperbranched polymers appeared in 1998 when Russo and Boulares described the preparation of hyperbranched aromatic polyamides using di- and tri-functional amines and acids [15–17]. In 1999, a similar work was published by Watanabe and his co-workers [18], and in 2001 by Kakimoto and Jikei [19], while Fréchet and his colleagues described aliphatic polyethers from diepoxides and tris(hydroxymethyl)ethane [20] and hyperbranched porphyrin from a triepoxy and di-hydroxyl functionalized porphyrin [21]. In 2000, Yan and Gao reported on polyaddition of 1-(2-aminoethyl) piperazine to divinyl sulfone [22], and Okamoto and his co-workers described hyperbranched polyimides from tris(4-aminophenyl)amine and a series of different dianhydrides [23]. In 2002 Kricheldorf and his co-workers reported on the preparation of poly(ether sulfone)s from 1,1,1-tris(4-hydroxyphenyl)ethane and 4,4′-difluorodiphenyl sulfone [24, 25], and Bruchmann and his colleagues described a polyurethane from isophorone diisocyanate and glycerol [26], poly(urethane urea)s from isophorone or toluene diisocyanate and aminoalkanediol [27], and aliphatic polyesters from diacids and triols [28]. In 2003, poly(arylene ether phosphine oxide)s from tris(4-fluorophenyl)phosphine oxide and a variety of bisphenols were reported by Fossum and Czupik [29], and a very interesting photosensitive fluorinated polyimide from 4,4′-(hexafluroisopropylidene) diphthalic anhydride and 1,3,5-tris(4-aminophenoxy)benzene was described by Chen and Yin [30]. In the same year, Kakimoto and co-workers published a very interesting comparison of hyperbranched polyimides prepared by either BMNLP $A_2 + B_3$ or AB_2 polymerization approaches [31]. Long et al. reported on a series of aromatic hyperbranched polymers, including polyesters [32–35], segmented poly(urethane urea)s utilizing isocyanate end-capped polyethers as oligomeric A_2 reagents and triamine monomers [34] and poly(ether esters) from poly(ethylene glycol) and 1,3,5-benzenetricarbonyl trichloride [35]. Most recently, In and Kim reported on nanoparticles with controlled end-functionality and topology by the polymerization of 4,4′-difluorophenylsulfone and 1,1,1-tris(hydroxyphenyl)ethane [36], while Perignon and co-workers described metal-containing nanoparticles [37], prepared from polyamidoamine (PAMAM) hyperbranched polymers previously developed by us [38].

In a series of patents issued between 2002 and 2004, we described a variety of different new hyperbranched polymers prepared by BMNLP, including the first silicon-containing ones: polycarbosiloxanes and polycarbosilanes [39, 40], and a variety of hyperbranched polyureas, polyamidoamines and polyamides with silicon-containing end-groups designed for subsequent crosslinking [41, 42]. We also reported on a unique new type of hyperbranched polymers with more than one type of reactive functional groups per molecule resulting directly from the polymerization. These unique polymers contain latent functionalities that remain dormant during the polymer formation but can become reactive under appropriate conditions [43]. They can be prepared from $A_xC_z + B_y$ or $A_xC_z + B_yD_z$ monomer systems, where, as usual, A and B are functional groups reactive in BMNLP, while C and D are the latent functionalities stable under A/B polymerization conditions.

16.3 Theory of $A_x + B_y$ Bimolecular Non-linear Polymerization

The theory of the reactions between two multi-functional monomers A_x and B_y, was developed by Flory [7, 10, 11] and Stockmayer [8, 9] as early as the 1940s–1950s, but the practical application of their findings was mostly overlooked until recent years. Other theoretical treatments of similar or more complex $A_x + B_y$ systems based on different approaches were later given by Macosko and Miller [44], Gordon [45], Case [46] and Charlesby [47, 48], and reviews have been published by Stauffer et al. [49] and Dušek and Dušková-Smrčková [50]. However, since for the systems of interest to this chapter the results of these other theories are equivalent to those of Flory and Stockmayer, only the latter will be discussed.

The Flory-Stockmayer theories are based on the following four simple assumptions:

1. Both A and B groups react only with each other (A + B) and not with themselves (neither A + A nor B + B).
2. All functional groups of the same type have equal reactivity (i.e., the reactivity of a functional group is independent of the size or the shape of the molecule to which it is attached).
3. Intra-molecular reactions do not occur in molecules of finite size (i.e., cyclization reactions are absent) (see also Chapter 15).
4. Functional groups act independently of one another.

Starting from these simple assumptions, and using combinatorial arguments, Flory [9, 10] predicted a number of properties of the branching reaction in an $A_x + B_2$ system ($x \geq 3$), which in contrast to an AB_x system can undergo gelation (the generation of infinite molecular weight species). He established the criterion for gelation of such a system in terms of the A/B stoichiometry and the extent of A + B reaction, and derived the molecular weight distribution of the system before gelation and for the sol fraction after gelation. His treatment also allowed for the presence of A_2 as well as A_x molecules in the reaction, but the reader is directed to the original references [10, 11] for details of this and other extensions of the theory. However, while in the original Flory's treatment, one of the reacting species was restricted to the functionality of two, this restriction was soon eliminated by Stockmayer [9] who generalized Flory's treatment to include arbitrary numbers of functional groups per molecule, and treated $A_x + B_y$ systems, in which $x \geq 2$ and $y \geq 3$. Ever since, the criteria derived by these theories for gelation in terms of the molar ratios of the reacting functional groups and the extent of reaction provide the bases for the design of synthetic parameters required for the preparation of hyperbranched polymers by this reaction route because they allow one to avoid gelation and obtain soluble reaction products.

Flory started by defining p_A as the probability that an A group has reacted, and, similarly, p_B as the probability that a B group has reacted at any given stage of the reaction between them. Furthermore, he designated r as the ratio of the number of A and B groups (not molecules) initially present in the system and α as a quantity representing the probability that a given chain ends in a branch unit:

$$\alpha = r\,p_A^2 = p_B^2/r \tag{E.16.1}$$

He then pointed out that gelation, or the formation of an infinite network, will occur when the probability that a randomly selected A-element is connected to $(x\text{-}1)$ other branching units, i.e., $\alpha(x\text{-}1)$, exceeds unity, so that the critical value α_c for the onset of gelation becomes $\alpha_c \geq 1/(x\text{-}1)$. Subsequently, Stockmayer generalized this critical condition for gelation to accommodate arbitrary functionalities of the reactive monomers, i.e., for an $A_x + B_y$ system:

$$rp_A^2 \geq 1/(x-1)(y-1) \tag{E.16.2}$$

This implies that soluble hyperbranched polymer product will be obtained if A is the minor component when

$$rp_A^2 > (x-1)(y-1) \tag{E.16.3a}$$

and if B is the minor component when

$$p_B^2/r < 1/(x-1)(y-1) \tag{E.16.3b}$$

For example, if all B groups present in a reaction system $aA_x + bB_y \rightarrow$ products, completely react (i.e., if $p_B = 1$), then a hyperbranched polymer having all A end-groups will be formed without gelation if $r > (x\text{-}1)(y\text{-}1)$, which requires (because of $r = ax/by$) that $ax/by > (x\text{-}1)(y\text{-}1)$. Conversely, if all A groups completely react, a soluble hyperbranched polymer with all B end-groups will form if $r < 1/(x\text{-}1)(y\text{-}1)$,

Table 16.2 Conditions for the formation and existence of soluble hyperbranched polymer product in selected $aA_x + bB_y$ polymerization systems in which $p_B = 1$

x	y	r	a/b
2	3	>2	>3
3	2	>2	>4/3
3	3	>4	>4
4	2	>3	>3/2
2	4	>3	>6
2	Large	>y	$>y^2/x$
Large	Large	>xy	$>y^2$

i.e., if ax/by < *1/(x-1)(y-1)*. Table 16.2 lists these conditions for several $aA_x + bB_y$ systems that are most commonly encountered in practice.

However, it is important to note that these results only provide the starting point, but not the complete answer to avoiding the gelation problem, since one of the fundamental assumptions made in their theoretical development was the absence of intra-molecular reactions in the finite species, i.e., the absence of cyclization. The failure of this assumption in practice (see also Chapter 15) is easily recognized by noting that with sufficient dilution a reaction that potentially could gel will proceed to completion without gelling. The reaction takes place preferentially intra-molecularly rather than inter-molecularly as is required for gelation.

This effect of dilution on the critical gelation conditions was investigated by Stockmayer and Weil [51] who studied critical conditions as a function of concentration. They reasoned that if cyclization was an increasing function of dilution, then at infinite concentration cyclization should be completely eliminated and the results would correspond to theory. They studied the reaction of equivalent numbers ($r = 1$) of OH groups (from pentaerythritol, x = 4) and COOH groups (from adipic acid, y = 2) in the presence of different amounts of a solvent, i.e., as a function of concentration. The theoretical prediction for the critical gelation point of this reaction with $r = 1$ is $p_A^2 = 1/(4–1)(2–1) = 0.33$ or $p_A = 0.577$. However, they experimentally found this value in the solvent-free system to be $p_A = 0.63$, and to increase with increasing amounts of solvent. On the other hand, by plotting the critical conditions vs. the reciprocal concentration of the reactants and then extrapolating back to 1/concentration = 0 (i.e., to infinite concentration), they found the critical intercept to be $p_A = 0.578$, which was in remarkable agreement with the value of 0.577 predicted by the theory. Clearly, this result gives confidence to other assumptions of the theory, namely to those of equal reactivity of all equivalent groups and the independence of the nature (i.e., size and shape) of the molecules to which they are attached.

The effect of the ratio of the reactive groups, *r*, on gelation can be seen from the data of Xu and Hu [52], shown in Table 16.3. They measured molecular weights and bulk reaction mixture viscosities as a function of *r* in the hydrosilylation reaction of divinyltetraethoxydisiloxane (x = 2), with tetrakis(dimethylsiloxy)silane, TKDMSOS (y = 4) (see Reaction Scheme 16.2), in which the vinylsilyl groups (component A) were completely reacted ($p_A = 1$).

Table 16.3 Monitoring of the course of BMNLP of Reaction Scheme 16.2 in the bulk by reaction mixture molecular weights and viscosity

Functionality ratio r = 4[B]/2[A]	M_n (GPC)	M_w (GPC)	Viscosity cSt
2.0	–	Gel	∞
2.4	3,500	26,000	186
2.6	3,400	17,000	119
2.8	3,000	5,000	58
3.0	2,700	4,500	50
5.0	1,800	2,000	13

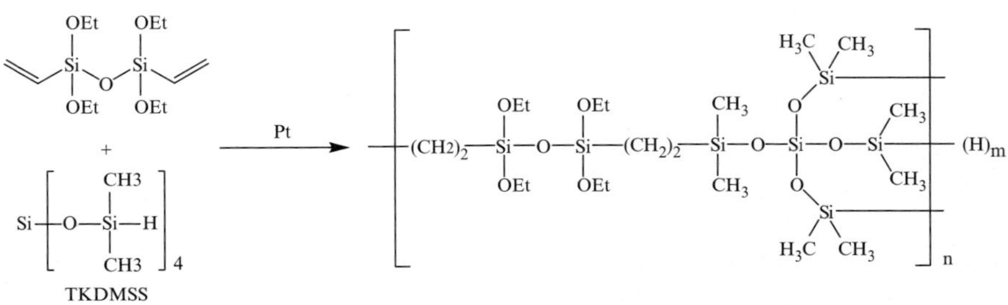

(16.2)

It was found that in contrast to the theoretical prediction for the critical ratio in this reaction system, r = (4–1)(2–1) = 3 (see Table 16.2), the experimentally observed value was near 2.0. Also, although this discrepancy was most likely due to cyclization reactions that are known to occur in such systems (see also Chapter 15) and the gel permeation chromatography (GPC) results seriously underestimated the true molecular weights of the resulting hyperbranched polymer products, the trend of the obtained data is strongly indicative of the approach to gelation as the ratio r decreases from 5.0 to 2.0 (with M_w tending to infinity while M_n remained finite).

Consequently, it can be concluded that because the above relationships are dependent on assumptions that are not exact for real systems, the actual degree of conversion that can be achieved for a given ratio of the reacting functional groups before gelation occurs tends to be somewhat higher than the theoretically predicted conversion. However, the theoretical value provides an excellent starting point from which to plan and conduct experiments to determine the allowed ranges for the ratio of the reacting functional groups and the allowable extent of conversion to avoid gelation problems.

16.4 Bimolecular Non-linear Polymerization by Hydrosilylation

Similar to the situation in linear polymers [53, 54], dendrimers (see Chapter 3) and hyperbranched polymers from AB_x monomers (see Chapter 13), the hydrosilylation reaction has also proved to be a very versatile method for the synthesis of hyperbranched polymers by BMNLP.

16.4.1 Hyperbranched Polycarbosilanes and Polycarbosiloxanes

A variety of different polycarbosilanes and polycarbosiloxanes have been prepared using a generalized synthetic approach shown in Reaction Scheme 16.3 [39, 40].

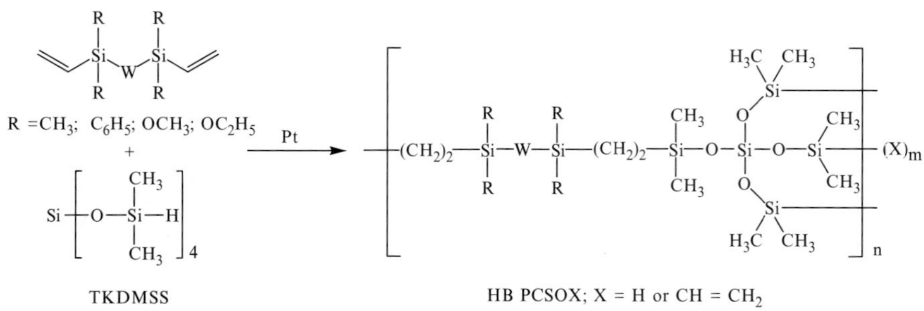

(16.3)

where: W = O or $-(CH_2)_2-$; R = CH_3; and TKDMSS can be replaced with its trifunctional derivative $Si(CH_3)[OSi(CH_3)_2H]_3$.

The reactions are best performed in bulk (to minimize cyclization; see Section 16.3) and in the presence of catalyst, such as Karstedt's catalyst (Pt-divinyltetramethyldisiloxane complex in toluene). A characteristic induction period is usually observed for about 30–60 min after which the reaction temperature starts increasing to about 90°C and the viscosity noticeably increases as the polymer molecular weight builds up, as exemplified by the data of Table 16.3. The obtained polymers are viscous oils, usually with broad molecular weight distributions (M_w/M_n ranging from 2 to 8), their molecular weights depend on how close to gelation one drives the reaction (see Table 16.3) and they can be obtained with Si-vinyl or Si-allyl groups, depending on which unsaturated monomer is used, or with Si–H end-groups, depending on which of the reacting monomers is used in excess (see Reaction Scheme 16.3 and Table 16.1).

16.4.2 Hyperbranched Polymer Networks

Because of their high molecular density of functionality, which can resemble that of dendrimers, hyperbranched polymers also represent extraordinarily useful building blocks for the creation of nano-domained networks, as shown in Fig. 16.1 (compare with Chapter 11, Section 11.4). In principle, there are five main types of such networks, including:

(a) Simple dendritic ("homopolymeric") networks having regularly distributed dendritic domains connected by the units resulting from self-condensation of single hyperbranched precursors (by, for example, hydrolysis of reactive Si–OR end-groups and subsequent silanol condensation; see Section 16.6).

(b) Multi-dendritic ("compositionally copolymeric") networks containing two or more compositionally different dendritic domains (A and B) connected by bonds resulting from the reactions of at least two compositionally different hyperbranched precursors (A and B) that have complementary reactive end-groups and react with each other under the crosslinking reaction conditions.
(c) "Architecturally copolymeric" networks with dendritic and linear polymer domains obtained by reactions of hyperbranched polymer precursors and linear (either α,ω-telechelic or with pendant reactive groups) polymer crosslinkers, which often have lower crosslinking densities than the networks of the types a and b above and by proper choice of the reagents and their relative amounts can be made either elastomeric or plastomeric.
(d) Dendritic networks obtained by crosslinking hyperbranched precursors with small molecular weight multifunctional crosslinkers (such as di-, tri- or tetra-functional reagents), which generally have very high crosslinking densities, and correspondingly high glass temperatures, rigidity and chemical resistance.
(e) Dendritic-linear interpenetrating polymer networks, which are the most complex members of this polymer networks family, generally show combinations (sometimes synergistic) of the properties of their component networks and are often the most economical to produce.

Although each of these five main types of dendritic polymer networks has its own specific properties, as a class they also have a number of common characteristics

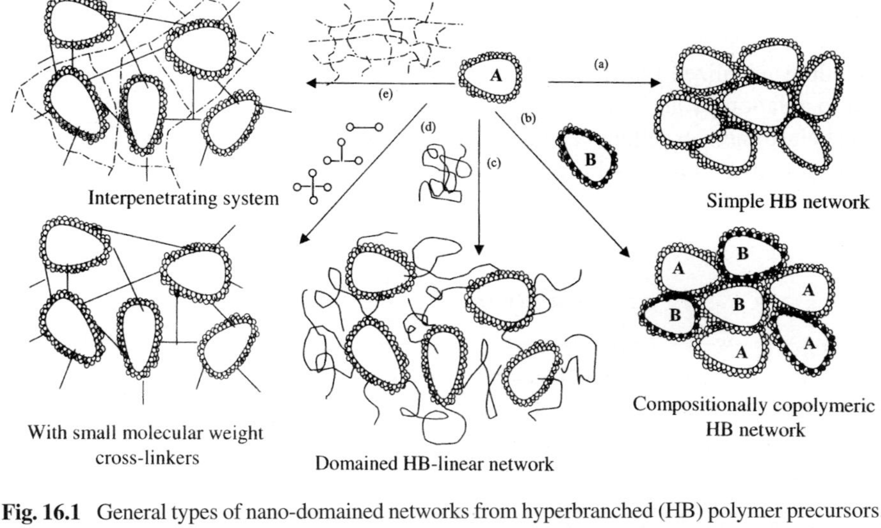

Fig. 16.1 General types of nano-domained networks from hyperbranched (HB) polymer precursors. Blank egg-shaped contours represent HB molecules/domains, while small circles in pearl necklace-type formation represent their reactive or reacted end-groups. Reagents in the clockwise direction include: (**a**) HB polymer of composition A; (**b**) HB polymer of composition B; (**c**) linear polymer with reactive chain ends; (**d**) di-, tri- and tetra-functional small molecule reagents; (**e**) interpenetrating polymer network comprising one HB and one linear polymer network.

that result from their basic fractal architecture. These include: (a) three-dimensional, nano-domained, honeycomb-like architecture that can be conveniently used (particularly when compositions of the dendritic cells and intercellular walls are sufficiently different) for sequestering (i.e., templating) a variety of different guest species to obtain unique host-guest nano-composites (see Section 16.6); (b) a very high degree of inter-domain crosslinking (resulting from hyperbranched polymer precursor multi-functionality and providing excellent mechanical properties and chemical resistance); (c) an ability to form elastomeric or plastomeric films, sheets, membranes or coatings that represent a completely new class of true nano-structured materials, and (d) a wealth of functional groups available for further chemical modification and convenient modulation of network properties, including their surface characteristics, such as reactivity, hydrophilicity/oleophilicity, nano-texture, etc. (see also Section 16.5) In addition, these hyperbranched polymer networks can be prepared in a variety of compositional combinations, and may exhibit pronounced robustness, durability, mechanical, chemical and weathering resistance.

For example, hyperbranched polycarbosilanes and polycarbosiloxanes of Reaction Schemes 16.2 and 16.3, can be easily and rapidly crosslinked, in the presence of Karstedt's catalyst, with α,ω-telechelic polydimethylsiloxanes, PDMS, bearing either vinylsilyl or silyl end-groups, depending on which type of hyperbranched polymer is used, as shown in Reaction Scheme 16.4.

The resulting elastomers have many desirable properties: they can be formed into films, sheets or coatings, or compounded with a variety of fillers such as silanol-functionalized silicas or polyhedral oligosilsesquioxanes (POSS), and exhibit low glass transition temperatures (T_g) (well into the −50°C to −70°C range and even below), high temperature stability and surface hydrophobicity that is typical of methylsilicones. They also exhibit pronounced abrasion resistance (e.g., in falling sand tests) and toughness, excellent adhesion to a variety of substrates including metalized plastic surfaces, resistance to solvents (such as isopropanol) and to ablation by an IR laser. This unique combination of properties makes these novel elastomers excellent candidates for a variety of applications, including very interesting nano-structured lithographic coatings.

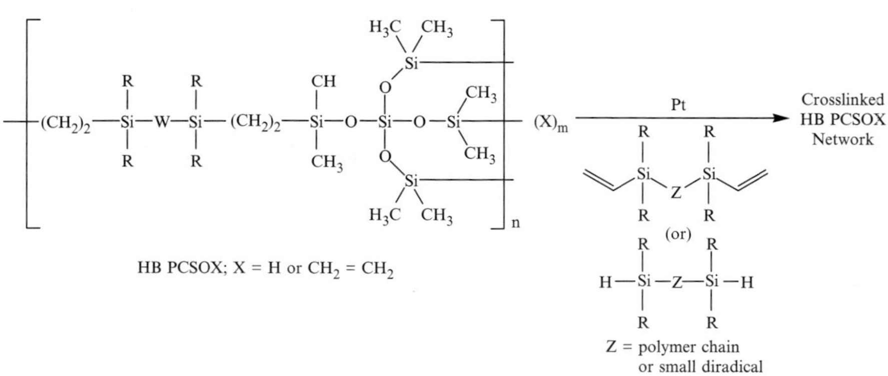

(16.4)

16.4.3 Perfluorinated Hyperbranched Polycarbosiloxanes

Using perfluorinated unsaturated monomers, such as diallyl- [52] or diallylether- [55] perfluoroalkanes in hydrosilylation polymerizations analogous to those shown in Reaction Schemes 16.2 and 16.3, copolymeric perfluorocarbosiloxanes can be obtained as shown in Reaction Scheme 16.5.

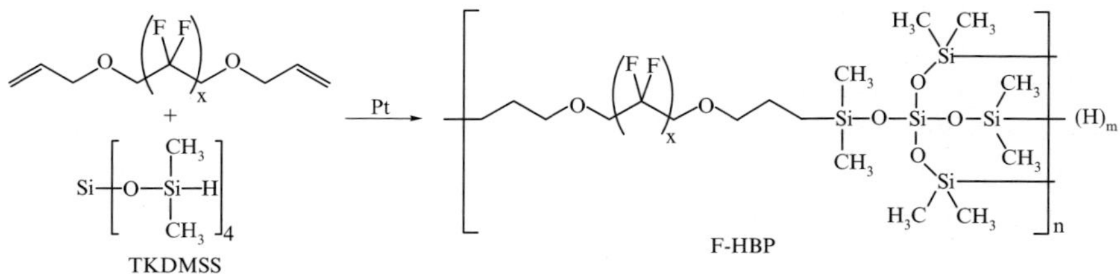

(16.5)

where x is an integer usually ranging from 4 to 8.

These polymers can be crosslinked, in ways similar to those shown in Reaction Scheme 16.4, into highly hydrophobic films or coatings with advancing contact angles of water (ACAW) ranging as high as 126°, or if their surfaces are specifically nano-textured to over 155° (see Fig. 16.2) [52].

However, these perfluorinated polymers can also be modified with hydrophilic end-groups, such as gluconolactone to create polysaccharide termini or poly(ethylene oxide) chains (see Reaction Schemes 16.6), to completely invert their nature from being water insoluble to becoming water soluble [55]. The resulting nano-scaled, covalently bonded perfluorinated micelles showed pronounced ability to dissolve relatively large quantities of oxygen in water and have been investigated as potential artificial blood plasma substitutes for emergency situations.

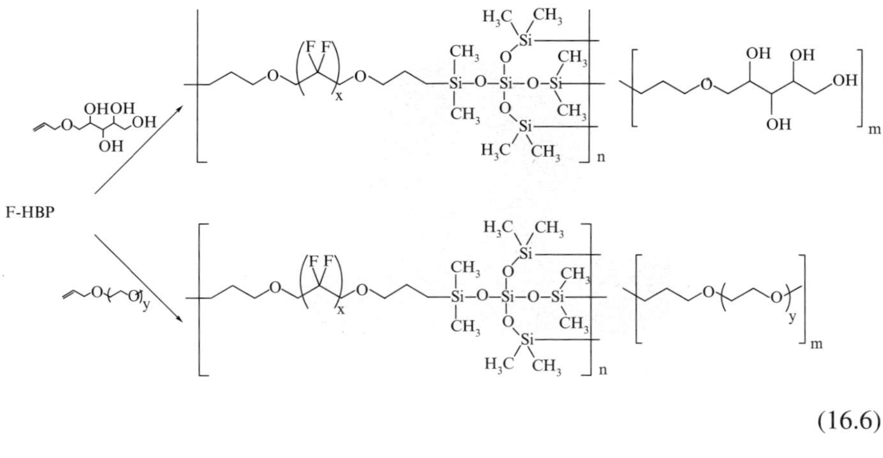

(16.6)

16.4.4 Hyperbranched Polycarbosiloxanes with Latent Functionalities

A unique type of hyperbranched polymer prepared by BMNLP is polycarbosiloxanes of Reaction Scheme 16.3 which in addition to their end-groups also contain ethoxysilyl, Si–OCH_2CH_3, functionalities uniformly distributed within each repeat unit throughout their molecular interiors [43]. These functionalities are introduced when tetraethoxydivinyldisiloxane monomer is used for the hyperbranched polymer preparation since they are inert under the hydrosilylation conditions, but represent potentially reactive, latent, functionalities that can be activated in the resulting products under different conditions. For example, as illustrated in Fig. 16.3, after the crosslinking via another hydrosilylation reaction (Fig. 16.3a: reacting polymer Si–H end-groups with difunctional divinyl crosslinker in the presence of Karstedt's catalyst, analogous to Reaction Scheme 16.4), the resulting hyperbranched network can undergo further crosslinking via condensation of the silanol groups resulting from the hydrolysis of Si–OCH_2CH_3 functionalities in the presence of a catalyst such as titanium di-*n*-butoxide (Fig. 16.3b), to create *in situ* silica "particles" that can act as nanoscopic fillers to self-reinforce the resulting crosslinked material [65].

Fig. 16.2 Superhydrophobic surfaces from crosslinked perfluorinated polycarbosiloxanes: ACAW = 158°.

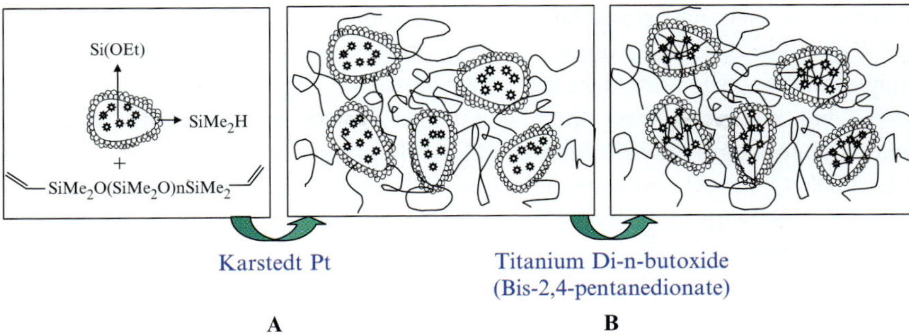

Fig. 16.3 Preparation of self-reinforcing networks from hyperbranched polycarbosiloxanes (egg-shaped structures) with latent ethoxysilyl-reactive groups (∗ shaped objects inside the egg-shaped structures). (**A**) crosslinking of an Si–H functionalized hyperbranched polycarbosiloxane via hydrosilylation; (**B**) hydrolysis of ethoxysilyl groups and their condensation into highly crosslinked silica particles inside the dendritic (egg-shaped) nano-domains.

In this figure, large egg-shaped contours represent molecules of hyperbranched polycarbosiloxanes (obtained as in Reaction Scheme 16.3) having a pearl-like necklace of $-Si(CH_3)_2H$ end-groups, and intramolecular ethoxysilyl units represented by star-like (∗) features. The latter crosslink to form nanoscopic silica particles (stars connected with short solid straight lines) within the hyperbranched domains of the already crosslinked network.

16.5 Bimolecular Non-linear Polymerization by Nucleophilic Substitution Reactions

16.5.1 Hyperbranched Silarylenesiloxanes

Hyperbranched silarylenesiloxane polymers were prepared via nucleophilic substitution BMNLP reaction of arylenedisilanols and tri-or tetra-functional chlorosilanes, as shown in Reaction Scheme 16.7 [56].

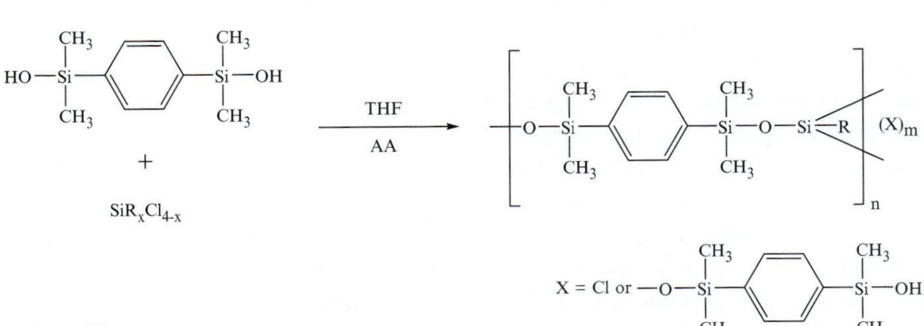

(16.7)

Similar to their hyperbranched counterpart obtained from an AB_3, CH_2=CH–C_6H_4–Si[–O–Si$(CH_3)_2$H$]_3$, monomer [57], these polymers with vinylsilyl end-groups from vinyltrichlorosilane monomer showed T_gs ranging between −10°C and −20°C, but relatively lower thermal and thermo-oxidative stabilities (knees in dynamic thermal gravimetric analysis (TGA) thermograms in air and in nitrogen at around 200°C) than their linear analogues [58]. The polymers were slightly opaque viscous liquids with molecular weights (M_w, via light scattering) ranging between 5,000 and 8,000.

16.5.2 Hyperbranched Polysiloxanes

Hyperbranched polysiloxanes can be prepared via the BMNLP method as shown in Reaction Scheme 16.8.

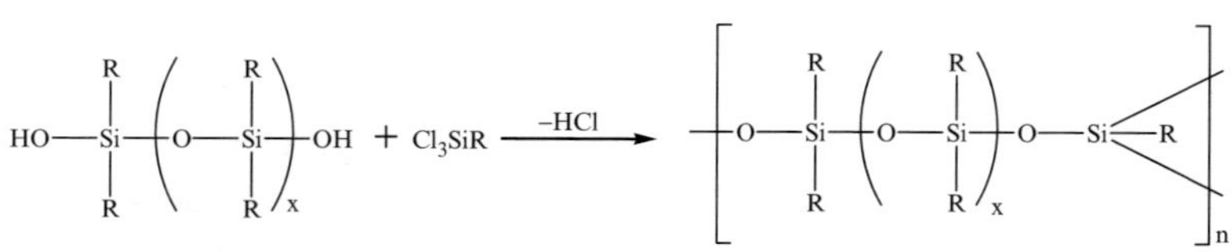

(16.8)

The key to this polymerization is successful synthesis of the disilanol monomer(s) which, for example, can be performed by the hydrolysis of dimethyldichlorosilane (DDS) with water under carefully controlled neutral conditions, as described by either Lucas and Martin in 1952 [59], or Kantor in 1953 [60]. (During the final revisions of the galley proofs of this book, an interesting paper on the synthesis and applications of dimethyl and diphenyl-disilanols was published by the Flinders University group of Adelaide, South Australia [64].) The former authors obtained the "dimeric" tetramethyldisiloxane-1,3-diol (TMDS) in about 60% yield by passing gaseous ammonia through the reaction mixture of DDS and ice-cold water to neutralize the evolving HCl byproduct. In contrast to this, Kantor showed that the "elusive" dimethylsilanediol monomer, Si$(CH_3)_2(OH)_2$, could be obtained in 65–80% yield from dimethyldimethoxysilane (which, in turn, was obtained by the alcoholysis of DDS with methanol in the presence of dimethylaniline as the acid acceptor) by the hydrolysis with boiling distilled water in the reactor that was carefully cleaned to completely neutralize its interior surfaces. In our group, we confirmed the former procedure and prepared TMDS which remained stable under nitrogen for several months [61].

It should be particularly noted that once the disilanol monomers are obtained, BMNLP provides a very simple and easy route to the hyperbranched polysiloxanes, including polydimethylsiloxane (PDMS) (R = CH_3 in Reaction Scheme 16.8), which could be considered the parent polymer of the entire silicon-containing hyperbranched polymer family and for which no alternative AB_x route has yet been proposed.

16.6 Siliconized Hyperbranched Polymers by BMNLP

Using the BMNLP principles a large number of other polymers that do not contain silicon has also been prepared, as outlined in Section 16.2. Of these, especially interesting for the topic of this book are the amine-terminated polyamides, polyamidoamines, polyureas and polyurethanes which have been used in a variety of post-polymerization siliconization reactions, and which can be prepared as illustrated for the polyurea formation by the diisocyanate addition to triamines shown in Reaction Scheme 16.9 [41, 42].

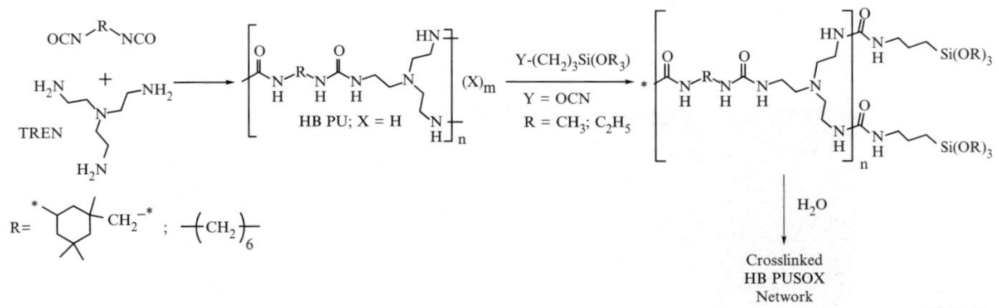

(16.9)

where R in RO = CH_3 or C_2H_5 and x = 6.

These polymers can be considered as hyperbranched analogues of PAMAMOS dendrimers (see Chapter 11) with hydrophilic and nucleophilic molecular "interiors" and organosilicon "exteriors" [62]. Also like their PAMAMOS dendrimer relatives, these siliconized hyperbranched polymers can be easily crosslinked by simple sol-gel processes through the water hydrolysis of alkoxysilyl groups and subsequent condensation of the resulting silanols [42, 62]. This crosslinking can be performed with water vapor (including moisture from air) or liquid, under controlled humidity or in open air, in the presence of suitable catalysts to regulate the reaction rate or without them, and to quite high degrees of crosslinking which directly predetermine mechanical properties and swelling of the resulting networks. The selection of catalysts depends on whether methoxysilyl or ethoxysilyl functionalized polymers are used, the crosslinking conditions and the desired rate of the reaction, and may vary from traditional tin-based catalysts, to various bases and acids, such as amines and acids, including acetic, benzoic, etc. In the case where the base polymer is nucleophilic enough, and very fast crosslinking is not of crucial importance, the use of catalyst may not even be required.

Crosslinking of the siliconized hyperbranched polymers described above results in self-supporting films, sheets or coatings, and the ensuing products can be used for various applications. The coatings usually have quite smooth surfaces which are often hydrophobic (ACAW ranging from 80° to 110°) indicating methylsilicone groups at the surfaces, and can exhibit very high pencil test hardnesses ranging to above 7 H on hard supports. Depending on the composition of the base polymers, these films and coatings can serve as very good matrices for a variety of different additives, offering possibilities of positive interactions with electrophiles and/or hydrophilic guest species, as shown in Fig. 16.4.

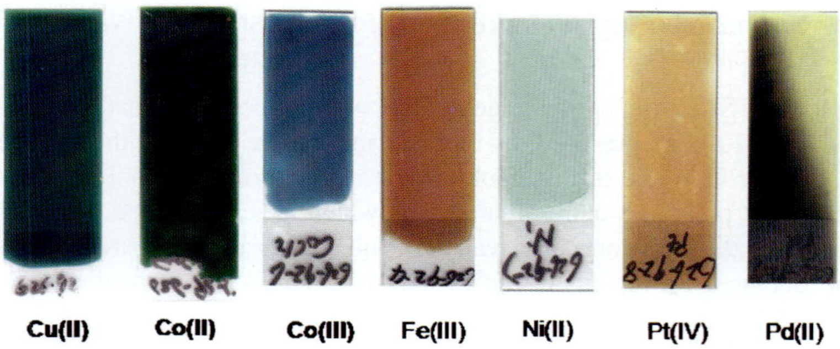

Fig. 16.4 Colorful world of nanocomplexes obtained from crosslinked coatings of siliconized hyperbranched polyureas of Reaction Scheme 16.9 and soluble salts of indicated metal cations.

Fig. 16.5 Antifouling effect of biocide-containing crosslinked hyperbranched polyurea coatings on zebra mussels in Lake Michigan after a year of exposure. Three clean rectangular objects are coated stainless steel plates attached to a rack made of polyvinylchloride pipes that are heavily encrusted with foulants.

One of the most recently studied applications of such systems was as antifouling coatings for use in marine and fresh water environments against the settlement of barnacles or zebra and quagga mussels, on man-made surfaces such as boat hulls, docks, pipe intakes, rigs, etc. Similar to PAMAMOS dendrimers (compare with Chapter 11) [62], it was found that the secondary bonding between the cations (such as Cu^{2+} or Zn^{2+}) used in biocidal nanocomplex components and these hyperbranched polymer matrices, together with strong steric hindrance imposed upon the mobility of the cations by the dendritic hyperbranched architecture, resulted in an outstanding reduction of biocide leaching into the aquatic environment (by orders of magnitude over a period of more than a year). This property, coupled with excellent antifouling activity [63] (see Fig. 16.5), makes these formulations excellent candidates for efficient, environmentally benign antifouling paints that are desperately needed by a number of prominent industries, including transportation, fishing, defense, telecommunications, water and energy production, etc., in these days when tin has already been banned from application and copper is about to suffer the same fate.

16.7 Conclusions and Future Outlook

Although the principles of BMNLP have been known for many decades and the reaction itself is routinely used in many standard crosslinking technologies, its application to the preparation of soluble hyperbranched polymers is only recent. In fact, it is beginning to emerge as one of the youngest major strategies for the construction of dendritic polymers. For example, at the recent 235th national meeting of the American Chemical Society in New Orleans, LU, in April 2008, almost one entire day of the symposium on branched polymers was packed with BMNLP-related presentations.

Until now, most of the work in this field has been directed at the synthesis of new hyperbranched compositions and among these the silicon-containing polymers have been relatively well represented. Very recently, however, the first studies aimed at better understanding of the fundamental nature of the process and how to control it have started to appear [1–4]. The process variables that are under investigation, and that will need further in-depth studies, include, but are not limited to, the effects of reaction mixture concentration, the rate and the order of monomer addition(s), the effects of monomer structures and reactivity, solvents and reactor shapes and designs. The effects of the process parameters on structural properties of the obtained hyperbranched polymers should focus on the polymer molecular weights (sizes), molecular weight distributions, degrees of branching, tendency to intramolecular cyclization, and ways and means how to control them.

Compared to the more traditional and much more developed AB_x monomolecular polymerization processes, BMNLPs appear to yield products that are smaller in sizes (i.e., lower in molecular weights, and most likely also in the degrees of branching) and narrower in the size (molecular weight) distributions. It seems from some very scant data that it can be estimated that their typical sizes may range between roughly 1 and 5 nm in hydrodynamic diameters. At the same time, however, BMNLP provides some extraordinary synthetic opportunities that are not available in the more traditional strategies, including the ability to use readily available monomers, no monomer shelf life problems, no need to focus almost exclusively on high temperature or catalyzed reactions, a high versatility towards varying polymer composition and the ability to produce polymers of the same compositions but with different end-groups as well as those with latent functionalities that can lead to very interesting post-synthetic chemical applications. All of these features offer some extraordinary application potentials if the required chemical engineering aspects of the necessary synthesis control can be provided. As a consequence, the future of BMNLP looks bright as more detailed theoretical-experimental studies, the preparation of new polymer compositions and new polymer applications appear. The grand prize that will further inspire and stimulate these efforts is the main strength of the BMNLP processes which includes their unique combination of versatility, simplicity and economical polymer production.

Acknowledgements The authors gratefully acknowledge financial support for a part of this work by the National Science Foundation (grants no. DMI-0419193 and DMI-0522183).

References

1. Unal S, Oguz C, Yilgor E, Gallivan M, Long TE, Yilgor I (2005) Polymer 46: 4533.
2. Oguz C, Unal S, Long TE, Gallivan MA (2007) Macromolecules 40: 6529.
3. Oguz C, Gallivan MA, Cakir S, Yilgor E, Yilgor I (2008) Polymer 49: 1414.
4. Oguz C, Cakir S, Yilgor E, Gallivan M, Yilgor I (2008) Polym Prepr 49(1): 48.
5. Kienle RH, Hovey AG (1929) J Am Chem Soc 51: 509.
6. Kienle RH, van der Meulen PA, Petke FE (1939) J Am Chem Soc 61: 2258.
7. Flory PJ (1941) J Am Chem Soc 63: 3083, 3091, 3096.
8. Stockmayer WH (1943) J Chem Phys 11: 45.
9. Stockmayer WH (1952) J Polym Sci 9: 45.
10. Flory PJ (1952) J Am Chem Soc 74: 2718.
11. Flory PJ (1953) Principles of Polymer Chemistry, Cornell University Press, Ithaca, NY, Chapter 9.
12. Sunder A, Heinemann J, Frey H (2000) Chem Eur J 6: 2499.
13. Voit B (2001) J Polym Sci, Part A: Polym Chem 38: 2505.
14. Jikei M, Kakimoto M (2001) Prog Polym Sci 26: 1233.
15. Russo S, Boulares A (1990) Macromol Symp 128: 13.
16. Russo S, Boulares A, da Rin A (1999) Macromol Symp 143: 309.
17. Monticelli O, Mariani A, Voit B, Komber H, Mendichi R, Pitto V, Tabuani D, Russo S (2001) High Perform Polym 13: S45.
18. Jikei M, Chon S-H, Kakimoto M, Kawauchi S, Imase T, Watanabe J (1999) Macromolecules 32: 2061.
19. Jikei M, Kakimoto M (2001) High Perform Polym 13: S33.
20. Emrick T, Chang H-T, Frechet JMJ (1999) Macromolecules 32: 6380.
21. Hecht S, Emrick T, Frechet JMJ (2000) Chem Commun 313–314.
22. Yan D, Gao C (2000) Macromolecules 33: 7693.
23. Fang J, Kita H, Okamoto K (2000) Macromolecules 33: 4639.
24. Kricheldorf HR, Vakhtangishvili L, Fritsch D (2002) J Polym Sci, Polym Chem 40: 2967.
25. Kricheldorf HR, Fritsch D, Vakhtangishvili L, Schwarz G (2003) Macromolecules 36: 4337.
26. Bruchmann B, Koniger R, Renz H (2002) Macromol Symp 187: 271.
27. Bruchmann B, Schrepp W (2003) e-Polym 014.
28. Bruchmann B, Stumbe J-F (2004) Macromol Rapid Commun 25: 921.
29. Czupik M, Fossum E (2003) J Polym Sci, Polym Chem 41: 3871.
30. Chen H, Yin J (2003) Polym Bull 49: 313.
31. Hao J, Jikei M, Kakimoto M (2003) Macromol Symp 199: 233.
32. Lin Q, Unal S, Long TE (2003) Polym Mater Sci Eng 88: 544.
33. Lin Q, Long TE (2003) Macromolecules 36: 9809.
34. Unal S, Yilgor I, Yilgor E, Sheth JP, Wilkes GL, Long TE (2004) Macromolecules 37: 7081.
35. Unal S, Lin Q, Mourey TH, Long TE (2005) Macromolecules 38: 3246.
36. In I, Kim SY (2004) Polym Mater Sci Eng 91: 779.
37. Perignon N, Mingotaud A-F, Marty J-D, Rico-Lattes I, Mingotaud C (2004) Chem Mater 16: 4856.

38. The procedure was taken from the US Patent application 0161113 A1, 2002, which was later granted as Reference 40.
39. Dvornic PR, Hu J, Meier DJ, Nowak RM (2004) Polym Preprints 45(1): 585.
40. Dvornic PR, Hu J, Meier DJ, Nowak RM (2002) US Pat. 6,384,172 B1.
41. Dvornic PR, Hu J, Meier DJ, Nowak RM, Parham PL (2003) US Pat. 6,534,600 B2.
42. Dvornic PR, Hu J, Meier DJ, Nowak RM, Parham PL (2004) US Pat. 6,812,298 B2.
43. Dvornic PR, Hu J, Meier DJ, Nowak RM (2003) US Pat. 6,646,089 B2.
44. Macosko CW, Miller DR (1976) Macromolecules 9: 199.
45. Gordon M (1962) Proc R Soc Lond, Ser A 268: 240.
46. Case LC (1957) J Polym Sci 26: 333.
47. Charlesby A (1954) Proc R Soc A 222: 542.
48. Charlesby A (1955) Proc R Soc A 231: 521.
49. Stauffer D, Coniglio A, Adams M (1982) Adv Polym Sci 44: 105.
50. Dušek K, Dušková-Smrčková M (2000) Prog Polym Sci 25: 1215.
51. Stockmayer WH, Weil LL (1945) Advancing Fronts in Chemistry, Twiss SB, Ed., Reinhold, New York, Chapter 6.
52. Xu Q and Hu J (2005) Unpublished work, Michigan Molecular Institute.
53. Dvornic PR, Gerov VV (1994) Macromolecules 27: 1068.
54. Dvornic PR, Gerov VV, Govedarica MN (1994) 27: 7575.
55. Kaganove SN, Satoh P, Dvornic PR (2006) US Pat. Appl 0147414 A1.
56. Dvornic PR, Zhang T, (2006) Unpublished results, Michigan Molecular Institute.
57. Rubinsztajn S, Stein J (1995) J Inorg Organomet Polym 5: 43.
58. Dvornic PR, Lenz RW (1990) Silarylene-Siloxane Polymers in High Temperature Siloxane Elastomers, Hüthig & Wepf Verlag, Basel, Chapter 3, pp. 85–214.
59. Lucas GR, Martin RW (1952) J Am Chem Soc 74: 5225.
60. Kantor SW (1953) J Am Chem Soc 75: 2712.
61. Hu J, Dvornic PR, (2004) Unpublished results, Michigan Molecular Institute.
62. Dvornic PR (2006) J Polym Sci Part A Polym Chem 44: 2755.
63. Scheide JE, Dvornic PR, (2007) Unpublished results, Michigan Molecular Institute.
64. Schamschurin A, Uhrig D, Fisher M, Clarke S, Matisons J (2008) Silicon Chem 3: 313.
65. Dvornic PR, Hu J, Meier DJ, Nowak RM (2008) US Pat. 7, 446, 155B2.

Index

D

L

M

N

O

P

Q

R